ANIMAL BEHAVIOUR
(ETHOLOGY)

ANIMAL BEHAVIOUR
(ETHOLOGY)

For B.Sc. and M.Sc. Students of all Indian Universities

V K Agarwal
MSc, PhD
Former Head of the Zoology Department
Meerut College
Meerut, Uttar Pradesh

S Chand And Company Limited
(ISO 9001 Certified Company)

S Chand And Company Limited

(ISO 9001 Certified Company)

Head Office: Block B-1, House No. D-1, Ground Floor, Mohan Co-operative Industrial Estate, New Delhi – 110 044 | Phone: 011-66672000

Registered Office: A-27, 2nd Floor, Mohan Co-operative Industrial Estate, New Delhi – 110 044 Phone: 011-49731800

www.**schandpublishing.com**; e-mail: **info@schandpublishing.com**

Branches

Chennai	:	Ph: 23632120; chennai@schandpublishing.com
Guwahati	:	Ph: 2738811, 2735640; guwahati@schandpublishing.com
Hyderabad	:	Ph: 40186018; hyderabad@schandpublishing.com
Jalandhar	:	Ph: 4645630; jalandhar@schandpublishing.com
Kolkata	:	Ph: 23357458, 23353914; kolkata@schandpublishing.com
Lucknow	:	Ph: 4003633; lucknow@schandpublishing.com
Mumbai	:	Ph: 25000297; mumbai@schandpublishing.com
Patna	:	Ph: 2260011; patna@schandpublishing.com

First Edition 2009
Reprints 2012, 2013 (Twice), 2014, 2015, 2016 (Twice), 2017, 2018 (Twice), 2019, 2020 (Twice)

Reprint 2021 (Twice)

ISBN: 978-81-219-3210-3 **Product Code:** H6ABR68ZOOL10ENAA09O

PRINTED IN INDIA

By Vikas Publishing House Private Limited, Plot 20/4, Site-IV, Industrial Area Sahibabad, Ghaziabad – 201 010 and Published by S Chand And Company Limited, A-27, 2nd Floor, Mohan Co-operative Industrial Estate, New Delhi – 110 044.

PREFACE

A honeybee visiting the freshly open rose flower, herds of cheetal deer grazing in a grassland, cubs of tiger resting quietly together in a shade in Corbett National Park and Tiger Reserve, fireflies beam light flashes to each other across a darkened landscape, a peacock dancing infront his harem of peahens and a mongoose swiftly and deftly bites its prey to death. The study of animal behaviour is about all these things and many others. It is about the chase of the hunter and the flight of the hunted. It is about the spinning of webs, the digging of burrows and the building of nests. It is about incubating eggs and suckling young. It is about the migration of a hundred thousand animals and the flick of a tail of one. It is about remaining motionless and concealed as well as about leaping and flying. In the words of **Manning** and **Dawkins** (1998)

"Behaviour involves the static postures and active movements, all the noises and smells and changes of colour and shape that characterise animal life."

This textbook of **Animal Behaviour** deals with various types of behaviours of the animals and even of human beings. This book covers the syllabuses of B.Sc. classes of All Indian Universities and has been prepared according to U.G.C. model curriculum.

This book includes 16 chapters, more than 400 pages and an exhaustive glossary. The language of the text of book is simple, lucid and illustrative. Most of the illustrations of this book have been redrawn by Mr. Trilok and Mr. Shimpy of Meerut.

Author has been teaching the subject of Animal Behaviour to students of M.Sc. and B.Sc. from last three decades. He has found this subject quite fascinating.

Author wishes to thank his wife Mrs. Usha Gupta and his three lovely daughters namely Dr. Surabhi Gupta, Dr. Shubhra Agarwal and Er. Anubha Agarwal for regulating his behaviour. At this juncture, author also wishes to thank all the faculty members of Zoology Department, Meerut College, Meerut; our animated discussions on every possible cognitive topic under the sun at tea often have generated entirely new ideas and helped the author to edit his writings.

Author expresses his sincere thanks to Mrs. Nirmala Gupta, CMD, Mr. Himanshu Gupta, JMD, Mr. Amit Gupta, CEO, Mr. Navin Joshi, Vice-President (Publishing) and Mr. R.S. Saxena of S. Chand & Company Ltd. for their wholehearted cooperation to bring out this new textbook.

Author also is indebted to Mr. Shishir Bhatnagar and his Editorial Section and Mr. Amit Goel for their various sort of help during author's visit of S. Chand & Company Ltd., New Delhi.

Author is open to constructive criticism of this book. Please do write to author about the book; your feedbacks would be helpful in further improvement of the book.

Dr. V. K. AGARWAL

'Neel Kanth'
566, Brahmpuri, Sharda Road
Meerut–250002
Uttar Pradesh
India

PREFACE

CONTENTS

Introduction

1.1. DEFINITIONS OF ETHOLOGY AND ANIMAL BEHAVIOUR

One of the basic aspects that separates living from non-living matter is the capacity of living organisms to respond to changes in the environment. The response of living organisms is adaptive, that is, it enhances the probability of the species to **survive** and **reproduce.** One of the most important ways of adapting to environmental change, at least for animal, is behaviour and its success depends to a great extent upon its degree of variability. Generally, the term behaviour is used to refer to all those responses to environmental changes that involve the integrated functioning of the entire organism. For instance, plants react to stimuli at the cell level, as do unicellular animals (*i.e.*, protozoans); in multicellular animals, genetically inherited sensory and nervous systems provide the framework for reaction, the way in which animal experiences its environment also influences its behaviour.

The study of animal behaviour from a biological point of view forms a new and young branch of biology called **ethology.** The term ethology is based upon two Greek words– *ethos* which means a characteristic disposition or habit or custom of an individual or group and *logos* which refers to study. **St. Hilaire** chose the term ethology in late eighteenth century to refer to the study of animals as living beings in their natural environment. Ethology is a biological discipline that involves the study of the total repertoir of the innate and learned behaviours that animals employ to resolve the problems of survival and reproduction.

Why do Animals have to Behave?

Animals are defined to a large extent by their behaviours. They are motile creatures that explore and manipulate their environments. Even the most sedentary animals are more active and manipulative than the most motile plants. For example, hydras search into the surrounding medium with tentacles; clams sweep food-laden currents of water into their mouths with rows of cilia and parasitic worms creep through the cavities and sway in the body fluid of their hosts. In fact, there is a sound ecological reason *why animals must be active?* Animals are the consumers of energy and not the primary producers, the role of which is filled by green plants and microorganisms (*e.g.*, cyanobacteria). Even when animals search out or prey on other animals, they merely act as channels for energy captured and packaged by primary producers. One of the fundamental laws of ecology states that consumers can get only a small fraction (10 per cent) of the energy that passes through plants. Consequently animals must literally collect a living from the environment, patrolling it or sucking part of it into their alimentary canal in order to secure a sufficient quantity of energy. Once such a option of **heterotrophy** was chosen in evolution by the earliest animals years ago, the effect spilled into every other aspect of their life. The body was compacted into precise symmetric shapes and powered by masses of muscles; alimentary tracts were assembled for the processing of food in packets; and entire new organs (*i.e.*, sensory organs) were invented to monitor odour, light and sound. Above all there was the nervous system to coordinate behaviour. However, behavioural responses are not absolutely dependent on a nervous system. The members (zooids) of colonies of siphonophores (*e.g.*, *Physalia*) communicate with one another partly through special cells in the epithelium. They are able to coordinate their movements in such an extent that the whole colony acts like a single organism or super-organism.

What is Behaviour?

In simple terms, what an animal does, how it acts is called **behaviour.** Behaviour is evolved in the success of the animal in caring for itself, in seeking appropriate shelter, in obtaining food, in escaping enemies, in courtship and mating and in the caring for the young. **Niko Tinbergen** (1966) has provided a very convincing definition of behaviour: *Behaviour is the movements animals make.* These involve more than running, swimming, crawling and other types of locomotion. They also comprise the movements animals make when feeding, when mating, even when breathing. Nor is this all, slight movements of parts of body, such as pricking the ears or making a sound are also parts of behaviour. And many animals do something analogous to our blushing — they change colour, sometimes as a way of concealing themselves from predators, while on other occasions when they are aroused to attack or are courting a female.

In fact, some actions of animals such as the honking of peacocks, which we should wish to count as behaviour, are not movements of the whole animal in the ordinary sense. The honking sound is produced as air is forced out of peacock's lungs by the contractions of muscles, which causes a region of throat to vibrate. Thus, there is movement due to the pulmonary musculature, just as there is muscular movements when an animal feeds or walks. In more accurate sense, therefore, animal behaviour consists of a series of muscular contractions.

In the words of **Aubrey Manning** (1979) behaviour includes all those processes by which an animal senses the external world and the internal state of its body, and responds to changes which it perceives. Many of such processes tend to take place 'inside' the nervous system and may not be directly observable. What the animal does may involve violent activity or complete inactivity but all are equally behaviour.

Raven *et al.* (2005) have defined behaviour as the way an animal responds to stimuli in its environment. A stimulus may be as simple as the odour of food. In this sense, a bacterial cell "behaves" by moving toward higher concentrations of the sugar. This behaviour is very simple and well suited to the life of bacteria, allowing these microoganisms to live and reproduce. As animals evolved, they occupied different environments and face diverse problems that affected their survival and reproduction. Their nervous systems and behaviour concomitantly became more complex. Nervous systems perceive and process information concerning environmental stimuli and trigger adaptive motor responses, which we see as patterns of behaviour.

1.2. OVERVIEW OF BEHAVIOUR

Mechanism of Behaviour

Behaviour is the tool with which an animal uses its environment. Through behaviour the animal manoeuvers itself in an organised and directed way and manipulates objects in the environment to suite its requirements. In order to behave, the animal must act as an integrated and coordinated unit. It must juggle a wide array of stimuli from inside its body and from the external environment and organise the information into a series of commands to its muscles. Most animals have evolved complex systems of cells and chemicals to detect, to transmit, to integrate and to store environmentally supplied information for later use in making decisions. This consists of following four components:

1. Various types of **sensory cells** which pick up different changes in the environment.
2. A more or less complex system of neurons or nerve cells (*i.e.*, **nervous system**) which transmits and integrates information from sensory receptors.

3. Chemical messengers (*i.e.*, **hormones**) which transmit information on a more leisurely time scale than the nervous system.

4. **Muscle cells** which transform information from the nervous system into actions (*e.g.*, movement).

Most animal behaviour is a sequence of responses related to some need of the species such as getting food, acquiring and maintaining living space, protection and reproduction. A precise sequence of behaviour is the characteristic of the species and is performed in the same manner by all members of the population throughout its geographical range. In fact, *no two species behave exactly alike.* Thus, behaviour adds another dimension to the biological study of animals besides the already employed aspects of morphology and physiology.

Proximate Versus Ultimate Causation

Even simple behaviours such as migration and overwintering of monarch butterflies are amazingly complex and difficult to study. Many animal behaviourists focus on "how" questions: those involving the ways in which behaviour is directly produced and regulated. We call these structures and mechanisms **proximate factors.** For example, if we ask, *How do male monarchs produce the "settling" pheromone ?* We could study hormonal and environmental cues that trigger the production and release of the pheromones. However, if we focus only on proximate factors, we may miss the important adaptive components of this behaviour.

Both behavioural biology and sociobiology are rooted in investigations using "why" rather than "how" questions. These are questions that have influenced the evolution of behavioural patterns. For example, why should there be both migratory and nonmigratory forms of the monarch butterflies ? or, what are the evolutionary adaptive value, of a female monarch mating more than once ? An evolutionary perspective generates important questions about the functions of behavioural patterns.

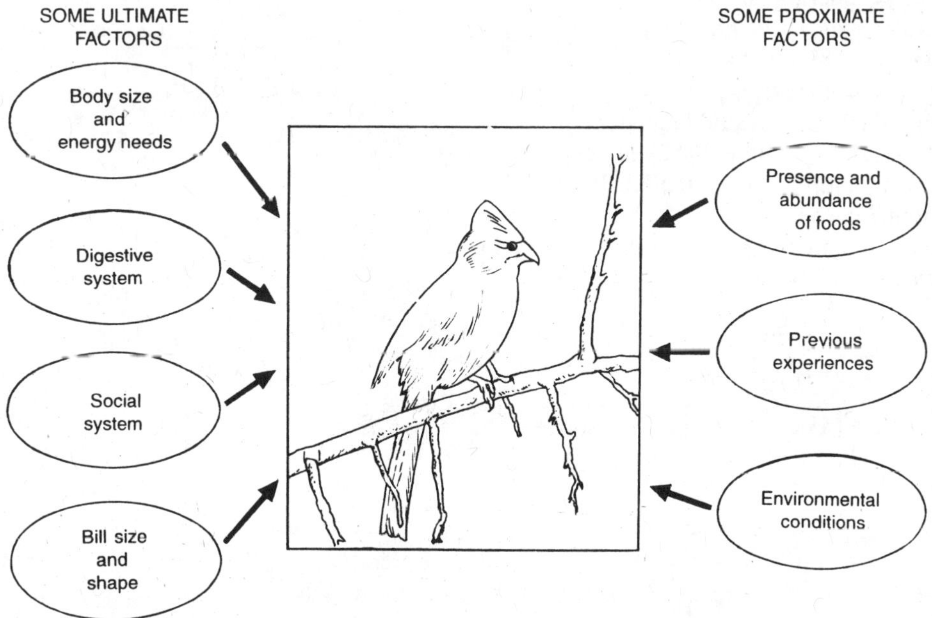

Fig. 1.1. Factors affecting the feeding behaviour of northern cardinal birds. Constraints that have arisen through evolution establish the limits on the dietary habits for the cardinal. Past experience and current environmental conditions influence the immediate choices made by the animal as it forages (after Drickamer, *et al.*, 2002).

Evolution by natural selection is the cornerstone of all subdisciplines of biology, including animal behaviour. However, natural selection works at the level of the mechanisms that control and integrate behaviour. To understand behaviour more completely, therefore, we must have a solid grasp of both the internal mechanisms by which behaviour is produced ("how" questions) and the external factors that influence whether those mechanisms will be passed on the future generations ("why" question).

Behavioural ecologists are concerned with ultimate and proximate questions about behaviour. Suppose we are interested in the feeding habits of the bird, called northern cardinal (*Cardinalis cardinalis*), living in a variety of places in the North American country side (Fig. 1.1). *Ultimate constraints* (factors) affecting the cardinal bird would include its body size and related energy needs; the type of bill, which affects the foods it can consume; the digestive system, with regard to what foods the bird can process; and the social system of the species, which could influence the partitioning of available food resources. *Proximate factors* influencing feeding would include the presence and relative abundance of specific foods; past experiences of the bird in searching for and handling particular foods; and the season of the year, with particular regard to variations in energy needs due, for example, to reproduction or cold winter weather. Ultimate factors establish the limits, and proximate factor, affect the behaviour of an animal within those limits.

Various Types of Behaviour

Animal behaviour can be divided into that which is shown by all members of a species (called **species characteristic behaviour**) and that which varies from one individual to another (called **individual characteristic behaviour**). The species-characteristic behaviour includes the stereotyped behaviour patterns distinctive of particular species, for example, courtship and copulation of many animals. The individual characteristic behaviour includes the behaviour learned by an animal during its life time, for example, 'tricks' performed by individual dogs.

Ethologists of 1940's and 1960's have broken down any behaviour in the following *nine* categories of functions or patterns:

1. Eating or ingestive behaviour;
2. Shelter-seeking behaviour;
3. Agonistic behaviour;
4. Reproductive or sexual behaviour;
5. Care-giving or epimeletic behaviour;
6. Care-soliciting or et-epimeletic behaviour;
7. Eliminative behaviour;
8. Mutual-mimicking or allelomimetic behaviour;
9. Investigative behaviour.

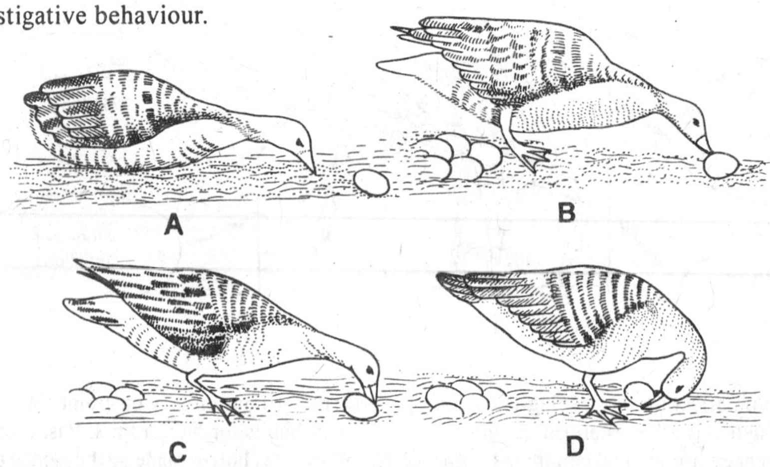

Fig. 1.2. Egg retrieval in greylag goose.

Modern ethologists tend to retain some of the above categories, to modify some and to discard others such as the following discussion may reveal.

1. Communication and releasers. Only by effective **communication** can one individual influence the activity of others. All sensory channels have been used by one species or another in communication—odour, sound, touch and sight. The behaviour pattern an animal first recognizes is a **sign stimulus** or **releaser** such as sound, colour, appropriate structure, odour or movement which is given by another animal or some object. Stimulus releases in the receiver a pattern of **motor response.** Frequently, the animal's responses are **stereotyped.** For example, a nesting greylag goose observing a displaced egg or a reasonable facsimile of one, will try to roll it back into the nest (Fig.1.2).

The signs an animal recognises and the responses it makes depends upon the nature of its receptors and the activities of appropriate effectors by the nervous and endocrine systems. Each animal has receptors attuned to signals of importance to it and a neuronal circuitry (*i.e.*, brain, spinal cord and nerves). Hormones affect the development of sexual and other responses and set the physiological stage.

2. Motivation and drive. It is a matter of common observation that an animal does not respond to a stimulus in the same way every time that stimulus is encountered. For instance, we can consider the case of a South African lion which is moving through the bush and coming upon a herd of wildebeest (African gnu). On some occasions such an encounter results in the lion stalking and perhaps killing one of the wildebeest. On others, the lion walks casually past the wildebeest apparently ignoring the presence of potential food. Thus, something about the lion has changed between one encounter with food and another. But what has changed? Since there is no difference in the stimulus itself, we are left with possibility of some internal change in the animal. That internal change is called **motivation.** Being a descriptive term, motivation defines some kind of internal variable (factor) which influences the relationship or **intervenes** between stimulus and response. Historically, such intervening variables were called **drives.**

A drive or "specific motivation" can be defined as an *urge to perform a particular activity.* Animals exhibit a variety of drives, such as *hunger drive, thirst drive, courtship drive, mating drive, migration drive* and so on. The expression of a drive is completed in the following three steps:

(*i*) Appetitive behaviour. It is the specific search for a releasing situation. It places an animal in a position to achieve a biological goal. For example, a hungry fox becomes restless and begins a characteristic hunting behaviour that is adapted to a situation.

(*ii*) Consumatory behaviour. The biological goal of appetitive behaviour is the consumatory act. Thus for the hungry fox, catching and eating a chicken is the climax of hunger drive. Likewise, in the mating drive, location of a mate by a songbird following a search (appetitive behaviour) releases courtship behaviour (consumatory act). Once the goal is achieved, the goal directed activity ceases.

(*iii*) Quiescent or refractory period. Following the consumatory act, appetitive behaviour does not immediately resume. This is a period of quiescence. The satiated animal does not hunt.

3. Biorhythms and biological clocks. Biorhythms means any cyclic pattern of activity in an organism. Most animals display a day-night rhythm or **circadian rhythm.** Many of their behaviour patterns occur only at certain times of day. The synchroniser ("**entraining agent**" or **zeitgeber**) for this daily periodicity or 24 hour cycle is the alternation of light and dark periods, which is dependent on sun. Some animals also have an annual rhythm (called **circannual**

rhythm). Such a rhythm occurs in migratory birds which fly south in the fall, return in the spring and then breed. The tendencies to migrate, to breed and to perform many other activities fluctuate with the length of day. When days become shorter in the fall, these birds get an urge to migrate.

Quite amazingly, some animals with daily or annual rhythms maintain this periodicity even when isolated for a long time from external synchronisers under constant conditions. Evidently they possess an "**internal biological clock**" or an "**inner calendar**" that regulates the temporal organisation of behaviour independently of external, rhythmically fluctuating stimuli or cues. However, external stimuli may accelerate or decelerate this rhythm. Under natural conditions, they ensure that the cycles correspond to the external year or day. The biological clocks of mammals appear to be based on groups of neurons in the **hypothalamus** that have an inherent oscillating activity.

4. Instinct and learning. The invariable and predictable nature of stereotyped behaviour suggests that it is inherited or innate behaviour or **instinctive behaviour.** Instinctive behaviour depends on the activation of preprogrammed neurone circuits. Appropriate **fixed action patterns (FAP)** are made the first time animal sees necessary release sign. Many kinds of unpracticed stereotyped behaviour appear suddenly in animals and are indistinguishable from similar behaviour performed by older, experienced individual. For example, newly emerged moths and butterflies fly perfectly as soon as their wings are dry. Orb-weaving spiders "know" how to build their webs without practice and without having watched other spiders build there. Newly hatched gull chicks crouch in the nest the first time they hear the alarm call of the parent.

Some instinctive behaviour, such as spider's web spinning and courtship behaviour of grasshoppers and spiders, are **closed** and not modified by experience. Other instincts are **open** and although functional when first performed, are modifiable and improve with practice. New born herring gull chicks peck at the beaks of the parents, which regurgitate partially digested food for the chick. **Jack Hailman** found that a herring gull will initially peck equally at a model of a herring gull adult, or of a laughing gull. After a few days in that nest, however, the chicks will peck more at the herring gull model. Thus, they experienced being fed by herring gulls and have modified their preference accordingly.

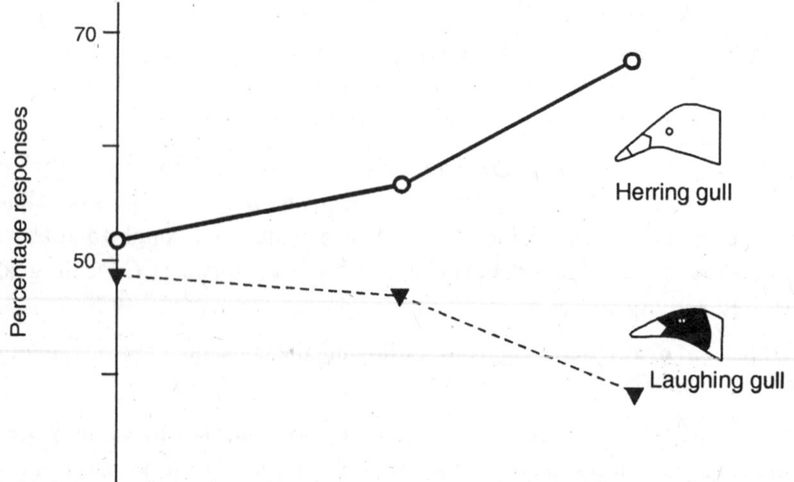

Fig. 1.3. At birth, herring gull chicks will peck equally at models of adult of either herring or laughing gulls, but after seven days they have developed a preference for the model of the adult of their own.

Learning is the modification of behaviour by experience. As a rule, behavioural modifications through learning are adaptive, since they improve the animal's fitness. Learning is of two types — associative and non-associative learning. In **associative learning**, the animal learns that different properties (stimuli) of the environment are associated and modifies its behavioural responses to one of them accordingly. For example, a chimp may learn an association between poking a stick into termite mounds and extracting a stick covered with edible termites. It will then learn to poke sticks in termite mounds when it is hungry. In **non-associative learning** the animal also learns to modify its behaviour but not because of any association of stimuli.

Types of Non-associative learning. Non-associative learning includes various kinds of learning such as imprinting, habituation and the development of bird song.

(i) Imprinting. It is a *restricted learning* which takes place early in life of an animal and is not subsequently modified. Many newly hatched birds (*e.g.*, goslings) **imprint** on the first moving objects they see which is normally their mother and follow only this object as they continue to develop.

(ii) Habituation. It is a simple form of learning found in almost all animal species. It consists of learning not to respond to a certain stimulus as a result of repeated presentation of that stimulus. For example, rustling leaves are worth reaction to (*e.g.*, by hiding) because they sometimes indicate the approach of predator. However, repeated rustling without the appearance of a predators is likely to be caused by the wind. In this case, the animal ceases to respond (*i.e.*, habituates) to the rustling.

Habituation is thus a type of *flexible learning* which continues to be modified as a result of continued interactions with the environment.

Types of associative learning. Associative learning is of three kinds such as classical conditioning, operant conditioning and latent learning.

1. Classical conditioning. It was first proposed by the Russian physiologist and pioneer learning theorist **Ivan Pavlov.** In a typical experiment **Pavlov** would sound a bell when bringing a dog its food. As the dog learned the association between the sound of the bell and being fed, it salivated on hearing the bell in expectation of its meal. Soon Pavlov could make his dog salivate just by sounding the bell, even without bringing its food. He claimed that the salivation of the dog had been conditioned. Thus a stimulus (bell ringing) which does not initially elicit a response (negative salivation) comes to do so (positive salivation) by association with a stimulus (food). The eliciting property of the stimulus (bell ringing) is conditional upon its association with an established stimulus–response relationship (dog's food and salivation).

2. Operant conditioning. It occurs during so called trial-and-error learning. It was first worked out at the turn of the century by the American psychologist, **E. L. Thorndike** and more recently by another American psychologist, **B. F. Skinner**, and many others (*i.e.*, **W. S. Small,** 1900). The experimental apparatuses they used have come to be called **puzzle box, maze** and **Skinner box.** The pigeon in the Skinner box has a choice of two coloured disks to peck at. If it pecks at one it receives a grain of food; if it pecks at the other it does not. Thus the bird learns the association between doing something (*operation*) and being fed (*reinforcer* or *reward*) and accordingly it more or less accurately pecks the correct disc when it is hungry.

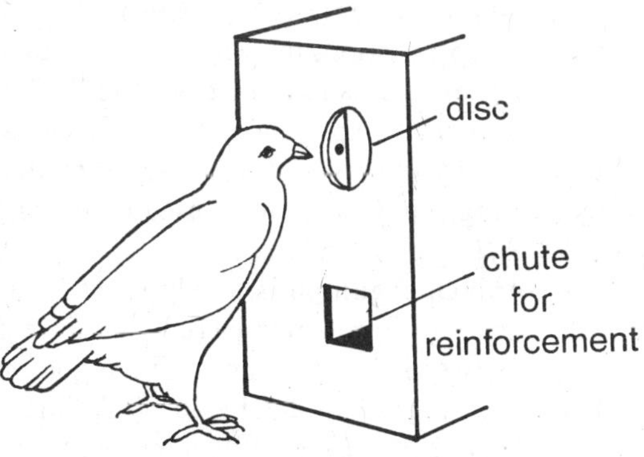

Fig. 1.4. A pigeon in a Skinner box demonstrating instrumental (operant) conditioning.

3. Latent learning. It is a special type of associative learning. It occurs in the absence of any obvious reward and the occurrence of learning may not be immediately apparent. Learning remains unexpressed or **latent** within the animal. An example of latent learning comes from the predatory wasp (*Philanthus triangulum*), which inhabits burrows in sandy soil. The wasp must leave their burrows in order to capture prey some distance away. When a wasp is departing from her nest, she flies in circle possibly learning features of the terrain. **Tinbergen** (1958) provided landmarks (pine cones) around the wasp's nest holes; upon emergence the wasp reconnoitered the area and then flew off. If Tinbergen removed or rearranged some landmaks, the returning waps become disoriented. A series of studies demonstrated that the entire configuration of landmark is used by the returning wasp as a guide to the location of its burrow (see **Drickamer** *et al.*, 2002). This type of learning of digger wasp was termed **spatial learning** by **Alcock** (1993).

4. Insight learning (reasoning). It is most advanced form of learning. Responses produced by insight are those resulting from a rapid appreciation of relationships in which animals solve problems too quickly to have gone through a trial-and-error processes. The animal seems to arrive at a solution by **reasoning.** Reasoning can be defined as the ability to combine spontaneously two or more separate or isolated experiences to form a new experience, which is effective for obtaining a desired end (**Manning,** 1979).

Between 1913 and 1917, **Wolfgang Kohler,** studied insight learning in chimpanzees on Island Tenerife. One of chimpanzee was given two bamboo poles, neither of which was long enough to reach the fruit placed outside the cage. However, the poles could be fitted together to make a longer pole. After many unsuccessful attempts to reach the fruit with one of the short poles, the chimpanzee gave up, started playing with the poles, and accidently joined them together by pushing the narrower pole inside hollow end of the other. The chimpanzee then jumped up and immediately ran to the bars of the cage to retrieve the fruit with the long pole. Kohler interpreted this as an example of insightful behaviour. Insight learning and other cognitive processes belong to Gestalt School of psychology (see **McFarland,** 1985).

Insight learning also includes various forms of **imitation.** Imitation covers an heterogeneous range of behaviours including social **facilitation** (*e.g.*, the performance of a behaviour already within an individual's repertoire as a result of its performance by another individual); **local enhancement** (*i.e.*, the increased tendency to respond to a particular part of the environment as a result of responses to it by another individual, *e.g.*, the spread of milk-bottle opening behaviour through tit population) and **true imitation** (*i.e.*, the copying of otherwise improbable utterances as acts, *e.g.*, vocal mimicry in birds, food washing in primates; **W. H. Thorpe,** 1963).

5. Cognition. Cognition is presently used in two somewhat different ways by animal behaviourists. In one sense, cognition has been defined as a general term for mental function, including perception, thinking and memory (**Immelmann** and **Beer,** 1989), or as the study of the minds of organisms (**Roitblat,** 1987). Currently, cognitive ethology has adopted a perspective that the mental experiences and conscionsness that characterize nonhuman animals can be explained by examining the impact of evolution on cognitive processes as well as ontogeny and underlying neural bases for these processes; this has been termed **ecological approach (Kamil,** 1998).

Animal cognition is nicely illustrated by **Tolman** and **Honzik's** (1930) experiment with rats. Three groups of rats were tested in a maze. (A maze is a network of pathways and blind alleys between a starting point and a goal, used chiefly to study problem solving or **cognitive behaviour** in animals and humans). Animals (rats) in a **conventionally reward group** received food when they successfully completed the maze. No **reward rats** were allowed to wander in the maze at will but were never presented with food in the goal box. A third, **delayed reward** group received no food reward for the first ten days of the experiment but was given food on successful completion of the maze on the eleventh and subsequent days.

Evidently the *no reward* group showed little evidence of learning (the number of errors remained high). The *conventionally reward* group showed a typical learning curve (Fig. 1.5.).

Delayed reward animals, however, learned at a rate comparable with the no reward group until days 10, but then showed a precipitous decrease in error–making. This sudden decrease when food was presented was far more dramatic than would be expected if animals only started to learn about the maze at that point. The data implies that the rats developed some kind of 'map' (called **cognitive map**) of the maze during the ten days in which they were wandering about without reward. This latent knowledge, however, was only put into practice when a reward was attainable.

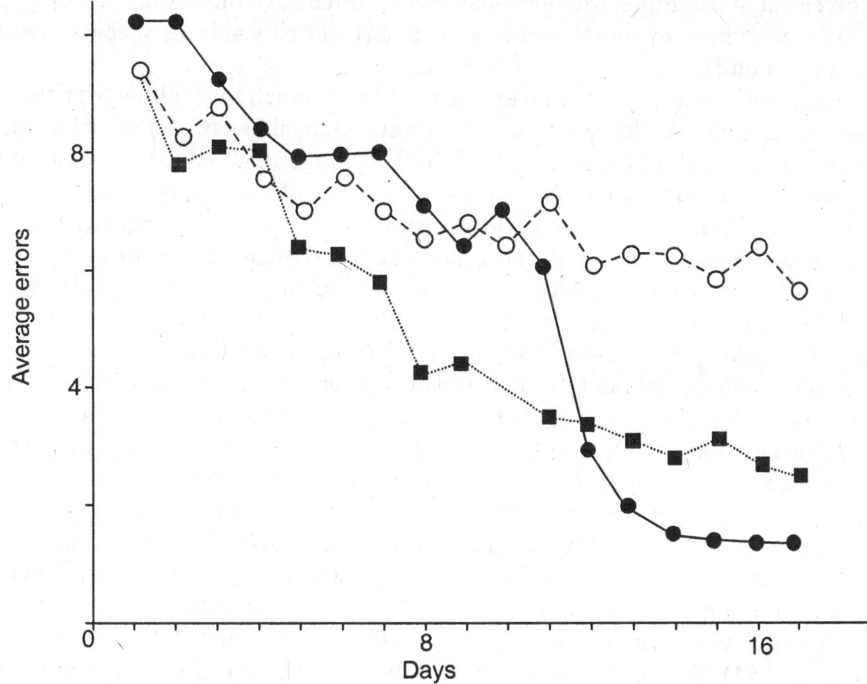

Fig. 1.5. Animal cognition by rats in a maze. *Open circles* show animals trained without food; *squares* indicate animals trained with food; *closed circles* represent animals trained without food until day 11, then provided with food in the goal.

6. Genetics and evolution of behaviour. Behaviour like other traits of organisms, is found to be coded by the genes in the breeding experiments. Given an underlying genetic basis, it is to expect that behaviour is subject to evolutionary change and is adaptive. **Behavioural ecologists** study behaviour in its ecological context to ascertain its evolution and adaptive nature.

7. Agonistic behaviour. Attacking, threat, submissive and fleeing behaviour form a complex, often referred to as **agonistic behaviour.** Agonistic behaviour or aggression represents the way members of the same species resolve conflicts for food, mates and other limited resources. Combats tend to be ritualized and end with the dominance of one individual and submission of another.

8. Reproductive behaviour and parental care. Behaviour increases reproductive success and maximizes an individual's genetic contribution for succeeding generations. Since there is a fixed amount of energy available for gamete production and eggs contain more energy reserves than sperm, female produces fewer eggs than male produces sperm. It is to the female's reproductive advantage to select superior males to fertilize her relatively few eggs, while it is to the male's advantage to fertilize as many eggs as possible. Sexual competition among males for access to females and epigamic (attractive to the opposite sex) selection by females for superior males, have frequently led to the development of large male size, ornaments and weapons to elaborate courtship rituals.

Parental care of the young increases the probability of their survival but reduces the number of young produced. In general, females have a greater investment in the young than the males and are more likely to assume the major care. Intense predation against the young and certain other factors may make it advantageous for the male to help in parental care. Whether the mating or breeding system is **polygyny** (*i.e.*, habitual mating of one male with several females), **monogamy** (*i.e.*, condition of marriage (mating) to one female or male) or **polyandry** (*i.e.*, mating of one female with several males), it depends to a large extent on the amount of male involvement in parenting. Monogamous species often have biparental care (*e.g.*, pigeon), polygynous species have maternal care (*e.g.*, baboon) and polyandrous species have paternal care (*e.g.*, jacana bird).

9. Social behaviour. In a broad sense, any kind of interaction resulting from the response of one animal to another of the same species represents social behaviour. A **social group** is an aggregation of individuals of the same species that come together because of mutual advantages such as a reduction in predation or an increase in foraging efficiency. Disadvantages of social group formation are increased competition within the group and greater risk of disease. Advantages must outweigh before the social groups form. Social groups range in complexity from simple aggregations of unrelated individuals (*e.g.*,school of fishes, large wildebeests and zebra herds, etc., these groups are formed largely by non-breeding individuals based on a mutual attraction; see **Mathur,** 2004) to complex societies of closely related individuals in which there is considerable division of labour (*e.g.*, honey bees, ant, termites, etc.). Social behaviour exhibits the phenomena of altruism, territorial behaviour and dominance hierarchy.

(i) Altruism. It is a selfless or self-sacrificing behaviour. It means the transfer of some benefits from the altruist to the recipient, at the cost of the altruist. For example, parental care is a type of altruistic behaviour where parents sacrifice their own interests to preserve their offspring. In most advanced form of altruism which is observed in social insects such as **termites** (Isoptera) and **ants, bees** and **wasps** (Hymenoptera), sterile "worker" castes work to increase the reproduction of another individual ("queen"), they do not reproduce themselves. Thus, an animal's fitness depends upon not only its own fitness but on fitness of kin or relatives (**B. Hamilton,** 1964). The term **fitness** is akin to *survival value* and is a measure of the ability of genetic material (gene) to perpetuate itself in the course of evolution. Such type of natural selection which takes account of other relatives as well as immediate descendants is called **kin selection.**

(ii) Territorial behaviour and dominance hierarchy. In territorial behaviour, an animal (often male) defends a specific area, the **territory** against others (often just against males) of the same species. The territory may change shape, position or be defended only at certain times of the day or year. Thus a territorial animal has almost exclusive use of an area for a period of time, *e.g.*, many fishes, lizards, birds and carnivores (dogs and cats). In the **social** or **dominance hierarchy** form of social organisation there is a rank order of individuals with a most dominant, or **alpha,** individual at the top of the rank order and other individuals increasingly subordinate to those above them in rank order. The hierarchy usually develops after much real or ritualized aggression when the animals first come together. The frequency of aggression decreases after a few days and the dominance order is maintained by occasional aggressive behaviours. In fish and reptiles the order is usually related to size and colour, and thus the alpha male is likely to be the largest and most brightly coloured individual. In birds such as chickens, dominance hierarchy is called "**peck order**". The upper levels of a dominance hierarchy are usually stable and clearly defined; the lower levels are often unstable and poorly defined.

Territorial behaviour is commonly found in animals when the habitat is fairly uniform, productive and stable and the population is small in relation to resources. A hierarchy often develops when food or other resources become limited or when animals are crowded. The transition from territoriality to a hierarchical social organisation under adverse conditions allows more animals to live in a given area. It also allows animals to concentrate around a localized place containing highly concentrated resource (such as food, water, basking or sleeping site).

1.3. HISTORY OF ANIMAL BEHAVIOUR

The modern study of animal behaviour has its roots in three different lines of scientific thoughts: **psychological, physiological** and **zoological.** These three traditions have given rise to **comparative** and biological psychology, to the evolutionary analysis of behaviour and to the modern field of ethology.

Animal psychology has a philosophical heritage. To understand the meaning of concepts of soul, mind and consciousness (Box 1.1) and thinking one has to go through the views of Greek philosophers:

Box 1.1.

What is Consciousness?

Consciousness is divided by psychologists into the following three major functions or processes:

1. Conation. It is the act of "willing". As such it is intimately tied up with such concepts as "mind", "soul" and "ego". It includes voluntary reaction. Now conation is recognised as "ego strength", "motivation", and "decision making".

2. Affect. It is a technical term for "feelings" or "emotions".

3. Cognition. (Latin = to know). It is the act or processes of knowing including both awareness and judgement. Roughly speaking, cognitive activities are those internal processes that involve **information processing, thinking, reasoning, remembering, perceiving** and so forth. It does not include emotions or feelings.

1. Concept of Soul in Greek Philosophy

The dichotomous view of humans and animals can be traced to Greek and other European philosophers who proposed that there had been two kinds of creation: humans and gods were the products of rational creation, while irrational brutes were a separate category of living creatures. They were believed to differ with respect to the number and types of souls they possessed. **Plato,** more than 2000 years ago, said that the human spirit is both rational and irrational. **Aristotle** and **John Locke** taught that human beings are born devoid of ideas — they are "blank slates" at birth. Aristotle also wrote that man is motivated by desire to avoid pain and seek pleasure. He while accepting the view that humans differed from non-humans in regard to their souls, attempted to place all species on a continuous scale, or *scala natura.* Humans were placed at the top of scale of living organisms.

In 13th century, human being was officialy removed from his place on the natural scale of animals by **Albertus Magnum (Big Albert). St. Thomas Aquinas,** a student of Albertus, supported the separation of humans and other animals based on the idea that human being has a "soul" divinely placed in the embryo, sometime before birth. This marked the beginning of a religious explanation of **instinct.** The reasoning was that there is a life after death and the part of human being that continues to live is called the **soul.** The precise conditions, good or bad under which the soul continues after death are determined by how person behaved during life on the earth. In order to make the "correct" decisions *reason* is necessary. Animals do not share in the hereafter and do not need reason. Instead, they are guided by blind instinct.

This philosophical notion of instinct was rejected by many people. For example, **Erasmus Darwin,** Charles Darwin's grandfather, believed that behaviour has no innate quality at all but that it is entirely learned.

2. Darwin's Contribution in Growth of Ethology

Charles Darwin (1809-1882) influenced the development of ethology in three main ways. First and foremost his theory of **natural selection** (forwarded in his book *Origin of Species,*

1859) set the stage for consideration of animal behaviour in evolutionary terms, a key aspect of modern ethology. Second, Darwin's views on instinct can be regarded as direct fore-runner of those of founders of classical ethology (*i.e.,* **Lorenz** and **Tinbergen**). Third, Darwin's observations on behaviour were important, especially those that stemmed from his belief in the evolutionary continuity of human being and other animals. For instance, in his book *The Descent of Man and Selection in Relation to Sex* (1871). Darwin wrote "we have seen that the senses and intuitions and various emotions and faculties such as love, memory, attention, curiosity, imitation, reason, etc., of which human being boasts may be found in an incipient or even sometimes a well developed condition in the lower animals ". In his another book *The Expression of the Emotions in Man and the Animals* (1872) **Darwin** elaborates on this theme : "with mankind some expressions, such as bristling of the hair (Box 1.2) under the influence of extreme terror, or the uncovering of the teeth under the furious rage can hardly be understood, except on the belief that man once existed in a much lower and animal condition".

Box 1.2.

Erection of body hair ("goose flesh") that sometimes occur when we are in a state of fear seem inexplainable until we think of it in evolutionary terms as the vestige of a behaviour pattern that was once serviceable. If our remote ancestors had more body hair than we now possess, it is possible to imagine that the erection of these hair might be sufficient to create the image of creature large to scare away any potential predator.

3. Growth of Ethology in 19th Century

The American behaviour psychologist **Jacques Loeb** (1859-1924), was an extreme advocate of the law of parsimony and attempted to account for nearly all behaviour in terms of **tropisms** which were viewed as "forced movements". According to him animal behaviour was composed of a matrix of reflexes, a simple example of which are the spinal reflexes. This left no room for spontaneous behaviour — actions not precipitated by an immediate external stimulus. Thus **Loeb** was of the view that animal behaviour could be understood as resulting from immediate physical and chemical effects of stimuli upon protoplasm.

According to **Loeb's tropism theory** the primary feature in the directed movements of lower organisms is the source of stimulation. This theory inspired others to explore further and many important investigations during this period such as **A. Alverdes, S. J. Holmes, A. Kuhn** and **S. O. Mast** (1911) made the orientation mechanisms of invertebrates a major target of their research.

Herbert Spencer Jennings (1868-1947) was an early behaviourist who stressed the descriptive study of the full range of behaviours of species under investigation. His book *Behaviour of Lower Organisms,* written in 1904 dealt mainly with protozoans.

Douglas A. Spalding (1840-1877) was a pioneer animal behaviourist who is best known for a series of experimental investigations of the development of behaviour in young chicks. For conducting the factors controlling behavioural ontogeny, he conducted deprivation experiments in which animals were reared in the absence of certain normal portion of their environment. **Spalding** (1873) also conducted early research on the phenomenon of imprinting in domestic chick.

Charles O. Whiteman (1842-1910) was an American Zoologist who studied the behaviour of pigeons and doves and is regarded as *founding father of ethology.* He declared in 1898 "*Instincts and organs are to be studied from the common viewpoint of phyletic descent*". Thus, he proposed that behaviour, like structure, can be studied from an evolutionary perspective, *i.e.,* one can study the evolution of behaviour just as one studies the evolution of organs. **Whiteman** (1919) also studied **fixed-action patterns (FAP)** in pigeons.

Ivan Pavlov (1849-1936), Russian digestive physiologist and Nobel laureate (1904) demonstrated the **conditional reflex.** A hungry dog salivates on seeing or smelling meat. **Pavlov** rang a bell each time the dog was fed. After a few repetitions, the ringing of the bell alone would cause the dog to salivate. Thus the dog had learned that the sound of the bell usually meant food. Any stimuli perceivable by the dog could be used in the place of the bell.

Edward L. Thorndike (1874-1949) Emphasized the need for systematic, replicable experiments in comparative animal psychology. **Thorndike** used the **puzzle box** (Fig. 1.6) to perform a series of task-learning experiments, using cats as test subjects. A cat was placed in the box, which was fastened shut; by manipulating a shuttle lever, the cat could open the door.

Fig. 1.6. Thorndike puzzle box. A cat inside the cage can already see the reward, in this case the fish, placed outside. In order to obtain the reward, the cat must learn to manipulate a shuttle-lever system that raises the door of the cage (after Drickamer, *et al.*, 2002).

and obtain a reward placed outside the box. From these experiments, **Thorndike,** concluded that much of animal learning take place by trial and error and that rewards are a critical component or learning process.

5. Growth of Ethology in 20th Century

Oskar Heinroth (1871-1954), a German zoologist, wrote his major papers in 1910 and 1911 on the ethology of ducks and geese. He first of all used the term **imprinting** (*pragung*). Like **Whiteman,** he drew conclusion that certain patterns of actions are just as fixed and characteristic of different species of birds as their plumage and form.

In 1918, **J. S. Szymanski** demonstrated the existence of **biological clocks** in animals. He showed that animals had some means of measuring time independently of such physical factors as light and temperature, for he discovered that 24-hour activity patterns were synchronized with the day-night cycle when animals were kept in constant darkness and temperature.

T. Schjelderup-Ebbe (1922) reported **social dominance-subdominance hierarchies** among birds. He observed that there existed certain types of social hierarchies among birds in which higher ranking individuals could peck those of lower rank without being pecked in return.

In 1925, **W. Rowan** proposed **photoperiodism hypothesis** of bird migration. In 1927, **G. E. Coghill** studied origin and growth of innate behaviour patterns of salamanders (amphibians) by following the sequence of coordinated movements and nervous connections through all stages of embryonic development.

Wallace Craig (1876-1954) an American student of **Whiteman,** developed theoretical models of the control of animal behaviour. He termed the stereotyped, species typical behavioural patterns studied by Whiteman and Heinroth as "**consummatory acts**". **Craig** also noted that not all behaviour is so rigidly fixed and invariant. He differentiated "appetitive behaviour" from consummatory acts and noted that appetitive behaviour is variable, agitated behaviour that is terminated by the occurrence of stimulus releasing the consummatory act. According to Craig, thus, a food-deprived animal engages in variable "searching" behaviour, which is terminated by the appearance of food and the elicitation of relatively stereotyped eating patterns.

Jacob von Vexkull, in 1934, emphasized the fact that only very few stimuli, out of the entire array that continuously bombard the animal, are capable of eliciting responses by the organism. He termed the unique perceptual world of each animal as its **umwelt.** Because of its unique sensory and neural structure, each animal is particularly sensitive to some particular stimuli (now are called **sign stimuli**) and insensitive to others. Approach for feeding in the tick, for example is elicited by the stimulus of butyric acid. Any object emitting butyric acid elicits approach behaviour from the tick, while items with no such odour will be ineffective.

For the method of stimulus filtering, **von Vexkull** proposed the **searching image** concept. We are all aware of the perceptual phenomenon of suddenly seeing something that we previously overlooked. For example, when we look at a photograph of camouflaged insects, we may not see at first, but then we suddenly may see one. Thereafter, the insects seem easy to pick out we have developed a searching image for the insect.

Fig. 1.7. Karl von Frisch.

Classical ethology reached its peak in the works of **Konrad Lorenz** (1903-1989) and **Niko Tinbergen** (1907-1988), two European ethologists who together with **Karl von Frisch** (1886-1983) shared the 1973 Nobel prize for Physiology or Medicine. The citation stated that they were the chief architects of the new science of ethology.

Lorenz, in 1937 and 1950, refocussed the attention of zoologists on the invariant component of behaviour — the **instinct** or **fixed action pattern (FAP).** Lorenz's method assumes that one can draw valid conclusions about the significance of an animal's behaviour only if its actions are observed while it is undisturbed and in its natural surroundings. For it is under these conditions that the behaviour evolved and presently has adaptive value in survival.

Fig. 1.8. Konrad Lorenz.

This viewpoint suggested a new method of studying the relationships and evolution of different species. Lorenz's students became known as **ethologists.** One of their methods was to assemble **ethograms,** catalogues of various behaviour patterns of the species belonging to certain **taxa** or groups of organisms. These ethograms were built up through field observations and then compared with each other in manner similar to that employed by museum workers and comparative anatomists in comparing the structure and form of different groups. Thus, an organism's behaviour became just as valuable in establishing its species relationships as the shape of its body or its colour patterns. For example, the bird species grouped and classified as pigeons vary so greatly in form and other respects that no one common structural characteristic could be found to label a bird a pigeon. Lorenz showed that the group can be distinguished from other birds by the way they drink water. *Pigeons all make sucking movements while other birds lift their heads and swallow.*

In addition to provide a new way of studying the relationships of organisms, the work of **Lorenz** identified innate components that are useful for the study of evolution even in the process of learning, by explaining the phenomenon of **imprinting** (or restrictive learning), in which young animals fixate on certain other individuals or objects during a brief 'critical period' in their development.

Niko Tinbergen was experimental behaviour biologist who found ways (1942, 1952) to analyze the behaviour patterns observed in free-ranging animals. He and his coworker devised numerous experiments to isolate and to identify the key stimuli, called **social releasers,** used by animals to communicate with members of their own species. A great deal of complex behaviour in fish (stickleback, *Gasterosteus aculeatus*) and birds (herring gulls, *Larus argentatus*) was found to be "released" by elementary *stimuli* that could be successfully imitated with crude models held in the experimenter's hands.

Fig. 1.9. Niko Tinbergen.

Karl von Frisch (1886-1982) had conducted research on animal sensory process and made important contributions to the study of bee behaviour and communication.

Later on, certain other important discoveries took place in ethology. In 1952, **G. Kramer** studied *orientation* of birds (pigeons and starlings) to positional changes of sun. In 1957, **F. Sauer** discovered *celestial navigation* by migratory birds (*e.g.*, old world warblers). In 1959, **A. F. J. Butenandt** chemically analyzed the first pheromone — the sex attractant substance of silk moth (*Bombyx mori*) and named it **bombykol.**

Daniel Lehrman (1955, 1964) studied *sexual behaviour* (*i.e.*, mating and nesting behaviours) in the ring dove and observed that sexual behaviour of ring dove is governed by interactions of outside or environmental stimuli, the hormones and the behaviour of each mate. In 1964, **W. D. Hamilton** studied social behaviour in social insects and proposed the concept of **kin selection** and inclusive fitness. In 1968, **J. Van Lawick Goodall** performed a long term complete field study of social behaviour of free living chimpanzees in the Gombe Stream Reserve. This work provided an impetus and stimuli for a large number of similar studies on other primates.

Jay Rosenblatt (1958-1979) of Rutgers University proposed **hormone "triggers" theory** to explain onset of maternal behaviour in rats, cats and other animals. According to him, the hormone levels of pregnant mammalian females changes dramatically just before labour starts. The amount of *progesterone* decreases sharply while the level of estrogen increases. **Rosenblatt** stated that the onset of maternal behaviour before delivery is based on this rise in estrogen. He called this effect a **hormonal trigger** that significantly changes both **motives** and **behaviour.**

Moltz and **Kilpatric** (1978) found that rat pups that are nearing weaning are greatly attracted to eating maternal pellets (**coprophagy**). Through ingestion of mother's faecal pellets they ingest **deoxycholic acid** (a maternal pheromone) which protects them against a highly lethal disease of the gastrointestinal tract, called **necrotizing enterocolitis.**

1.4. PROFILES OF SOME PIONEER BEHAVIOUR BIOLOGISTS

1. Ivan P. Pavlov

Ivan Petrovich Pavlov(1849-1936) was a son of a priest. He was born in the Russian town Ryszan, where he attended the religious school and seminary. Pavlov entered St. Petersburg

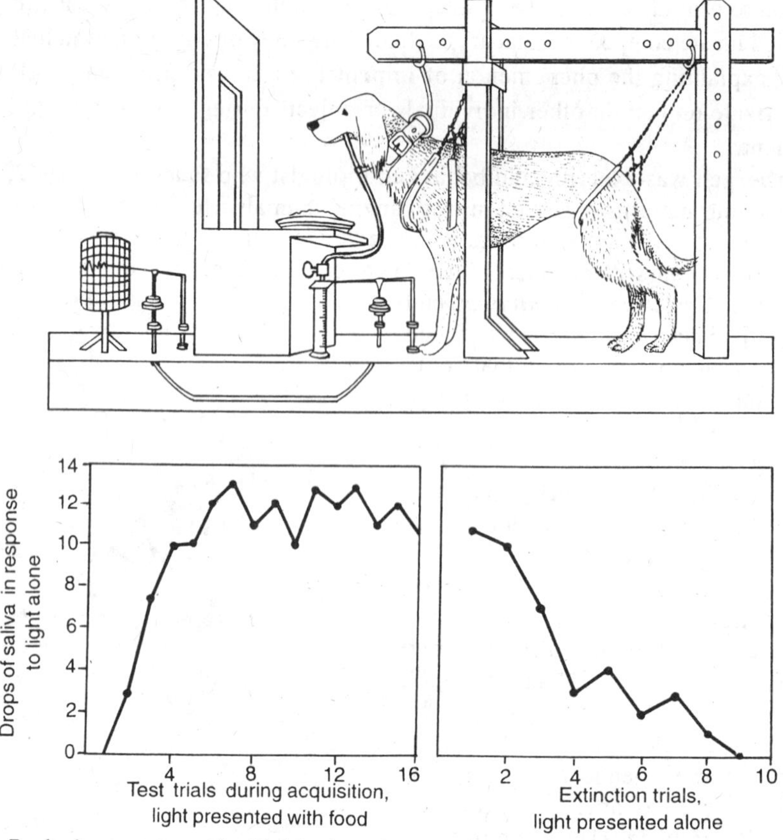

Fig. 1.10. Pavlov's apparatus with which he demonstrated classical conditioning in the dog. Note the graphs at the bottom.

University in 1870 and graduated in natural science in 1875. After obtaining his doctorate at the Military Medical Academy in 1883, he studied in Germany. He became Professor of Pharmacology at the Military Medical Academy in 1890, and Professor of Physiology in 1895. This great Russian physiologist received the Nobel Prize for Medicine or Physiology in 1904 for his work on the physiology of digestion.

Pavlov's early work was on the physiology of circulation, especially the mechanisms that regulate blood pressure. He discovered that the vagus nerve controls blood pressure, and investigated nervous control of the rhythm and depth of the heartbeat. In 1879 Pavlov began work on the physiology of digestion, which culminated in his book the *The Work of the Digestive Glands*, published in 1879. He investigated the mechanism involved in the secretion of various digestive glands and came to the conclusion that they were controlled exclusively by nervous mechanisms.

During the course of his work on the physiology of digestion, **Pavlov** noticed that salivation could be induced in his experimental dog by the sight of food, or other stimuli that normally preceded feeding. This led him to the discovery of **conditioned reflex** or **conditional reflex,** now regarded as fundamental aspect of learning (Fig. 1.10). From 1902 until his death in 1936, **Pavlov** concentrated his researches on the phenomenon of conditioning. He suggested that the cells of central nervous system changed structurally and chemically during conditioning, Pavlov's major work, translated into English as *Conditional reflexes: an investigation of the logical activity of the cerebral cortex* (1927), had a major impact on the development of psychology.

2. Konrad Lorenz

Konrad Zacharius Lorenz (1903-1989) was born in Austria. He studied medicine in Vienna and also studied comparative anatomy, philosophy and psychology. He became demonstrator and then lecturer in comparative anatomy and animal psychology. At the same time he studied animal behaviour at his family home in Altenberg. In 1940 he was appointed Professor of Philosophy at the University of Konigsberg, but in 1943 he was drafted into the army medical service. In 1944 he was taken prisoner of war by the Russians. He was released in 1948, became attached to the University of Munster and then moved to Seewiesen with the founding of *Max Planck Institute for Behavioural Physiology,* where he remained until his retirement in 1973.

Lorenz is regarded as **founder of ethology.** He, like **Tinbergen,** emphasized the importance of straightforward observation of animal behaviour under natural conditions. Lorenz's approach was somewhat more philosophical and his numerous theories became quite influential. One of Lorenz's main contributions to ethology is his work on the development of social relationships, especially the phenomenon of **imprinting** (1935). He confirmed **Heinroth's** observations on imprinting in goslings and also studied imprinting in mallard ducklings, pigeons, jackdews and many other birds. In 1935, Lorenz recognised the importance of **sign stimuli.** He (1937) maintained that much of animal behaviour was made up of a number of **instincts** or **fixed action patterns** or **FAP,** which are characteristics of the species and largely genetically determined. In 1950, Lorenz studied IRM or **innate releasing mechainsms** by which sign stimuli were recognised. To understand pattern of evolutionary development, Lorenz (1958) studied courtship and other kinds of social behaviour in twenty species of ducks and geese. Some well known publications of Lorenz are **King Solomon's Ring** (1952); **Evolution and Modification of Behaviour** (1966); and **On Aggression** (1966). Lorenz's classic studies on grey-lag goose first came to the attention of English speaking scientists during 1950's, nearly after 20 years. Lorenz was awarded for his work the 1973 Nobel prize for Physiology or Medicine along with Frisch and Tinbergen.

3. Nikolaas Tinbergen

Tinbergen (1907-1988) was born in The Hague, the Netherlands, and studied biology at the University of Leiden. In 1930 he went on an expedition to Greenland, and in 1938 he visited Lorenz at Alterberg. During World War II Tinbergen was interned in a hostage camp in the Netherlands, afterwards he became Professor of Zoology at the University of Leiden. In 1949, he was invited to become a Lecturer in Zoology at the University of Oxford, where he founded the **Animal Behaviour Research Group.** He retired in 1974.

Tinbergen was an experimentalist who found ways to analyze the behaviour patterns observed in free ranging animals. He and his coworkers (1942, 1952) devised numerous experiments to isolate and to identify the key stimuli, called **social releasers** (= sign stimuli), used by animals to communicate with the members of their own species. A great deal of complex behaviour in fish (*e.g.,* stickleback) and birds (*e.g.,* herring gull) was found to be "released" by elementary stimuli that could be successfully imitated with crude models held in the experimenters hands. His book '*The Study of Instinct*' (1951) integrating ethology and relating it to the study of the nervous system, has been possibly the single most influential synthesis of animal behaviour published in last century.

Tinbergen's Four Questions. Tinbergen was a gifted field biologist who carried out many elegant experiments in the natural environment. A significant feature of the work of **Lorenz** and **Tinbergen** was their attempt to combine evolutionary or functional explanations of behaviour with casual or mechanistic explanations. For example, in his 1963 paper *On aims and methods in ethology*, Tinbergen posed four questions he thought necessary for a full account of any aspect of animal behaviour. The ethologists, he believed, should aim to answer the following four questions of the causation, development, survival value and evolution of any behaviour pattern under study.

1. What makes behaviour happen at any given movement? How does its machinery work? **(Causation)**
2. How does the behaviour machinery develop as the individual grows up? **(Development)**
3. Have the behaviour systems of each species evolved until they became what they are now? **(Evolution)**
4. In what ways does this phenomenon (*i.e,* behaviour) influence the survival, the success of the animal? **(Function or Survival)**

Psychologist **Thomas McGill** has noted that the first six letters of the alphabet can be used to remember these questions: Animal Behaviour : Causation, Development, Evolution, Function (See **Drickamer** *et al.*, 2002).

4. Karl von Frisch

Karl von Frisch (1886-1982) was born in Vienna and educated at the University of Munich and Vienna. He became professor of Zoology at the University of **Rostok** in 1921, at Breslam in 1923 and at Munich in 1925. As a child, von Frisch was interested in natural history, and while still at school he published some of his observations, including experiments on the light sensitivity of sea anemones. Most of his subsequent research on animal behaviour was concerned with how animals obtain information about their environment. For most of his life Von Frisch spent winters in his laboratory, studying fishes and summers at his family home at Brunnwinkl, studying honeybees. He discovered that fishes are capable of colour vision and of discriminating underwater sound waves. Both of these discoveries were contrary to the prevailing scientific opinion and thus aroused opposition. Von Frisch also discovered that when the skin of a minnow (= a small European cyprinoid fish, *Phoxinus phoxinus*) is damaged, a pheromone is released that causes other minnows to flee from the area. In these studies, success of von Frisch was based upon careful behavioural observation and a profound understanding of biological function. This was also exhibited by his work on honeybees.

In 1921 Karl von Frisch started his experiments on honeybees. In his celebrated studies of honeybees, stretching unbroken over sixty years, von Frisch used elementary training techniques to demonstrate which environmental stimuli the bees can detect. Methodically and painstakingly, von Frisch and his students plotted the sensory world of this one species of insect. They proved, among other things that bees have **colour vision** and have the capacity to perceive **polarized light** and **magnetic fields.** In 1948, von Frisch discovered the communication mechanisms of honeybees. After many years of patient work, he deciphered the meaning of **bee "dances"** behaviour patterns of individual bees returning to the hive. The dances communicate information to the other bees regarding location of food and water, location and suitability of sites for a new hive and other information. With **Tinbergen** and **Lorenz**, von Frisch was awarded the 1973 Nobel Prize for Physiology or Medicine for his important research.

5. Edward C. Tolman

E. C. Tolman (1886-1959) was an American psychologist working at University of California, Berkeley. He is considered as **Father of modern congnitive approach to animal**

behaviour. His major publication was the book "*Purposive Behaviour in Animals and Men*" (1932), though he also published many papers. Unlike other congnitive theorist of his time, such as **G. Romanes** and **W. Kohler**, Tolman's view was not mentalistic. His system was **purposive** but not anthropomorphic. He believed that animals behaved in a purposive manner, but did not assume they had mental images of their goals.

Tolman regarded himself as a behaviourist, but he advocated a molar rather than a molecular behaviourism. In the **molar behaviourism**, acts have distinctive properties of their own, which can be described independently of the particular physiological processes responsible for behaviour. **Molecular behaviourism,** on the other hand, is reductionist in that it seeks to account for behaviour in terms of underlying physics and physiology. About Pavlovian conditioning he believed that the animal learns about the **reinforcement** (A stimulus capable of increasing the rate of performance of conditioned response in instrumental or operant conditioning), and not simply because of the reinforcer. He attacked the stimulus-response theory prevalent in his time and viewed that the conditional stimulus is a sign that some other event is to follow.

Tolman also pioneered the idea of **cognitive maps.** His view was that animals acquire items of knowledge or **cognitions** which are organised so they can be utilised when needed. Thus while **Thorndike** and **Pavlov** believed that animals learned each section of a maze by trial and error, Tolman thought that animals acquire a "mental map" of the maze and can "think" their way through to the goal — rather than respond reflexively (and unconsciously) to the specific stimuli in each section. Now it is believed that animals such as the rat can do a little of both (**McConnell,** 1986).

6. B. F. Skinner

Burrhus Frederic Skinner (1904-1990) of Harward University is widely known as the world's leading behaviourist and his classical book "*The Behaviour of Organisms*" (1938) introduced a distinction between operant behaviour and respondent behaviour. (Note: Skinner preferred the term operant to instrumental). He defined **operant behaviour** as spontaneous action without any obvious stimulus. Thus classical conditioning is based on respondent or **elicited behaviour.** Food is placed in a dog's mouth and this elicits a salivary response. On the other hand, instrumental or operant conditioning is based on **emitted behaviour,** behaviour that seems to spring from an unknown source within the organism. According to **Skinner,** the emitted behaviour is just as conditionable and potentially just as predictable as elicited or respondent behaviour. He demonstrated this view through the use of the **instrumental-conditioning apparatus** or **Skinner box** (as it is more popularly known). Skinner's behaviourist philosophy brought about a revolution in experimental techniques that has persisted to this day.

Thus in place of the trial and error procedure characteristic of classical conditioning and experiments using puzzle boxes or mazes, Skinner devised the free-operant procedure in which the animal (pigeon or rat or human child) is allowed to indulge freely in various activities while the experimenter attempts to manipulate the consequences. Operant conditioning consists essentially of training animal to perform a task to obtain a reward. A rat may be required to press a bar, or pigeon to peck an illuminated disk, called a "key". The typical method of training is called **shaping.**

Skinner's best known books are *Science and Human Behaviour* (1953) and *Beyond Freedom and Dignity* (1971).

7. Edward L. Thorndike

A pivotal publication in the development of animal psychology was **Edward L. Thorndike's** report (1898) of **trial-and-error learning** in cats and dogs escaping from puzzle boxes (Now a days trial-and-error learning is called **instrumental learning** — the correct response being "instrumental" in providing access to reward). **Thorndike** stated that trial and error learning was influenced by two laws:

1. His **law of exercise** (1911) states that the more frequently an animal repeats an **stimulus-response** or **S-R-bond,** the stronger it becomes.
2. His **law of effect** (1913) states that stimulus-response (S-R) bonds are strengthened if the effect of the response is rewarding or satisfying.

Thorndike believed that punishment 'broke' S-R bonds or suppresses responses temporarily. He concluded that principles governing learning were essentially the same for all species. So comparative research was deemed unnecessary, expensive and inconvenient.

1.5. SOME RIDDLES OF ANIMAL BEHAVIOUR

1. Gestalt Concept

Gestalt is a mental or perceptual pattern. It is a German word meaning "*figure*", "*form*", "*pattern*" or "*configuration*". The ability to see something "as a whole" — rather than just see its parts — is the ability to "form a gestalt". Thus gestalt means the tendency to see things as "whole rather than a jumbled bits and pieces". The basic viewpoint of gestalt psychology is **holism** which is a doctrine that organised wholes cannot be explained in terms of parts.

2. Nature and Nurture Controversy

It is one of the most persistent and difficult arguments in the study of behaviour. The term **nature** refers to a person's or animal's inborn endowment, instinct or heredity. The term **nurture** refers to the way in which a person or animal is raised, the quality of his or its environment during the developmental years.

Ethologists questioned: Is an animal endowed with nerve connections enacted by its genetic makeup and accordance with ethology, does this determine the way it behaves? Or, is its nervous system first a *tabula rasa*—a blank page—on which, in accordance with behaviouristic psychology, each individual experience registers the nerve connections that favour the animals's survival?

Thus the question centres around whether we can treat behavioural pattern inherited variable only within limits and subject to natural selection? If heritable, is it "species-specific" (characteristic of a single species)? Can it be used to describe species and relationships between species? As we know there are some behavioural patterns which persist in populations, so it was concluded that behaviour is transmitted via genes through the generations. Therefore to most ecologists the species-typical behaviour was a prime concern. They therefore tended to concentrate on patterns which were highly stereotyped and similar in form throughout a species and they are often referred to as **innate** or **instinctive.** The assumption here was that nature was all important and nurture was of little consequence. Indeed **Konard Lorenz** once remarked that the developmental origin (ontogeny) of behaviour was a subject of more interest to embryologist than to ethologists.

Once the nature-nurture controversy was a bitter one, but it was also fruitful, for each side had much to gain from the other, and its resolution brought them closer. Psychologists came to recognise that evolution had led animal species to be different from each other and placed constraints on what each could learn. For their part, ethologists came to reject the idea that any behaviour was fixed and inflexible and to realise that the acts they studied no matter how stereotyped, may have been influenced by learning and by other environmental influences. They also came to appreciate the merits of a carefully controlled experimental approach. So today many ethologists work in the laboratory and some of them even use the sorts of equipment, developed by psychologists for the study of learning, adapted to shed light on ethological questions. In the battle over nature or nurture, the result has been a compromise: both sides have gained from the realisation that neither nature nor nurture can be ignored in the development of any behaviour pattern.

1.6. IMPORTANCE OF BEHAVIOUR FOR THE ANIMAL ITSELF

All animals behave for their benefit and survival. For example, the male stickleback fish habitually builds a tubular nest of grass and weeds on the bed of the river. After inducing one or more females to spawn in it, the male fertilized the eggs and guards them by swimming around the nest and intermittently would be seen **fanning the eggs** with the help of pectoral fins. By doing this, it directs water towards the fertilized eggs to *ensure fresh supply of adequate dissolved oxygen*. As the eggs grow they require more oxgen, so the male increases its intensity of fanning and forms a number of extra openings in the roof of nest to make ventilation more efficient. If the male is removed the eggs will die. This is simple and direct example to indicate how behaviour is necessary for survival.

1.7. SIGNIFICANCE OF ANIMAL BEHAVIOUR

The study of behaviour of animals is the final objective of all other branches of biology. For instance, if one looks closely at any one of the elemental actions of the animals such as to avoid being eaten, to gather food and to reproduce, he/she finds that it depends on an extraordinarily complex and beautifully synchronised mechanism of nerve cells, glands and muscles, plus a supporting skeleton. Thus the conduction velocity in a nerve fibre saves an earthworm from a robin; the speed of a muscle contraction determines whether a predator will feed or go hungry; a single molecule of pheromone falling on olfactory receptor cell determines whether two silk moths will mate. These structures are only parts of complex mechanisms that must combine properly through the inherited constitution of each individual. If they do not, animals would not be here to behave. Therefore the students of behaviour have to be familiar with the structure (anatomy) and function (physiology) of these elegant mechanisms of the animals.

All animals have a variety of complex relationships with members of their own species, with members of other species, and with the physical environment. The survival of a species depends on its individual member's ability to obtain food and shelter, to find mates and reproduce, and protect themselves from parasites and predators. In many ways, adaptive behaviour ensures survival, and survival ensures evolutionary success. By studying the behaviour of animals, we learn much about the relationships between them and their environments, the physiological processes that determine their behaviour, and some of the reasons for their abundance and distribution.

Another purpose of the study of animal behaviour is to apply behavioural information to the care and management of both wild as well as domestic animals. So most intelligent, humane and cost-efficient approach to the breeding, care and use of domestic animals must be based on an understanding of their natural behavioural tendencies. Some animals are maintained as pets primarily because of their behaviour; this is true of cats, dogs and horses. The more we know about their behaviours, the happier we will be with them as pets. Likewise, with the livestock that are kept for economic reasons, we can use behaviour to our advantage or disadvantage. Studies have shown for example that behavioural stress may take an economic toll. Animals that display inadequate sexual performance or abnormal maternal care can also cause economic losses.

The study of animal behaviour also helps in understanding human nature (behaviour) and has added value in zoo keeping, economic zoology (sericulture), apiary (honey production), agriculture and biological control of pests. A knowledge of the migratory routes of an endangered whale or shorebird may enable conservationist to design adequate reserves to save the animal from extinction. Even if there were absolutely no practical benefits to be gained from learning about animal behaviour, the subject would still be worth exploring because it is so fascinating. Who would have guessed that preying mantis can detect the ultrasonic cries of predators bats, while Belding's ground squirrels treat their full siblings differently from their half-siblings and male black winged damselflies use their penises to scrub other male's sperm out of their mates sperm storage organ before transferring their own sperm (**Alcock**, 1998).

QUESTIONS

Long Answer Questions

1. Define the term ethology ? Explain with suitable examples the various types of behaviours in animals. (*Allahabad 1992, 94*)
2. Define the term behaviour ? Explain the mechanism of behaviour. (*Lucknow 1994*)
3. Write an essay on "Animal Behaviour" (*Lucknow 1996*)
4. Write on different patterns of behaviour giving examples of each. (*Lucknow 1992*)
5. Give a brief account of history of animal behaviour.
6. Describe the significance/purpose of study of animal behaviour.

Very Short Answer Questions

1. Who first studied the social behaviour in honeybee ?
   (**Ans.** Karl von Frisch) (*CCSU Meerut 2006*)
2. Write the name of two scientists who got Nobel prize for working on animal behaviour.
   (**Ans.** Niko Tinbergen and Konrad Lorenz) (*CCSU Meerut 2006, 07*)

Multiple Choice Questions

Choose the correct answer from the four alternatives given.

1. The biologist who discovered the meaning of the dances performed by returning honeybee foragers was
   (*a*) Karl von Frisch (*b*) Niko Tinbergen
   (*c*) Konard Lorenz (*d*) Ivan Pavlov
2. Which of the following biologists is given credit for developing the concept of imprinting in 1937?
   (*a*) Niko Tinbergen (*b*) Karl von Frisch
   (*c*) Konard Lorenz (*d*) Oscar Heinroth
3. In classical conditioning, the unconditional stimulus is transferred to a conditioned stimulus. The scientist who did the first systematic research on this phenomenon was
   (*a*) Harry Harlow (*b*) Ivan Pavlov
   (*c*) B. F. Skinner (*d*) Niko Tinbergen
4. Which of the following is true of imprinting?
   (*a*) it occurs only in birds
   (*b*) its effects last only for short time
   (*c*) species-identity and sexual-identity imprinting always occur at the same time
   (*d*) it can occur in both young and mature animals.
5. In the nature versus nurture debate, nature focuses on an individual's
   (*a*) training (*b*) tendency to eat unprocessed food
   (*c*) genes (*d*) early influences
6. Jane Goodall studied
   (*a*) gorillas in smoky mountain (*b*) chimpanzees in Gombe National Park
   (*c*) baboons in Africa (*d*) tigers and Himalayan fauna

Answers (MCQs)

1. (*a*); **2.** (*c*); **3.** (*b*); **4.** (*c*); **5.** (*c*); **6.** (*b*).

2

Approaches to the Study of Behaviour

The ideas, methods and theories established during the last half of the nineteenth century form the foundation of today's experimental approaches to the study of animal behaviour. Various investigations in animal behaviour took place mainly in following three directions:

1. Comparative psychological approach. Comparative psychologists and physiologists have attempted to determine the underlying causes of behaviour – the control mechanisms.

2. Ethological approach. Classical ethologists have been concerned primarily with the functional significance and evolution of behaviour patterns but have also developed explanation of mechanisms, including drives, innate releasing mechanisms (IRMs) and similar concepts.

3. Behavioural ecological and sociobiological approach. Behavioural ecologists and sociobiologists have explored the ways in which animals interact with their living and nonliving environments and have applied the principles of evolutionary biology to the study of social behaviour and organisation in animals.

We can now briefly examine the historical development of each of these approaches to the study of animal behaviour.

2.1. HISTORICAL DEVELOPMENT OF VARIOUS APPROACHES OF ANIMAL BEHAVIOUR

A. Studies of Mechanisms

Comparative psychology is the study of different animal's behaviour patterns in order to determine the general principles that explain their actions. Comparative psychology can best be understood by looking at the variety of approaches to behaviour studies taken over the past century, which eventually led to development of comparative psychology. Currently, comparative psychology has melded into the larger discipline that we call animal behaviour or ethology (*Meld* means inclusion in one's total score during card games).

1. Perceptual psychology. During the mid-nineteenth century were emerged several distinct approaches for the discovery of mechanisms underlying behaviour. Researchers who were concerned with mind/body dichotomy studied the relationship between physical and mental processes. Investigators were interested in separating the processes of sensation (body) and perception (mind). This usually involved the objective measurement of sensation (*i.e.,* the reception of stimuli through the senses, such as sight and hearing) and the comparison of this direct measurement to objective interpretation (perception) of the sensation. The newly emerging **psychophysics** is an outgrowth of these early studied. These types of studies make clear to us that what an animal makes of its world, both regard to sensory systems and with respect to the animal's interpretation of its sensation.

2. Physiological psychology. Modern physiological psychology developed from early attempts to relate behaviour with the internal physiological properties and events of the organism. For example, **Marie-Jean-Pierre Flourens** (1794-1867) surgically removed portions of brains of pigeons and recorded the resulting changes in the behaviour of the bird. **Hermann von**

Helmhotz (1821-1894) studied the *conduction speed of nerve impulses,* and later, the *physiology of vision.* He simply measured the speed of nerve conduction by experimenting on the frog motor neuron that triggers muscle contractions. First he stimulated the nerve at one point near the muscle and then at a second point farther away from the muscle. The difference in amount of time elapsed (*i.e.,* passed) between stimulus and muscular contraction in the two measurements is the conduction time for the distance between the two stimulus points. From this information, he calculated the speed of conduction.

In another classical study, **Roger W. Sperry** (1913–1994) and his coworkers surgically manipulated the position of the eyes in newt (*Notophtalmus viridescens*). He removed the eyes of newt and replaced them so that they were upside down. Newts treated in this way behaved as if they saw the world upside down: they moved their eyes upward in response to the movement of an object downward in response to the movement of an object downward in their visual field. This effect persisted even after several years. Sperry's work showed that in the visual system of the newt, the neurons in the optic nerve traveling from the retina to the brain are labeled for spatial orientation. Thus, even though the eye has been rotated, the message sent to the brain along the nerve remains the same as if the eyes were in the correct normal orientation. For his work on the nervous system, **Sperry** shared the Nobel Prize in 1981.

During late 1800s and early 1900s, two major new theoretical and experimental points of view arose during this period: functionalism and behaviourism.

3. Functionalism. The **functionalists,** such as **John Deway** (1859–1952) studied the functions of the mind and how the mind operates, in contrast to studying how the mind is structured. Functionalists tried to ask three major questions: 1. How does mental activity occur ? 2. What does mental activity accomplish ? and 3. Why does mental activity take place ? Functionalism applied objective observation rather than introspection as its primary method.

The approach of functionalists introduced into the psychology the concept of **adaptive behaviour** which is a idea prevalent in biology that *behaviour function in the survival of animal in its natural habitat.* To these early psychologists, the concept of adaptive behaviour implied that the response to a stimulus changes the sensory situation in such a way that the original conditions that produced the response are altered. For example, pain disappears when a sharp splinter is removed from the hand, and the original condition – the existence of a splinter – is also changed.

4. Behaviourism. John B. Watson (1878 – 1958) was the principal founder of a new approach to the study of behaviour, called **behaviourism.** The basic view of behaviourist is that animal behaviour consists of an animal's responses, reactions, or adjustments to stimuli or complexes of stimuli. Thus, most activities of an organism are products of its past expriences. Thus, behaviour, rather than the mind, became the primary focus of study. To what degree can we predict and control behaviour based on a knowledge of an animal's previous experiences ? The methods utilised by **Watson** and his followers, for example, **B.F. Skinner** (1904 – 1990) were mainly objective. Reports of subjective feelings or emotions were, by definition, not acceptable as scientific data. This restriction forced the behaviourists to study human behaviour in much the same way they studied the behaviour of any other animal, without benefiting from their subjects verbal judgements or reports of feelings and perceptions.

5. Animal psychology. Simultaneous with the development of three viewpoints was the emphasis by **Edward L. Thorndike** (1874 – 1949) on the need for systematic, replicable experiments in comparative animal psychology. By experimentating with cats in the puzzle-boxes (Fig. 1.6), he concluded that much of learning of animals take place by trial and error and that rewards are a critical (*i.e.,* deciding) component of the learning process.

Quite correctly, **Frank Beach** in 1950 streesed that the subject of comparative psychology was devoting too much attention to the white rat as a test subject, while ignoring many other types of available vertebrate organisms. Other workers, notable **Lockard** (1971) and **Hodos** and **Campbell** (1969), drew our attention to the lack of an evolutionary perspective in comparative psychology and *to the incorrect use of the rat as a model for other organisms, especially humans.* These reviewers initiated investigations of correct comparative nature, for example, the experiments of **Dewsbury** (1972, 1975) on reproductive behaviour of rodents. More attention has also been given to the natural context and actual field investigation of animals (**Lockard** 1971; **Barash** 1973, 1974).

Modern animal psychology is a diverse mixture of subdisciplines, both new and old. The study of comparative learning and learning theory is still quite important, as the works of **Bitterman** (1975), **Seligman** (1970) and **Roitblat** and **Meyer** (1995) exemplify. Ecological aspects of learning in a variety of species of animals have been investigated by workers such as **Kamil** and **Sargent** (1981), **Davey** (1989), **Balda** *et al.,* (1998) and **Dukas** (1999). The development of behaviour was studied by **Oppenheim** and colleagues, who studied the development of feeding behaviour in reptiles such as garter snakes (*Thamnophis sirtalis*) (**Burghardt** and **Krause**, 1999). **Suomi** and **Harlow** (1977) studied primate behaviour development. The study of the physiological processes underlying behaviour has also diverged into several pathways: the relationship of hormones and behaviour (**Lehrman,** 1965; **Crews,** 1980; **Goy** *et al.,* 1988; **Knappet** *et al.,* 1999; **Stirier** *et al.,* 1999); neural correlation of behaviour (**Hubel** and **Wiesel** 1965, **Brown** *et al.,* 1988; **Glendinning** *et al.,* 1999) and brain chemistry and psychopharmacology and behaviour (**Kelley** *et al.,* 1999; **Ferris** *et al.,* 1999). Behaviour genetics developed by the efforts of workers such as **Hirsch,** 1967; **Oliverio,** 1983; **Miklosi** *et al.,* 1997; **Kim** and **Ehrman,** 1998).

B. Studies of Function and Evloution

1. Growth of Ethology. The systematic study of the function and evolution of behaviour is called **ethology.** This branch of biology is now a little over a century old. One of its most important principles is that *behavioural traits, like anatomical and physiological traits, can be studied from the evolutionary viewpoint.* For example, **C.O. Whitman** (1842 – 1910) made extensive observations of display patterns, which he termed **instincts,** in various species of pigeons. Whiteman found that he could use **displays** (*i.e.,* patterns of behaviour exhibited by animals that function as communication signals) to classify animals according to similarities and differences in behaviour. From its early beginnings, ethology developed into a separate science, with its own concepts and terminology, much as comparative psychology did. Today, as we noted previously, those working as ethologists are conducting the same sorts of studies as all others who study animal behaviour.

Ethogram is an inventory of the behaviour of a species (**Drickamer** *et al.,* 2002). The ethogram has been a starting point for many ethological studies. After making observations of an organisms behaviour, ethologists then formulated specific questions about the adaptiveness and function of particular behaviour patterns. A student of Whitman's, **Wallace Craig** (1876–1954), defined two key categoreis of behaviour patterns from his work with doves and pigeons. The first category includes the variable actions of an animal, such as its searching behaviour to find food, a nest site, or a mate; these are called **appetitive behaviour.** The second category includes stereotyped actions that are repeated without variation, such as the act of mating of killing of prey; these are called **consumatory behaviour.**

The ethological approach is utilised in another major area of inquiry: the determination of how key stimuli trigger specific behaviour patterns. **J. von Uexkull** (1864 – 1944) demonstrated that *animals perceive only limited portions of the total environment with their sense organs and central nervous systems.* This sensory-perceptual world was termed the **Umwelt** by **von Vexkull.** Among the stimuli recorded by the sense organs, certain specific cues that ethologists called **sign stimuli** trigger particular stereotyped responses called **fixed action patterns (FAPs).** For example, the enlarged belly of female three-spined stickleback fish triggers courtship behaviour in male stickleback (Fig. 2.1.).

Fig. 2.1. Courtship of male and female three-spined stickleback fish. The enlarged belly of the female three-spined stickleback fish (top) is a sign stimulus for male of the species (bottom) to court and to induce the female to enter the nest he has built (after Drickammer *et al.*, 2002).

The synthesis of these early findings and the further development of modern ethology depended heavily upon two ethologists **Konrad Lorenz** (1903–1989) and **Niko Tinbergen** (1907–1988). **Lorenz** pioneered studies of genetically programmed behaviour and investigated the importance of specific types of stimulation for young animals during critical periods of early development (*i.e.*, imprinting). Modern ethology's concern with four areas of inquiry – **causation, development, evolution** and **function** of behaviour – developed from a scheme proposed by **Tinbergen** (1963). Recognition of animal behaviour as an independent discipline came in fall of 1973, when the Nobel Prize of Psysiology or Medicine was awarded to three ethologists: Konrad Lorenz, Niko Tinbergen and **Karl von Frisch** (1886–1982). Von Frisch had conducted research on animal processes and made important contributions to the study of bee behaviour and communication.

In modern times, ethology is characterized by varied types of investigations ranging from traditional observational studies in natural environments, **Geist,** 1971; **Joerman** *et al.*, 1988; **Millesi** *et al.*, 1998) to experiments on the physiological foundations of behaviour (**Bentley** and **Hoy,** 1974). Some ethologists work primarily with behaviour genetics and the evolution of behaviour (**Mannings,** 1971; **Gerhardt,** 1979; **Ukegbu** and **Huntingford,** 1988) or analyse the relationship between hormones and behaviour (**Hinde,** 1965; **Truman, Fallon** and **Wyatt,** 1976) or the nervous system and behaviour (**Nottenbohm,** 1981; **Rose** *et al.*, 1988; **Oliveira** and **Almada,** 1988). Other ethologists work on research problems in the field or in the laboratory setting that resembles the natural habitat.

2. Ethological approach. By exercising experimental manipulation to test specific hypotheses, **Kummer** (1971) investigated the effects that transplantation of individuals from troop to troop had on the social behaviour of baboons, **Wickler** (1972) studied the significance of colour patterns in fish; **Gowaty** and **Wagner** (1988) tested the aggressive behaviour of eastern blue-birds and **Panhuis** and **Wilkinson** (1999) investigated the effect of male eye span on contest outcome in stalk-eyed flies.

Since the mid-1950s, the distinctions between ethology and comparative psychology have been disappearing. Several events have opened communication between scientists of two approaches. These events include the biennial meeting of the International Ethological

Conference, many cross-visitations between researchers in Europe and America, and the publication of a number of international animal behaviour journals. A common approach which started with the notion of species-typical behaviour, has emerged from this exchange of information. **Species-typical behaviour** involves actions and displays that are broadly characteristic of a species and that are performed in a similar manner by all its members. **Dewsbury** (1985) produced the autobiographical sketches of many leaders in animal behaviour. This work included a variety of angles and provided notable insight into the way the various approaches have developed independently, and how they have recently blended into a unified and integrated approach to the study of behaviour.

In recent years, one major, long-standing controversy (*i.e.,* nature versus nurture) has been also resolved. Early ethologists believed that much of animal's behaviour was instinctive or preprogrammed and was not affected to any great extent by experience. Many psychologists claimed that learning and experience were the major determinants of behaviour. Today, most animal behaviourists believe that neither of these viewpoints is entirely correct. Instead, the correct focus is on the interaction of genotype, physiology and experience as the determinants of behaviour, and on how the relative contributions of genetic and environment effects differ among animal species.

Comparison Between Ethology and Comparative Psychology

Ethology was developed, largely in Europe, by researchers, trained in biology. Ethologists traditionally observed a wide variety of animals in nature and conducted experiments under conditions that mirrored the natural setting as closely as possible. They concentrated their efforts on exploring questions of ultimate causation – the "why" question of the evolution and function of behaviour.

Comparative psychology orginated primarily in America. Until the past several decades, most psychologists generally worked under controlled laboratory conditions. Much of their research work was carried out on small rodents, particularly the domesticated rat. Comparative psychology placed main emphasis on proximate issues–the "how" questions of the physiological and developmental mechanisms underlying observed behaviour patterns. **Dewsbury** (1984) defines comparative psychology as *the attempts to make comparisons across species in order to develop principles of generality regarding animal behaviour.*

Animal behaviour is now a unified discipline with a broad synthetic approach: much of the research conducted by animals or comparative psychologists today is indistinguishable from that of other animal behaviourists with different backgrounds. These research experiments include explorations of the genetic aspects of food-searching behaviour in blowflies (**McGuire** and **Tully,** 1986), the effects of distasteful conditioning on learning behaviour of honeybees (**Adramson,** 1986), the role of hormonal factors in infanticidal behaviour in rats (**R. E. Brown,** 1986), and the role of the brain in budgeriger's interpretive acousstic information from contact calls (**S.D. Brown** *et al.,* 1988).

C. Behavioural Ecology and Sociobiology

In the past 57 years, third approach to the study of animal behaviour has emerged. **Behavioural ecology** and **sociobiology,** with origins in zoology examine the ways in which *animals interact with their environments and the survival value of* behaviour (**Morse,** 1980; **Krebs** and **Davies,** 1993, 1997). "Environment" as used here includes animals of the same species (conspecifics), other animals within the same ecological community, plants and inorganic physical features of habitat.

Behaviour ecologists are concerned with both animals and proximate questions about behaviour (Fig. 1.1.). They are trained primarily in zoology, ecology and related fields, are also greatly influenced by the methods of comparative animal psychology. Behavioural ecologists often begin a field investigation and define questions about, for example, population regulation or predator-prey relations. Does the predator maximize its energy intake by utilizing some form of optimal foraging strategy ? What are the most important features of the prey for predator recognition and detection ? Certain aspects of the overall investigation (such as just mentioned latter question) may require experiments more systematic than those that can be done in the field setting. Thus, as behavioural ecologists we might bring specific, testable hypotheses into the laboratory findings to what is known about the animal in its natural field setting.

For an example, here we can consider the prey-catching behaviour of the shore crab (*Carcinus maenas*). When these animals are given their choice of what sized mussel to consume (Fig. 2.2), they selected the size that provides them with the highest rate of energy return (**Elnes** and **Hughes,** 1978). Notice that although the crabs do not eat mussels in a variety of sizes, they may avoid the larger mussels because of the extra time and energy needed to crack open the shells. The wide size range that the crabs eat may represent a compromise: lots of time spent searching for just the right sized mussel would be inefficienet. Thus, we see that both the time it takes to find food and the ability of the predator to handle the prey can influence prey selection.

Certain noted investigations in the field of behaviour ecology include **Emlen** (1952) *On bird behaviour and energy budgets,* **Davis** (1951) *On population biology of rats*, and **King** (1955), *On the relationship of prairie dog social behaviour to habitat.*

Recently topics that have received particular attention by investigators include foraging strategy (**Stephens** and **Krebs,** 1985; **Vander Wall,** 1990; **Bell,** 1991; **Ydenberg** and **Hurd,** 1998); proximate mechanisms in behavioural ecology (**Real** 1994; **Braude** *et al.,* 1999); predator-prey systems and predation risk (**Haskins** *et al.,* 1997); **Randall** and **Matocq,** 1997), the ecology of sex and strategies of reproduction (**Askenmo,** 1984; **Clutton – Brock,** 1988); and social systems in relation to ecology (**Thornhill** and **Alcock,** 1983; **Christenson,** 1984; **Hill,** 1998).

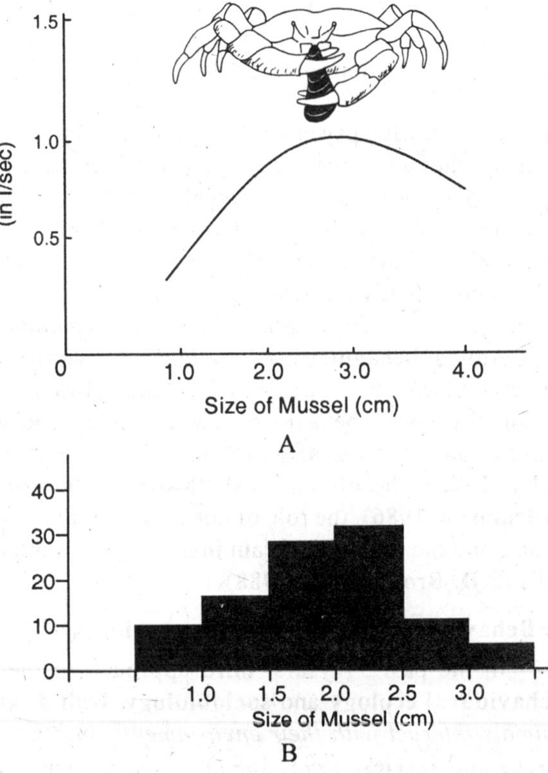

Fig. 2.2. Shore crabs select mussels for food. A–shore crabs select those sizes of mussels that provide the best rate of energy return more often than other sizes. B–shore crabs, in fact, do eat mussels of a variety of sizes (after Drickamer *et al.,* 2002).

Sociobiology born in 1975 with the publication of **E.O. Wilson's** *Sociobiology: The New Synthesis.* Helping behaviour in nest of birds were studied in Florida scrub jays (**Woolfenden,** 1975), African white-fronted bee-eaters (**Emlen,** 1984), acorn woodpeckers in the western United States (**Koenig** *et al.,* 1984) and seychelles warblers (**Komdeur,** 1992; **Komdeur** *et al.,* 1995). The investigations of these and similar social systems in birds reveal that nests with helpers are more successful, *i.e.*, the number of young fledglings is higher.

In fact, the various theories and concepts that constitute sociobiology have their roots in many earlier works. Among the most significant are the writings of **Williams** (1966) on *natural selection and concept of adaptation*, **Trivers** (1971, 1972) on *the evolutionary aspects of altruism and parental behaviour* and **Hamilton** (1964, 1971) on *the genetic theory underlying the evolution of social behaviour.* Studies conducted under the general heading of sociobiology include, for example, those on altruism in ground squirrels (**Sherman,** 1977); on strategies for reproduction in damselflies for reproduction in damselflies and other insects (**Waage,** 1979, 1997); on parental investment in water bugs (**Smith,** 1997) and on mate slection in American kestrels or European falcons (**Duncan** and **Bird,** 1989). In recent years, a major topic for investigators using the sociobiological approach has involved sexual selection and various factors influencing mate choice (**Anderson,** 1994; **Cowaty,** 1995; **Eberhard,** 1996). **M. Petrie** *et al.*, (1991) reported that peahens prefer peacocks with elaborate trains (*i.e.,* greater number of eyespots in their tail feathers). **Prasad** and **Chaturvedi** (2006) studied 12-hour temporal relationship of circadian serotonergic and dopaminergic activity influences seasonal testicular growth and secondary sex characters in Indian weaver bird, *Ploceus phillippinus.*

2.2. TOOLS AND TECHNIQUES OF ANIMAL BEHAVIOUR

The discipline of ethology may be approached from different viewpoints depending upon the questions asked and the inclination of the investigator. This is in part a reflection of the diversity of the animals, the tools available for study, the diversity in behaviour and an accident of history. The following three disciplines have converged to produce its current complexion:

1. Psychological techniques. Psychologists have been traditionally interested in questions involved with how an animal learns — its motivation and drive to perform certain acts. Thus they devise puzzle-boxes, skinner-boxes and mazes and training tasks for animals as their methodological approach.

2. Physiological techniques. Physiologists have been traditionally interested in describing the sense organ response, the function of nervous system and effect of hormones and pheromones on the animal behaviour. They record the sensitivity of these systems, using sophisticated electronic equipments. For example, neurophysiology of cricket song was studied with the aid of implanted **micro-electrodes.** Likewise, to understand the magnetic sense of pigeons used during homing and their migration can be studied by the help of **Helmholtz coils** which may be set above the pigeon's head and around its neck.

The psychological techniques and physiological techniques are performed in the laboratory (see Section 2.3).

3. Field biological techniques. Field biologists have traditionally been interested in the animal's relationship with its environment — the social behaviour and behavioural adaptations that permit individual and population survival. They often use observational and recording techniques perhaps aided by **camera, radar, radio transmitters, oscilloscope, stroboscopic-flash lamp, ultrasonic loudspeakers, multipen event recorder** and healthy amount of patience. Field biologists compare behavioural patterns between individuals and between species. They may expose the animal to different environments to know how behaviour is modified, all is done to understand the relevance of the behaviour to the animal's past and present survival.

Certain significant field-biological tools and techniques of ethology are the following:

1. Study of animal's behaviour in its natural surrounding. The first task is simply to observe the behaviour of an animal in general in as many contexts as possible. Since we know that the surroundings can distort and modify behaviour, the study should be made under conditions which are as natural as possible. Ideally we would follow the animal around on its daily routines in the wild state. In practice we usually have to compromise. A common solution is to cage the animal as comfortable as possible and allow it to behave in as natural fashion as it will when left without disturbance.

The bulk of behaviour data is gathered in this way, and it is supplemented by whatever field observations are possible. Gradually a picture is constructed of the motor patterns which the animal uses in its daily life, the stimuli (both physical and social) to which it seems to be responsive and the ways in which the behaviour changes with a shift in the physiological condition of the animal.

A **camouflage tent** is also an aid in observing the natural behaviour of animals which soon get used to it. In this way the scientist can make his observations without being noticed by the subject (Fig. 2.3).

2. Selection of experimental animal. The success of a behavioural investigation depends very much on the selection of an experimental animal (subject) which is appropriate to the area of interest. Obviously a study of birds is likely to be a poor introduction to the use of olfaction in social communication.

3. Quantification of behaviour. To gain full picture of an animal's behaviour it is often necessary to record its activities, which are frequently repetitive, over a long period of time. To do this by direct observation would drive all investigators insane; therefore a variety of activity recorders are used. For example, **multi-pen recorders** allow continuous recording of an animal's behaviour. Each pen of the recorder is wired to a separate switch of the manual keyboard. When a behaviour occurs, the appropriate key is depressed to actuate (to move to action) the recording pen and the key is held down until the behaviour terminates. The duration of each behaviour is also recorded by stop-watch or other means. The recordings provide a permanent record of the observations.

A time-sampling method may improve the accuracy of behavioural observation. This method uses a timing device to pace the observer's recording. For example, a tape-recorded signal delivered through an earphone to the observer marks off fifteen-second intervals and on hearing each signal the observer notes the behaviours presently being displayed.

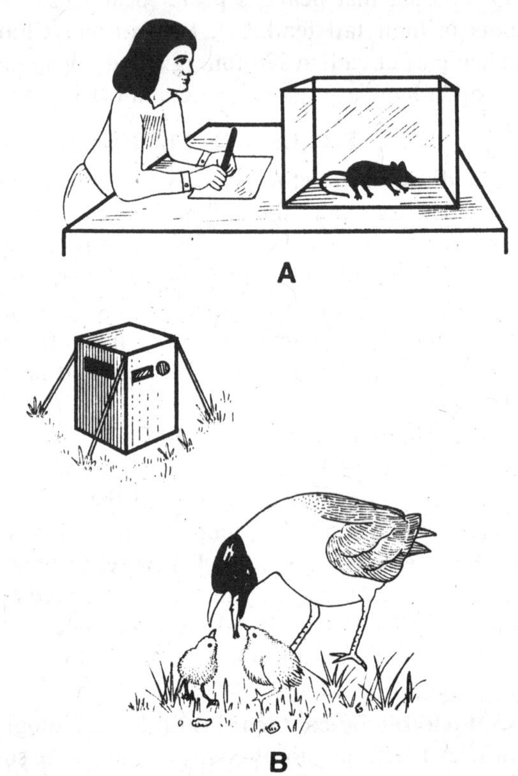

Fig. 2.3. A traditional view of the distinction between ethology and psychology. The psychologist puts his animal in a small box and peers in to see what is is doing, while the ethologist puts himself in the box and looks out at what the animals are doing.

4. Use of kymographs and cinematography in recording animal behaviours. There are many other recording techniques for slow, fast or repetitive behaviour patterns of animals. For example, **kymographs** may be used to make a graphic record of the muscular responses of simple animals such as worms and sea anemones. **Slow-motion cinematography** is used to record the activities of fast-moving animals such as birds. The **time-lapse photography** is used for slow-moving animals such as sea anemones. For extremely rapid movements, such as jumping **in salticid spiders, multiple-flash photography** may be used.

5. Use of video-tape in analysis of behaviour. The current use of audio-visual media has revolutionized the science of animal behaviour. Video-taping fills the need for a visual record of behaviour as it actually occurred. By repeated analysis of the same behavioural sequence the investigator often makes critical observations not seen at the first time. In addition, repeated data collection on the same behavioural sequence provides the observer reliability of the data estimation.

Video-taping has several advantages over filming: 1. The visual record is available for use immediately after taping, because there is no processing delay, 2. The video monitor enables one to view the visual record while taping is actually in progress, thus enabling mechanical errors to be rectified immediately, 3. Taping may be done under normal lighting conditions, thereby avoiding bright lights and intense heat from interfering with the subjects's behaviour, 4. Because video-tape may be reused, it is less expensive means of recording behaviour once the initial equipment has been purchased, 5. Video-tapes are more suitable than film for recording long sequences of behaviour since they are capable of longer continuous recording, 6. Video-tape provides a more convenient means of linking vocalisation (auditory signals with behavioural patterns).

Aside from the initial cost, there are two major disadvantages to video-tape equipment. It is generally less portable than movie cameras, thus, confining taping techniques largely to laboratory studies. Unlike movie film, video-tape is not easily sectioned into single-frame visual records to be reproduced as slides and photos, or made into line drawings.

6. Use of tranquilizers and radio-telemetry. To study social behaviour, observations of natural behaviour in individuals (wild animals) over long periods of time become important. The tranquilizer gun enables us to capture animals, mark them and release them unharmed. For this a shell is filled with an anesthetic drug, then loaded into an air rifle and shot at the animal and then the shell with its barbed hook is removed. The scientists now have upto an hour to take measurements and to mark the subject.

One method of marking which allows us to track the individual animal more easily is **radio-telemetry**. In this method, an anesthized animal (*e.g.*, lions) is fitted with a collar containing batteries and a radio transmitter. A receiver to make the transmitter signals audible and a directional antenna enable us to locate the animal at any time, even in dense under bush.

2.3. STUDY OF BEHAVIOUR IN LABORATORY

The first major goal of brain researchers or psychologists was the localization of brain functions, to know which areas or structures of the brain were involved in controlling particular behaviour. These studies include following three types of techniques.

A. Neuroanatomical or Lesion or Ablation technique

B. Neurophysiological technique

C. Neurochemical technique

A. Neuroanatomical or Lesion or Ablation Technique

This is oldest and most crude method for studying the relationship between particular neural structure and behaviour. For this purpose, certain regions of the brain are destroyed and their functions is deduced from the abnormal behaviour it causes in the animal. The larger **ablation** (*i.e.*, removal of tissue) may be carried out free hand by means of knife cuts, whereas the localised and small **lesions** (*i.e.*, pathological changes in tissue) are produced by passing electric currents through platinum, irridium electrode to cauterize the area. Small lesions can also be created by neurotoxic **kainic acid.** Lesion methods are different from **stimulation** method, in the latter case, neurons are activated while in the former they are inhibited or destroyed. Neuroanatomical technique helped many workers to form brain atlases or brain maps, which are known **stereotaxic atlases** that are presently available for many mammalian species such as rat (**De Groot,** 1959), cat (**Jasper** and **Marsan,** 1954) and dog (**Lin** *et al.*, 1961).

The first indications that brain strcutures such as amygdala influence aggressive and defence behaviour in cats was obtained by lesion experiments. Bilateral lesions in amygdala caused enhanced sexual behaviour.

Pierre-Paul Broca (1824–1880) was a French doctor who in 1861 came across a patient with head injury whose speech was defective. This correlation of brain lesion and speech fascinated him and he started observing different patients with similar injuries. After few years, Broca concluded that there is a definite area in the brain cortex which is responsible the articulate speech. Later using the same technique another scientist, a German neurologist, **Carl Wernicke** (1848-1905), in 1880, established another area in the brain responsible for speech. These two areas are known as **Broca's** and **Wernicke's areas** (Fig. 2.4), respectively. Broca's area controls

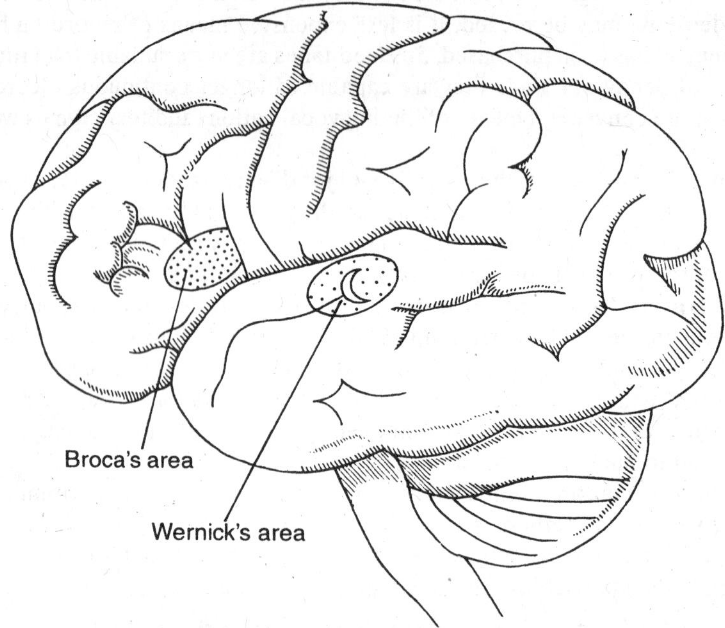

Fig. 2.4. Human brain showing Broca's and Wernicke's areas for speech.

the infrastructure for speaking a word, this means it has control over neck muscles and larynx. Wernicke's area is word retriever or dictionary. A lesion in Broca's area causes slurring in speech and there is difficulty in understanding the words spoken by such a person though the person would talk sensibly. Opposite to it, if there is a lesion in Wernicke's area then the person can talk clearly but the speech will not be articulated or coherent.

The toads (amphibians) have been the major objects for ablation techniques; the experimental finding that there are zones in different areas of the toad brain responsible for the triggering of prey catching **(Optic tectum)** and avoidance behaviour **(TP-region)**, is supported by ablation experiments.

Different nuclei of hypothalamus (Fig. 2.5) and their role in controlling vital activities such as feeding, drinking, mating, aggression and parental behaviour have been discovered by lesion technique. Lesion in hypothalamus can lead to **adipsia** (*i.e.*, loss of the thirst), **hyperdipsia** (*i.e.*, abnormally increased thirst), **aphagia** (*i.e.*, loss of the ability to swallow) and **hyperphagia** (*i.e.*, abnormally increased appetite for food). It has been identified that lesion in ventromedial hypothalamus increases hunger in the experimental animals. This indicates that this area of hypothalamus has **satiety centre.** Similary, a lesion in the lateral hypothalamus reduces hunger and animal stop eating. It means **feeding centre** lies in the lateral hypothalamus.

By further findings, it has also been established that there are specific nuclei in the anterior and ventro-medial parts of hypothalamus that start and stop the sexual behaviour.

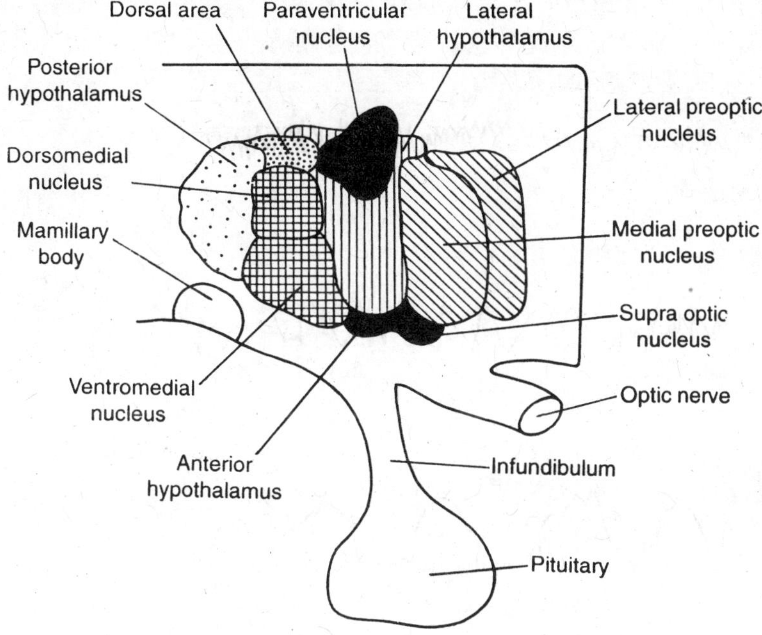

Fig. 2.5. Nuclei of hypothalamus of human brain.

2. Neurophysiological Technique

The discovery of **animal electricity** or **bioelectricity** came by chance. The Italian scientist **Luigi Galvani** (1737–1798), in 1786, hung some frog legs on his balcony railing one day and noticed that the frog's legs twitched when they touched the metal fence. **Galvani** at first misinterpreted this phenomenon attributing it to muscle alone. By 1793, Galvani discovered that twitches occurred in the nerve muscle preparations even without any metal; it indicated that the nerve and muscle were producing electricity by themselves. This finding laid the foundation of **electro-physiology.**

The discovery that the messages in nervous system travel in the form of electric current replaced the neuroanatomical lesion technique by a less destructive and more precise neurophysiological technique for the study of behaviour. This technique involves two ways.

1. Recording of electrical activities from brain while animal is performing specific action patterns.

2. Stimulating the brain region for eleciting behaviour.

1. Recording of electric impulses. With the help of stereotaxic machine the electric impulses in the brain are recorded and their correlation is traced with the behaviour. Following four basic types of brian waves have been recorded in human beings.

(*i*) Alpha waves. These waves are recorded from the parietal and occipital lobes when the person is at rest with eyes closed. The brain is alert but unoccupied. These waves are also produced when we are awake but with a peaceful relaxed mind and not in excited state.

(*ii*) Beta waves. These are generated from the frontal lobe when brain is stimulated by sensory inputs. It means these waves indicate the active state of brain when a person is mentally alert and is thinking or concentrating on some issue.

(*iii*) Theta waves. These waves are produced from the temporal and occipital lobes under emotional stress. They are associated with earliest stages of sleep, hallucination and creativity.

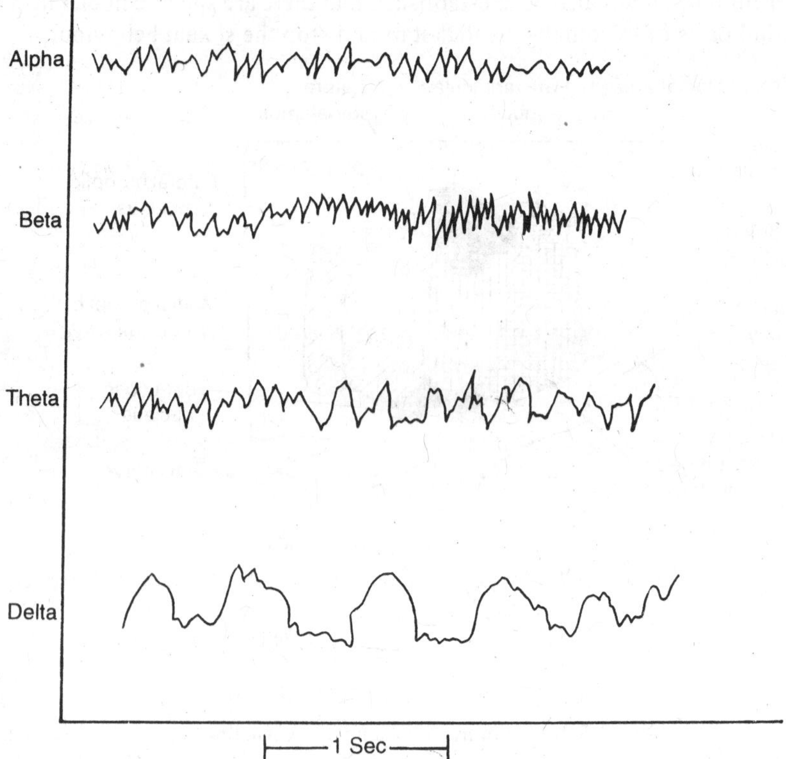

Fig. 2.6. Types of electric waves generated in different parts of the human brain.

(iv) **Delta waves.** These waves are produced during deep sleep.

2. Initiation of behavioural patterns by electric stimulation. David Ferrier (1873) and **Bartholow** (1874) identified that particular motor patterns of muscular movement can be elicited by stimulating defined regions of the cerebral cortex of human brain. At the end of the 19th century, **Ferrier** was able to roughly localise the sensory centres in the cerebral cortex by vision, hearing (auditiory), taste (gustatory) and smell (olfactory). Electric stimulation method is the study of behavioural patterns that are initiated by electrical stimulation in specific brain regions by implanted electrodes. For example, **J. Flynn** (1929) stimulated a particular locus in the diencephalon of a cat and elicited prey catching behaviour. **A. Zukerman** (1965) was able to induce behaviours such as fleeing, courtship and threat by electrically stimulating different regions of a pigeon's brain. **E. von Hoist** and **Von Saint Paul** (1960) elicited sleeping behaviour in a cock by stimulating the brain stem. By using the same technique of electric stimulation by electrodes **pleasure centres** in human brain have been identified by **Jacobson** (1967).

3. Neurochemical technique: This technique involves local application of drugs, neurotransmitters, some blocking agents or hormones to stimulate neurons selectively. It is also called **chemical brain stimulation technique.**

Psychoactive drugs (such as transquilisers, sedatives), alcohol, opium, hashish and bhang, etc., are used for inducing mental relaxation and sleep (Box 2.1). These can be administrated via many routes such as *intraperitonially, intramuscularly* or *subcutaneously* by injection or orally by ingestion. **Fisher** (1956) demonstrated that sexual and maternal behaviour can be induced by injecting minute quantities of sex hormones (estrogen or testosterone) into

BOX 2.1.

The chemicals can affect behaviour in different ways. Depending upon their nature and chemical composition they has been classified as follows:

1. Physiological agents. Adrenalin, histamine, serotonin and glutamic acid are known as physiological agents. They are present normally in the body but they are considered **drugs** when administered in concentrated dose.

2. Foreign agents. Reserpine, barbiturates and lysergic acid fall under this category. They are not normally present in the body.

Neuroactive drugs are also identified as tranquilizers, energizers or hypnotizers. **Tranquilizers** reduce agitation, excitement and emotion; **energizers** are stimulants such as caffine, amphetamine, etc. They tend to produce hyperexcitability and also facilitate sensory and motor perfomance. They also act as anti-depressants. **Hypnotizers** such as barbiturats are the drugs that produce sensory and motor depression and deep sleep. In large doses they cause unconciousness and even death.

hypothalamus. **R.P. Mitchell** (1968) found that administering the estrogen in anterior hypothalamus of brain of female cats when not in heat produced mating behaviour. On the contrary, if the same hormone was injected in some other parts of brain, similar behaviour was produced.

Most recently the brain is being, explored for its various functions using PET (*i.e.,* Positron Emission Tomography) technique.

2.4. CAPTURING, MARKING AND TRACKING ANIMALS AND ANIMAL SIGNS

Before discussing specific methods used for data collection in field settings, let us depart a moment to examine several related problems– issues that generally arise when we are conducting field studies, but that may also apply to some laboratory work in animal behaviour. These issues concern the methods of capturing animals, ways of marking animals for individual animals, and signs left by animals that may assist us with our interpretation of their behaviour.

1. Capturing Animals

To study animals properly, it is often necessary to capture them. We may require to mark the animals for identification. We may require to make measurements of their size, sex or age. Or, we may wish to bring them into the laboratory for certain portion of our research work. There are a variety of methods of capturing the animals that we can employ (**Eltringham,** 1978). Exact capture technique we use in a particular situation will depend upon the species and the nature of its habitat. The most common methods involved **nets** or **traps.** For many insect species we would use either **sweep nets** (Fig. 2.8) or **dip nets.** For birds we often use **mist nest,** nets of fine mesh set up using two poles, some what like a volleyball net. Mist nets are often set up to intersect the flight of the birds, for example, across a clearing in the woods or along the path of a stream.

Larger nets can be used with small cannon to capture flocks of birds at a feeding station or mammals baited to come to a particular site. **Live traps** for vertebrates and invertebrates are available in an enormous variety of shapes and forms and can be used in the capture of almost all species of animals. **Seine nets** (Fig. 2.7) are often used to obtain samples of fish populations from a stream or lake. **Pitfall traps** can be placed into ground so their top rims are flush with the ground surface, are used for capturing a variety of invertebrates, reptiles, amphibians and mammals. Insects can be removed from tree trunks via the use of brushes or small vacuums (*e.g.,* Bulb aspirator), and they can be removed from vegetation by shaking or beating the plant while holding a tray underneath (Box 2.2).

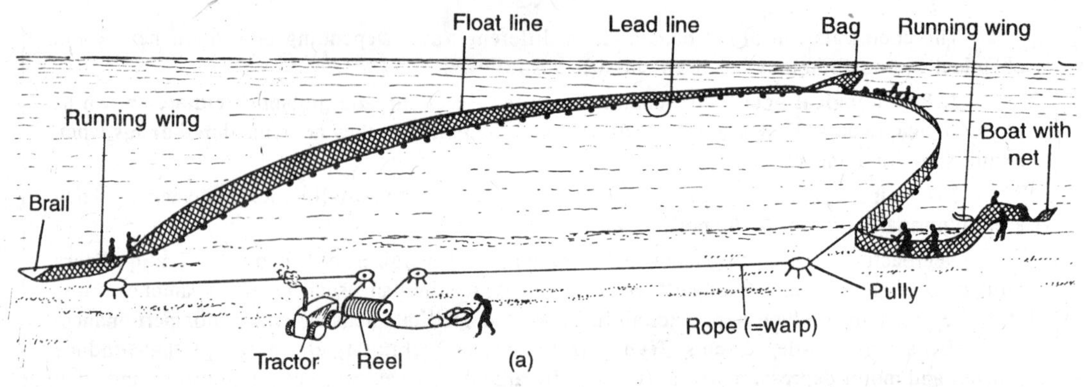

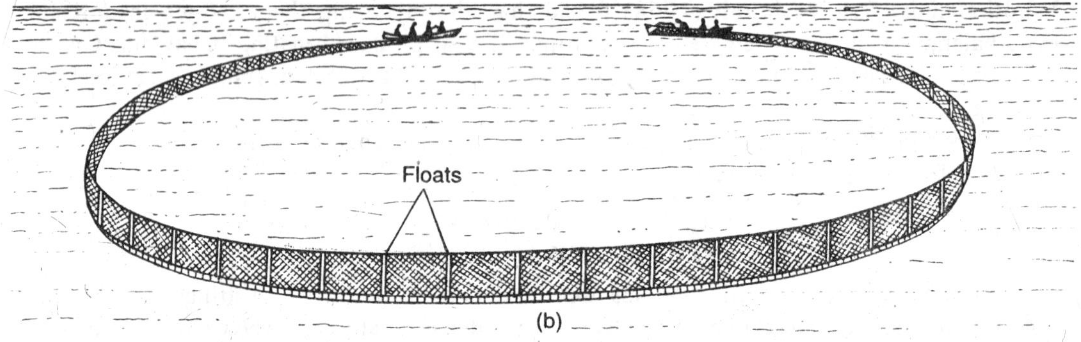

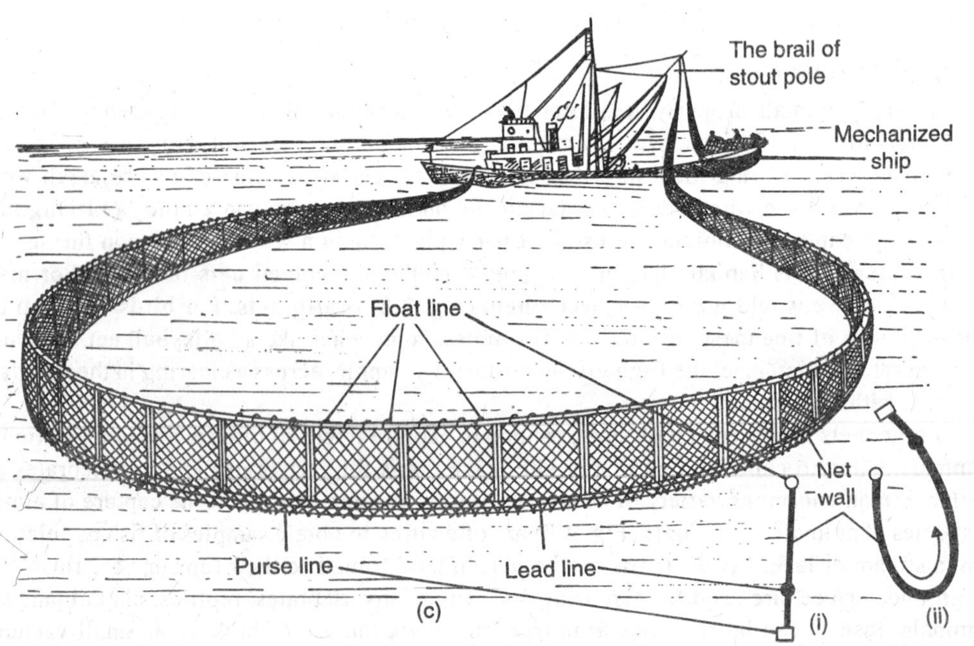

Fig. 2.7. The seines. A – Beach seine; B–Ring or Danish seine; C– Purse seine.

BOX 2.2.

Sweeping and *beating* are most productive methods of collecting large number of insects.

1. Sweeping. The purpose of sweeping is to dislodge insects from the vegetation by means of an insect net, *e.g.*, sweep net, **sweep net** (Fig. 2.8) contains a stout and short handle attached

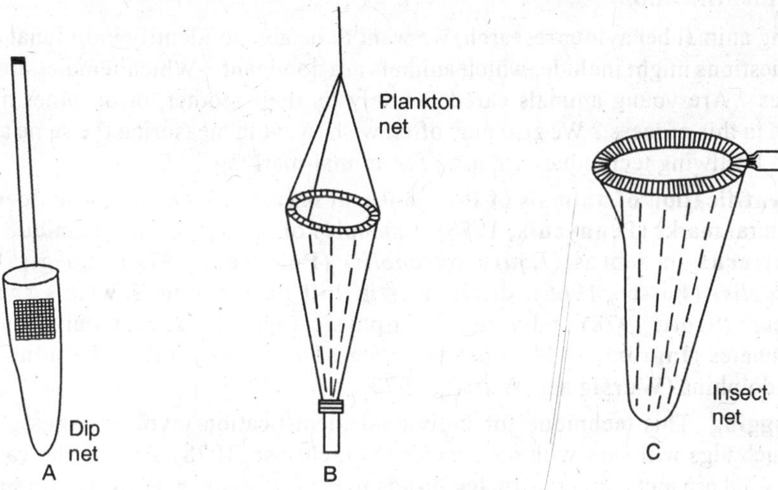

Fig. 2.8. Sweep nets.

to a metallic ring having bag of muslin or canvas. The collector sweeps the net back and forth though the grass or bushes as he walks. During use, the contents of net should be frequently checked and the specimens sorted and placed in a proper vial.

2. Beating. The principle of beating is to hit a branch of a tree or shrub hard enough with a heavy stick in such a way that the insects and other arthropods fall on a tray or white sheet placed below from where they can be captured and collected.

3. Collection by aspirators. The aspirator is a simple suction apparatus used for collecting small fragile insects, mites and spiders. In **bulb aspirator** suctions is applied through the suction

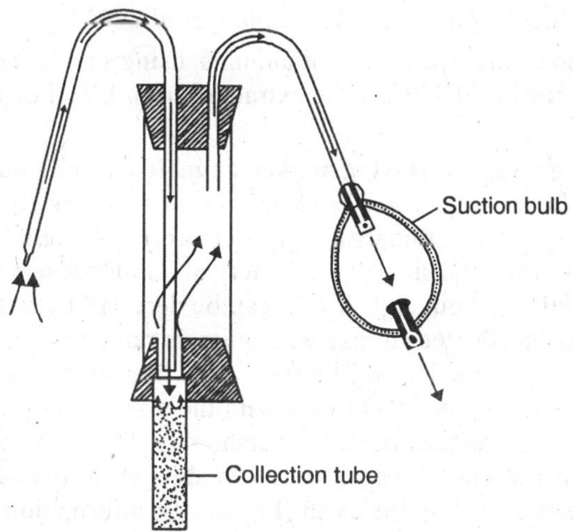

Fig. 2.9. A bulb-aspirator.

bulb and the desired specimen sucked in through the intake tube directly into the collecting vial.

In the aquatic habitats, electric shock can be used to stun large numbers of fish, which then float to the surface; they soon recover from their stunning. Finally, for many mammals, particularly large beasts such as bears or ungulates, we employ a **CO_2 cartridge gun** and a **tranquilizer dart.**

Animals shot in this manner are anesthetized for varying periods, and they are given an antidote when the investigator has finished tagging or measuring them.

2. Animal Identification

During animal behaviour research, we want to be able to identify individual animals. Our research questions might include: which animals are dominant ? Which females are mating with which males ? Are young animals cared for only by their mother, or do other members of a group share in this process ? We also may often wish to avoid measuring the same animals twice by mistake. Folllwing techniques are used for animal-marking:

1. Identification of animals of their natural marks. We can primarily identify animals by their natural marks (**Pennycuik,** 1978). Examples of the use of this technique include **coat colour patterns** in zebras (*Equus burchelli*) (**Petersen,** 1972) and giraffe (*Giraffa camelopardalis*) (**Foster,** 1966); differences in **bill patterns** in Bewick's swans (*Cygnus Columbianus*) (**Scott,** 1978) and variation in physical charcteristics, natural mutilations and scars in primates (**Ingram,** 1978), lions (*Panthera leo*) (**Pennycuik** and **Rudnai,** 1970), and bottlenose dolphins (**Wursig** and **Wursig,** 1977).

2. Tagging. This technique for individual identification involves tagging animals; the nature of such tags will vary with the species (**Stonehouse,** 1978). Among the various tagging techniques used are metal and plastic **leg bands** in birds (**Spencer,** 1978; **Patterson,** 1978), **fin clips** or **punches** and metal **fingerling tags** for fish (**Laired,** 1978); **toe clips, ear punches, dye marks,** and **tatoos** for a variety of mammal species (**Lane-Petter,** 1978); and **small dots** of paint or dye on invertebrates (**Southwood,** 1994).

Another animal-marking technique involves **radioisotopes,** either in collars or bands or in subcutaneous implants. Because of potential problems with radioactivity and possible effects both on the animal bearing the radioisotope label and on the environment, the use of this technique has been curtailed considerably. More recently, **Sheridan** and **Tamarin** (1988) have developed the use of radio-nuclides for identification of individual meadow voles (*Microtus pennsylvanicus*) in order to study longevity and reproductive success; radionuclides are safer and not harmful to the animal carrying the label (see **Drickamer** *et al.*, 2002).

We can also examine the DNA of animals, using several techniques; the most widely used is **DNA fingerprinting.** DNA can be extracted from blood or tissue samples or, simply, from hair or faeces.

3. Tracking animals. Having marked animals for individual identification, we now might ask: *How can the animal be followed over time to study its behaviour?* The most popular technique of tracking the animals is **radiotelemetry.** An animal is captured and fitted with a collar containing a radiotransmitter and battery or implanted with a similar device. Once released, the transmitter will send out signals that can be detected by a radio receiver and antenna. This technology has been adapted for use with animals ranging in size from small birds, lizards and rodents, to whales and elephants. The size of the transmitter package and the range over which the receiver can pick up the signal vary with the size of the animal and the requirement of the investigator. By following an animal for many days, it is possible to obtain an accurate picture of the area that it uses and the portions of the day when it is active. Radio signals can now be detected by **satellite** circling the earth (Fig. 2.10); information from the satellite is then fed to a **computer** at a ground station and animal movements can be analysed. This technique has been used effectively with caribou (*Rangifer tarandus*) in Alaska (**Miller** *et al.*, 1975) and migration of olive Ridley's turtles (see **Mohanty,** 2009).

For some animals, particularly large mammals or birds in flight, we may use vehicles moving along the ground or air planes from overhead to follow movement patterns. For many aquatic animals, including many vertebrates (especially fish and whales) and a variety of invertebrates, we can use boats and diving gear to observe both from the water's surface and from below. In all of these situations, **photographs** or **videotapes** may aid in gathering information and enhance the process of identifying the individual and scoring the behaviour patterns.

Fig. 2.10. Satellite telemetry system.

Shake-and-Bake Method. This technique has been developed for studying rodent movement by **Kaufman** in 1989. The procedure, somtimes called the "Shake-and -Bake" method, involves using a finely powdered fluorescent dye that comes in a variety of colours (*e.g.*, magenta, lime green, blue, etc.). The animal to be tracked is captured and placed into a plastic bag with the powder. After gently shaking the bag for a few seconds, the pelage (*viz.*, the coat or covering of a mammal, as of fur, wool, etc.) of the animal becomes covered with the dye. Once the animal is released, it will deposit a trail of dye for four to eight hours wherever it goes. On the night after the animal has been allowed to put down the dye trail, we go out with a special black light that causes light to fluoresce, and the trail of the animal can be followed quite easily. In this way, we obtain a record of the travels of the animal; the technique can be used with facility on both nocturnal and diurnal animals.

4. Animal signs. In fact, any trace, sign or construction which animals leave, can help us interpreting their behaviour. Examples of animal signs that we might expect to find include **tracks, faecal material, egg shell,** and animal remains. Almost all animals have tracks of some kind, except those in aquatic habitats. Impressions of feet can be found in mud, sand or other types of surfaces. We may use tracks to determine the direction of movement of an individual or group, the composition and size of a group, and depending upon whether there are age or sex differences in the nature of the tracks, something about the age and sex of the individuals. Examination of fecal material can tell us something of the diet of an animal. Egg shells may help us identify nest sites and indicate which species are nesting in a particular habitat. The remains of animals might hold clues to the cause of death or could even tell the tale of an act of predation. Other, possibly more active signs we might discover would be **nest, egg cases** and **spider webs.** Nests of birds can provide important information about the site selection and construction of the nest, and when eggs or young are present, about the development of nest.

Sampling Methods

Workers of animals behaviour use following sampling techniques:

1. Focal animal sampling. In recent years, field studies of many animal species have focused much attention on research stractegies and methods of collecting data under field conditions. Focal animal sampling (**Altmann,** 1974) involves recording all of the actions and interactions of one particular animal during a prescribed time period. Using this method, an observer may watch a large number of animals, recording the behaviour of each for a short period (*e.g.,* five or ten minutes per animals), or the observer may record the behaviour of fewer animals, each being watched over a longer time period (hours per animal). As an example of the use of this technique we can consider here the study conducted by **Bercovitch** (1986) on dominance rank and mating activity in male olive baboons (*Papio anubis*). The question tested concerning the relationship between male dominance rank and access to sexually receptive females for mating. Intense observations were made on only one or two baboons each day. The females that were the subjects or the focal sampling were in behavioural estrus, exhibiting consort (companion) relationships with males. Using focal sampling, it was possible to record all of the male partners of the sexually receptive females. When the data were analysed using information from only adult males that interacted with the receptive females, there was no clear relatioship between dominace status and mating activity. However, if adolescent males were included in the analysis, then a relationship emerged, with dominant males engaging in more sexual activity with receptive females. One advantage of this observation technique is that intense samples can be gathered with a focus on particular animals.

2. Scan sampling or Instantaneous sampling. An observer using this technique watches each animal for only a few seconds at periodic intervals and records the activity (*ies*) that the animal is performing only at the specific time marks indicated by the sampling scheme. The intervals between samples of the behaviour of each individual can vary, but generally, they are a few minutes to a half hour. As an example of the use of scan sampling method, consider a study of feeding behaviour of two captive groups of lemurs (**Ganzhorn,** 1986): the ring-tailed lemur (*Lemur catta*) and the ruffed lemur (*Varecia variegata*) housed in seminatural environment. Scan sampling was conducted on the groups of lemurs, using a thirty-minute interval between records of individual activity. The species and part of the plant being eaten were recorded when the lemur was eating – at the time the data were to be recorded. By recording the data in this fashion, the observer was able to make many records on all the lemurs in both groups during all seasons of year. The analyses of these feeding data indicated that roughed lemurs spent less time eating during the day, had longer feeding bouts, and consumed more fruits than ringtails.

BOX 2.3.

Bout. The term bout is generally applied to (*i*) a repetitive occurrence of the same behaviour (*e.g.*, a bout of pecking) or (*ii*) a relatively stereotyped sequence of behaviours that occur in a burst (*e.g.*, courtship display).

One advantage of this technique is that it allows the observer to sample widely across all animals in the group and across a wide range of behaviour patterns.

3. One-zero sampling. It involves recording the occurrence or nonoccurrence of each of a set of behaviour patterns within a series of time periods (**Renner** and **Rosenzweig,** 1986). This scheme is seen by some investigators as the best way to record a wide range of activities encompassing solitary actions, object-directed behaviour, and social interactions. This may also be a useful method for capturing the occurrence of behaviour patterns that either occur with very low frequency or are of brief duration.

4. Sequence sampling. This technique was used by **Altamann** (1974). In sequence sampling the focus is on the chain of behaviours. These may be performed by a single individual, *e.g.,* courtship displays in male ducks, parturition (*viz.,* action or process of giving birth to offspring) in female, courtship display in stickleback fishes. The initiation of a sample period is usually determined by the beginning of sequence and the sample period terminates where the observed sequence terminates.

Some other sampling technique **ad libitum** (*viz.,* without restraint or imposed limit, or as much or as often as is wanted by the animal) and all occurrence sampling.

2.5. ROLE OF STATISTICS IN THE STUDY OF BEHAVIOUR

Statistical data regarding different aspects of behaviour is important in deriving conclusions from various experiments. It is used to summarise and analyze data and to describe variability, differences and similarities and to determine the appropriate taxonomic classification of ambiguous species and classification of whole groups of species.

1. Discrimination of different species. The animal species that are morphologically so similar that they can not be discriminated are defined on the basis of their breeding patterns and reproductive isolation. For example, a number of species of fireflies are identified by their characteristic courtship pattern and they do not interbreed. This judgement is possible only by maintaining their statistical records.

2. Classification of ambiguous species. Data from animal behaviour (ethology) can be used in determining appropriate taxonomic classification for ambiguous species and determining their similarities and derivation (ancestry). **Bekoff, Hill** and **Milton** (1975) determined the proper taxonomic position of New England canid relatives in relation to other canids by using behavioural data. It supported the anatomical data that New England canids are more closely related to coyotes then to wolves.

3. Classification of whole groups of species. Behavioural data is also useful in the calssification of groups. For example, different species of ducks and geese display different behavioural patterns. These have been used in revising their taxonomic classification.

Analysis of variation and variance. Animals exhibit variation in appearance or performance for most morphological, physiological and behavioural traits. For example, some male red-winged blackbirds (An American bird, *Agelaius phoeniceus*) perform territorial boundary displays repeatedly in the face of an intruder attempting to enter the territory. The wingspread display is often associated with the conkaree (conch-like) call. While others display much less frequently under the same circumstances. Similarly, nests constructed by redwings may vary in their height above ground and the types of plant material used in construction. How can we design experimental procedures to account for this variation ?

Variation may be either discrete or continuous. **Discrete** variation involves those measures that can take on only certain values. For example, clutch size (number of eggs layed in a nest) in birds is a discrete measure; a species may lay 1, 2, 3 or upto 6 or more eggs per clutch, but they do not lay 1.5 or 3.6 eggs. Other measures are **continuous;** they can take on any value between some lower an upper limit. The nesting territories of red-winged black birds may vary from a few square meters to thousands of square meters (Fig. 2.11). Values assigned for different birds can be anywhere within this range and are limited only by how refined we like to make our measurement.

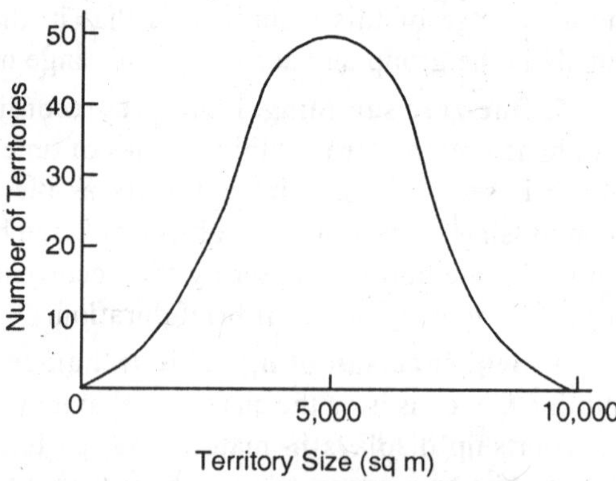

Fig. 2.11. Territory size in red-winged blackbird. Many measurable behavioural traits show some form of continuous variation. Territory size in redwings (black birds) provides such an example. This graph illustrates territory size for a large sample of over 500 birds (after Drickamer *et al.*, 2002).

To assess the amount of variation that occurs for a particular behaviour or trait in a sample, we use two principal measures, the *mean* and the *variance.* We can calculate the mean and variance for any group of values for a trait regardless of whether the variable is discrete or continuous. The **mean** ($\underline{x}$) or **central tendency,** is the arithmetic average. Thus, from a sample of four clutches of eggs from yellow-shafted flickers we might get values of 2, 4, 6 and 8 – with a mean of 5 eggs per clutch. In a second sample of four clutches, the values of 4, 5.5, and 6 eggs would also produce a mean of 5 eggs per clutch.

An estimate of the amount of deviation of the values in a sample around the mean of that sample is termed as **variance.** By definition, variance is the average squared deviation of the values from the mean of the sample. We calculate the variance S as,

$$S = \frac{\sum (x - x_1)^2}{{}^{n}N - 1}$$

Where each x_1 represents an individual value in the sample, x is the mean and N is the total number of values in the sample. So, for our first sample of clutches the variance S is 6.7, whereas for the second sample, the variance would be 0.7. The means for the two samples are identical, but the variance are quite different; there is much more variation around the mean in the first sample.

Two other measures of the variation around the mean, the *standard deviation* and the *standard error* of the mean, are also often encountered in the animal behaviour literature. The **standard deviation** of a sample is simply the square root of the variance and, is, therefore, a more useful number for describing variation around the mean. The **standard error of the mean** is an estimate of the standard deviation of a group of means from sample taken from a large population.

QUESTIONS

Long Answer Questions

1. Explain ethology and its importance. What are the main approaches of its study ? (*CCSU Meerut 2008*)
2. Describe briefly the methods used for study of animal behaviour. (*Allahabad 1992, 96*)
3. How the animals can be studies in wild or in their natural habitat ?
4. Compare ethology with comparative psychology.

5. Describe growth of sociobiology.
6. What is sampling ? Discuss various sampling methods.

Short Answer Questions

1. Give an account of neurophysiological techniques of studying the behaviour.
2. Write about the following.
   1. Neurochemical techniques.
   2. Role of statistics in the study of animal behaviour.

   (*Kanpur* 2001; *Bundelkhand* 2002)

   3. Satellite telemetry system.

Very Short Answer Questions

1. Which part of brain is assoicated with aggression and defence activities ?
2. What is Broca's area ?
3. Define Wernicke's area.

Multiple Choice Questions

Choose the correct answer from the four alternatives given.

1. Which method you will use to take notes on mother-infant relationship ?
   (*a*) focal sampling (*b*) scan sampling
   (*c*) ad libitum (*d*) sequence sampling
2. Methodology used to study behaviour in wild includes
   (*a*) sampling (*b*) marking technique
   (*c*) satellite-telemetry technique (*d*) all of them
3. Sleeping behaviour in cats was introduced by stimulating
   (*a*) cortex (*b*) pones
   (*c*) brain stem (*d*) hypothalamus
4. Neuroanatomical technique involves
   (*a*) use of chemical (*b*) use of electric electrodes
   (*c*) lesion and ablation (*d*) low energy cyclotron
5. In the human brain, dictionary of learned words is located in
   (*a*) broca's area (*b*) amygdala area
   (*c*) wernicke's area (*d*) none
6. Adrenalin and histamine are
   (*a*) physiological agent (*b*) psychosomatic
   (*c*) foreign agent (*d*) hormone

Answer (MCQs)

1. (*a*); **2.** (*b*); **3.** (*c*); **4.** (*c*); **5.** (*c*); **6.** (*a*).

Behaviour Patterns

Behaviours may be broadly classified into two main types: 1. **Group behaviours** which include behaviour of the group or species. It is subdivided into two types: (i) **stereotyped behaviour** and (ii) **acquired behaviour.** 2. **Individual behaviours** which are characterised by the individuals.

In fact, different behavioural patterns in animals develop as adaptive ends, and range from simple, brief and stereotyped acts to complex and highly variable acts — comprising a chain of events. A behaviour is stimulus-bound but modified later during the life history of an individual in response to the surroundings. Therefore, a behavioural pattern seems to be first innate (endogenous) and then acquired.

3.1. STEREOTYPED BEHAVIOUR

A behaviour is called stereotyped when an individual repeats the same pattern of behaviour again and again. In case of stereotyped behaviour, the animal is to a large extent **stimulus bound,** where a pattern of stimuli trigger a sequence of responses. Since this kind of behaviour is essentially the outcome of *inherited properties* of the nervous system of the animal, it is also known as **innate** or **inborn** or **inherent behaviour.**

Characteristics of Stereotyped Patterns (SPs)

1. They are highly complex consisting of a chain of acts in a definite sequence.
2. They are complex in origin and are predictable.
3. They are species-specific or sex-specific.
4. They are initially stimulus-dependent.
5. They do not result as a consequence of experience.

Pattern of Movement in Stereotyped Behaviours — Fixed Action Patterns (FAPs)

Stereotyped behaviours involve a highly intricate pattern of movements and such patterns sometimes become so established that the animal ignores other acts coming in between and continues the one and the same pattern. Then stereotyped acts are considered to be fixated and the behaviour patterns are called **fixed action patterns (FAPs).** A FAP involves a series of events in a well ordered sequence. Such action patterns in animals were first described by **K. Lorenz.** He attributed FAPs to those which were "*innate and invariant*", *i.e.*, being endogenous they did not vary between individuals or between repetition by the same individual.

All stereotyped actions are not fixed, *i.e.*, some FAPs are stereotyped, others are not. The species have a series of stereotyped and species-specific patterns. Some patterns are inhibited by the nuptial males, some only by the young ones and still some only by lactating mothers. But all individuals in a category exhibit patterns of behaviour more or less in the same fashion.

Example of FAPs

Lorenz, while characterizing FAPs enumerated that FAPs were triggered by the external stimuli. He stressed that once triggered, the FAPs had become independent of the external world. **Lorenz** and **Tinbergen** (1939) exemplified this hypothesis from *egg-rolling behaviour of graylag goose* (which is a ground nesting bird).

It has been observed that when an egg rolls out from the nest, the graylag goose attempts to take it back into the nest. This is done by placing the beak ahead of the egg and drawing it back toward the chest. The movement of beak towards the chest is in fixed fashion and described as an action of FAP. A side to side movement of beak which might occur in this retrieval process is described as a variable component. The nature of FAP in this case was confirmed by removal of egg in the middle of retrieval process; as expected it did not stop the movement of beak in these individuals. It is to be noted that the goose continued repeated beak movements towards its chest despite the fact that triggering stimulus, the egg, had been removed.

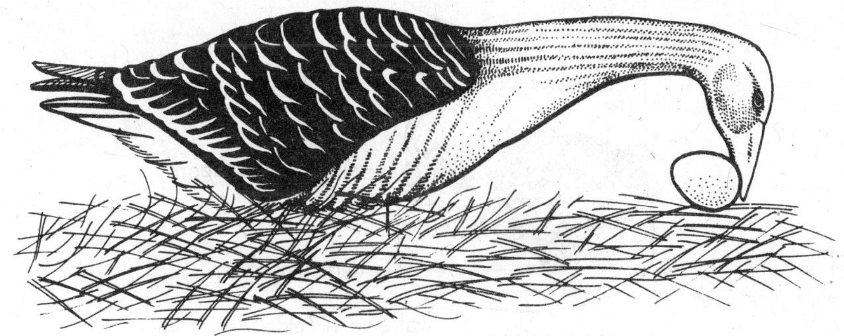

Fig. 3.1. The egg-rolling response of graylag goose shows FAP.

Modes of Stereotyped Behavioural Patterns

Stereotyped behaviour includes the following types of behaviours :

1. Spatial orientation (Kineses and Taxes)
2. Reflexes
3. Instincts
4. Motivation

1. Spatial Orientation (Kineses and Taxes)

Orientation means the act of placing an animal in a definite relation to the point of the compass or other fixed direction or stimulus. It includes ways in which animals direct appropriate behavioural responses to environmental stimuli; an orientation response describes a steering reaction and may or may not be accompanied by movement. Because orientation precedes the behavioural response, it is regarded as one of the simplest component of behaviour. Like the reflex it can be modified by experience.

Orientation involves single animals and being a type of pre-programmed, instinctive or stereotyped behaviour, it is built into the nervous systems of animals. Such a built in behaviour is a necessity for organisms that live solitary lives and that do not have parental care, thereby lacking any opportunity to learn by imitation how to carry out certain actions. **Spatial orientation** or **orientation** on a local scale (over a few centimeters or so) to easily defined stimuli such as a light source or humidity gradient, can be seen in many invertebrates. It falls into the following two main types : (1) movements in which animals attain their preferred habitat without orienting with respect to the source of the stimulus (**kineses**) and (2) movements which are made at some fixed angles to the stimulus source (**taxes**).

(i) Kineses

The simplest form of spatial orientation is **kineses,** in which the animal's response is proportional to the intensity of stimulation but is independent of the spatial properties of the stimulus. As a result of kineses, an animal may arrive by chance in a favourable environment, by which the intensity of stimulus is reduced or entirely eliminated. Kineses are of following two types:

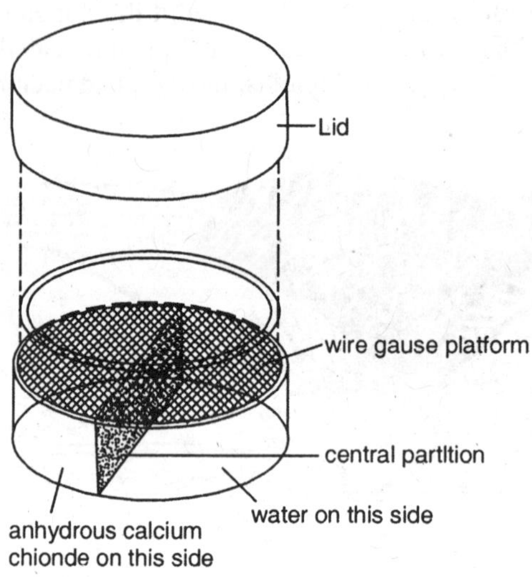

Fig. 3.2. A choiçe-chamber for experiments on orthokinesis of woodlice (*Oniscus*).

(a) Orthokinesis. In an orthokinetic response, the animal alters its rate of movement, according to the intensity of the stimulus. When the stimulus, which may be light or moisture, is of the right intensity animal slows down and thus spends more of its time under these conditions. For example, when woodlice (*Oniscus porcellio*) are allowed to move about on gause, one half of which is maintained at high humidity (Fig. 3.2) and the other half at low humidity, they scuttle rapidly over the dry half but slow down on the damp half. After a while a cluster of animals builds up on the damp part of the gause.

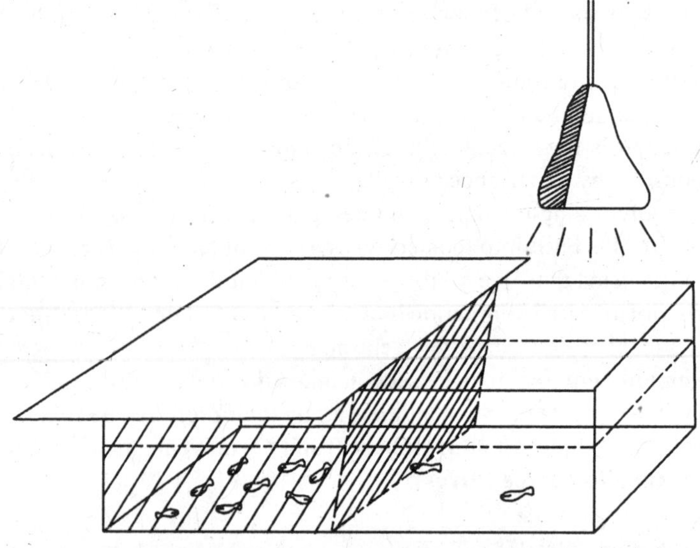

Fig. 3.3. Flatworms (*Dendrocoelum* sp.) showing klinokinesis.

(b) Klinokinesis. When the rate of change of direction of the animal increases as the intensity of a stimulus increases, it is called **klinokinesis.** An example of klinokinesis is provided by the flatworms *Dendrocoelum lacteum*. If one side of a dish of water with some flatworms in it is covered with a dark cloth, the animals will be found to accumulate in the darkened area (Fig. 3.3). In fact, they do this because they are escaping from the light: it is because they move more slowly and turn more often in the dark.

Another example of klinokinesis is provided by the unicellular organism, *Paramecium,* which exhibits klinoketic response with respect to the local concentration of carbon dioxide (Fig. 3.4).

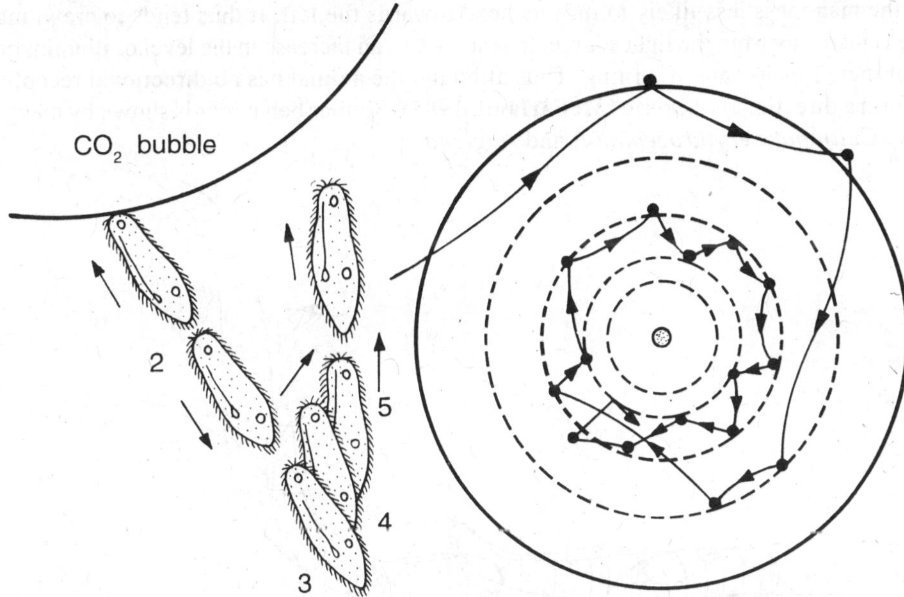

Fig. 3.4. Klinokinetic response of *Paramecium* in the vicinity of a CO_2 bubble. When it senses a high carbon dioxide concentration, *Paramecium* withdraws, turns through a certain angle and advances again; but its new direction is unrelated to the direction of the stimulus.

(*ii*) Taxes

Locomotor responses by which an animal orients itself toward or away from particular stimuli are called **taxes.** A taxic response is directional. If the response is movement towards the stimulus, it is a **positive response**; an avoiding reaction is a **negative response**. With respect to the type of stimulus, a taxis might be classified as one of the following: **thermotaxis,** response to heat; **phototaxis,** response to light rays; **thigmotaxis,** response to contact; **chemotaxis,** response to chemical substances; **hydrotaxis,** response to moisture; **rheotaxis,** response to currents of air or water; **galvanotaxis,** response to constant electric current; or **geotaxis,** response to gravity, and so on.

Distinguishing features of taxis

1. It is a fixed action as well as stereotyped pattern.
2. It involves spatial orientation.
3. There must be orientation of the whole body.
4. The direction of the movement should be continuously guided by the external stimulus.
5. The orientation movement is directly proportional to the stimulus strength.

(a) Phototaxis. The mechanisms of phototaxis — movement in relation to light source – have been studied extensively, mainly because most animals readily react to light and because light is so easily controlled in quality, direction and intensity. For example, when the larva of the housefly (*Musca domestica*) has finished feeding, it seeks out a dark place to pupate. At this stage it will crawl directly away from a light source, *i.e.*, it shows **negative phototaxis.** The maggot has primitive eyes on its head that are capable of registering changes in light intensity but are not able to provide information about the direction of light source. As the maggot crawls it moves its head from side to side (Fig. 3.5). When the light on the left is brighter than that on the right, the maggot is less likely to turn its head towards the left. It thus tends to crawl more toward the right, away from the light source. In response to an increase in the level of illumination, the maggot increases its rate of turning. Thus although the animal has no directional receptors, it can perform a directional response (**McFarland,** 1985). Similar behaviour is shown by maggots of blowfly (*Calliphora erythrocephala*) and *Euglena.*

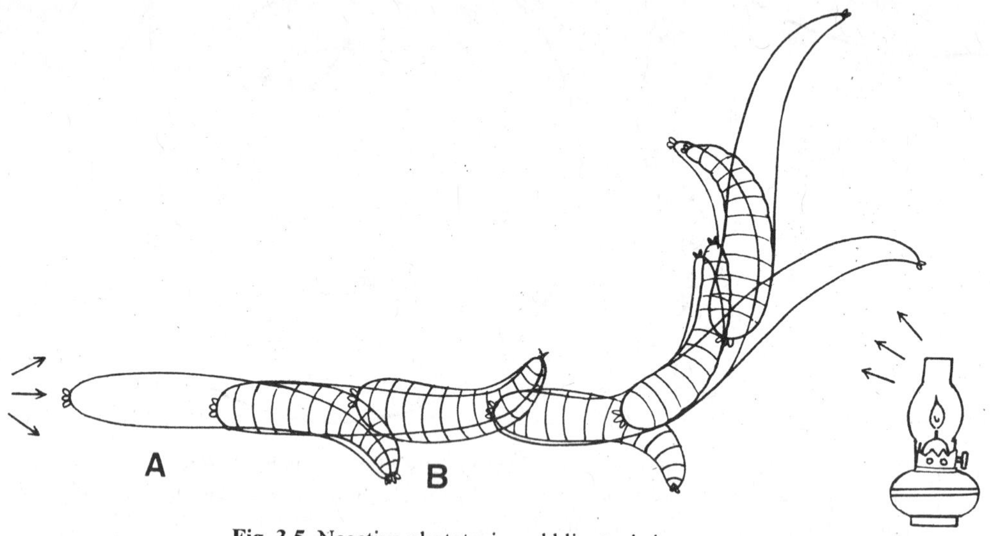

Fig. 3.5. Negative phototaxis and klinotaxis in a maggot.

The unicellular flagellate *Euglena* requires light to carry on its photosynthesis. It moves towards a dim light, *i.e.*, away from a very bright light (since strong sunlight will disrupt its chlorophyll molecules). Thus in *Euglena,* the taxic response is dependent on the intensity as well as on the direction of the stimulus (**Wallace,** 1979).

Klinotaxis, tropotaxis and telotaxis. Orientation by successive comparison of stimulus intensity requires turning movements. Usually it is called **klinotaxis.** For example, klinotaxis is shown by maggots of housefly and blowfly during negative phototaxis (*i.e.*, by moving away from a directional light source). As the larva moves, it swings its head from side to side alternately exposing left and right photoreceptors to the light behind. When the left receptor is stimulated, the animal responds by bending to the right and vice versa, thus moving away from the light. Many animals show klinotaxis in response to gradients of chemical stimulation.

Klinotaxis is a key feature of those animals in which the photoreceptor is placed asymmetrically in the body. However, an animal having a pair of shaded eye spots pointed in slightly different directions can make simultaneous comparisons centrally of the light intensity falling on the two sides of the body. It can then achieve **tropotaxis** which enables the animal to steer a course directly toward or away from the source of stimulation. An example of tropotaxis is the *negative phototaxis* shown by planarian worms which instead of bending from side to side take a fairly direct course away from the light. Another example of tropotaxis is provided by the pill woodlouse (*Armadillidium vulgare*) which lives under stones or fallen trees and shows a *positive phototaxis* after periods of dessication or starvation with its two compound eyes on the head, animal is able to move directly toward a light source. If one eye of woodlouse is blacked out or masked, *the animal moves in a tight circle* keeping the good receptor towards (**positive stimulation**) or away from (**negative stimulation**) the stimulus source. This shows that the two eyes normally provide a balance of stimulation. Likewise, to escape from its pursuing predators, the grayling butterfly (*Eumenis semele*) flies upwards toward the sun. If blinded in one eye, it will "escape" in circles as it attempts to equalise the stimulus.

In contrast to tropotaxis, **telotaxis** is a form of directional orientation that does not depend upon simultaneous comparison of the stimulation from two receptors. It involves orientation to a directional stimulus when only one of a symmetrical set of sense organs is operating. For example, honey bees with one eye masked will walk up a beam of light. Orientation is achieved by bending the body until the good eye is maximally stimulated. By reducing the amount of turning as stimulation of the eye increases, the bee maintains a steady course up the beam. Likewise, in case of hermit crab, when there are two sources of stimulation (*i.e.*, light), the hermit crabs move toward one and never in a median direction, showing that the influence of one of the stimuli is inhibited.

Pharotaxis, menotaxis and mnemotaxis. Pharotaxis is a sophisticated type of phototactic orientation that resembles with a ship that navigates at night by the flashes from a distant lighthouse (*pharos* means lighthouse). Pharotaxis can often lead an animal badly astray. If a naive sailor kept a constant angle between his ship's course and a distant lighthouse, he would travel in a closing spiral and end up on the rocks at its base. This seems to be what happens to moths and other animals when they are attracted to a candle flame or other point light source at night. They tend to approach the light source slightly from one side and then spiral in on a collision course. The usual explanation of this maladaptive, often fatal behaviour is that insects are indeed naive navigators with respect to earth-bound light source. Their pharotaxis has evolved to operate with respect to very distant (*i.e.*, astronomical or celestial) light sources whose rays are practically parallel (polarised light). This enables them to hold to a substantially straight course during a journey of short length and duration.

There is another type of telotaxis which resembles with pharotaxis, called **menotaxis** or **compass orientation (Griffin,** 1955). Menotaxis involves orientation of an animal at an angle to the direction of stimulation. An example of menotaxis is the light compass response shown by homing ants. These animals are guided, in part, by the direction of sun. If the apparent direction of the sun is changed slowly by means of a mirror, then the ants change course accordingly. Once it was thought (**Brun,** 1910's) that if ants (*Lasius niger*) were confined in a dark box for a few hours in the middle of their homeward journey, they would maintain the same angle to the sun when released. But it later came to light (**Jander,** 1950s) that the ants compensated for the movement of the sun and continued in the same direction upon being released. Such time-compensated compass reactions have been shown to occur also in the beetle (*Geotrupes sylvaticus*), the pond skater (*Velia currens*) and the honeybee (*Apis mellifera*).

Some invertebrate animals (*e.g.*, digger wasp *Philanthus triangulum*) are capable of much more complex orientation called **mnemotaxis (Kuhn,** 1919) or **piloting (Griffin,** 1955). Mnemotaxis is the ability to find a goal by referring to familiar landmarks. The animal may either search randomly or systematically for the landmarks. It is called mnemotaxis in part because of the configurational nature of the stimuli involved and in part because these stimuli must be learnt (**Hinde,** 1970).

(*b*) Geotaxis. Geotaxis, the orientation with respect to gravity is practically universal in animals. Some animals including vertebrates have special gravity-sensing organs, *e.g.*, statocysts, semicircular canals. Insects depend on sense cells that detect the direction of gravity on long appendages such as the antennae as these are held outspread from the head. Most animals can detect up from down with great precision, and may show **positive geotaxis** (downward movement) under one set of circumstances and **negative geotaxis** (upward movement) under other. For example, caterpillars move down the stems of their food plant when about to pupate in the ground; moths emerging from the pupae show a marked negative geotaxis as they climb to the top of the nearest stem, where they spread and dry their wings.

(*c*) Phonotaxis. Orientation of animals in relation to a sound source, called **phonotaxis,** may be almost as precise as phototaxis (tropotaxis and pharotaxis). Some moths hear the ultrasonic cries made by a hunting bat as it uses its sonar to find its way and its prey in the dark. Two simple ears of moths that are located on their thorax, enable them to take a flight path directly away from their potential predator. The ears of crickets and grasshoppers are located on their front legs, which can be spread quite wide apart so as to give them an added information about the direction of sound source. Birds and mammals locate and orient themselves in relation to a sound source. Mostly this depends on comparing the signals reaching the two ears together with "search movements" such as turning the head back and forth. Animals such as cats and horses with their directional and highly mobile external ear flaps undoubtedly are better at acoustic localisation than the human beings.

(*d*) Chemotaxis. Orientation with respect to the source of some chemical diffusing in air or water, called **chemotaxis,** is almost universal. It is less precise than phototaxis or phonotaxis because of the way in which chemical stimuli become distributed through the medium. Nevertheless, chemotaxis makes possible some behavioural performances that seem amazing to "non-olfactory" creatures like ourselves. For example, *Paramecium* back-and-turn if they happen to swim into a region containing a noxious chemical. Such type of chemotaxis is called **phobotaxis (Jenning,** 1900s). Further, chemical stimuli may cause an animal to become active and then to engage in a trial-and-error search pattern, like a dog following the scent of game animal. Orientation to a chemical source is often accomplished by a combination of search activity and **anemotaxis.** Anemotaxis is the tendency of animals to move upwind if the signal is received on land. In contrast to it, **rheotaxis** is the tendency of aquatic animals to swim upstream, if water is the medium.

The nose of mammals seems to have an intrinsic directional capability as the eyes and ears do. Also owing to rapid adaptation of the olfactory receptors, it is unlikely that animals can perceive gradual concentration differences in order to discover whether they are approaching or retreating from a source.

2. Reflexes

Reflex behaviour is the simplest form of reaction to stimulation. It describes the rapid autonomic response of the body or part of body to a simple stimulus. Reflexes can be identified by following features:

1. Reflexes are the simplest units of complex behaviour.
2. They are normally automatic, involuntary and stereotyped. (*Stereotype* means to repeat an action without a variation).
3. Unlike taxes, reflexes usually involve the movement of a part of body.
4. They are not necessarily to be continuously guided by the stimulus.
5. Reflexes are the outcome of the neural mechanism; therefore, they are innate in origin. (*Innate* means existing in an individual from birth).

6. A reflex action is directly proportional to the stimulus strength; stronger the stimulus shorter the latent period, weaker the stimulus longer the latent period. (*Latent period* is defined as the time lag between the application of the stimulus and the observations of its effects).

Types of Reflexes. Different reflexes (Box 3.1) can be grouped into following two types:

1. Tonic reflexes. These are slow, *i.e.*, long lasting responses. They are involved in maintenance of the muscular tone, posture and equilibrium adjustment.

2. Phasic reflexes. These are fast, *i.e.*, short lived responses. They occur during flexion response of the body.

Box 3.1.
Classification of Reflexes

A reflex may be nutritional, defensive, locomotor, etc. Besides tonic and phasic reflexes, there are many types of reflexes:

1. Exteroreceptor reflex. It is due to stimulation of receptor on the outer surface of body.

2. Enteroreceptor reflex. It is due to stimulation of receptors of the internal organs.

3. Proprioreceptor reflex. It is due to stimulation of receptors on the skeletal muscle joint and tendons.

4. Spinal or cranial reflex. It depends on the part of central nervous system involved.

5. Conditional reflexes. These are those habits or voluntary activities which in due course of time become reflex actions, *e.g.*, cycling, salivation, driving, swimming, etc.

6. Unconditional reflexes. These are due to receptors involved and stimulus. These are normal reflex actions of the body, *e.g.*, contraction of the pupil of the eye in bright light, closure of eyelids, etc.

Example of Reflexes

1. The reflex action can be easily understood by a simple knee jerk reflex (Fig. 3.6). Simply a stimulus below knee cap (patella) provokes spinal cord which relays it through motor nerves to extensor muscles, causing them to contract. A reflex action, thus, is established with an overall effect of a kicking action.

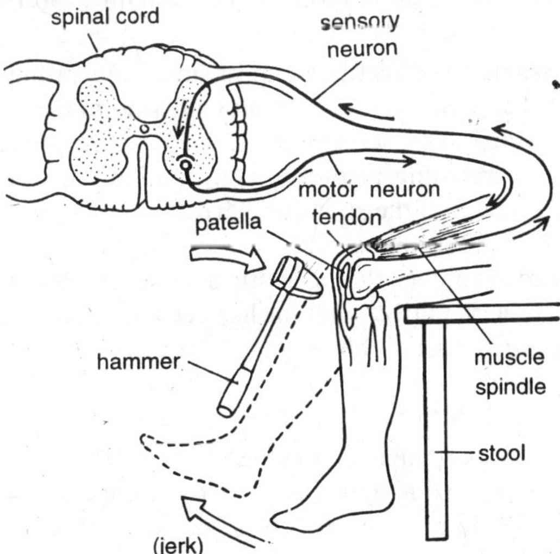

Fig. 3.6. Diagram of a simple reflex — the knee jerk reflex. A tap just below the pattelar cap below knee cap stimulates receptors in the tendon which send messages to the spinal cord. These are relayed to the motor nerves serving the extensor muscles which reflexively contract causing a kicking action.

2. Sometimes, the whole animal may be involved in the reflex response. For example, the **startle response** of humans require reflex coordinations of many muscles. Similarly, the **reflex withdrawal responses** of invertebrates such as polychaete worms and various molluscs involve the whole body. Such reflexes of these invertebrates are triggered by giant nerve fibres that carry very fast messages to all muscles involved so that they contract suddenly and simultaneously.

Reflex Arc and its Components

The nerve path traced by a reflex is known as a **reflex arc.** A reflex arc includes the *five* neural components: 1. a cell or group of cells acting as a **sensory receptor**; 2. an **afferent** or **sensory neuron** carrying impulses from the sense cells; 3. an **efferent** or **motor neuron** carrying impulses to effector cells; 4. an **interneuron** linking sensory and motor neurons; and 5. **effector organ** (muscle or gland). The reflex arc for the knee jerk is rather special (Fig. 3.6). Instead of connecting with an interneuron in the spinal cord, the sensory neuron connects directly with a motor neuron. The knee jerk reflex therefore forms a **monosynaptic arc** (it contains a single synapse), a type more typical of invertebrate nervous system. When a limb is pulled away from a painful stimulus, it involves somewhat more complex reflex arc. In this case, the passage of impulses from sensory to motor neurons is mediated by a mass of interneurons (called **association neurons**) in the spinal cord. Such a reflex arc is **polysynaptic.**

Physiological Characteristics of Reflex Arcs

1. Latency. In a reflex arc, there is usually a **latency** in response. For example, when a dog withdraws its paw from a painful stimulus, the response should occur within 27 seconds if the only limiting factor is the speed of impulse conduction through appropriate neurons. In fact, the response does not appear for 60-200 seconds. The reason for the delay is the number of synaptic junctions the impulse has to traverse. However, the latency of reflexes decreases with increased strengths of stimulus. This is important because reflexes are designed to act in emergencies.

2. After-contraction of muscles. Normally the muscle relaxes a few seconds after stimulation ceases. If contraction is elicited through a reflex arc, it may persist for several seconds after stimulus cessation. The duration and strength of this '**after-contraction**' of muscle is positively related to the strength of the stimulus.

3. Summation and warm-up effect. Among its other integrating properties, the CNS is able to accumulate repeated stimuli over time (**temporal summation**) from different parts of the body (**spatial summation**). The scratch reflex of the dog provides good example of summation. A dog shows a scratching response with its hind leg if an irritating stimulus is applied to certain areas of its back. If the stimulus is weak, scratching may not occur until after 20 or more applications. The CNS summates the irritations over the time they are applied. In this way summation may lead to another characteristic of reflexes, the **warm-up effect.** Reflexes often do not occur in full strength until a stimulus has been applied several times. Such warm-up effects are well known in various mobbing and courtship activities. In reflexes where the warm-up effect occurs, it seems that successive stimulation brings more motor neurons into play thus producing a stronger reaction. This is known as **motor recruitment.**

4. Fatigue. Reflexes show heightened effect of **fatigue.** Normally, a muscle stimulated to contract remains responsive for several hours. But if the muscle is stimulated via reflex arc, its response declines very rapidly. In some cases, as the scratch reflex in the dog, responses last only about 20 seconds. What appears to happen is that, with repeated stimulation, the interneurons in the CNS begin to block impulse transmission by increasing the resistance of their synaptic junctions. A stronger or qualitatively different stimulus, however, will soon re-establish a fatigued reflex.

Neurophysiology

Reflexes are organized at different neural levels — from the spinal cord to the brain. For example, simple flexion or extension reflexes are achieved at spinal cord and the locomotion and other mechanisms in our body are the functions of brain.

In general, the reflexes involve two control processes: one at central level and other at peripheral level. In the case of the **central control**, precise instructions are issued by the brain and are obeyed by all the muscles/organs involved. The coordination of swallowing movements in mammal is under central control. Central control is important in the coordination of many skilled movements, which requires rapid muscular activity (*e.g.*, movement of eye ball, Fig. 3.7 A). Central control is also operative in prey catching by cuttlefish (*Sepia*).

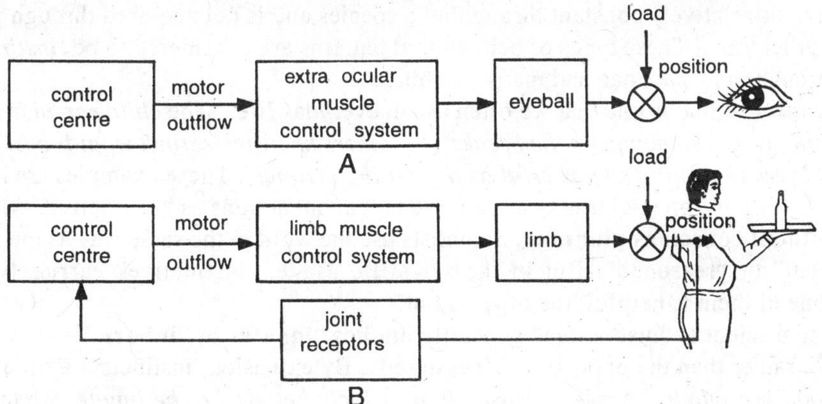

Fig. 3.7. Schematic representation of central and peripheral control of reflexes. A — The movement of eyeball is under central control; B — The movement of limb is under peripheral control.

The **peripheral control** of reflexes is attained through sensory and muscular system (Fig. 3.7B). For example, peripheral control is important in the movement of limbs, because these are often subjected to loads. Displacement of limb position is monitored by sensory receptors and feedback to the central control centre and the displacement is corrected.

Open-loop and closed-loop models of reflexes. Manning and **Dawkins** (1998) describes two models of the control mechanism – the open loop and closed loop models (Fig. 3.8).

1. Open-loop model. It has no monitor and input coming to behavioural system is interacted and we get the output (response). Even if output is subjected to some disturbance, it is not going to affect the behavioural system. Thus, in this case, output and input are not closely interlinked.

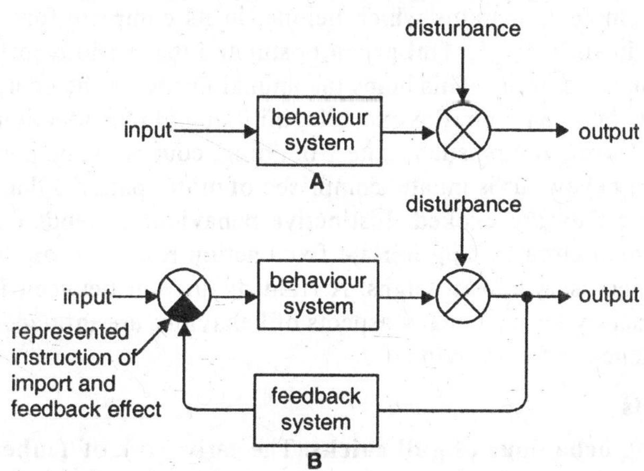

Fig. 3.8. Two models for reflex mechanism. A — open-loop system; B — closed-loop system.

2. Closed-loop model. In this case any alteration in the output feeds back to the input to affect the behaviour system and thus changes the output. Thus the change is proportional to the change done in behaviour system by disturbance. Open loop system seems to operate when responses are rapid, while closed-loop system works during slow and precise movements.

When two reflexes target the same organ, in this situation both become incompatible neurologically. One of them tries to inhibit the performance of the other. This is known as **inhibition.** Inhibition brings a greater coordination between the different responses in animals.

3. Instincts

Among animals, there are many examples of elaborate behaviour that show adherence to a 'plan' that is relatively constant throughout a species and is not acquired through previous experience or learning. These types of behavioural patterns are considered to be '*instinctive*', a term that through the years, has had many meanings.

The word 'instinct' is one that we often use in everyday life. "The *child ran in front of my car and instinctively I slammed on the breaks*", "*Mothers know by instinct what is best for their children*", "*There is nothing so powerful as the mating instinct*". These examples are all rather different but they have one point in common. We talk about actions as "instinctive" **when they are done without conscious planning.** Scientists use the word in the same way as the opposite of "intelligent" or "reasoned". But in the scientific usage, "instinctive" carries two extra meanings, one of them is helpful, the other is not.

First, to a scientist "instinctive" generally implies "**innate**" or "**inborn**" — the opposite of "learned", rather than the opposite of "**reasoned**". By extension, instincts are often used to refer *to broad behavioural tendencies which are widely believed to be innate:* we talk about 'the mating instinct', 'the maternal instinct', 'the gregarious instinct' and so forth. In other context we might call them "drives" or "motivations". This specialized scientific meaning of "instinct" is quite reasonable. What is not reasonable, however, is the tendency to see the '*description*' of behaviour as in some way an '*explanation*'. This is obviously wrong. To say that some behaviour takes place without conscious calculation, or even without individual learning, is no way to explain it; it is simply to label it as being, so far, unexplained.

The early ethologists considered innate behaviour to be behaviour that is determined by heredity and that is part of animal's original makeup and therefore that is independent of experience of the individual. Thus, **Lorenz** (1939) writes about the characteristics of innate behaviour as being **hereditary, individually fixed** and thus open to **evolutionary analysis. Tinbergen** (1942) refers similarly to instinctive act as being highly stereotyped coordinated movements, the neuromotor apparatus which belong, in its complete form, to the hereditary constitution of the animal. In 1951, **Tinbergen** postulated that various instinct centres in the body were arranged in a hierarchy. This helps the animal in saving the energy required for the stimulation of a particular behaviour. For example, initiation of reproduction in animals lead to a number of associated behaviours, such as nest-building, courtship and parental care.

Thus, instinctive behaviour is innate, comprised of motor patterns that can be carried out correctly the first time they are evoked. Instinctive behaviour depends on the activation of preprogrammed neuronal circuits. Appropriate fixed action responses are made the first time the animal sees the necessary release signs. A rigid distinction between innate and learned behaviour is unsatisfactory because many aspects of behaviour are influenced both by genetic factors and by experience of the individual.

Examples of Instincts

1. Food-begging behaviour of gull chicks. The early work of **Tinbergen** and **Perdeck** (1950) has shown that shortly after emergence from the egg the herring gull chick begins to peck at the tip of its parents beak. The pecks, in turn, induce the adult gull, to regurgitate a mass

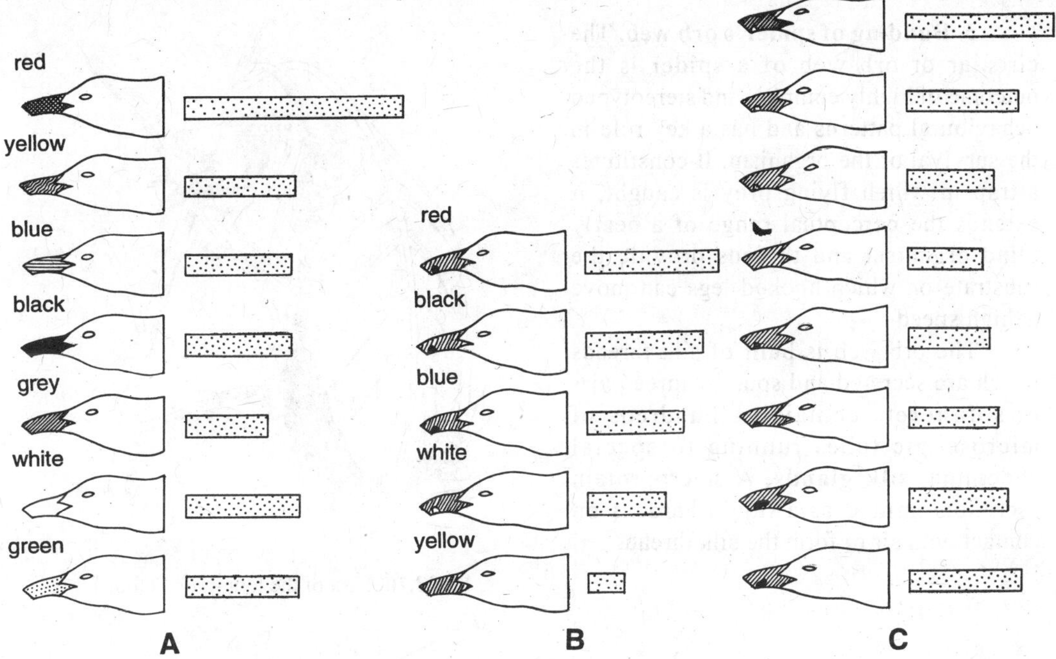

Fig. 3.9. Three series of model heads used in the pecking tests: A — Measures the effect of bill colour; B — That of patch colour (all the bills were yellow) and C — The effect of varying the contrast between patch and bill colour (all the bills were grey). The length of bar besides each model is proportional to the number of pecks it received.

of half-digested fish, clams or refuse from the local garbage dump. This serves as first meal for the baby bird. **Tinbergen** considered the chick's pecking response to be an innate or instinctive behaviour because it was characteristic of all baby herring gulls, performed in apparently stereotyped manner right from the start of their lives outside the egg.

Tinbergen used cardboard models of beaks and heads to determine exactly what caused the gull chicks to beg. He and his colleagues went to gull colonies, took newborn chicks to a test site and presented them with various models. These tests showed that only some of the model's characteristic were significant in affecting the behaviour of the young birds. The relevant stimulus properties were the *shape of bill* (long, pointed objects were preferred over short, stubby one), the *colour of patch* (red preferred over all other colours) and the *contrast of the patch with its background* (the more the merrier). In addition, a moving-beak model attracted many more pecks than a stationary one and a beak that is pointed down was more effective than one oriented in any other direction.

2. Nest building behaviour in tailor birds. A young long-tailed tailor-bird has managed to acquire a mate and is in the process of building its first nest. The quick, little bird first finds two large, hanging leaves and then proceeds to sew them together in a most curious fashion, first by punching holes in their margins with its sharp beak and then threading strands of spider webs, cocoons or bits of string through the holes. It manages to tie stopknots in each piece of thread and to pull it tight so that the leaves form a short of pouch. The bird then searches the area for twigs and grass, which will make the floor of nest, and on this cushion it deposits its eggs.

The tailor-bird has not witnessed nest building by older birds and it has had no previous experience in such a task. But on its first try it build a nest — a nest that may not be perfect, or ever as good as those of older birds, but it is adequate and the tailor-bird is able to raise its young (Fig. 3.10).

3. Building of spider's orb web. The circular or orb web of a spider is the outcome of highly complex and stereotyped behavioural patterns and has a key role in the survival of the organism. It constitutes a trap in which flying prey is caught, it extends the perceptual range of a nearly blind organism and threads provide the substrate on which hooked legs can move at high speed.

The orb web is built of silk threads which are secreted and spun by three pairs of **spinnerets** containing hundreds of microscopic tubes running to special abdominal **silk glands.** A scleroprotein secretion emitted as a liquid hardens on contact with air to form the silk thread.

Fig. 3.10. Nest of the long-tailed tailor-bird.

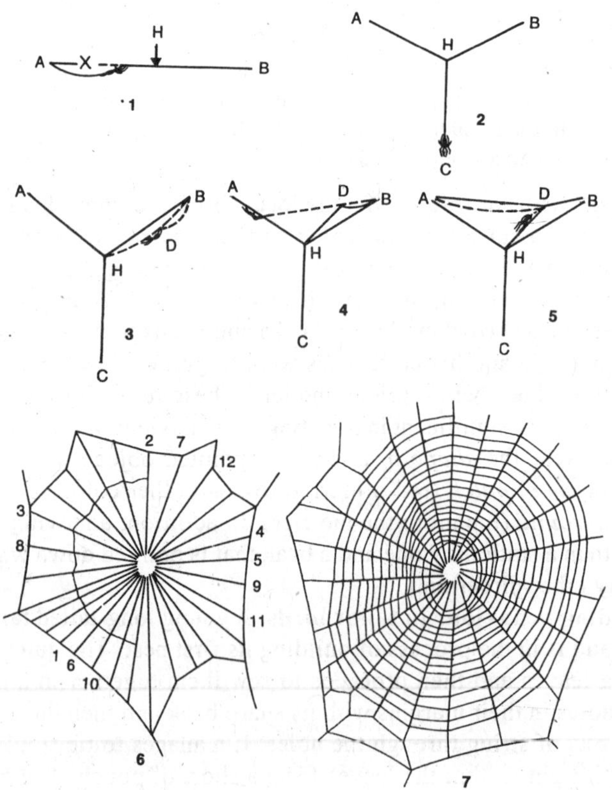

Fig. 3.11. Various stages in the construction of an orb web. A-E — Formation of frame threads, radii and central hub (H). F — An orb web with most of its radial spokes but no catching spiral. The numbers indicate the order in which the radii are laid down. Note that each radius is built in a different sector from the previous one. G — A completed orb web with viscid or sticky catching spiral. The spider lays down a tightly coiled viscid spiral, starting at the outside and working inward, removing and eating the temporary scaffolding as it goes. Spider sits in the centre or hub of the orb web.

Spider's silk threads are stronger than steel threads of the same diameter and are said to be second in strength only to fused quartz fibres. The threads will stretch one-fifth of their length before breaking.

The orb webs are spun by common spiders all over the world and the whole operation takes less than half an hour. Most spiders in the field tend to build their webs early morning at sunrise. Orb web is a regular structure made up of **frame, radial spokes** and catching **viscid spirals.** The stages of construction of a orb web were first worked out by **Hans Peters** in 1939 (Fig. 3.11).

The main rules of the orb web spider are first to build a Y-shaped scaffold, then in a set order, the frame and radial spokes, and finally the auxiliary (temporary and non-sticky) and stick (viscid) spirals (Fig. 3.11). Each main rule contains a set of subrules for measuring angles and walking set distances up certain threads. By following the programme of elementary rules the spider can build a complex structure without having a plan of it in her head.

Fig. 3.12. A newly hatched and blind honey guide's nestling which is a brood parasite kills a barbet, the host nestling, with special mandibular hooks on its bill.

4. All young newly hatched cuckoos of the brood parasitic European species will maneuver the eggs of their host parents on to their backs and out of the nest. In certain species of brood parasites, such as honey guide *Indicator indicator*, the newly hatched and blind nestling may even kill the host nestlings (the barbet, *Lybius bidentatus*) or push them out of the nest (Fig. 3.12).

Characteristics of Fixed Action Patterns

Ideally, a pattern of behaviour should have all the following characteristics if it is to be described as **fixed action pattern (FAP) (Lorenz,** 1932; **Tinbergen,** 1970; **Lea,** 1984).

1. Stereotypy. This behaviour always occurs in the same form.

2. Universality. The behaviour occurs in all members of a species. By this we should not expect ganders (male geese) or goslings (young geese) or even mother geese outside the nesting season to show the egg rolling behaviour. But we do expect all mother geese with nests to show it.

3. Independence of individual experience. Fixed action patterns should occur regardless of the individual animal's past history. This leads directly to the ethologist's famous isolation experiment: an animal is reared by hand in isolation from all other members of its species.

4. Ballisticness. A ballistic response is one that cannot be varied if circumstances change after the response has been launched. **Lorenz** and **Tinbergen** illustrated this fact for the graylag goose's egg rolling by offering a goose a giant egg, in fact a Cardboard Eastern egg painted with the same sort of markings as a real goose egg (Fig. 3.13). The goose responded to this giant egg as if it was a real egg, and when it was just put down outside the goose's nest, she stretched out her neck, stepped towards the Easter egg, hooked her beak over it and began to roll it in.

Teal-capturing behaviour of falcons. Hunting of birds and other animals involves instinctive behaviour. As a hunting roaming falcon begins searching for prey, its behaviour becomes highly variable. It is hard to predict which way it might bank, or whether it will rise lazily on air columns or suddenly drop toward the ground. At such times, the bird is hungry and might be equally pleased at the sight of a flying bird or a scampering mouse. If such a falcon sees a group of teal flying below, its behaviour suddenly becomes less random as it swoops toward them in the first stage of **"teal-hunting behaviour"**. The swoop is performed in somewhat the same way throughout the species, but it may vary a little bit from one individual to the next. The falcon might hit at the front or the back of the group of teals and might strike at birds to the right of the left of the flock as it passes. In fact, this is likely to be a sham pass to scatter the flock so that an individual bird (teal) may be selected for pursuit, and there is less variation in its performance than in the random searching. Once an individual teal is selected, the falcon goes through a series of maneuvers to keep the hapless bird separated from the flock as it closes in. The falcon's behaviour becomes increasingly less variable as it enters a final stage of high-speed dive.

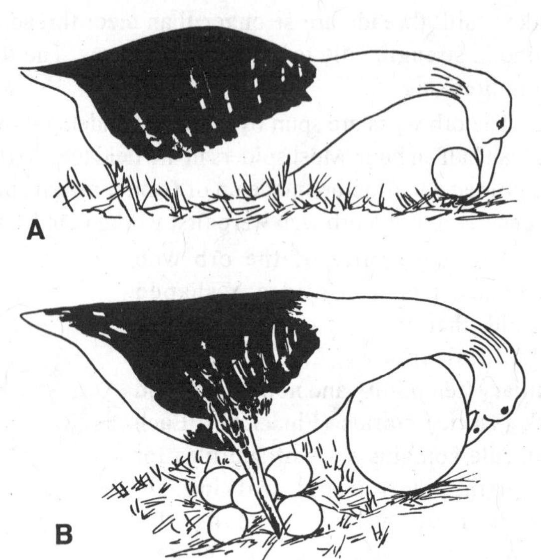

Fig. 3.13. Greylag goose retrieving an egg which is outside the nest. This movement is very stereotyped in form and used by many ground-nesting birds. The goose attempts to retrieve a giant egg in precisely the same fashion.

Fig. 3.14. When a roaming falcon spots its prey, it may show increasingly stereotyped behaviour until finally it ends the instinctive sequence with the final consumatory act.

The early part of falcon's dive is still variable to a degree and can be altered in response to change in the teal's escape behaviour; if the teal changes direction, so does the falcon. However, there is a point at which the falcon's behaviour can no longer be altered. It is committed to a specific pattern. Falcon's actions now are performed in very much the same way as they have been at this stage in previous hunts. This is the brief moment at which falcon's feet are tightly clenched into a fist as it swoops at almost unbelievable speed. Unless the teal makes a last second change in its pattern, it will be knocked from the sky, and the falcon's hunt will be successful (Fig. 3.14).

The performance of the strike (a FAP) ends that part of the instinctive sequence, but is usually triggers another one that also begins with a variable pattern. The falcon may follow its stunned prey to the ground through a number of routes and it may glide, flap or walk to the body. Once there, it grasps the body of the teal in a number of ways but its behaviour is again becoming increasingly stereotyped. It tears food from the body in much the same way as it has in previous kills and in much the same way other falcons treat their kills. Finally when falcon swallows, a series of muscles are called into play that contracts for the same duration and in the same sequence under all conditions. Swallowing is also a fixed action pattern.

5. Singleness of purpose. Fixed action patterns have only one function. Hooking an egg in with the beak is not the most efficient way of recovering an egg. But it is a method that works and one imagines that it could be useful in other contexts, for example, in nest-building, herding goslings or even perhaps in feeding on the surface of water. Yet Lorenz suggested that the behaviour pattern is never shown in any other context except that of errant eggs.

6. The existence of known trigger stimuli. We can find same stimulus or set of stimuli, which will reliably trigger the response. Experimental proof of this came from Lorenz's experiment with graylag goose in egg-retrieval and in **food begging response** of the herring gull chick as studied by **Tinbergen** in Netherland. Herring gull nests on the ground. The parents go off to feed at sea, on a local scrapheap or at some other distant source of food. When they return, they land on their small nesting territory and stand near (often over the chick), pointing their beaks with red mark at the ground. The chick then pecks at the parent's beak, which stimulates the parent to regurgitate the food it has collected, allowing the chick to feed (Fig. 3.15).

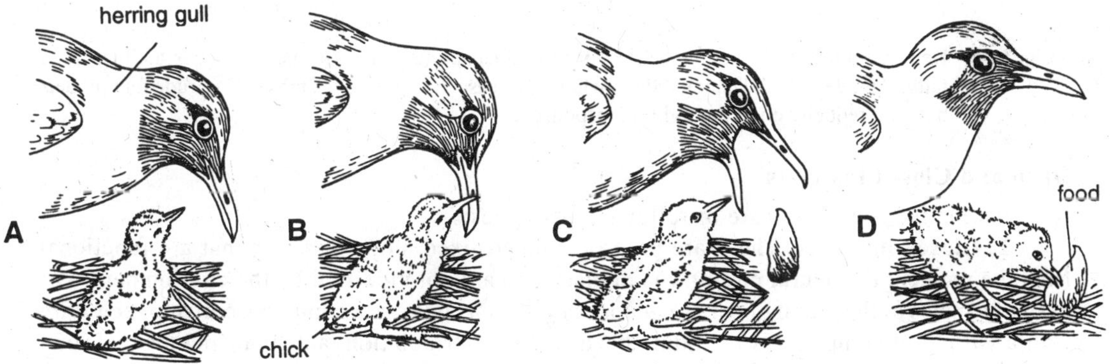

Fig. 3.15. Response of herring gull chicks towards the red spot on the beak of the parent. A — The parent arrives, the chick looks at the parent bill, B — Chick pecks at the red spot, C — Parent regurgitates the food, D — Chick eats the food.

The stimuli that trigger off fixed action pattern is referred to as either **"releasers"** or **"releasing stimuli"**. **Lorenz** (1937, 1970) used both terms, *.i.e.,* sigh stimulus and releaser. **Tinbergen** demonstrated that the territorial defense in male three-spined stickleback fish was stimulated by red belly of intruding male. If a dummy model even not figured as stickleback but with red belly was placed near the nesting male, it would arouse the fighting (instinct) behaviour in a fish. But a true model without red belly fails to evoke or release the fighting response (Fig. 3.16).

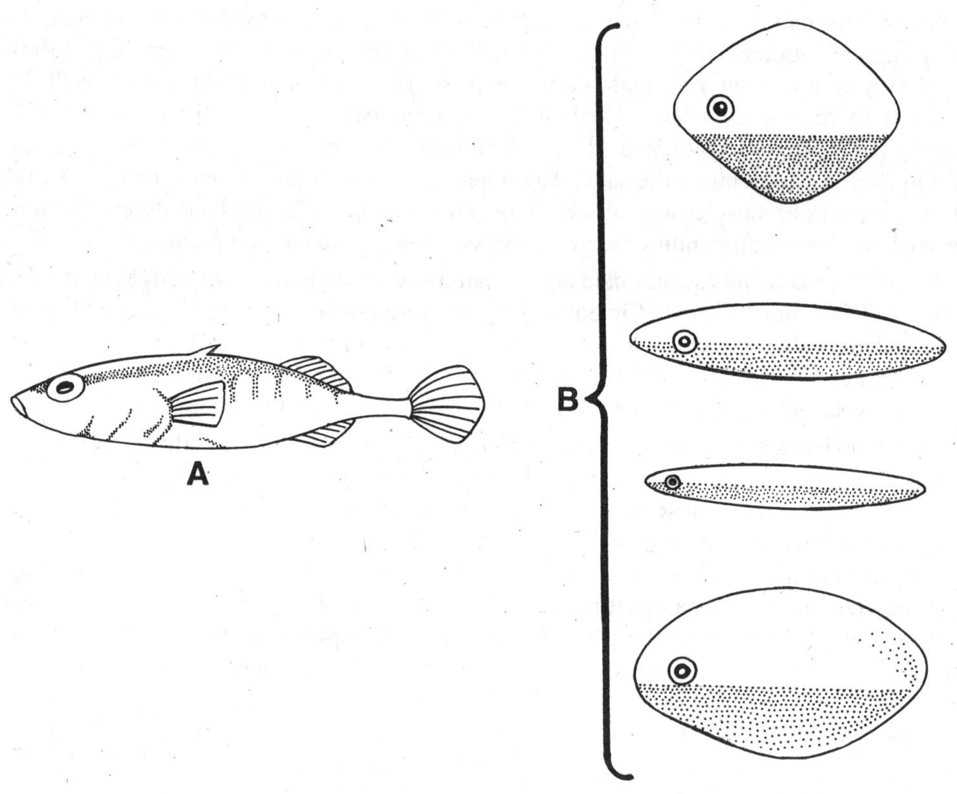

Fig. 3.16. Demonstration of instinctive behaviour in stickleback fish in territorial defence. Note that (A) a faithful model of fish lacking red belly fails to release the fighting response, (B) Crude four models but with redbelly elicit the fighting response.

Open and Closed Instincts

Instinctive behaviours are of following two types:

1. Closed instincts. These are preprogrammed fixed motor patterns that are functional from the moment the neural circuitry is in place and are not modified by the environment. For example, males of the grasshopper *Gomphocerippus rufus* possess a complex courtship repertoire with a variety of components including head rocking, stridulation, antennal flicking and loud song. These elements are organised in an invariant sequence; the entire courtship pattern is performed for the female over and over again until either she accepts the male and copulates or the male wearies and goes elsewhere (**Elsner,** 1973). **Hoyle** (1976) has found that there is no indication that the courtship sequence is ever altered in any way; instead they are the product of a closed "motor tape" that can be played back time and again.

Closed instincts are quite common in courtship signals, where there is a premium on clear-cut messages and little benefit associated with modification of the message. For example, a mature male spider that locates a receptive female of his species mechanically performs a series of behaviour patterns in response to sign stimuli. If he is fortunate, copulation follows. Depending on the species, courtship signals may include an elaborate set of leg-waving movements or the use of silk to tie down the female in a special way or the presentation of wrapped prey in a particular manner (**Bristowe,** 1958). In fact, in some species of spiders, the male runs the risk of being eaten prior to copulation if he does not provide the correct courtship messages. This act as selective force that favours males whose nervous systems permit them to generate a complete and accurate courtship sequence the first time they attempt to mate.

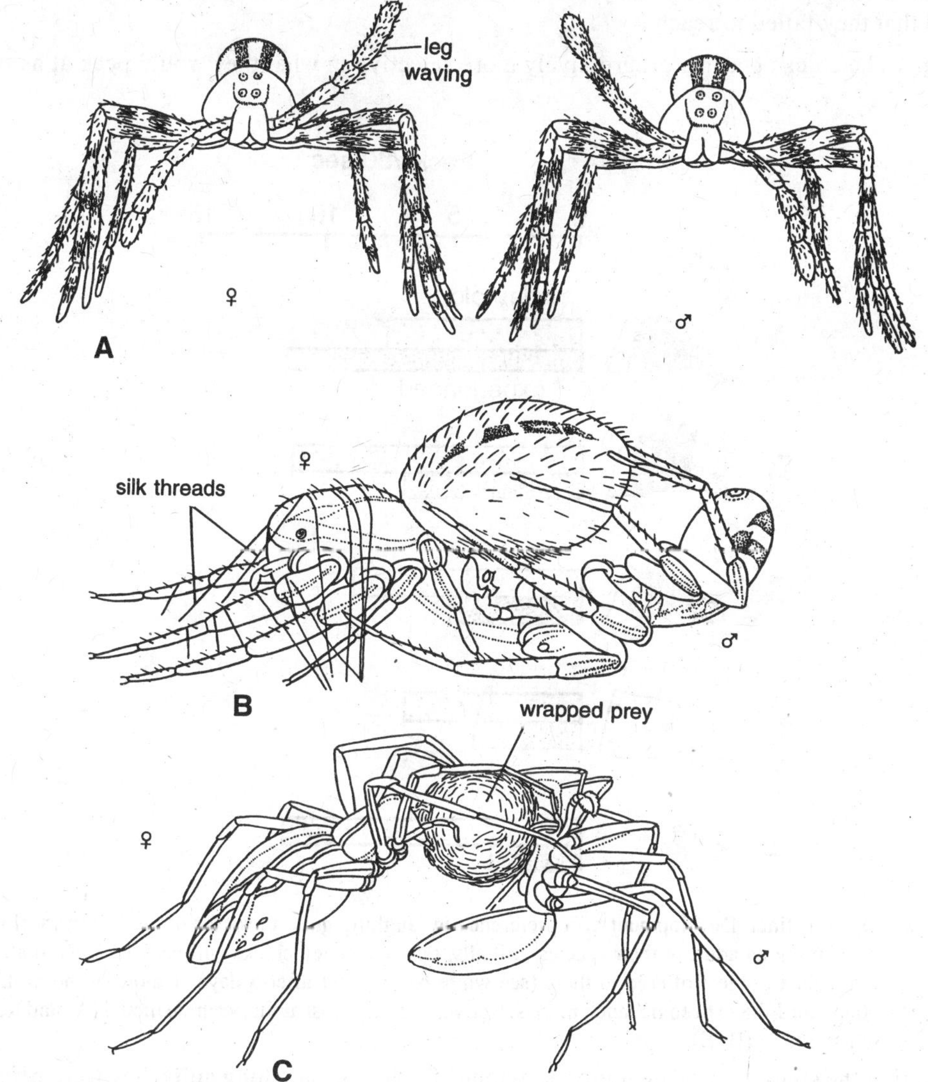

Fig. 3.17. Closed instinct. Courtship and mating in three species of spiders. A — A courtship signal of a male wolf spider (*Lycosa amentata*) given before approaching the female, B — The male of *Xysticus cristatus*, having fastened his mate to the ground with silk threads, is about to deposit sperm in her genital aperture, C — The male of *Pisaura mirabilis* (on the right) presenting wrapped prey to the female before mating.

2. Open instinct. In many species, behaviour that is functional when first performed, is capable of modification as a result of interaction with the environment. For example, by presenting cardboard models to chicks of the laughing gull of different ages, **Hailman** (1967, 1969) found the following responses (Fig. 3.18):

(*i*) Older birds were better able to place most of their pecks on the bill of a model and not off to one side or the others.

(*ii*) Pecking efficiency improved as well, with the chicks growing better at judging the distance between themselves and the beak. Initially they might be right on target, but would strike the bill so powerfully that they would be knocked head over heels or be so far away from the bill that they failed to reach it.

(*iii*) The chicks became progressively more selective in what they would peck at as time passed.

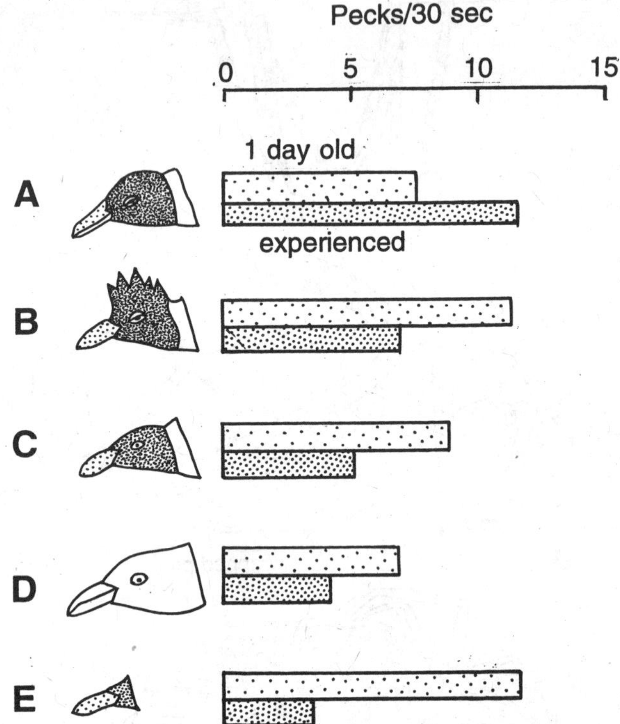

Fig. 3.18. Open instinct. Development of a preference in laughing gull chicks for the model most closely resembling an adult of their species. Initially, newly hatched chicks will peck actively at almost any pointed object offered to them (see white bars). Later after 3 days or more in the nest, the young birds have become much more selective, pecking most at the accurate model (A) and less at other models (B -E).

During the study of food begging behaviour of chicks of laughing gulls they were offered a series of different models, some close in appearance to a laughing gull adult, others not remotely similar to a laughing gull. Chicks that initially pecked at almost anything would later refuse to beg from models other than those that were fairly accurate representations of the head of a laughing gull (Fig. 3.18).

Adaptive significance of instincts. Instincts or FAPs are helpful for animals. An instinct renders the animals to acquire a prehand knowledge of aversive (hostile) situations in the environment and thus the animal is always ready to meet the challenges.

4. Motivation

It is a matter of common observation that an animal does not respond to a stimulus in the same way every time that stimulus is encountered. For example, we can consider the case of African lion moving through the bush and coming upon a herd of wildebeest. On some occasions such an encounter results in lion stalking and perhaps killing one of the wildebeest. On other occasions, the lion walks casually past the herd apparently ignoring the presence of potential food. Thus something has changed between one encounter with food and another. But what has changed? Since there is no difference in the stimulus itself, we are left with the possibility of some internal change in the animal. That internal change is usually labelled as **motivation.** The word motivation comes from the Latin word *motivare* meaning "to move".

Thus, by motivation we mean *fluctuations in the physiological state of an animal that result in the animal responding in different ways to the same stimulus at different times* (**Barlow,** 1977). One may substitute the term motivation with certain more familiar terms such as **drive, mood** or **tendency.** When we say that an animal or a person, is motivated to do something, we generally imply that its behaviour is driven or directed by some internal force or urge. Essentially, an animal's motivational state is a "**black box**" by which we may know a lot about the inputs (external stimuli) and the output (behaviour), in those whose internal mechanisms are not observable. The task that faces those studying motivation is to reduce their ignorance about what goes on inside the box.

Distinguishing Features of Motivation

Motivated behaviour is defined as response (drive) that is directed towards a specific goal leading situation. It can be recognised by the following features:

1. Motivation is a goal oriented behaviour. The goal may be defined as a condition which leads to satiation.
2. Drive towards the goal is done by a stimulus.
3. A motivated behaviour includes the following three components:
   (*i*) A goal-searching phase (*e.g.*, food for feeding, water for drinking, etc.);
   (*ii*) A phase of consummating acts (feeding and drinking);
   (*iii*) A quiescent period (a period after satiation).
4. These components of motivation are FAPs and stereotyped in nature.

Examples of Motivated Behaviour

There occur in animals a number of daily or seasonal behaviours which are goal directed. For instance, an animal eats or drinks in response to food and water to satiate its hunger or thirst. A bird builds a nest to complete its life's most vital activity — the reproduction. These activities, *i.e.*, hunger, thirst or desire to build a nest, are described as **drives** (Box 3.2). A drive is **biogenic** (*e.g.*, sex drive) if it is for an urgent biological need.

Box 3.2.

Drive Concept

The traditional view of motivation is built upon a simple feedback principle. A change in the animal's internal state is sensed by brain and leads to buildup of **drive** to perform the appropriate behaviour. The drive gives rise to appetitive and consumatory behaviour. **Appetitive behaviour** involves a search for suitable external stimuli (the particular requirement or goal); when these are encountered, **consummatory activity** or behaviour, such as eating or drinking, takes place. The consequences of the consummatory behaviour reduce the drive, either directly or by diminishing the internal or external stimuli that led to the drive. The consummatory behaviour then ceases. Thus, animal can rarely just reach out and grab food. They have to spend time and energy foraging for and processing suitable items. Following consumatory behaviour, the animal may enter a temporary period of quiescence with respect to that particular goal (called **quiescent** or **refractory period**), although it may still show appetitive behaviour for other goals.

The term *drive* was introduced by **Robert Woodworth** (1918) as an alternative to **William McDougall's** (1908) concept of instinct. **Woodworth** distinguished between the *energising* (drive) aspects of motivation and the *directing* aspects of motivation. **Primary drives** result from tissue needs and other **secondary drives** were derived from learned habits. **Lorenz** (1950) developed similar notion of drive: 1. accumulation of action-specific energy giving rise to appetitive behaviour; 2. appetitive behaviour striving for and attaining the stimulus situation activating the innate releasing mechanisms; and 3. setting off of the releasing mechanisms and discharge of endogenous activity in a consummatory action.

Neurophysiology of Motivation

Hypothalamus and cerebral cortex of vertebrate brain play significant roles in the motivation behaviour:

1. Role of hypothalamus in behaviour. The hypothalamus which lies on the floor of the diencephalon, is considered as control centre of motivated behaviour for it is connected with the brain as well as with the hypophysis. Specific areas of the hypothalamus are sensitive to the concentrations of salts in the circulating body fluid (blood). It acts via two circuits: 1. via the neurohypophysis by the secretion of ADH (antidiuretic hormone) which enhances water reabsorption in kidneys; 2. by directing the animal for searching and drinking the water. In fact, the *stimulation* of ventromedial nucleus of the hypothalamus (VHM) suppresses, while the lateral hypothalamus (LH) increases the feeding activity. Likewise, some other types of behaviour, such as **sexual behaviour, sleep, emotional** and **maternal behaviour** are associated with specific areas of the hypothalamus.

The hypothalamus seems to contain two regulatory mechanisms which work antagonistically. If one way is excitatory, the other one is inhibitory. For example, for feeding ventromedial area of the hypothalamus provides an **inhibitory mechanism** and lateral area produces **excitatory influences.** Thus, drive is associated with excitatory but satiation with the inhibitory mechanism of hypothalamus. These two mechanisms are also influenced by the environmental stimuli, changes in **internal milieu** and cerebral cortex. Hypothalamus is known to readily acknowledge the changes in environmental temperature and fluctuations in concentration of body fluid, for it contains the thermoreceptors and osmoreceptors.

2. Role of cerebral cortex in motivation. Apart from the hypothalamus, other areas of brain also appear to subordinate the motivated behaviour. For example, basal regions of the cerebral cortex, frontal parts of the hemisphere and rhinencephalon (*i.e.,* small brain– a part of forebrain) are connected to the hypothalamus and exert an influence on the motivated behaviour. **Neocortex** is the part associated with the arousal of aggression.

Theories of Motivation

It is quite difficult to understand about the factors which motivate animals from inside. A two-fold approach is usually made for its investigation. One is through stimulating different nerves, tracing their pathways to sensory systems and effector organs and subsequently analysing the physiological factors which bring a change in the overt behaviour. The other approach is through altering the inputs to animal, for example, by food deprivation and then directly observing its effects on an overt behaviour rather than observing its internal mechanisms. Following classical models (now obsolete) have been proposed to explain the motivation:

1. Lorenz's hydraulic model of motivation. Konrad Lorenz (1950) proposed a clear-cut model of motivation by which he explained how internal and external causal factors interact to elicit behaviour. This model is variously called **hydraulic model, psychohydraulic model** or **Lorenz's water closet,** since it was designed by using analogy of an hydraulic flow system (Fig. 3.19).

A **causal factor** (stimulus) is any event, process or change in some condition which can be shown to activate, sustain or inhibit a particular behaviour pattern. Causal factors may be

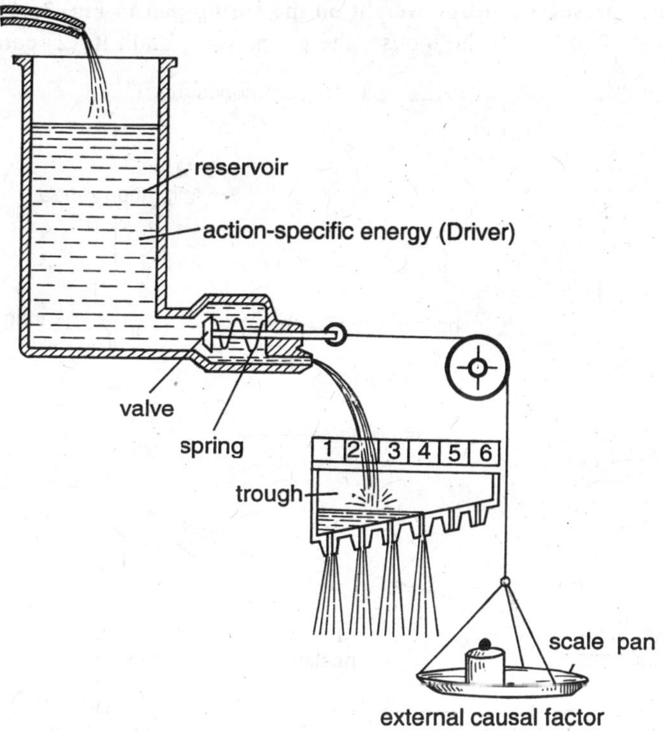

Fig. 3.19. Lorenz's hydraulic model of motivation. Action-specific energy is represented by water, which accumulates progressively in a reservoir when the behaviour concerned is not being expressed. The behaviour pattern occurs when water passes out of the reservoir (representing animal's drive level) into the trough beneath, higher threshold aspects of the behaviour (numbered 4, 5, 6) only being shown when a lot of water is passing into the trough. The valve in the reservoir is so arranged that it is opened by the combined effect of water in the reservoir (action-specific energy or internal causal factor) and of weights on a scale pan which represent the adequacy of external causal factors.

external stimuli, such as the appearance of a predator, the presentation of food or the behaviour of a mate, or **internal states,** such as the level in the blood of a hormone or of glucose. Internal and external causal factors interact to elicit a behaviour, a point graphically made by **Lorenz** in his hydraulic model. The water (or any fluid) in the system is analogous to **action-specific energy** which accumulates spontaneously over time. The longer an animal has not performed a behaviour, the more energy-specific to that behaviour accumulates in the animal's system and the more likely the behaviour is to be performed. This indicates that the animal's drive for that behaviour is increased. Energy is released through the valve when the animal encounters an appropriate stimulus (the external causal factors). The strength of the stimulus is related to the weight on the spring pan (Fig. 3.19). The ease with which the valve opens is a joint function of the characteristics of the stimulus (weights) and the amount of energy (fluid or water) in the reservoir. Depending on how far the fluid shoots out of the valve, different numbers of outlets in the trough come into operation. In behaviour terms, each outlet represents one stage of expression of the animals's drive.

For explaining Lorenz's hydraulic model of motivation, **Alcock** (1975) has used the example of a cat which has not eaten for some time. The cat's **feeding drive** has been building up like the water in the reservoir when suddenly a mouse appears. The mouse provides a strong stimulus to feed (it represents a heavy weight on the spring pan in Fig. 3.19) because it is a preferred prey of cats. On seeing the mouse, the cat may (1) stalk it, (2) catch it, (3) kill it,

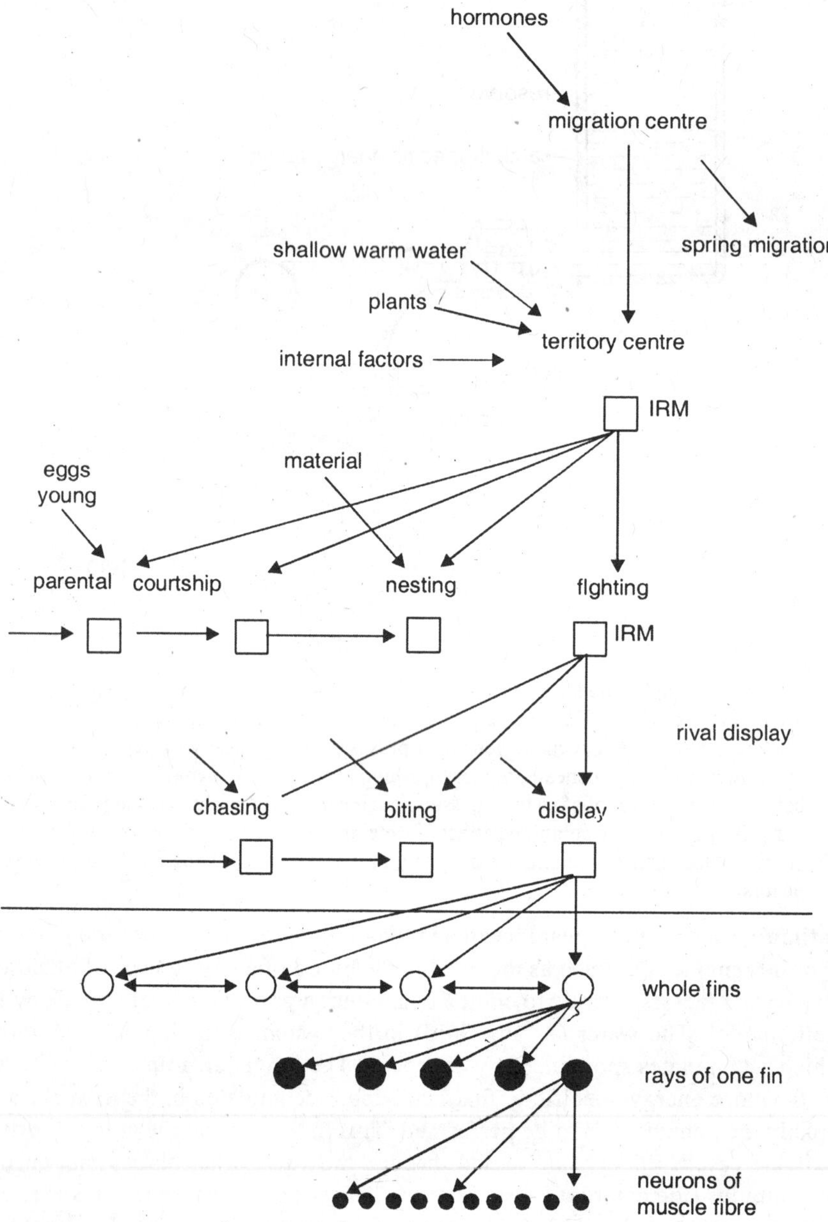

Fig. 3.20. Tinbergen's hierarchical model of the reproductive instinct of the male three-spined stickleback. Motivational impulses (arrows) 'load' behavioural centres. Impulses may come from the external environment, superordinated centres or spontaneously from within the centre itself. Innate releasing mechanisms or IRMs (open squares) inhibit the discharge of centres until released by an appropriate stimulus.

(4) eat it. It is as if the mouse released all the action-specific energy available for feeding which poured into the trough and flowed out as the sequential performance of acts 1-4, making up the cat's feeding behaviour. But what would happen if the cat has not been without food for very long time? Now it is as if only a little water has built up in the reservoir. When the water is released it shoots out only a short way and only brings into operation the first and second outlets in the trough. The cat is then observed to stalk and catch, rather than follow through to kill and eat it.

2. Tinbergen's hierarchical model of motivation. Tinbergen (1951) suggested a more complex model, called **hierarchical model of motivation** to explain the organization of behaviour over longer periods of time. In this model energy is not completely specific to individual's behaviour but is held in higher level behavioural centres such as those for **reproductive behaviour** or **feeding behaviour.** Tinbergen visualized motivational energy as flowing down through various inhibitory blocks to progressively finer levels of appetitive behaviour or consummatory acts through to muscular activities and finally indivisible motor units (Fig. 3.20).

In Tinbergen's hierarchical model, the observed patterns of behaviour reflect an order of functional organization (an **instinct**) within the central nervous system. Thus in reproductive instinct (Fig. 3.20) **hormones** are assumed to affect the highest centre, *i.e.,* the **migratory centre**, controlling reproduction in the three-spined stickleback (*Gastrosteus aculeatus*). This results in appetitive behaviour in the form of migration. Appetitive migration ceases when the fish encounters a particular type of habitat. The stimuli from the well-suited habitat (the **key stimuli**) excite a specific innate releasing mechanisms (IRMs). In its resting state, however, the IRM blocks the next centre in the hierarchy and inhibits the performance of behaviours beyond that point. When it is excited, the block is removed and the centre is freed for propagation. Impulses can now travel down to lower centres controlling, for example, *brood care, resting* and *fighting*, but each of these centres is blocked until their appropriate key stimuli appears.

Tinbergen's hierarchical model of motivation is an ingenious idea, and unlike "as if" model Lorenz proposed, was based on the ideas of physiologists. Tinbergen hoped it might have some neurophysiological reality, with various centres in the brain devoted to different instincts. Unfortunately it turns out not to be that simple — the nervous system does not store and use up energy as these models suggest. Nor this energy is neatly compartmentalised into centres and pathways with clear and distinct behavioural functions. Furthermore, though to some extent behavioural systems such as those controlling feeding, drinking or sexual behaviour, can be thought of as distinct, the factors affecting them overlap and they often influence each other. For example, the hormone **estrogen** is an important internal factor making female rats receptive to the male, it also makes them very active so that they are more likely to come across a male, and it makes them less interested in food so that, when receptive, they spend less time in eating and more time in looking for mates. This hormone thus influences the systems concerned with feeding, with activity and with sexual behaviour.

There are certain other (modern) models of motivation such as **feedback models** (Deutsch's model 1960; **Control models** (1997) to explain newt courtship) and **state-space approach.**

3.2. ACQUIRED BEHAVIOUR: LEARNING

W. H. Thorpe (1963) has defined learning as " that process which manifests itself by adaptive changes in individual behaviour as a result of experience". In a way learning is the acquision of new behavioural patterns based upon past experiences (Box 3-3). **Lorenz (1969)** has defined learning as an adaptive change in behaviour that results from experience.

Box 3.3.

Following two criteria are used to distinguish learning from other modifications of stereotyped behaviour:

1. Learning must be permanent and not the result of fatigue or fluctuations in motivation.
2. Learning must not be simply a permanent change in behaviour resulting from maturation.

For example, swimming behaviour of salamanders has been proved to be a stereotyped behaviour appearing as a result of maturation and not learning.

For example, if a baby toad encounters a tiny moving bug for the first time after losing its tadpole tail and hopping of onto land, it will be able to perform the stereotyped prey-capture behaviour of its species. It may orient toward the object, open its mouth, flip out its tongue, strike the prey, withdraw the tongue and the creature stuck to it and finally swallow its food. Some would argue that this is an innate response to a certain class of stimuli.

If the toad is taken into a laboratory after having matured further, it can be offered various insects under controlled conditions. A hungry, cooperative toad will go through the routine of snaping up flies, meal worms and other edible creatures presented to it. If the experimenter then places a toxic millipede in the amphibian's enclosure, the toad may take this bait as well. The millipede responds by exuding a violently nauseating substance from pores on its body, whereupon the toad will spit and push the prey from its mouth. Later on, *this toad will refuse to attack this millipede species even though it is hungry and will take edible insects eagerly.* The toad's behaviour has changed. The change continues for some time and is adaptive (*i.e.*, it makes biological sense for the toad not to waste its time trying to eat an animal that is poisonous) . Its altered behaviour can be traced to a discrete event in the toad's lifetime. The animal has **learned** to avoid this millipede after having tasted just once. Toads also can learn with equal facility not to attack other dangerous prey, such as stinging bumblebees and honeybees. Thus, although it is difficult to imagine how the toad could learn the technique of capturing its food, selected experiences can lead this predator to modify a presumably innate response.

Types of Learning

Learning is often classified into the following *five* major categories:

1. Non-associative learning
   (*i*) Habituation; (*ii*) Sensitization
2. Associative learning
   (*i*) Pavlovian learning (Classical conditioning)
   (*ii*) Operant conditioning (Trial-and-error learning)
3. Latent learning
4. Insight learning
   (*i*) Reasoning; (*ii*) Intelligence; (*iii*) Cognitive thinking
5. Phase-specific learning
   (*i*) Imprinting; (*ii*) Avian song learning; (*iii*) Language learning

1. Non-associative Learning

The mode of learning which develops in the absence of its association with any reinforcement (reward or punishment) is called **non-associative learning**. It is of following two types:

(i) Habituation

Habituation is a simple learning not to respond to repeated stimuli which tend to be without significance in the life of the animal (**Wood-Gush** 1983). Unlike the other forms of learning, habituation involves not only the acquisition of new responses but the loss of old ones. If an animal is repeatedly given a stimulus which is not associated with any reward or punishment, it ceases to respond. Thus, the phenomenon in which repeated applications of stimulus result in decreased responsiveness is called **habituation. Razran** has defined habituation as *learning what not to do.*

Animals are constantly bombarded by a host of different stimuli emanating from the environment. The time and energy costs of responding to every one stimulus would clearly be prohibitive. Furthermore, only a small proportion of the stimuli require a response (key stimuli). Habituation is a way of eliminating responses to stimuli which are sometimes important but, which in a particular case, are irrelevant. Rustling leaves, for instance, are worth reacting to (*e.g.*, by hiding) because they sometimes indicate the approach of a predator. However, repeated rustling without the appearance of a predator is likely to be caused by the wind. In this case, the animal ceases to respond (*i.e.*, habituates) to the rustling.

Likewise, the escape response of fish to a shadow passing overhead diminishes progressively if the stimulus is repeated every few minutes, until the fish cease to react at all. Similarly, the orientation response of the toad (*Bufo bufo*) toward potential prey progressively declines of non-edible prey-like objects are presented repeatedly. In fact, it is well known to fruit growers that scarecrows erected to deter birds are effective for a short time, and the birds soon become habituated to them. Attempts to scare birds from airfields by broadcasting alarm calls have run into similar problems of habituation. We all also know how one ceases to notice sounds, like that of a ventilator fan (exhanst fan) running continuously.

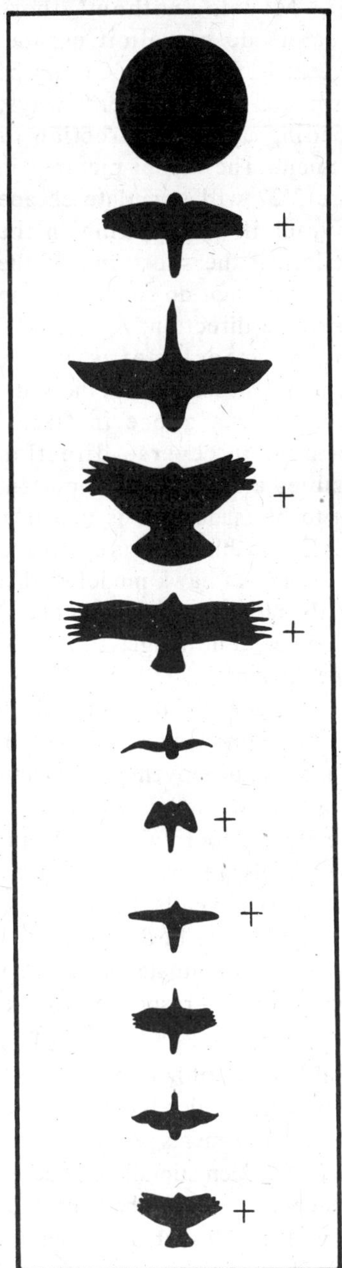

Fig. 3.21. Habituation: Models used to test birds in their reactions to silhouettes of birds of prey. Those models marked with a + stimulated escape response.

Classical Examples of Habituation

1. Habituation was first reported in 1887 by investigators testing the reaction of spiders to vibrating tuning forks. Initially, when the fork was vibrated, a spider would drop from its web by a thread to a distance of half metre. It would remain there for a time before returning to the web. With repeated tests the spider gradually reduced the distance to which it dropped and shortened the time of its return. After one month's training one spider remained on the web in spite of vibrations.

2. Various birds are preyed upon by hawks. **Tinbergen** has shown that these birds will flee if a hawk silhouette (*i.e.*, portrait, picture cut from black paper) is displayed overhead.

Silhouettes of other shapes do not evoke escape behaviour (Fig. 3.21). The characteristics that are important in the bird's identification of hawks are the wing shape, the long tail and short neck. Models without these characters do not elicit escape. Interestingly, the same model may be interpreted in different ways, depending upon its direction of movement. The model pictured in Figure 3.22 will stimulate escape behaviour if it is moving in the direction of the short end of the body; it does not do so it moves in the reverse direction. Apparently, when the model moves in the direction of the long end of the body, it resembles a goose in flight; movement in reverse direction resembles a hawk. The important point to be made here is that it is possible to habituate a prey bird to the presence of hawk model so that when the model is moved over head many times, it no longer stimulates escape response.

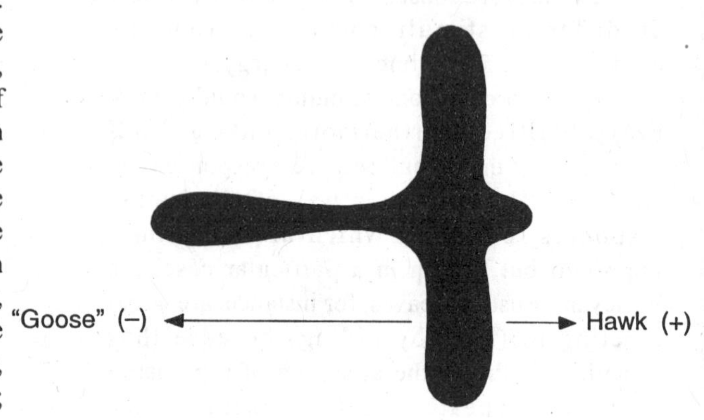

Fig. 3.22. Habituation. Model used to test birds in their reaction to the direction of movement of a silhouette. when the model is moved to the right it evokes an escape response, indicated by a (+) sign. Moving the model to the left does not stimulate escape, indicated by a (–) sign.

How does the distinction between silhouettes develop? Experiments with young turkeys have shown that they will initially respond with escape behaviour to all silhouettes of a certain size and rate of movement. Thus they will initially give alarm calls to discs and rectangles as well as to hawk models. After a few days these responses decline as habituation occurs. As long as the models are routinely presented to the turkeys, they show little reactions, however if any of the models are absent for several days and then presented to birds, they will stimulate a great number of alarm calls. Based on these tests it seems likely that birds in the wild do not normally respond to various birds as they fly overhead, because they are relatively abundant in the habitat. They become habituated to certain shapes. In contrast, hawks are relatively rarer and when they appear, the alarm response is immediately given.

3. A study by **Clark** (1960) on the polychaete *Nereis* (the ragworm) illustrates some of the typical features of habituations. *Nereis* is an annelidan marine worm which normally lives in a burrow or tube which it constructs in the mud at the bottom of brackish estuaries. The worm's head and anterior segments protrude from the tube whilst it feeds from the surface of mud. At such a time variety of sudden stimuli will cause the worm to jerk back rapidly into its tube. In the laboratory, **Clark** could easily get the worms to live in glass tubes in shallow basins of water. He found that a variety of stimuli such as jarring the basin (mechanical shock), touching the head of the worm, a sudden shadow passing over, etc., would all cause rapid retraction into the tube, but the majority of worms emerged again within a minute. If these stimuli were repeated at 1 minute intervals the proportion of worms responding fell off until none of them were retracting. **Clark** found that habituation occurs more rapidly if stimuli were given close together (**Manning,** 1979).

4. Sea hare *Aplysia* breathes through its gills, which are situated in a region called the **mantle cavity**; the gill's enclosure opens to the outside through an opening called the **siphon.** If an experimenter prods the siphon, the *Aplysia* withdraws siphon and gills, and folds them up

within the mantle cavity. This is called the **gill withdrawal reflex** and is simply a protective reaction. After a while, if undisturbed, the *Aplysia* puts its siphon out again, and if it is then prodded a second time, it will show the same withdrawl reflex. However, it will not do so indefinite number of times. If it is repeatedly prodded, it comes to ignore the stimulus, and leaves its siphon and gills out. Such a behavioural change is the habituation, *i.e.*, the *Aplysia* has learned not to respond to an apparently harmless stimulus (**Alcock,** 1986).

Certain characteristics of habituation. Habituation exhibits the following characteristics:

1. Stimulus specificity. In habituation, the decrease in responsiveness occurs only with reference to the **habituating stimulus.** This stimulus specificity is shown clearly by habituation in territorial **sticklebacks. Peeke** and **Veno** (1973) observed decrease in the level of aggression between neighbouring territory owners. Fish were aggressive to one another when they first establish territories but quickly ceased to respond. The adaptive advantage of this habituation is clear; there is no point wasting time and energy chasing fish which are neighbours rather than intruders.

2. Length of the interstimulus interval. Another important factor influencing habituation is the length of the **interstimulus interval (ISI).** The longer the ISI, the less habituation we expect. **Davis** (1970) investigated the effects of increasing the ISI in rats. He trained animals to expect different ISIs for a 50ms alarm tone of 120 db, then compared their tendencies to habituate under identical conditions. He found that long ISIs were less effective in producing habituation but had longer-lasting effects overall.

3. Dishabituation. If a novel stimulus is presented during the process of habituation, then there is an increase in responsiveness. This is called **dishabituation.** It is thought to be due to changes in the animal's level of arousal and is very similar to Pavlovian disinhibition (**McFarland,** 1985).

4. Sensitisation. Sensitisation is the opposite kind of change. While habituation means to become less sensitive to a stimulus, sensitisation means to become more sensitive to a stimulus. For example, if an *Aplysia* receives an alarming stimulus such as an electric shock on the tail, it then respond more readily to other stimuli (such as prods to the siphon) to which it would otherwise have been less responsive. It has become more sensitive, for if an *Aplysia* receives a dangerous stimulus naturally, it probably means some hazardous entity nearby, and it will pay to be careful.

Box 3.4.

Neurophysiology of Habituation and Sensitisation in *Aplysia*

To understand habituation and sensitisation of *Aplysia* more clearly, we can consider its neurophysiology. The nervous control of the gill withdrawal reflex is a simple unit of one sensory neuron and one motor neuron (Fig. 3.23). The siphon contains the sensitive end of the sensory neuron, which at its other end is directly connected by a synapse with a motor neuron controlling the muscles of the mantle cavity. When the sensory neuron is stimulated, it fires the motor neuron and the siphon and gills are withdrawn. How does the system habituate? There are two possible mechanisms for the regulation of such simple system: either a change in the amount of neurotransmitter released by the sensory neuron, or a change in the sensitivity of the motor neuron to constant doses of neurotransmitter. In the case of *Aplysia,* the former possibility is the real one. It is as if repeated activity in the sensory neuron has exhausted the supply of neurotransmitter, making the system as a whole less responsive.

For sensitization, two more kinds of neurons are needed. The dangerous stimulus is sensed by another sensory neuron, whose synapse connects with one or more interneurons that eventually run into the same synapse connected with the motor neuron that controls the mantle muscles. When the sensory neuron in the tail becomes active, it fires the interneurons which in turn causes a series of chemical changes in the motor neuron. The effects of these chemical changes is to make the motor neuron fire more readily. It requires less of stimulus to become depolarised. Hence the system is sensitized.

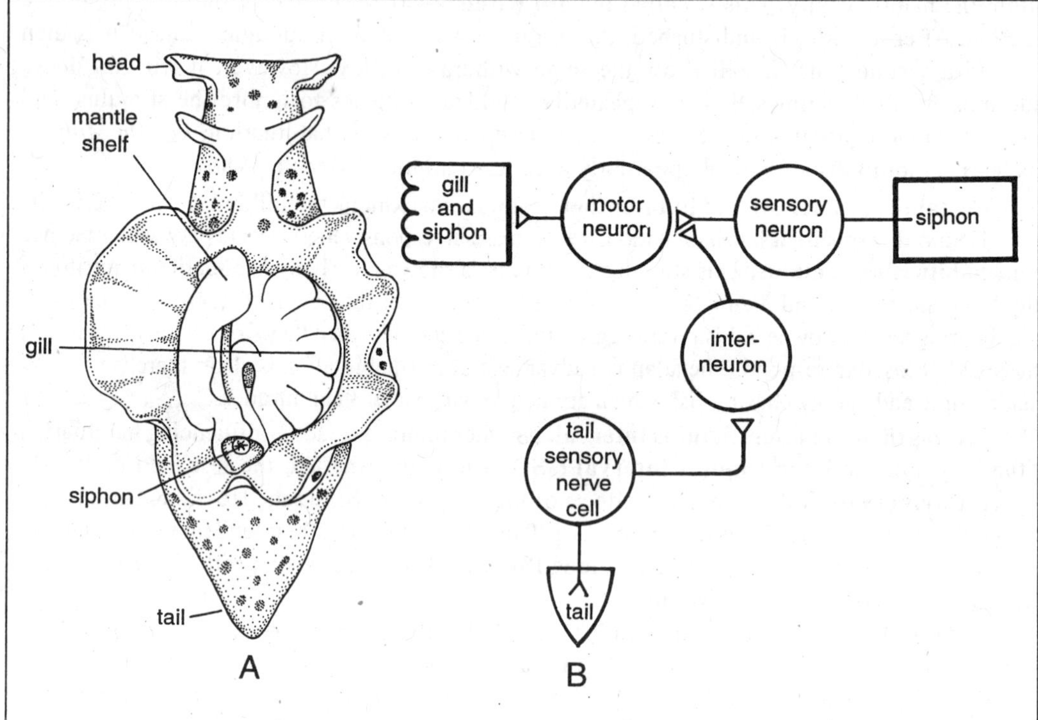

Fig. 3.23. The nervous control of siphon withdrawal in the sea hare *Aplysia.* When an *Aplysia* is tapped on the siphon, it withdraws its siphon and gills into its mantle chamber; but if it is repeatedly tapped, it ceases to respond. This habituation is controlled in the synapses of the siphon's sensory neuron and motor neuron.

2. Associative Learning

We tend to think of learning as a process whereby we acquire new responses and new capacities. In associative learning, a previous neutral stimulus or action has sufficiently important consequences to be singled out from other such events. After some repetitions followed by same consequences, a long-term association is built up between the event and its result and the animal's response changes accordingly.

Associative learning is of following two types:

(i) Classical Conditioning (Pavlovian Learning)

Most people have heard of the Russian physiologist, **Ivan Petrovich Pavlov** (1849-1936), who discovered that it is possible to train a dog to salivate at the sound of a bell. This type of learning is called **classical conditioning.**

Pavlov's classical experiment with dogs often involved the "**salivary reflex**". Dogs salivate when food is put into their mouths and Pavlov could measure the strength of their response by arranging a fistula through the cheek from the salivary duct, so that the drops of saliva fell from a funnel and could be counted (Fig. 3.24). A hungry dog was placed on a stand, restrained by a harness and every precaution was taken to exclude disturbances. In this position, it could be given various controlled stimuli such as lights, sounds or touch, and meat powder could be puffed into the mouth through a tube. In Figure 3.24, the device for blowing a controlled amount of powdered meat into dog's mouth to reward the salivation response has not been shown. A standard quantity of meat powder caused the secretion of a certain amount of saliva. Now Pavlov preceded each ration of powder by the sound of a metronome ticking (Box 3.5). At first, this stimulus caused no response, *viz.*, the dog pricked up its ears momentarily. However, after five

Fig. 3.24. A typical experimental set up in Pavlov's laboratory to show classical conditioning.

or six pairings of metronome followed by food, saliva began to drip from dogs fistula soon after the metronome started and before the meat powder arrived. Eventually the amount of saliva produced to the metromone alone was the same as that which was given by the meat powder.

Thus, the dog had learnt to respond to a new stimulus, previously neutral, which Pavlov called the **conditioned stimulus (CS** or **S_2).** The salivation response to the CS is the **conditioned response (CR** or **R_2).** Prior to learning, only the meat powder, called **unconditioned stimulus (UCS** or **S_1)** produced salivation as an **unconditioned response (UCR** or **R_2;** Fig. 3.26).

Box 3.5.
Metronome

A metromone is an instrument for indicating and marking exact time in music, consisting usually of a reversed pendulum whose period of vibration is regulated by a shifting weight.

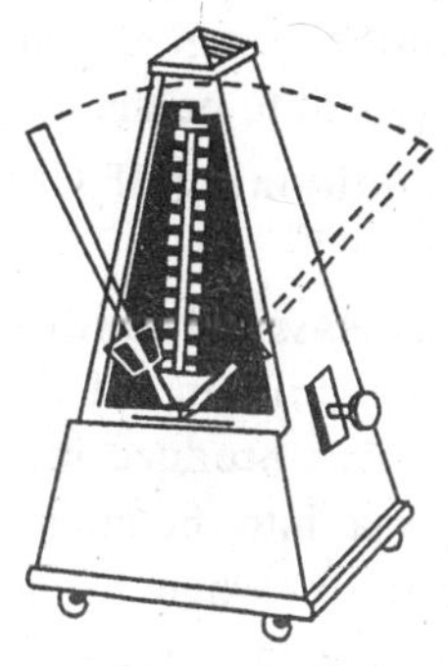

Fig. 3.25. Metronome.

Laws of classical conditioning. The phenomenon of classical conditioning is regulated by following four laws:

1. Law of contiguity. It states that the stimuli to be associated must occur together in time and space. For example, if the meat powder stimulus precedes the bell-ringing, conditioning will not occur. If the bell ringing precedes the meat stimulus by as much as three seconds, conditioning is hard to establish.

2. Law of repetition. It means that the conditioned response becomes progressively stronger and more probable in occurrence with progressively greater number of training trials.

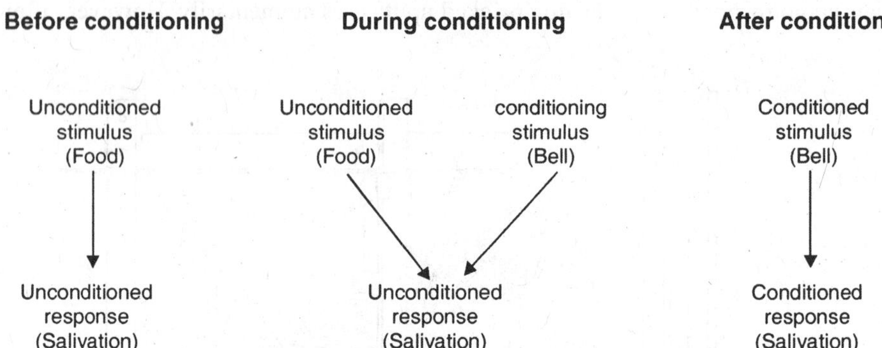

Fig. 3.26. Classical conditioning involves the formation of new connections. Before it takes place, a stimulus leads to a response. As a result of that stimulus being associated with a different one (the conditioning stimulus), the second stimulus comes to lead to the response. In the case studied by Pavlov, the bell (*i.e,* metronome) came to lead to salivation without the need for food to be present.

3. Law of reinforcement. It states that if the conditioning stimulus (such as the ringing bell) is presented to a conditioned animal without being associated with the normal stimulus (the meat powder) for a large number of times, the conditioned response (salivation) will gradually disappear (*i.e.,* the behaviour becomes **extinguished**).

4. Law of interference. It states that conditioning (learning) may disappear (forgotten) by new conditioning that interferes with the original conditioning.

Types of classical conditioning. Classical conditioning is of following types:

1. First order conditioning. If unconditioned stimulus (UCS) is paired with conditioned stimulus (CS) for a number of times, CS alone elicits a response. This direct pairing of CS with UCS is called **first order conditioning.**

2. Second order conditioning. If pairing and presentation of CS (sound of bell) with a second CS (light) reconditions the dog to second CS (light), it is called **second order conditioning.** In this type of conditioning, the animal is conditioned to a second CS presented along with first CS.

3. Positive conditioning. If conditioned response (CR) is beneficial as unconditioned response (UCR), it is said to be the **positive conditioning.**

4. Negative conditioning. IF CR becomes a negative reinforcer, it is said to be **negative conditioning.**

Significance of classical conditioning. Classical conditioning is useful to an ethologist in understanding of conditioning of animals to the environmental stimuli. For example, during reproduction, ring doves produce in their crop a specialised material, called "**milk**". They regurgitate this material into the mouth of their young. Doves that are parent for the first time initially regurgitate "milk" when they are touched on their breast over the engorged crop by the young birds. After a time, the parent becomes conditioned so that other stimuli (*e.g.,* sound and sight of brood) will produce regurgitation.

Many human's psychosomatic disorders (*i.e.,* physical disorders attributable to psychological causes) are thought to be due to the classical conditioning of automatic functions such as changes in blood pressure and dilation and constriction of blood vessels. The Russians have provided many elegant demonstrations of the conditioning of many responses usually classed as involuntary in both animals and humans. Around year 1970s, interest has centred on conditioning faster or slower heart rates. The conditioning model has also been applied to explain certain types of **asthma, skin allergies** and **gastric ulcers** (**Roderick** *et al.*, 1973). Bed-wetting (medically known as **enuresis**) is cured in children through classical conditioning (**Freyman,** 1963).

(ii) Instrumental or operant conditioning. Instrumental or operant conditioning is a type of learning where the animal has some control over the stimuli it receives and often over the responses it produces; its behaviour will influence its situation. In the 1930s, **B. F. Skinner** developed an apparatus that made it possible to demonstrate operant conditioning. This device, now called a **Skinner box,** typically is a sound-dampened and constantly illuminated box with a lever that activates a food magazine. Once inside the Skinner box, an animal had to press a small bar in order to receive a pellet of food from an automatic dispenser. When the experimental hungry animal (usually a rat, hamster or pigeon) was placed in the box, it ordinarily responded to hunger with random investigation of its surroundings. When it accidentally pressed the bar, a food pellet was delivered. The animal did not immediately show any signs of associating the two events, bar pressing and food, but in time its searching behaviour became less random. It began to press the bar more frequently. Eventually, it came to spend most of its time just sitting and pressing the bar. This sort of learning was called **operant conditioning** by Skinner.

In operant conditioning experiments, a number of reward procedures have been used with an equal number of varying results. For instance, if the animal is rewarded every time it presses the lever, we speak of a **continuous reinforcement schedule.** In addition, there are several **partial reinforcement schedules.** If only a certain constant proportion of the lever pressing are rewarded, we speak of a **ratio reinforcement schedule.**

It is possible to build "stimulus" control into operant conditioning situation. For example, a pigeon in a Skinner box may be required to peck a bar in order to receive food. The apparatus is designed so that food is delivered only when the bar is pecked (Fig. 3.27). Within a certain period a sign lights up that says "peck", the pigeon will come to peck the bar mostly at that time. It will largely ignore the bar until the "peck" sign is lighted. If the pigeon pecks after a "don't peck" sign lights up, and that peck is not followed by food, the pigeon will come to ignore that sign through **discrimination learning.**

Fig. 3.27. A pigeon in a Skinner box demonstrating instrumental conditioning. Every peck pigeon makes at the disc is rewarded by the reinforcement.

(iii) Avoidance conditioning. It is a special class of instrumental conditioning. In this procedure an animal is presented with a stimulus that precedes some "punishing" stimulus (Fig. 3.28). The animal can avoid the "punishment" if it responds in a particular way. Many experiments have used avoidance conditioning of goldfish. The fish is placed into a test box with a low barrier across its centre. At each end of the box there is light. The procedure is to turn on the light at the end of the box where the fish is located. Five seconds later a shock is delivered to the fish. The fish can stop the shock or avoid it altogether if it swims over the barrier. After several dozen tests, the fish will swim over the barrier as soon as light goes on. Thus repeated

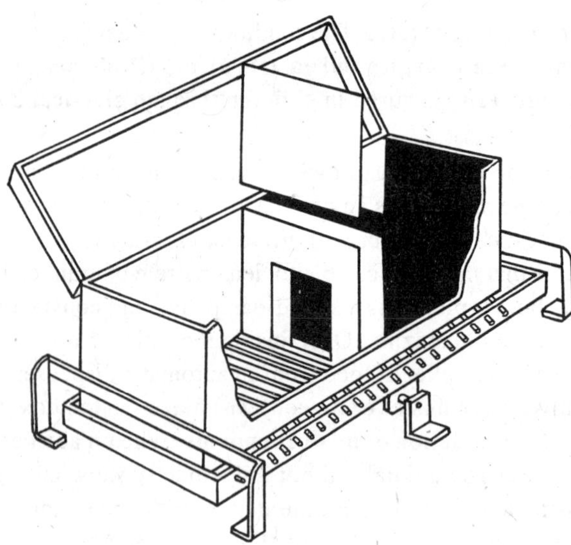

Fig. 3.28. A sketch of shuttle-box used for demonstration of instrumental conditioning based on avoidance.

presentation of the test situation reinforces the behaviour; however, in some cases if the punishment is absent, the behaviour is gradually becomes extinguished.

Comparison of Classical Conditioning and Operant Conditioning

Classical conditioning and instrumental (operant) conditioning are two types of learning and show the following differences and similarities between the two:

Difference between classical and instrumental conditioning.

1. The sequence of events is dependent upon the responses of the organism in instrumental, but not in classical conditioning.
2. The classical conditioning is the respondent conditioning, but operant conditioning is the characteristic of the instrumental conditioning.
3. In classical conditioning the reward comes regardless of the behaviour of the animal. The reward in instrumental conditioning, however, is linked to a definite behaviour of the animals.

Similarities between classical and instrumental conditioning. Both types of conditioning share the following features:

1. Acquisition. An acquisition of a response in conditioning is measured by following ways: amplitude or extent of response, latency of response, number of reinforcements or probability of response in a give time and condition.

2. Extinction. Learning through both types of conditioning temporarily vanishes if a conditioned response remains unrewarded for a number of times.

3. Generalization. If an animal is conditioned to a specific stimulus, it will respond to other closely related stimuli.

4. Discrimination. If a stimulus among a battery of stimuli is selectively rewarded, the animal becomes able to discriminate it from other similar stimuli.

5. Contiguity. The pairing of conditioned stimulus (CS) and unconditioned stimulus (UCS) in a conditioning process must be continuous.

6. Reinforecement. For any type of conditioning, reinforcement (reward) is necessary. Removal of reinforcements or its improper placements results in gradual extinction of learn responses.

7. Interference. A new learning, in between, usually interferes with a conditioned response.

8. Repetition. A maximum learning in both types of conditioning is directly related to repetitive rewards.

Trial-and-Error Learning

Animals are constantly confronted with situation in which they have a choice of stimuli to which they can respond. A chicken can peck either a black lever (for which it receives food) or a white lever. It learns this task by a procedure of trials. Some correct and others incorrect, and its performance improves over time. Soon the bird pecks the black lever almost exclusively. This type of learning is called **trial-and-error learning.** It is a type of operant conditioning involving maze-running.

Mazes have been constructed to measure the rate of trial-and-error learning of many animals (*e.g.*, rats). Typically, an animal will be placed into a maze with one or more choice points leading to a food source. Improvement in its learning is measured by noting the decrease in the time required in the maze or by noting the reduction of errors in touring the maze. This behaviour is commonly seen in vertebrates as they learn the sources of food, water and danger spots in their environment. The simplest maze is a T-maze in which the individual has to make one directional choice; by rewarding rats that turn one way rather than by the other (or by giving them an electrical shock if they turn in the opposite direction), the experimenter can teach them to make a consistent and predictable choice.

3. Latent Learning

When acquisition of a response is not displayed during learning but remains hidden (latent) and is expressed later, it is called **latent learning** (see chapter 1). For instance, most animals will have knowledge of their surrounding because they explore them out of their curiosity. Insects and birds once fix a home territory always have flight oriented toward their home.

Latent learning is a peculiar form of learning which remains unrewarded at the time of learning. Wild animals judiciously utilize it in avoiding their predators and in searching their food.

4. Insight Learning

Insight learning is the highest form of learning in which the animals solve their problems too rapidly without a normal trial-error approach. For example, a male chimpanzee in sight of a bunch of banana in his cage, which were out of his reach, uses unsuccessfully a short stick. But failing to get fruits, he rakes a long stick with the short stick present outside of his cage beyond his bodily approach. Having grasped the long stick he uses it to get the fruit. This demonstration of **Kohler** (1927) shows that chimpanzees used first latent learning of playing with sticks, but they applied it in some unusual context which speaks for insight learning.

(i) Reasoning

Reasoning is the ability to solve complex problems by behaving according to general principles rather than simply responding to the situation with simple trial-and-error behaviour or modification of stimulus-response behaviour. The animal should be able to put together elements from its past experience into new arrangements to meet different situations.

Several types of tests have been devised to test an animal's reasoning ability. For instance, a **detour problem** consists of placing the animal in an environment where it must follow a circuitous route to a food source (or a escape path). A direct pathway to the food is blocked and the animal must go away from the food to succeed. The question is : can the animal successfully solve the problem on the first try, or must it go through a series of trial-and-error procedures? Only the higher primates are good at this type of problem, although other animals may display the ability in rudimentary form. Dogs, rats and racoons use trial-and-error learning to solve the detour problem.

Another example of a reasoning problem is the **discrimination learning test** in which animal must choose between two or more responses depending upon the conditions. We can set up such a test using two doors, one black and the other white. The animal must learn to choose the white door if the light is on, but it must choose the black door if light is off. This is an **if-then reaction:** if the light is on, then choose the white door; if the light is out, then choose the black. Higher primates are good at this type of reasoning problem.

Other types of reasoning tests include the **oddity principle** where an animal is presented with several objects and it must choose the one that is different from the others; the **delayed reaction test**, in which the animal must find an object that has been hidden from view for a specified time; the **triple-plate problem**, in which the animal must learn a prescribed series of steps (such as pressing a number of levers in a correct order) in order to receive a reward or escape punishment.

(ii) Intelligence

According to Merriam Webster's Medical Desk Dictionary (1996) **intelligence** means (1) the ability to learn or understand or to deal with new or trying situations; (2) the ability to apply knowledge to manipulate one's environment; (3) to think abstractly as measured by objective criteria (as intelligence tests).

Darwin believed in the evolutionary continuity of mental capabilities and he opposed the widely held view that animals were merely automatons far inferior to humans. In his *Descent of Man* (1871) Darwin argued that "*animals possess some power of reasoning*" and that "*the difference in mind between man and higher animals, great as it is, certainly is one of degree and not kind*".

The intelligence of animals was exaggerated by Darwin's disciple **George Romanes,** whose publication of *Animal Intelligence* (1882) was the first attempt of scientific analysis of animal intelligence. **Romanes** defined intelligence as *the capacity to adjust behaviour in accordance with changing conditions.* In his book *Introduction to Comparative Psychology* (1894) **Conway Lloyd Morgan** contradicted the ideas of Romanes and suggested that *higher faculties evolved from lower ones.* He proposed a psychological scale of mental ability. Although the idea of a ladder-like evolutionary scale of abilities had a considerable influence upon animal psychology, it is not an acceptable view today. Current studies have made it clear that **different species in different circumstances exhibit a wide variety of types of intelligence.**

Methods of Assessment of Intelligence

There are two main ways of assessing the intelligence of animals: one is to make a behavioural assessment and the other is to study the brain. In the past, both approaches have been dominated by the idea that *there is a linear progression from lower, unintelligent animals with complex brains.* A survey of the animal kingdom as a whole tends to confirm this impression. Since **intelligence tests** for people were introduced by **Albert Binet** in 1905 to intelligence behaviourally, considerable progress has been made in improving and refining them (Box 3.6). Following tests have been designed to know the **speed of learning** which is an important measure of animal's intelligence:

Box 3.6.

1. Intelligence test. A test designed to determine the relative mental capacity of a person to learn, *e.g.*, achievement test, aptitude test, etc.

2. Intelligence quotient (IQ). It is a number used to express the apparent relative intelligence of a person that is the ratio multiplied by 100 of the mental age as reported on a standardized test to the chronological age.

***(i)* Concept of triangularity. Gellerman** (1933) described in detail experiments in which two chimpanzees and 2 year old children were learning that a food reward was associated with a white triangle on a black square and not with a plain black square. One child learnt in a single trial, but the other took 200 trials and both chimpanzees took over 800 trials to reach the criterion of 19 correct trials out of 20. On a comparable test most rats would learn in 20 to 16 trials, though admittedly they would usually be mildly punished for wrong responses as well as rewarded for correct ones. Discrepancies of this kind occurs both within and between species and we have no reliable evidence that speed of initial learning for simple associative problems varies within the vertebrates.

However, *speed is only one aspect of learning;* we might also ask *what is learnt?* For example, Fig. 3.29 shows that, although the chimpanzee may take longer to learn the *triangle discrimination* out lined above, it learn more about "triangularity" than does the rat. One aspect of "intelligence" is *the ability to strike a reasonable balance between generalization and discrimination in tests of this type.* Similarly, if we consider more complex forms of learning then we can at least note some gradation of ability within the vertebrates.

***(ii)* Harlow's learning sets.** Testing for what have been called **learning sets** has been useful in measuring other aspects of intelligence. Thus if an animal can form a learning set *it*

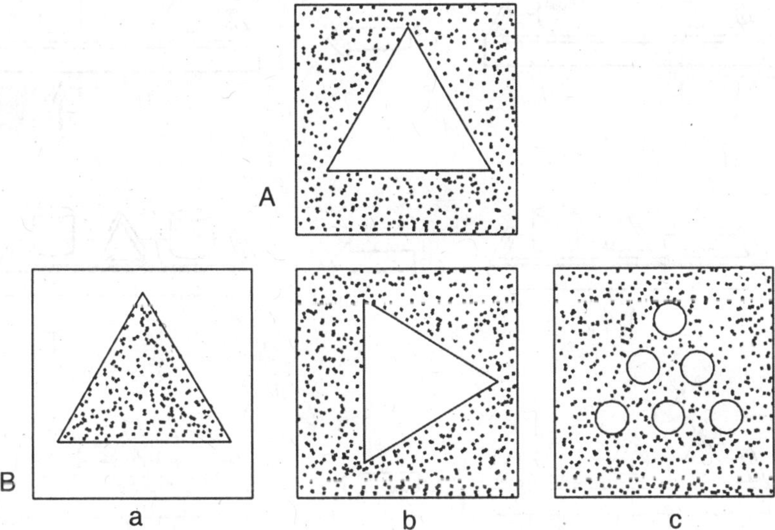

Fig. 3.29. The concept of triangularity. Trained to respond to the top figure (A) A rat makes random responses to any of the lower figures (B) a chimpanzee responds to (a) and (b), but makes random responses to (c). A 2 year old human child recognizes a triangle in (a), (b) and (c).

means that it can learn not just a problem, but something about the principle behind it and then can steadily increase its learning speed when given a series of similar problems (Fig 3.30). **Harlow** (1949) has described the basic technique with primates. A monkey is presented with a pair of dissimilar objects — a matchbox and egg cup, for example. The matchbox, no matter where it is placed always covers a small food reward, the egg cup never has a reward. After a number of trials, the monkey picks up the matchbox straight away. Now the objects are changed: a child's building block is rewarded, a half tennis ball unrewarded. The monkey takes about the same time to learn this, again the objects are changed, and so on. After some dozens of such *discrimination tests* the monkey learn each discrimination much more rapidly, although viewed as an individual problem it is just as difficult as the first one. Eventually after 100 or so tests the monkey presented with any pairs or objects lifts one and if yields a reward chooses it for all subsequent traits. It has learnt the principle of the problem or in Harlow's terminology, it has formed a **learning** set.

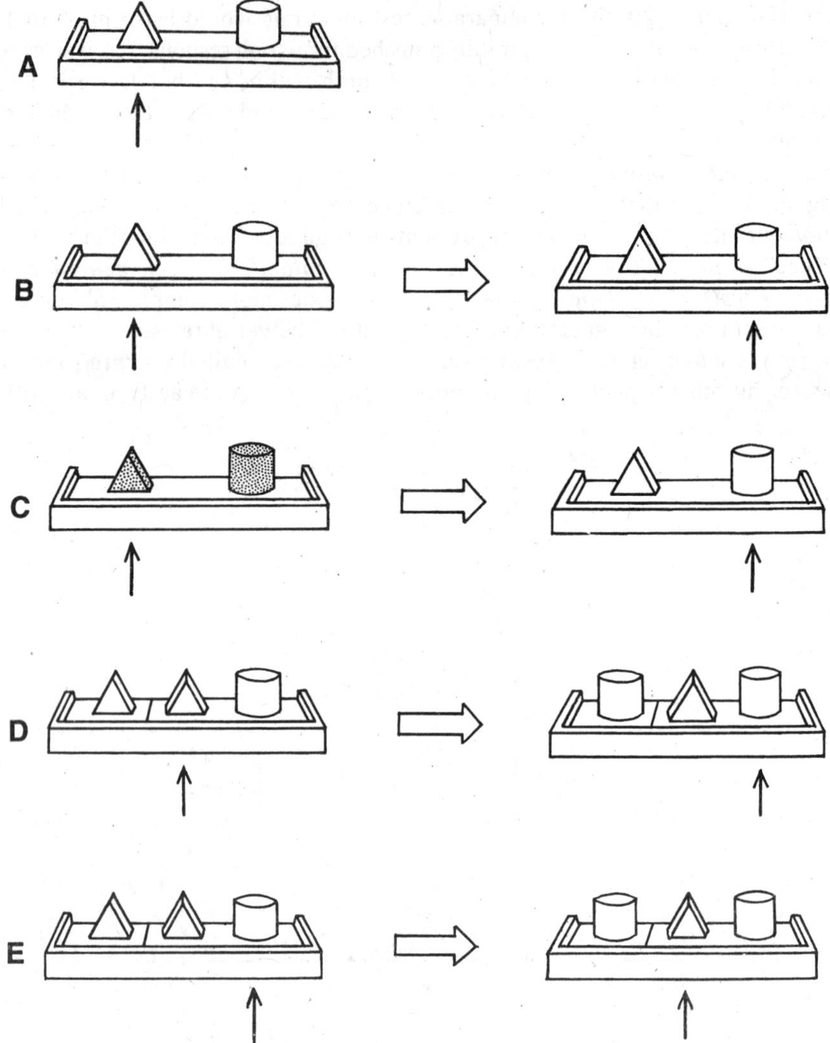

Fig. 3.30. Series of discrimination problems used to test learning sets. A — **Simple discrimination** (the arrow indicates the correct choice); B — **Reverse task** (the animal has to reverse its originally correct choice); C — **Conditional task** (one object is correct when both are gray, the other when both are white); D — **Matching task** (the animal has to match the sample at left on tray); E — **Oddity task** (the animal has to choose the odd one out).

Despite the criticism that the ability of different species to master learning sets depends largely upon the way the tests are set up, there do seem to be genuine differences among species (Fig. 3.31).

The ability of learning sets was once regarded as a capability of the more advanced mammals only but it is known that all the vertebrates with exception of fish have shared this ability. But goldfish and octopus are known to form repeated reversal sets (another type of learning sets).

The modern **I. Q. tests** include various subtest designed to assess a person's memory, arithmetic and reasoning power, language ability and ability to form concepts. Pigeons appear to have a prodigious ability to form concepts such as water, tree and human beings. Are we to take this as sign of great intelligence? In discussing language ability in animals, ethologists came to conclusion that human ability, in this respect, far exceed those of any other animal, however well trained. Does this show that humans have greater general intelligence or merely that human intelligence is highly specialized in the use of language ? These problems are still unresolved.

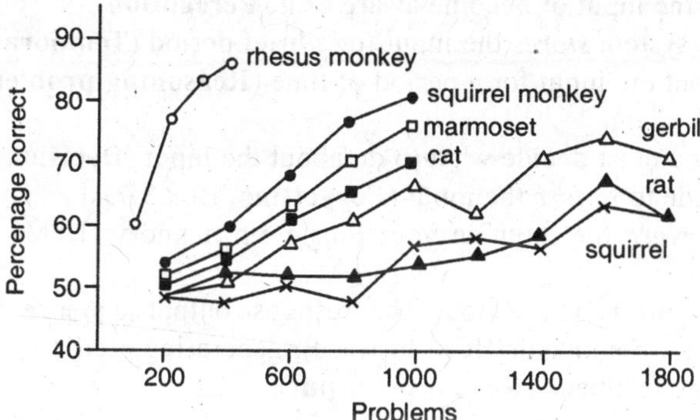

Fig. 3.31. Visual discrimination learning set in mammals. The percentage correct on the second trial of each problem is plotted against the number of problems given.

The Gestalt approach

Gestalt means mental or perceptual pattern. Term Gestalt originates from the German word meaning "figure", "pattern" or "good form". The ability to see something "as a whole" — rather than just seeing its parts — is the ability to "form a Gestalt".

The Gestalt movement began in Germany around 1912. It was started by several psychologists who believed that the perception of *shape* was inborn. To the Gestalt theorists, the circle was a "perfect form". Thus, to the Gestalt psychologists, you perceive the coin as "round" or "circular" because your brain has built-in concept of "roundness" or "circularity". When you look at the coin, the "perfect form" of roundness is called into your mind automatically. So you impose the innate concept of "circularity" on your sensory inputs. True, you *learn* to call this particular example a "coin" through experience. But according to Gestalt theory, your perception of roundness is a psychological experience that is a gift of nature.

But, **J. J. Gibson** (1950) took the opposite point of view. He stated that your brain is "hard-wired" to *see the world as it really is.* The coin is round and the light rays coming from the coin are rich in sensory cues describing its roundness. The pattern of excitation the light rays make on your retina is round. Thus, how could you perceive the coin as anything other than circular in shape? Gibson said we can explain all perceptual experience in terms of information to be found in the stimuli itself.

***(iii)* Cognitive Behaviour**

Cognitive behaviour involves **cognition** which is defined as the act or process of knowing including both awareness and judgement. Roughly speaking, cognitive activities are those "internal processes" that involve **information processing, thinking, reasoning, remembering, perceiving** and so forth. According to **James V. McConnel** (1986), cognition in humans is a complex set of operations:

1. Cognition begins with **attention,** the process by which you *detect* and *orient* toward sensory inputs **(Input detection)**.
2. Then you must *receive* these inputs at the level of your sensory receptors **(Input reception)**.
3. Your receptor organs "*process*" the inputs by *converting* physical energy (such as light rays and sound waves) into patterns of neural energy **(Input transduction)**.
4. Lower centres in your nervous system detect *critical features* and send information about these features upto higher centres **(Critical feature selection)**.

5. The lower centres also *screen out* certain types of information (**Input screening**).
6. Your cortex (of brain) attempts to match the input with items stored in memory (**Pattern recognition**).
7. At this point, the input *registers* on your consciousness (**Registration**).
8. You *perceive* the input or become aware of it (**Perception**).
9. Your nervous system *stores* the input for a brief period (**Temporary memory storage**).
10. You think about the input for a period of time (**Reasoning problem solving** and **Mental imagery**).
11. Eventually, you must decide what to do about the input (**Decision making**).
12. You may decide to *ignore* the input (**Forgetting,** Box 3.7).
13. Or, you may *store* the input in your long term memory (LTM) (**Permanent memory storage**).
14. You may also *respond* to the input. The 'response output sequence' involves such activities as issuing motor commands (feed forward), generating predictions about consequences, and monitoring feedback (**Response output**).
15. The *environment* then *responds* to your output and sequence begins all over again.

Box 3.7.

There are two main kinds of theory of why animals forget: the decay theory and the interference theory. According to **decay theory**, the memory of some events fades with time unless continually upgraded. According to the **interference theory**, animals forget things not because memory fades but because other memories displace them. The memory of more recent things interferes with the recall of things memorized longer ago.

5. Phase Specific Learning

(*i*) Imprinting

A more complex process than instinctive interaction between parent and young, occurs in species where the young are able to move around from the moment they are born or hatched. When such animals live in groups, both parents and young are faced with the problem of recognizing each other. There is a selective pressure on parents to care only for their own young and this creates a selective pressure on the offspring to approach only their own parents. Even when such animals are solitary, the young need to keep in touch with their parents.

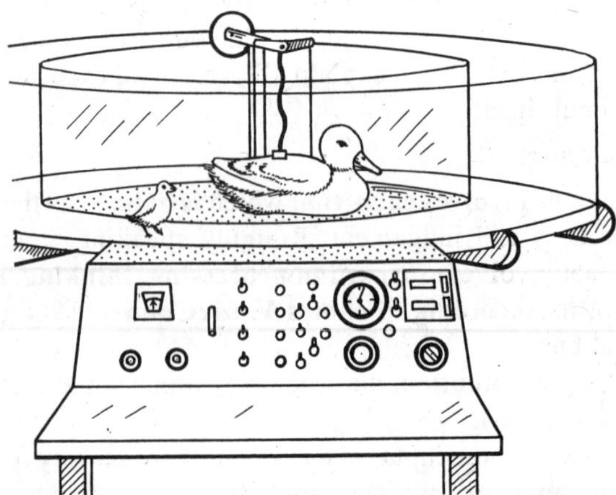

Fig. 3.32. Apparatus used in the study of imprinting. A decoy duck is moved around the centre of the ring while a duckling follows along a circular runway. The control panel for the apparatus is in the foreground.

In such species, there is a **rapid kind of learning** that is predisposed to happen at the time of birth or hatching. **Lorenz** first discovered this phenomenon in greylag geese. A gander (male goose) which he had hand-reared came to follow him everywhere, and when it reached sexual maturity it attempted to mate with him (*i.e.,* Lorenz). In other words, it treated him in all respects as though he was a mother goose. Lorenz called this kind of learning as **imprinting.**

History of the Idea of Imprinting

In 1910, **Oscar Heinroth** first of all described that goslings tend to follow a large moving object soon after hatching and to relate in specific ways to that kind of object later in life. He is often given the credit for being the first to use the term *imprinting* (Pragung), however, prior to him **Splanding** had conducted extensive studies on imprinting in domestic chicken many year earlier (1872-1875). **Lorenz** (1935) confirmed Heinroth's observations on goslings and also studied imprinting in mallard ducklings, pigeons, jackdews and many other birds. He argued that imprinting, unlike ordinary learning, took place at a particular stage of development and was irreversible.

Mechanism of Imprinting

What seems to happen, at least in ducks, geese and chickens is this: the young bird is predisposed to learn the characteristics of any object that it sees and hears during a fairly short period soon after hatching, called **critical period** by **Lorenz,** but now-a-days referred to as the "**sensitive period**". The exact timing of the sensitive period differs between species. Domestic chicks, for example, only follow objects they have seen during the first three days after hatching, whereas for mallard ducklings, the phase lasts for 10-15 days after hatching. The chick will not imprint on objects seen after that time.

Normally what the young bird will see and hear during the sensitive period is its mother. But if we arrange things so that what it sees is either **Konrad Lorenz** or a red watering can or a blue flashing light borrowed from a police car, then it will learn the characteristics of that object instead. Bright, coloured, noisy, moving objects are most effective stimuli for imprinting experiments. But if no such stimuli are available, then chick will even imprint on their own feet.

Although imprinting is most often studied in ducks and chickens, it occurs in many other species. For example, an imprinting process has been held to underlie mutual recognisation by many goats and their kids. Unless a goat is able to see and smell her kid very shortly after birth, she will subsequently refuse to suckle it (**Klopfer** *et al.,* 1964). Although later on it was suggested that the mother may take an active role by marking the kid in someway (**Gubernick** *et al.,* 1979), there is obviously a sensitive period here, whatever the mechanism. It has also been suggested that imprinting process may be involved in the formation of bonds between human mother and their babies.

Imprinting, thus, *is a method of rapid learning* of the mother's appearance by newly hatched chicks. This type of learning is preprogrammed to take place as part of the normal process of development and in whatever circumstance pertaining at the time. **Lorenz** (1935, 1970) argued that imprinting must have involved a special process, distinct from ordinary learning, since *it was exceptionally fast, was irreversible and could only occur within a sharply defined critical period.* But the concept of a critical period has been replaced by that of a more loosely specified sensitive period: *imprinting is most likely to occur within the sensitive period, but can occur outside it.* Recent experiments have shown too that sometimes imprinting can be reversed. And though imprinting is certainly a rapid kind of learning, this can be reasonably explained by the fact that the animal concerned has had virtually no previous learning experience and so there is no scope for what learning psychologist call "**proactive interference**" from previous tasks.

But if imprinting probably does not involve any special process, its occurrence in specific contexts does seem to involve specific adaptations to each species particular mode of life.

Thunberg (1971) studied imprinting in the alewife (a type of salmon fish), which breeds in the rivers that flow to the Atlantic coast of North America. Juvenile alewives (young salmon smolt) leave their natal rivers and live in the sea until the time comes for them to breed, when they always return to the river in which they were themselves spawned. **Thunberg** showed that alewives discriminate their natal streams by the traces of chemicals present in water, and they must learn this during the relatively brief period they spend in the rivers after hatching. Thus, here, the migratory response is imprinted in salmon.

Types of Imprinting

Imprinting is of following two types:

1. Filial imprinting. It takes place in many species of birds and mammals. These are the kinds of animals with most extensive parental care. Imprinting is adaptive because it enables the young to recognise and follow their parents. They will grow up in a world of many hostile enemies and one or two protective parents. If they are to survive, it is important that the young should choose the right animals to follow. The kind of imprinting that we have been considering so far is called **filial imprinting.** It is the imprinting of the following response which young animals make to their parents.

2. Sexual imprinting. It concerns the species to which the animal will direct its sexual behaviour. In geese, for example, the sexual imprinting and filial imprinting are two different processes. A goose which follows a human being as if it were its parent does not have its sexual behaviour anything like so disturbed; when it grows up it will court other geese.

Sexual imprinting also occurs early in life. Most experiments on sexual imprinting have been conducted on birds. It has been found that birds are most easily sexually imprinted on their own species, fairly easily on closely related species, and only with difficulty on very different species. Herring gulls and lesser black-backed gulls are similar species, and black-backed gull reared by herring gull parents (because some ethologist moves eggs between nests) become sexually imprinted on the herring gull *i.e.,* the adult lesser black-backed gulls so produced will try to mate with herring gulls. They are sometimes successful. Most of the gulls which are hybrids between the two species around the British Islands are the result of these experiments. *Sexual imprinting normally functions in the wild to ensure that the animal will, when it grows up, choose a mate of the correct species.* In nature, to look at your parents is a good method of learning the characteristics of your own species.

Current Status of Imprinting Idea

Lorenz (1935) believed that imprinting is fundamentally different from other forms of learning; but this is not a popular view today. Imprinting involves a narrowing of pre-existing preferences, a process that has much in common with other forms of perceptual learning. It has much in common with ordinary conditioning.

Although imprinting may provide an alternative to innate recognition of members of ones own species, it probably serve other functions. For example, it may be important in the development of kin recognition and in preventing inter-breeding among close relatives (**McFarland,** 1985).

3.3. MEMORY

Learning is nothing without memory. We must be able to store the results of experience and recall them to our benefit later. Indeed one of the most remarkable properties of the nervous system is that it can retain some representation of past events for almost a life-time — tens of years in some cases.

According to **W. H. Thorpe** (1963), memory has the following parts:

1. Learning. The process of acquiring some activity or knowledge.

2. Remembering. The process where by the effects of past learning manifest themselves in the present.

3. Forgetting. Not directly observable, but manifested by a failure to remember.

4. Retention. The storage of learnt material in the brain.

Nature of Memory

The nervous system operates by transmitting electrical impulses along defined pathways. The process of learning involves heightened activity in those channels that record sensory impressions and their outcomes. It seems unlikely, however, that heightened activity *per se* (*i.e.*, by itself) could constitute memory, *i.e.*, a memory could be stored in the form of a continuous train of nerve impulses running for years around the same pathways. Such ideas of "self-reverberating circuits" (for memory) were entertained at times in the history of psychology, but they are abandoned now. This idea of memory was discarded by the following experimental proof:

Andjus and his coworkers (1955) managed to cool rats down to 0°C for periods upto an hour, at which temperature all electrical activities in the nervous system ceases. When these rats were warmed up again, these rats were found to retain their memory of events prior to cooling as well as of normal animals.

On the other hand, it seems likely that the establishment of memory involves **structural changes** of some kind in the nervous system so that some channels are facilitated, then we can readily understand why function is restored when electrical activity begins again. It is now generally accepted that *memory storage must be represented in a chemical/physical form* (Box 3.8) and some insight have been gained regarding how such changes are brought about and what is the nature of the store.

Box 3.8.
Memory Traces

Physiologically, memories are caused by changes in the capability of synaptic transmission from one neuron to the next as a result of previous neural activity. These changes in turn cause new pathways or facilitated pathways to develop for transmission of signals through the neural circuit of brain. The new or facilitated pathways are called **memory traces.** Memory traces are important because once they are established, they can be activated by the thinking mind to reproduce the memories.

Experiment in lower animals (*e.g.*, *Aplysia*) have demonstrated that memory traces can occur at all levels of the nervous system. Even spinal cord reflexes can change at least slightly in response to repetitive cord activation which is part of the memory process. Also some long term memories result from changed synaptic conduction in the lower brain centres. Evidently most of the memory that we associate with intellectual processes is based on memory traces mainly in the cerebral cortex.

Obviously the study of memory mechanism will involve neurophysiological and biochemical methods, for we must investigate the fine-scale operation of neurons and the synapses between them. However, most crucial evidence still has to come from behavioural observations (**Manning** and **Dawkins,** 1998). In fact, it is what the animal actually does which provides our knowledge of what *it has learnt, stored and now recalls* and from this knowledge we go on to deduce mechanisms and then check them physiologically and biochemically (**Dudai,** 1989).

The study of memory mechanisms, such as learning, has benefited from the comparison of different types of animals. Valuable evidence has come from certain molluscs (*e.g.*, *Aplysia, Octopus,* etc.), the honeybee (Box 3.9), a few bird species and a few mammalian species. Human memory itself has been a rich source. Although convincing experiments are usually impossible in case of humans, we have various conclusive cases where the effects of brain damage arising from accidents can be studied. Of course we can get detailed evidence from humans who, unlike animals, can tell us verbally what they can or cannot recall.

Box 3.9.

Honeybees show very persistent recall after a single experience of a colour combined with a food reward. Like vertebrates, their memory goes through a *labile phase* (lasting about 3 minutes) when it is susceptible to destruction by cooling or anaesthesia, before becoming incorporated into a much more robust *long-term store.*

Positive and Negative Memory

Human brain is flooded with sensory information from all our senses. If human mind attempted to remember all this information the memory capacity of the brain would be exceeded within minutes. Fortunately, the brain has the peculiar capacity *to learn to ignore information that is of no consequence.* This results from **inhibition** of the synaptic pathways for this type of information. The resulting effect is called **habituation.** This is, in a sense, a type of **negative memory.**

On the other hand, for those types of incoming informations that cause important consequences, such as pain or pleasure, the brain has automatic capability of enhancing and storing the memory traces. This is **positive memory.** This results from facilitation of the synaptic pathways, and the process is called **memory sensitization.** Special areas in the basal limbic regions of brain determine whether information is important or unimportant and make the subconscious decision whether to store the thought as an enhanced memory trace or suppress it.

Three types of Memory

In chick and rat, the following three types (or stages) of memory (Fig. 3.33), each with different time course have been recognized (**Andrew,** 1985):

1. Short-term memory (STM);
2. Intermediate-term memory (ITM); and
3. Long-term memory (LTM).

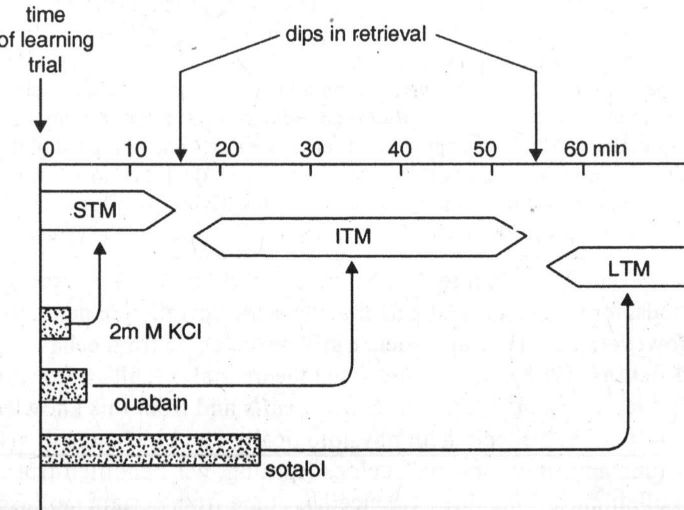

Fig. 3.33. A scheme for the three stages of memory storage constructed from experimental evidence in chicks. At time since the learning trial increases the chick retrieves memory of the event first from short-term memory (STM) then intermediate-term memory (ITM) and finally long-term memory (LTM). The dips in retrieval at 15 to 55 minutes represent periods when the chick is switching from one store to the next. The three chemical agents listed each affect memory of one type only and lead to subsequent amnesia when administered during the period indicated by the extent of bars at the left. This is supportive evidence for a three-stage memory system.

The evidence for three types of memory comes from two sources: *Firstly,* there are drugs (*e.g.*, methyl anthranilate or 2m MKcl, ouabain, sotalol; colour beads being coated by each of these drugs) which affect memory specifically at each of three time phases, and are ineffective earlier or later. *Secondly,* corresponding to these times of drug sensitivity, there are remarkable precise timed fluctuations in the ability to retrieve the memory of an event after it has occurred.

Such precision in timing of the ability to recall events suggests that animals retrieve their memories from different stores in turn. It is the process of shifting from one store which is now fading, to the next, now forming, which manifests itself as a dip in retrieval. It is still not clear whether — Is each memory is necessary precursor of the next, playing a part in its formation, or does each form and fade in parallel perhaps interacting and passing information across as it fades and the next comes into play? **Dudai** (1985) has suggested that there is a likelihood of a "**multi-step**", "**multi-channel**" nature of memory in different species.

The three types of memory interact, each phase of memory must be carrying a representation of the events which have been learnt. Long-term memory can be shown to be in place after about an hour in most animals studied. Establishment of LTM involves structural changes in the brain which subsequently facilitate neural transmission along new pathways. The process leading to the setting up of short and intermediate term memories probably involve persistent activity in some neural circuits. This is why they eventually decay and why they were so vulnerable to physical shock and to drugs.

Box 3.10.
Consolidation of Memory

For short-term memory to be converted into long-term memory that can be recalled weeks or years later, it must become "consolidated". That is, the memory must in some way initiate the chemical, physical and anatomical changes in the synapses that are responsible for the long-term memory. This process required 5 to 10 minutes for minimal consolidation and 1 hour or more for strong consolidation. Consolidation involves following processes:

1. Role of rehearsal in transference of short-term memory into long term memory. Psychological studies have shown that rehearsal of the same information again and again in the brain (mind) **accelerates** and **potentiates** the degree of transfer of short-term memory into long-term memory and therefore accelerates and potentiates consolidation. The brain has natural tendency to rehearse newfound information especially newfound information that catches its attention.

2. Codifying of memories during consolidation. One of the most important features of consolidation is that memories are codified into different classes of information. During this process, similar information is recalled from the memory storage bins and used to help process the new information. The new and old informations are compared for similarities and differences and part of the storage process is to store the information about these similarities and differences rather than to store the information unprocessed.

The Anatomy of Memory

We can now identify regions of mammalian brain involved in three stages of memory formation. The original evidence came from human subjects with particular type of brain damage. Over a century ago the Russian neurologist **Korsokov** described patients who were able to recall distant events normally but has lost permanently the ability to form new memories following a head injury, a stroke or severe alcoholism, the so called **anterograde amnesia** (Box 3.11). A common feature of such patients was damage to neural structures at the base of forebrain and adjacent to the hypothalamus, especially the mamillary bodies and the thalamus (Fig. 3.34 A). Since Korsokov's original descriptions, damage to other closely associated brain regions, notably the **hippocampus,** has been shown to have the same effect on the inner side of the base of the cerebral hemispheres, and the **amygdala** which lies anterior to it (Fig. 3.34). Bilateral damage to any of these structures can lead to a complete inability to store new memories, and recall beyond a few seconds may be impossible. **Sacks** (1986) in his essay '*The lost mariner*' has given a brilliant account of the extraordinary and tragic existence of a person (nicknamed 'HM') who has lost all ability to memorize events and has only a distant past to which he can relate. By working with 'HM', it was concluded that not all learning is processed along the same channels before it enters the memory store.

Box 3.11.
Some Definitions

Amnesia. Loss of memory; it sometimes includes the loss of the memory of personal identity due to brain injury, shock, fatigue, repression or illness or sometimes induced by anesthesia.

Anterograde amnesia. Permanent loss of the ability to form new memories following a head injury, a stroke, a shock, or severe alcoholism. It becomes effective for a period immediately following a shock or severe alcoholism.

Retrograde amnesia. It means inability to recall memories from the past, *i.e.*, from the long-term memory storage bins, even though the memories are known to be still there.

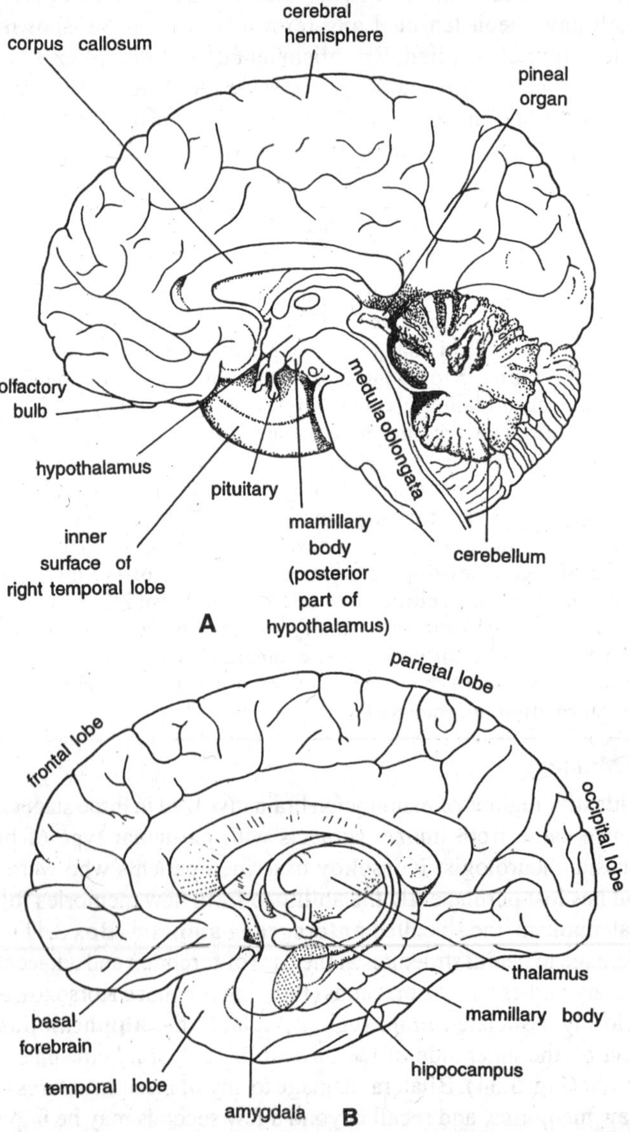

Fig. 3.34. The human brain viewed (A) in straight saggital section and (B) from the same aspect but with brain stem and cerebellum removed so as to reveal more of the structures involved in memory formation.

The human evidence has been supported by work with rats and primates (Fig 3.35). Numerous studies using specific brain lesions and drugs have build up a picture of several circuits for different types of memory (**Mishkin** and **Appenzeller,** 1987).

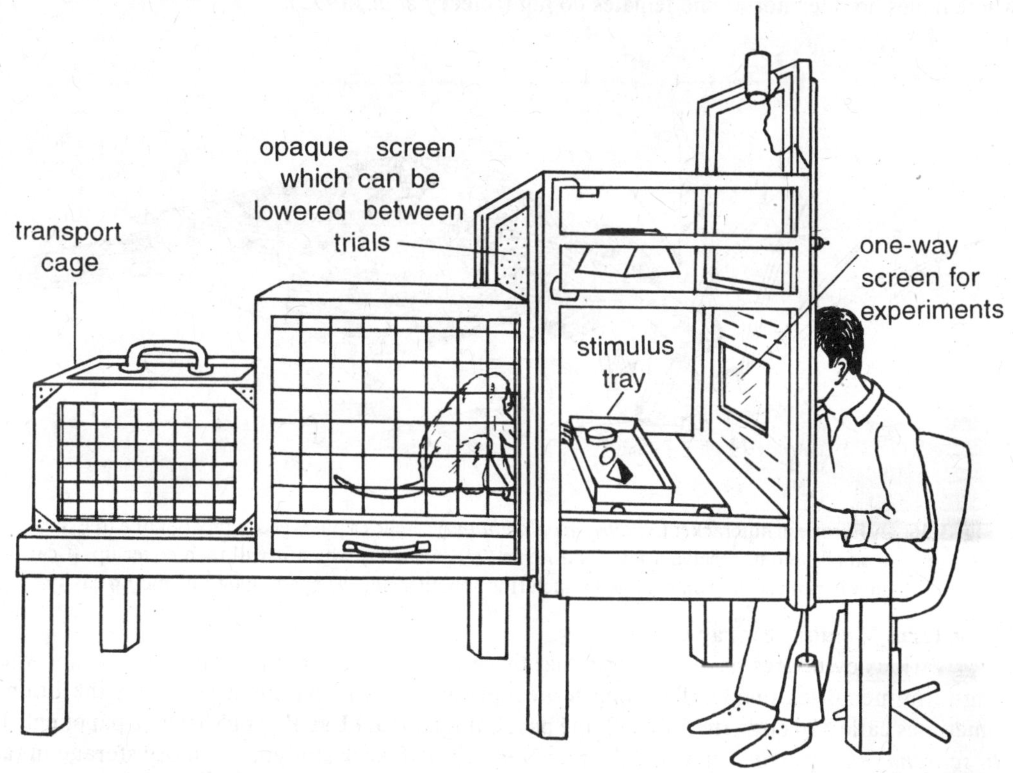

Fig. 3.35. The 'Wisconsin General Testing Apparatus' is now a familiar feature of laboratories where primate learning is being investigated.

By a new imaging technique, called **positron emission tomography (PET)** which has been described by **Ungerleider** (1995), it has become possible to "see" patterns of blood flow in different parts of the brain while a person actually learning or recalling some event or procedure. Active, *i.e.,* rapidly firing, nerve cells require more oxygen and energy than in their resting state, which means that blood circulation is switched toward such areas. For example, when a subject learned a spatial task such as navigating along an urban route, shows in PET 'active areas' in the region of **hippocampus.** Prior to PET scanning, the subject's blood was made very slightly radioactive by the injection of labelled water and the whole brain scanned to reveal which parts have enhanced blood flow.

Role of hippocampus in memory. For procedural type of learning sensory and motor areas of the cerebral cortex are involved, both for processing and for eventual long-term storage. The hippocampus has been focus of a great deal of memory research, particularly concerning spatial memory. In rats, hippocampus is involved not just in short-term memory formation but in the organization of spatial memory generally. In pigeons, too, hippocampus is involved in spatial memory. **Bingmann** *et al.,* (1988) have shown that homing pigeons with hippocampal lesions have great difficulty in homing to their loft. In birds, specialized spatial memory of food storing behaviour (Fig. 3.36) is also related with hippocampus of brain. **Krebs** *et al.,* (1989) have examined a wide range of birds and found that food-storing species have a hippocampus which is significantly larger in proportion than that of close relatives which do not store food. Their spatial learning specializations have apparently required neural specializations to match. Similar differences in size of the hippocampus have been found between mammals which hold territories and hence require a good spatial sense and those which do not. Such difference also exists between the sexes of one species where males hold territories and females do not (**Sherry** *et al.,* 1992).

Fig. 3.36. A European nutcracker (*Nucifraga columbiana*) catches a pine seed at the foot of a tree. It has dug a small hole to receive a few food items (2-12 usually) which it will then cover up. It carries a number of seeds in an enlarged pouch (the subgular pouch) opening into its mouth cavity.

Long-term Memory Storage

Various evidences have indicated that there is no special area in mammalian brain where long-term memories reside. The long-term representation of an event stored in the brain is sometimes called an **engram.** In 1950, the psychologist **Karl Lashley** published a paper entitled '*In search of the engram*', in which he received years of work studying memory storage in rats; he found the engram elusive. In one set of experiments, rats learnt a fairly complex maze and then had parts of their cerebral hemispheres removed. By the conclusion, there was no portion of hemispheres which had not been removed from one or another group of subjects. The overall result suggested that *the degree to which rats lost their memory depended not on which parts of the brain were removed, but only on how much.*

Imprinting and LTM. In fact, learning a maze is a complex task for the rat, since it involves essentially *visual, spatial* and *tactile* cues. The processing of these cues and subsequent storage probably involves several pathways and interactions between them. Consequently we might expect memory of a maze to be disrupted by lesions at several different sites.

Recently more simpler techniques have been used to learn regarding long-term memory storage. For example, **Horn** (1985, 1990) and **Rose** (1992) have used **visual imprinting** in newly hatched domestic chicks as a learning and memory system for behavioural, anatomical and biochemical studies. Such imprinting studies have the following advantages: 1. The rapidity with which young chicks learn; 2. The fact that they are innately biased to approach conspicuous objects and require no period of deprivation followed by reward (as with rats being trained to run a maze); 3. That imprinting occurs very early in life. Thus, it may be possible to study localized effects of learning on a brain which has been little "marked" (*i.e.,* imprinted) by previous experience. It is ultimately, structural effects that ethologists looking for. They have emphasized that long-term memory storage must involve *permanent (or at least persistent) changes to the structure of synapses.* These are growth processes and they are blocked by protein synthesis inhibitors such as **puromycin.**

In chick forebrain, the **intermediate hyperstriatum ventrale (IMHV),** has been identified which is crucial for storing a representation of the imprinting experience and mediating discrimination.

Other parts of the forebrain are not so involved. Indeed differences have been found in the way the left and right hemispheres and their components such as IMHVs operate. **Andrew** (1985) has described other evidence that the two sides of the chick brain act complementarily during memory formation and the same is likely to be true for mammals.

Long-term potentiation (LTP). In mammals, as already described, hippocampus is closely involved in memory processes and its neurons are known to have special property called **long-term potentiation (LTP).** If hippocampal neurons are stimulated with rapid trains of impulses, that is, they are artificially induced to become active, their synapses with other neurons become strengthened and facilitated (potentiated) and thus more ready to transmit excitation the next time a similar pattern of stimulation occurs. This changed state persists for some hours, even if it was stimulated by only a single burst of stimuli. If a second or multiple bursts of stimulation are given then LTP can persist at least in intact animals, for days or weeks. LTP can be induced in other parts of brain also.

QUESTIONS

Long Answer Questions

1. Define behaviour. Differentiate between innate and acquired behaviour. (*Lucknow 1993*)
2. Describe spatial orientation, kinesis and taxes in animals.
3. Explain taxes as an adaptive behaviour in animals. (*Kerala 1997*)
4. What do you mean by taxis ? Describe with the help of suitable diagrams the photo-thermo, chemo-and geo-taxis met with the animals. (*Purvanchal 1994, 96*)
5. Define taxis. Explain with examples phototaxis, geotaxis and chemotaxis shown by animals. (*Purvanchal 1998*)
6. Explain reflex behaviour in animals with suitable examples. (*Bangalore 1995*)
7. Give an account on the types of reflexes. (*Punjab 1996*)
8. With suitable examples explain instinctive behaviour. (*Kerala 1997*)
9. What is innate behaviour ? Discuss the features of innate behaviour using suitable examples. (*Allahabad 1995*)
10. What is learning? Describe the different types of learning behaviour with the help of suitable examples. (*Allahabad 1991*)
11. Write an essay on learning and memory. (*Lucknow 1998*)
12. Give a critical account on learning behaviour in animals. (*Bharathiar 1997*)
13. Describe the neural mechanism of learning. (*Kerala 1997*)
14. What is memory ? Discuss various theories of memory. (*Lucknow 1993, 95*)
15. Write a short note on five types of learning. (*Punjab 1996*)
16. What is the difference between learned and innate behaviour ? (*Kerala 1997*)
17. Define imprinting and give its characteristics. (*Punjab 1996*)
18. Write an essay on learning and memory. (*Lucknow 1998*)
19. Write an essay on motivation, learning and imprinting. (*Kochi 1999*)
20. What is animal behaviour ? Write only the names of the types of behaviour pattern in animals. Describe any one of them. (*CCSU, Meerut 2008*)
21. What is behaviour ? Describe the various types of behaviour patterns in animals. (*CCSU, Meerut 2006*)

Short Answer Questions

1. Write short notes on the following:
   1. Taxes; (*Allahabad 1991; Purvanchal 93, 97, 99*)
   2. Phototropism in *Paramecium*. (*Bangalore 1994*)
2. Differentiate between the following:
   (*i*) Thermotaxis and phototaxis (*Punjab 1995*)
   (*ii*) Phototaxis and phonotaxis (*Punjab 1995; Purvanchal 99*)
   (*iii*) Chemotaxis and phototaxis (*Purvanchal 1995*)
   (*iv*) Thermotaxis and geotaxis (*Purvanchal 1995*)
3. Write short notes on the following:
   (1) Types of reflexes (*Punjab 1996*)
   (2) Pavlov's experiment (*Allahabad 1997*)
   (3) Conditioned behaviour (*Allahabad 1996; Punjab 99*)
   (4) Classical conditioning (*Lucknow 1994*)
   (5) Learning (*Allahabad 1995*)
   (6) Motivated behaviour (*Lucknow 1995*)
   (7) Stereotyped behaviour (*Lucknow 1994; Bangalore 95*)
   (8) Maze experiments (*Allahabad 1995*)
   (9) Trial and error method. (*Allahabad 1990, 94, 96*)
4. Define instinctive behaviour and explain it with the help of three examples. (*Allahabad 1990*)
5. Define the term behaviour. How instinctive behaviour differs from learning behaviour ? Give suitable examples. (*Allahabad 1993*)
6. Write short notes on the following:
   1. Innate behaviour (*Allahabad 1995*)
   2. Instinct (*Allahabad 1991; 97; Lucknow 94*)
   3. Sequential instincts (*Punjab 1995*)
   4. Fixed action plan (*Allahabad 1991, 94*)
   5. Intelligence. (*Lucknow 1993, 96, 97*)
   6. I.Q. (*Lucknow 1994, 96*)
   7. Conditioned reflex (*Punjab 1995*)
   8. Imprinting. (*Kochi 1999*)
7. Explain physiological basis of motivation. (*Kerala 1999*)
8. Explain selective learning with a suitable example. (*Bharathiar 1997*)
9. Write a short note on imprinting. (*Bangalore 1995; Allahabad 97*)
10. Write a short note on memory. (*Lucknow 1994, 99, 97*)
11. Do comparison of classical and operant conditioning.
12. Differentiate between innate (stereotypic) behaviour and acquired (conditioned) behaviour. Give example of each. (*CCSU, Meerut 2007, 2008*)
13. What do you understand from learning in animals ? Describe the various types of learning in animals. (*CCSU, Meerut 2006*)

Very Short Answer Questions

1. Define taxis. Give its examples. (*Punjab 1995*)
2. Name three animals which show positive phototaxis. (*Purvanchal 1999*)

Multiple Choice Questions

Choose the correct answer from the four alternatives given.

1. Insight learning is commonly seen in (*Bharathiar 1997*)
(*a*) invertebrate (*b*) reptiles
(*c*) birds (*d*) mammals

2. The area of brain that monitors hunger and thirst is located in
(*a*) thalamus (*b*) hypothalamus
(*c*) cerebrum (*d*) medulla

3. Learning is related to
(*a*) hypothalamus (*b*) cerebellum
(*c*) cerebrum (*d*) medulla

4. Loss of memory can be done by destruction of
(*a*) cerebrum (*b*) cerebellum
(*c*) medulla (*d*) diencephalon

5. Which part of your brain is involved in planning and decision making ?
(*a*) broca's area (*b*) wernicke's area
(*c*) frontal lobe (*d*) temporal lobe

6. **Ivan Pavlov** carried out experiments which were mainly concerned with
(*a*) cardian reflexes (*b*) simple reflexes
(*c*) conditional reflexes (*d*) origin of life

7. Reflex action is controlled by
(*a*) brain (*b*) spinal cord
(*c*) peripheral nervous system (*d*) automomous nervous system

8. The nerves leading to the central nervous system are called
(*a*) efferent (*b*) afferent
(*c*) motor (*d*) none

Answers (MCQs)

1. (*d*); **2.** (*b*); **3.** (*d*); **4.** (*a*); **5.** (*c*); **6.** (*c*); **7.** (*b*); **8.** (*b*).

Behaviour Ecology : Habitat Selection

The **habitat** of an organism is the place where it lives (**Odum** and **Barrett,** 2005). For example, the habitat of the water backswimmer (*Notonecta*) and water boatman (*Corixa*) is the shallow, vegetation-choked area (littoral region) of ponds and lakes. Habitat is the "address" of the organism. Habitat also refers to the place occupied by entire community. For example, the habitat of the sand sage grassland community is the series of ridges of sandy soil occurring along the north sides of rivers in the southern great plains of United States. Habitat in this case consists mostly of physical or abiotic, complexes; whereas habitat for water bugs mentioned previously includes living and nonliving objects. Thus, the habitat of an organism or group of organism (population) includes other organisms and the abiotic environment.

4.1. DEFINITION OF HABITAT SELECTION

From the prehistoric times, human beings were aware that different kinds of plants and animals live in different habitats. The surface of our planet (Earth) consists of a mosaic or patchwork, of habitat types, and the distribution of organisms is neither uniform nor random. Let us see what cause these differences.

Plants are generally dependent on natural agents for **dispersal,** such as water or air currents or other organisms. The result is an essentially random dissemination of plant individuals ; few ever reach environments favourable to survival and reproduction. In contrast to the plants, animals have well developed locomotive abilities, at least during some point in the life cycle, and they play more active roles in finding places to live. According to **Rosenzwig** (1990), **habitat selection** is defined as *the choosing of a place in which to live.* This definition does not imply that the choice is necessarily a conscious one or that individuals make a critical evaluation of the entire constellation of factors confronting them. More often the choice of animal is an "autonomic" reaction to certain key aspects of the environment.

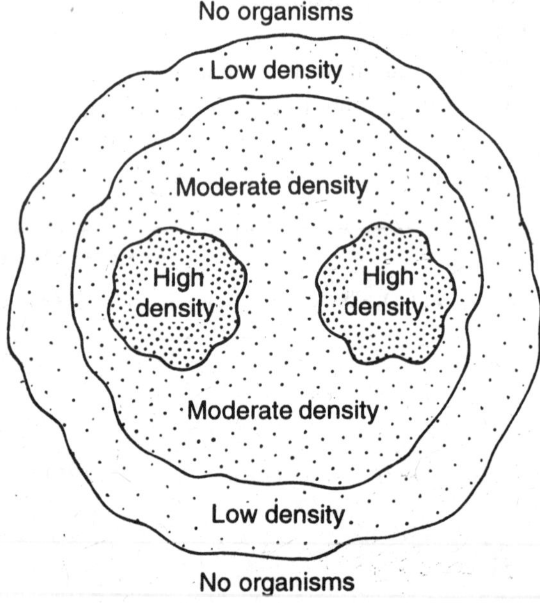

Fig.4.1. Typical relationship between distribution in space and density for a species. Area of optimal habitat support the highest density and are surrounded by suboptimal habitats with lower densities (after Drickamer *et al.,* 2002)

In this chapter will be described the distribution of species (*i.e.,* the presence or absence of a species in a particular habitat) and the dispersal of the individuals of a species within the habitat. When examining the distribution and abundance of most species, we often find areas of high population density near the geographic center of the species range; abundance decrease outwardly (Fig. 4.1).

The Niche Concept

If the habitat is the "address" of the organism, **niche** is its "profession", its trophic position in food webs, how it lives and interacts with the physical environment and with other organisms in the community.

The concept of ecological niche is not so generally understood outside the field of ecology. Terms such as niche are difficult to define and quantify; the best approach is to consider the component concepts historically. **Joseph Grinnell** (1917, 1928) used the word niche "*to stand for the concept of the ultimate distributional unit, within which each species is held by its structural and instinctive limitations ....no two species in the same general territory can occupy for long identically the same ecological niche*" (Incidently, the latter statement predates Gause's experimental demonstration of the competitive exclusion principle). Thus, **Grinnell** thought of niche mostly in terms of the **microhabitat,** or what is now called the **spatial niche. Charles Elton** (1927) was one of the first to begin using the term niche in the sense of the "*functional status of an organism in its community*". Because of Elton's great influence on ecological thinking, it has become generally accepted that niche is by no means a synonym for habitat. Because Elton emphasized the importance of energy relations, his version of the concept is designated the **trophic niche.**

G.E. Hutchinson (1957) suggested that the niche could be visualized as a **multidimensional space** or **hypervolume** within which the environment permits an individual or species to survive indefinitely. Hutchinson's niche can be measured and mathematically manipulated. Hutchinson (1965) also distinguished between the **fundamental niche** (*i.e.,* the multidimensional space that a species occupies under ideal conditions with no competition) and the **realised niche** (the space occupied under real world conditions that involve competitors, predators, and disease). The fundamental niche is often much larger than the realised niche.

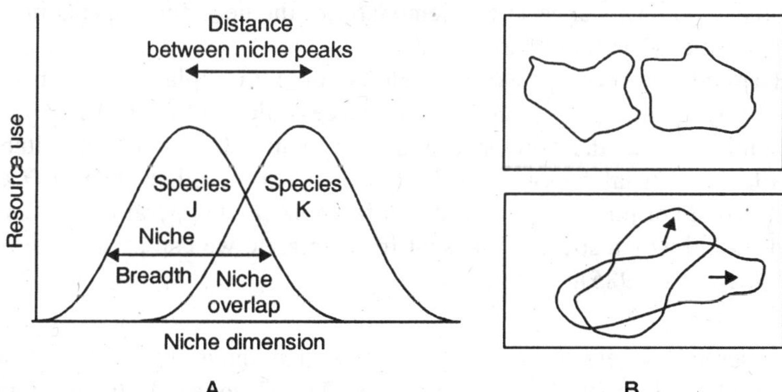

Fig. 4.2. Schematic representations of the niche concept. A–Activity curves for two species along a single resource dimension illustrates the concept of niche breadth and niche overlap. B–In the upper diagram, two species occupy nonoverlapping niches, whereas in the lower diagram, niches overlap so much that severe competition results in divergence, as indicated by the arrows (after Odum and Barrett, 2005).

4.2. PRESENCE OR ABSENCE OF SPECIES: FACTORS RESTRICTING HABITAT USE

If a species occupies an area and reproduces there, we know that all its needs are met and that it can compete with other species successfully. A convenient way to identify the factors affecting the distribution of a species is to determine why it is absent from a place. We can examine possible reasons one by one, by using transplant experiments, by giving individuals a choice of artificial habitats in the laboratory, or by building enclosures in the field that encompasses different habitats.

In an experiment, called **transplant experiment,** organisms are moved to a new environment, and their survival and reproduction are observed. Because organisms sometimes survive but fail to reproduce, a long-term project covering several generations is necessary. To study the distribution of animals, **G.W. Elmes** (1971) dug up colonies of heathland (*Heath* is an area of open uncultivated land, usually covered with heather (shrub), gorse (shrub) and coarse grasses) ants (*Lasius niger*) of England and moved them to sites that differed in temperature and in moisture content. He moved 18 colonies of this ant which normally inhabits low, wet heathlands to higher, drier sites, and moved 6 colonies of this ant to other, low, wet areas as controls. **Elmes** monitored their survival over a 5 year period and found that the higher the colony was moved, the less likely it was to survive (Table 4.1).

Table 4.1. Result of transplant experiments of heathland ant colonies in England (Source Drickamer *et al.*, 2002).

Location of transplant	Years survived						Total number of colonies transplanted
	0	1	2	3	4	5^+	
Control *Lasius niger* zone							
11–m level	1		1			4	6
Dry heath 13–m level			1		1	3	5
17–m level	6	3				1	10
25–m level	1	1	1				3

Those ants which were at 13m level did as well as the controls at 11m. One of the main factors that affected survival was competition with other species that normally inhabit the higher, drier part of the heath. This experiment demonstrates the need for controls in research and the need for long-term observation of transplants.

If a transplant is successful, two possible factors may explain why a species is not found naturally in the transplant area: 1. the area is inaccessible because the dispersal ability of the organism is limited, or 2. the organism fails to recognize the area as a suitable habitat. If a transplant fails, the causal factors may be the presence of other species (*i.e.*, competitors, parasites, pathogens) or physical and chemical factors (*i.e.*, temperature, pH and so on). When trying to understand why a species is absent from an area, we can proceed through a series of steps (Fig. 4.3; **Krebs,** 1985).

1. Dispersal Ability

Many cases of successful introduction by humans have demonstrated that **locomotive abilities** adequately explain a species absence. The European starling (*Sturnus vulgaris*) originally was found in most of the Europe and Asia. After several unsuccessful introductions of small numbers of starlings into the United States, 80 pairs were released in New York City's Central Park in 1890 by **Eugene Scheifflin** of the Acclimatization Society. The "goal" of this group was to familiarize Americans with the all the birds in Shakespeare's plays (**Miller,** 1975). Within fifty years, starlings had reached the west coast, and today they are probably the most

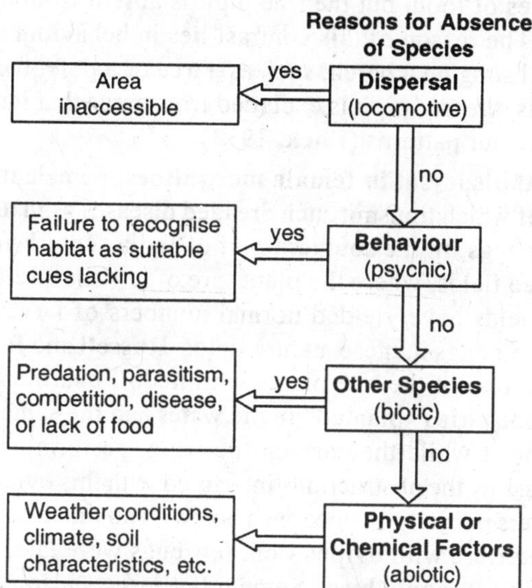

Fig. 4.3. Methodological approach for studying the geographical distribution of a species. A useful way to proceed in the analysis of a species' distribution is to determine why that species is *not* in a particular place. Four basic factors are listed : behaviour may be involved in each and the factors may interact (after Drickamer *et al.*, 2002)

numerous bird species in the U.S.A. Starlings nest in tree cavities and are more aggressive than most native species of birds – they will even evict larger species such as flickers (*Colaptes auratus*) from nest holes. The eastern bluebird (*Sialia sialis*) is one native bird species that now occupies only a fraction of its former range partly because of its unsuccessful competition with the starling. Although they are insectivorous during the summer, starlings are generalists and switch to seeds in the winter; many insectivorous species must migrate south to find insects. Other aspects of the starling's habits have led to their success in modern industrialized countries : a tolerance of loud noises and air pollution ; a willingness to roost and perch in various places, from trees to bridge supports ; and a preference feeding in grassy areas.

Sometimes, ecologists characterize species as being *r*– or *k*–selected, *r* refers to the rate of population increase, and *k* refers to the number in the population at the upper limit, or carrying capacity of the environment. The ***r*-selected species** have high-reproductive rates, rapid development, and great powers of dispersal (**MacArthur** and **Wilson,** 1967); they tend to live in ephemeral environments in which recolonization of areas is necessary. **Diamond** (1974) studied the birds of New Guinea and nearby islands and found that certain species were always the first to recolonize islands that had lost their fauna due to volcanic explosions or tidal waves; he referred to these *r*-selected species as **"super-tramps". K-selected species** have low reproductive rates, slow development, and limited powers of dispersal; characteristically inhabitants of stable environments, these species may fail to colonise new areas separated by relatively small barriers. One island Diamond studied was separated from New Guinea by only 10 m of water, yet it had only half the New Guinea species expected on the basis of the availability of comparable habitats in the two areas.

2. Behaviour

Sometimes behaviour patterns keep species from occupying apparently suitable habitat.

***(i)* Behaviour based dispersal in tree pipit birds.** Two closely related species of British birds, the tree pipit (*Anthus trivialis*) and the meadow pipit (*Anthus pratensis*), are both ground nesters and eat similar types of food, but the tree pipit is absent from many treeless areas that the meadow pipit inhabit. The reason of this contrast lies in behaviour associated with song—the tree pipit ends its aerial song on a perch, such as a tree or a pole; the meadow pipit ends its aerial song on ground. Thus, the tree pipit is excluded from areas that it could otherwise occupy because of a specific behaviour patterns (**Lack,** 1933).

***(ii)* Behaviour based dispersal in female mosquitoes.** Female mosquitoes of the genus *Anopheles* (many species of which transmit such dreaded diseases as malaria) are very particular about where they lay their eggs. In the southern part of India, *Anopheles caulifacies* eggs and larvae are found only in rice fields, where the plants are of less than 12 inches height; however, eggs transplanted to old fields have yielded normal numbers of larvae and adults, and other species of *Anopheles* lay their eggs in these mature fields. **Russell** and **Rao** (1942) demonstrated that the mechanical obstruction of the rice plants inhibited *A. caulifacies* females from laying eggs. Glass rods and bamboo strips" planted" in the water had the same effect; shade was not a factor. *A. caulifacies* oviposit while they are on the wing, performing a hovering dance 2–4 inches above the water; possibly the obstructions interfered with this oviposition dance. However, in one experiment researchers partially submerged a box with no lid in the shallow water; although the box did not directly interfere with oviposition, few eggs were laid.

***(iii)* Nest site selection in honeybees.** Some other insects also actively choose where to live. In the spring, a honeybee colony that has grown sufficiently large will split in two, with the old queen and half her workers force flying off in a swarm, deserting the old hive and the remaining workers to a daughter, who will become a new queen (see **Alcock,** 1993). The departing swarm settles temporarily in a tree, where the workers hang from a limb in a mass around their queen. Over the next few days, **scout workers** fly out from the swarm in search of small openings that lead to chambers in the ground, in cliffs, and in hollow trees. There are often many such sites within range of the waiting swarm, but only some motivate a scout worker to perform a dance back at the swarm, a dance that communicate information about the distance, direction and quality of the potential new home (see Chapter 9 of Communication). Other workers attend to a dancing scout and may be sufficiently stimulated to fly out to the spot themselves. If site is attractive to them, they too will dance and send still more workers to the area. Eventually most scouts will be advertising one location, and then the swarm leave its temporary perch and flies to the most popular nest site.

Thomos Seeley (1977) discovered that scouts are enthusiastic only about chambers with a volume between 30 to 60 litres. But the size of the chamber is not only factor assessed by scout bees. In Bavaria, Germany, where **Martin Lindauer** did his work, the bees generally choose holes in ground, with wooden structures and strawbasket hives the second and third choices, respectively. Moreover, if a Bavarian swarm is given a choice between two apparently equivalent hives, one close to the original hive (*i.e.,* about 50 meters away) and another farther away (*i.e.,* about 200 meters off), the bees will choose the more distant of the two (**Lindauer,** 1961). This is true even though the added distance means that the queen may become exhausted on the trip (since queens are wonderful egg layers but only marginal fliers).

Ethologists have tested Lindauer's hypothesis that *cold winters and feeding competition have influenced the evolution of the habitat preference of Bavarian honeybees* by predicting that honeybees that do not confront these special ecological pressures will differ in their behaviour. Italian honeybees offer a chance to test this prediction. They are members of the same species, *Apis mellifera,* but they live in southern Europe where the winters are much milder than those of northern Germany (**Jaycox,** 1980; **Jacox** and **Parise,** 1981). As expected, Italian bees do not prefer chambers in the ground to aboveground hives, nor do they favour sheltered over unsheltered hives. Moreover, they accepted smaller cavities and sites closer to a daughter's colony than do northern bees.

All of these differences are correlated with the reduced danger of winter mortality for Italian bees. Because they live in southern Europe, heavily protected sites are not especially valuable (**Gould,** 1982). There is also reduced selection pressure for large colony size, which appears to be adaptive primarily under cold climatic conditions. In cold regions, large numbers of colony members create many layers of insulation around the mass of overwintering bees, enabling the queen and a sufficient worker force to make it through winter. During the mild winters of southern Europe, large colonies gain no special survival advantage. Smaller colonies not only can occupy smaller cavities, they also can support themselves with food resources from smaller areas. Therefore, Italian bee swarms are not under pressure to disperse great distances from the home hive (**Gould,** 1982).

(*iv*) Ascidian larva as a site selector. The pelagic larvae of ascidians are very selective when it comes to invading a new site. The site must be of the right consistency (*i.e.,* mud or sand for molguloid ascidians and rock for other ascidians) : otherwise they cannot attach and undergo metamorphosis. For many ascidain larvae (*e.g., Herdmania, Ascidia,* etc.) vertical surfaces such as pilings are favoured sites. These larvae are adapted to select a preferred site, a role often superseding all other roles. Such larvae are locomotory but do not feed. They are dispersed by currents, but their pelagic lives are too short for extensive dissemination. Because it cannot feed, a larva must find a site and metamorphose within a few hours or days; otherwise it dies.

Thus, the ascidian larva has evolved as a most efficient site selectors. This tadpole-like larva with neural tube, myotomes, pharyngeal gill-slits, eye and otolith resembles a chordate body plan as it probably is the ancestral type for all vertebrates.

Further, the success of the ascidian species which are ecologically adapted for hard vertical surfaces, depends on the facility with which these larvae find and colonize a new site. However, some ascidians such as the molguloids, colonize sand or mud bottoms. These exposed surfaces are so widely distributed that the organisms do not need to be precise site selectors. Their larvae are structurally simple, without eyes and tails, and descend to the bottom in response to gravity, attach indiscriminately and metamorphose. In fact, the existence of tailless larvae among ascidians that do not require preferred sites is additional confirmation that site selection is the principal adaptation for the typical tadpole larva.

Why should a species not take advantage of a suitable habitat ? **Drickamer** *et al.*. (2002) have provided following two possible reasons for this question. 1. One possibility is that the habitat is not actually suitable, perhaps because of competition, predation, or other factors the scientists may failed to detect. 2. On the other hand, such habitats may not have been suitable in the past : if organisms responding to certain environmental cues in previous optimal habitats left more offspring, their genetically influenced behaviours would become widespread and persist. New environments, although suitable, may not contain those cues and therefore are not utilized.

3. Interaction with other Organisms

Even if an individual can and "wants" to get to a place, other factors may prevent its becoming established. These factors could be predators, parasites, disease agents, allelopathic agents (poison or antibiotics), or competitors. Demostrating conclusively that one species prevents an area from being colonized by another is difficult, but it has been indicated with experiments and observational data.

(*i*) Evidence of influence of predator on distribution of mussels. A careful series of experiments by **Kitching** and **Ebling** (1967) showed how mussels (*Mytilus edulis*) were kept out of protected bays along the coast of Ireland by three species of crabs and one species of starfish. Where the coasts was unprotected, crabs were restricted by wave action and small mussels could survive. In sheltered waters, mussels survived only in areas such as steep rock faces that the predators cannot reach. Kitching and Ebling proposed that four criteria must be met before we can conclude that a predator restrict the habitat of its prey :

1. If they are protected from predators, prey will survive when transplanted to a site where they normally do not occur.
2. If the distributions of prey and predator are negatively correlated.
3. If the predator is observed eating the prey; and
4. If the predator can be shown to destroy prey in transplanted experiment.

***(ii)* Influence of competition on distribution of warblers.** The more similar two species are, the more likely they are to compete intensely, and thereby restrict each others distribution. Although a species is usually less numerous at the edge of its distribution than at the center (see Fig. 4.1), it is sometimes abruptly replaced by a close relative, with both species at maximum density at the interface (Fig. 4.4). If interspecific competition restricts distribution, we would expect one species to extend its range in the absence of the other. This phenomenon was demostrated by **Diamond** (1978) with bird distribution on mountain tops in New Guinea, where a second, closely related species may be absent due to its inability to disperse (Fig. 4.5). In these cases, the single species occupied a much a larger altitude range than it did on islands where both were present. Competitive exclusion was actually witnessed by **Orians** and **Collier** (1963) when colonial tricoloured blackbirds (*Agelaius tricolor*) moved into a marsh already occupied by redwinged blackbirds (*Agelaius phoeniceus*). After the invasion, the red-wing territories were restricted to the periphery.

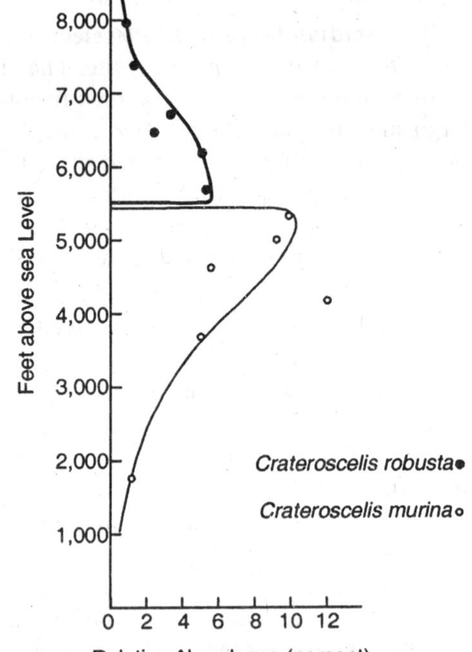

Fig. 4.4. Influence of other species on distribution of warblers. Abundance is measured as a percentage of all bird individuals observed. As one sample from the side of a mountain in New Guinea shows, *Crateroscelis murina* reaches maximum abundance at 5,400 feet and is abruptly replaced by *Crateroscelis robusta* (after Drickamer *et al.*, 2002).

The studies mentioned above, however, suggestive, they do not clearly show that competition by one species excludes the other. The responses are correlational, and experiments are needed to demonstrate cause and effect. One such experiment is to remove one species and note changes in nearby competitors, or to introduce a closely related species and monitor the success of each, as **Vaughan** and **Hansen** (1964) did with two species of pocket gophers (*Thomomys bottae* and *Thomomys talpoides*). Slight differences between these rodent species such as dispersal powers and environmental tolerances led to one or the other species, winning out.

4. Physical and Chemical Factors

If a transplant experiment fails, and no evidence exists that biotic factors have eliminated the species, some combination of physical and chemical factors may be involved. Each organism has a range of tolerance for these factors, and much of its behaviour is directed toward staying within these limits. Temperature and moisture are the main factors that limit the distribution of life on Earth, but physical factors (such as light, soil nutrients, salts and pH) are important as well.

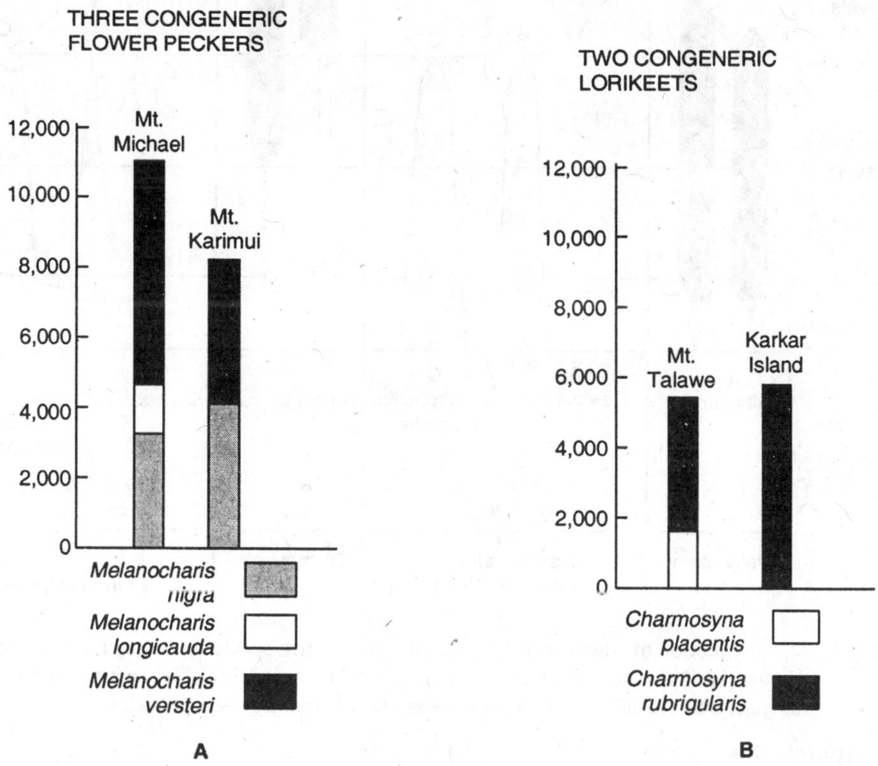

Fig. 4.5. Altitudinal range of species of birds on New Guinea mountains and surrounding islands. A–Three similar cogeneric flower peckers, *Melanocharis nigra, M. longicauda* and *M. versteri,* occupy nonoverlapping areas up to about 11,000 feet on Mt. Michael. On Mt. Karimui which is smaller and more isolated than Mt. Michael, *M. longicauda* is absent. B–Two cogeneric lorikeets, *Charmosyna placentis* and *C. rubrigularis* occupy an area on Mt. Talawe, but only *C. rubrigularis* colonized Karkar Island. Altitudinal ranges in all cases are nonoverlapping and are larger in the absence of other species (after Drickamer *et al.*, 2002).

The classic studies of **Connell** (1961) show that a combination of factors influences the distribution of organisms. He studied two species of barnacles (*Balanus balanoides* and *Chthamalus stellataus*) that live in the intertidal zone of rocky coastlines of Britain. Physical factors, especially tolerance to desiccation and high temperatures, predation by snails (genus *Thais*), and interspecific competition for space all interacted to determine distribution and abundance (Fig. 4.6) of barnacles.

4.3. DISPERSAL FROM THE PLACE OF BIRTH

Animals may also make decisions about whether to remain at (or return to) the natal site or to disperse to other breeding locations. **Natal dispersal** means leaving the site of birth or social group (**emigration**), traversing unfamiliar habitat, and settling into a new area or social group (**immigration**). Moving away from familiar ground is risky because the individual is strange with the location of food and shelter and is no longer in the presence of familiar neighbours and relatives.

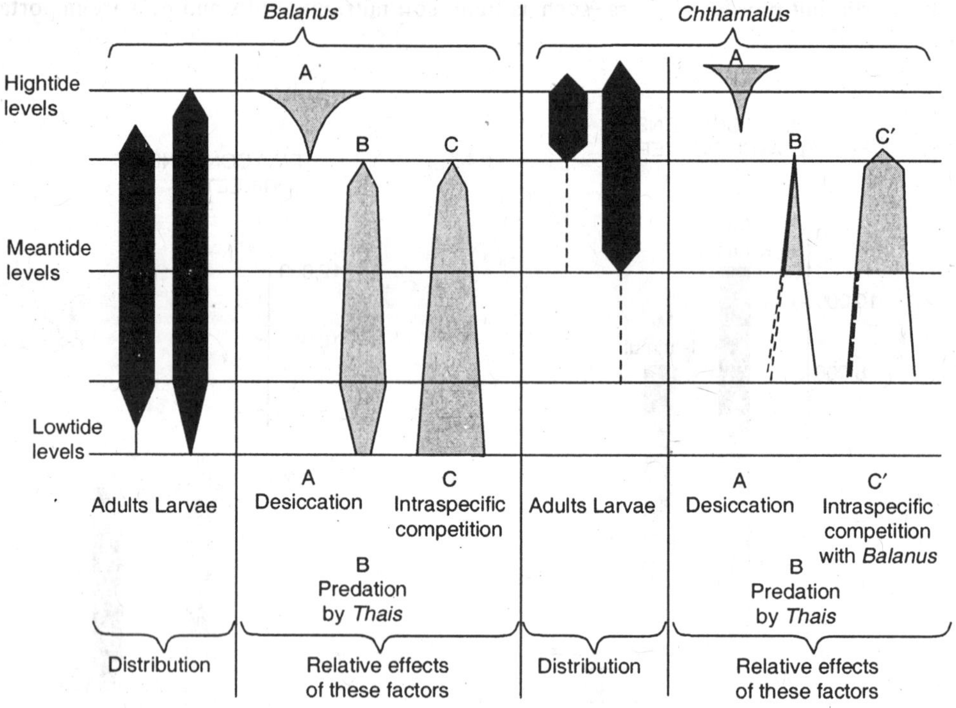

Fig. 4.6. The distribution of adult and newly settled larvae of two species of barnacles. The upper limit to the distribution is set by desiccation, while the lower limit is set by a combination of competition for space and predation by a species of snail (after Drickamer *et al.,* 2002).

1. Philopatric Sex

In most species of birds and mammals, members of one sex tend to disperse, while members of the other sex are **philopatric**, breeding near the place where they were born. Among mammals, it is usually the males that disperse, while among most species of birds (perching birds of the order Passeriformes), the opposite is true (*i.e.,* the females that disperse) (**Greenwood,** 1980). (*Note.* Among mammals, there are some exceptions. In the chimpanzee, *Pan troglodytes,* and humans, females are the despersing sex; see **Drickamer** *et al.,* 2002). The reason for such difference in mammals and birds may be that most bird species are monogamous, and the male defends a territory that contains resources vital to him and his mate. It is probably easier for a male to establish such a territory and attract a mate in or near his natal site, where he is familiar with the location of resources and / or predators. On the other hand, many species of mammals are **polygynous.** The females form the stable nucleus, and the males attempt to maximize their access to them, frequently moving from one group to another (**Greenwood,** 1980).

2. Causes of Dispersal

The causes of dispersal can be understood at several different levels. At the *proximate level* ethologists wish to know the immediate reasons why an individual leaves the natal area. For instance, it might be forced out by its parents or other residents, or it might respond involuntarily to increase in *testosterone* (hormone) level associated with sexual maturation. At the *ultimate level* ethologists wish to know the long-term, evolutionary causes of dispersal. For instance, individuals that fail to disperse may have lower reproductive success because their offspring are inbred and therefore less viable. Natural selection would then favour dispersal.

Following two hypotheses have been forwarded by ethologists for explaining the causes of dispersal of animals.

1. Inbreeding avoidance hypothesis. The ultimate cause of dispersal from the natal site has been argued by many to be inbreeding avoidance. The costs of inbreeding, referred to as **inbreeding depression** (depression means decline), have been documented in many laboratory and zoo populations (**Ralls** *et al.*, 1979). Recently inbreeding depression has also been studied in natural populations. Inbreeding depression demonstrates itself through reduced reproductive success and survival of offspring from closely related parents compared to offspring of unrelated parents. Inbreeding depression is caused by increased homozygosity of the inbred offspring and the resulting expression of deleterious recessive alleles. For example, in African lions (*Panthera leo*), males from a small, inbred population showed lower testosterone levels and more abnormal sperm than did males from a large , outbred population (**Wildt** *et al.*, 1987) (Fig. 4.7).

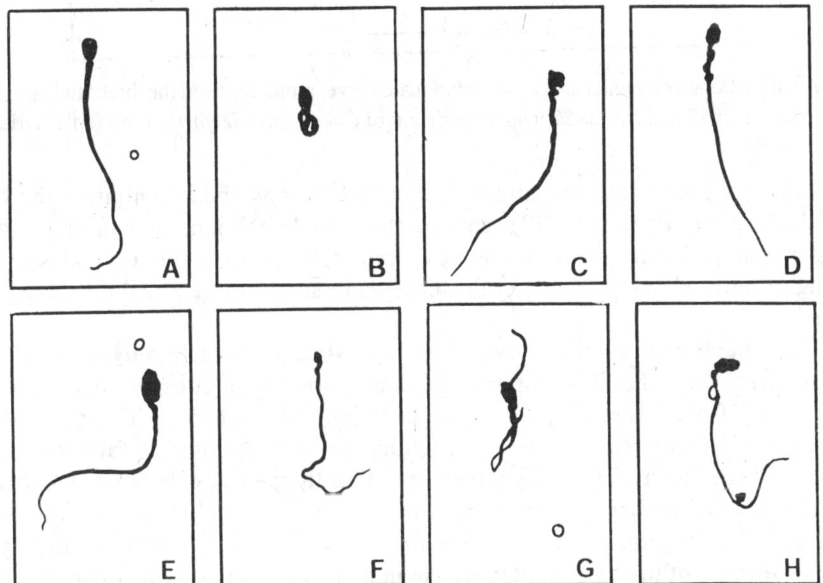

Fig. 4.7. Abnormal sperm from an inbred lion population in Africa worked out by Dave Wildth. A. Normal;B. Tightly coiled flagellum; C. Missing mitochondrial sheath; D. Abnormal acrosome and disordered midpiece. E. Macrocephalic with abnormal acrosome. F. Microcephalic with missing mitochondrial sheath. G. Bent flagellum; H. Bent neck with residual cytoplasmic droplet.

In an experiment when both inbred and outbred white-footed mice (*Peromyscus leucopus*) were released back into natural habitat, the inbred stock survived less well than the outbred stock, although differences between the two stocks were not great in the laboratory environment (**Jimenez** *et al.*, 1994).

When one or other sex disperses, there will be less chance of matings between related individuals. Among black-tailed prairie dogs (*Cynomys ludovicianus*), young males leave the family group before breeding; females remain. Also adult males usually leave groups before their daughters mature (**Hoogland,** 1982). Among primates such as vervet monkeys (*Chlorocebus aethiops*) and baboons (*Papio anubis*), males usually leave the natal group at sexual maturation or shortly after. They usually transfer to a neighbouring group with age peers (fellows) or brothers (Fig. 4.8). Several years later they may again transfer alone to a third group. **Cheney** and

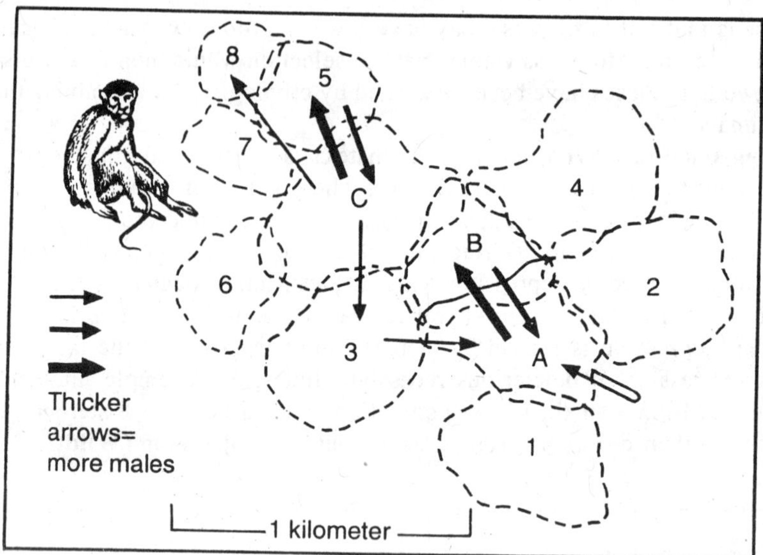

Fig. 4. 8. Group transfer by natal and young adult male vervet monkeys from the three study groups between March 1977 and July 1982 in an experiment of *Cheney* and *Seyfarth,* 1983 (After Drickamer *et al.* 2002).

Seyfarth (1983) argued that this pattern of nonrandom movement minimizes the chances of mating with close kin. **Packer** (1979) reported that a male baboon, after failing to disperse at sexual maturity and after mating with relatives, sired offspring with low survival rates compared to offspring of outbred males. Thus, in this case there seems to be a real cost associated with inbreeding.

(*ii*) Intraspecific competition hypothesis (or Mate competition hypothesis). A second cause of dispersal from the natal site may be competition with conspecifics (**Dobson,** 1982; **Moore** and **Ali,** 1984). Most species of mammals are polygynous , with males mating with more than one female. Males may be forced to disperse as they compete for access to females. Although the inbreeding avoidance hypothesis predicts that one sex should disperse, it does not predict which sex should disperse; the competition hypothesis predicts that males should be the dispersing sex in polygynous species. According to **Greenwood** (1980), in such systems the reproductive success of males is limited by the number of females with which they can mate, and males are likely to range farther than females as they search for mates. Females, on the other hand, are limited by resources (food and nesting sites) that can best be obtained and defended by staying at home. Among group- living mammals, females typically form the stable nucleus, and the males attempt to maximize their access to them, frequently moving from one group to another.

According to a slightly different competition model (**Hamilton** and **May,** 1977), animals disperse so as to avoid local resource competition with close relatives and thus avoid lowering their indirect fitness. In a new habitat they are likely to be competing with nonrelatives and therefore would suffer no cost.

5. Examples of Dispersal

1. Dispersal in lions. Dispersal patterns in lions follow the typical mammalian pattern: females usually remain in or near their natal pride. Whereas males always leave, usually before 4 years of age, to become nomads and/or take over new prides (Fig. 4.9) (**Pusey** and **Packer,** 1987). Competition with other males seems to be an important factor, because departures most often occur when a new association or coalition of males takes over the pride.However, some males appear to leave voluntarily, either in search of mating opportunities or to avoid breeding with kin . Once a coalition takes over a new pride, it is usually ousted by another coalition within a few years, or it leaves to take another pride. In all cases, males leave the pride before their daughters start mating. An additional factor is that new coalitions of males kill the young cubs in the new pride. Thus, breeding males must remain in a new pride long enough to ensure the survival of their cubs. **Pusey** and **Packer** (1987) concluded that male-male competition, mate acquisition, protection of young cubs and inbreeding avoidance all play roles in the evolution of dispersal patterns of lions.

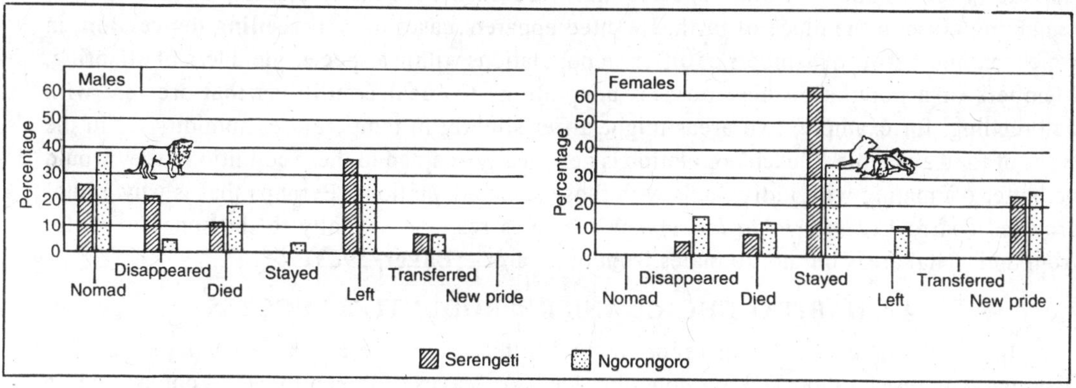

Fig. 4.9. The fate of subadult lions by 4 years of age at two African sites. One can note, here, the differences between the sexes in dispersal patterns.

2. Belding's ground squirrels. Belding's ground squirrels (*Spermophilus belding*) follow typical mammalian patterns: females remain in the natal area for life (*i.e.*, females are philopatric), while males disperse (**Holecamp** and **Sherman**, 1989). The proximate cause of male dispersal seem to be the effects of prenatal exposure to testosterone (organizational effects) and the attainment of a critical body weight. Effects of testosterone later in life (activational effects) seem less important, since castration of males just prior to natal dispersal did not prevent dispersal. **Holecamp** and **Sherman** were not able to test the inbreeding avoidance and competition hypothesis directly, but conclude that inbreeding avoidance was the more feasible means by which dispersal increased fitness (Table 4.2.)

Table 4.2. Why juvenile male Belding's ground squirrels disperse ? Answers have been found at each of four levels of analysis (Source : Drickamer *et al.*, 2002).

Levels of analysis	Summary of findings
1. Physiological mechanisms	Dispersal by juvenile males is apparently caused by organizational effects of male gonadal steroid hormones. As a result, juvenile males are more curious, less fearful and more active than juvenile females.
2. Ontogenetic processes	Dispersal is triggered by attainment of a particular body mass (or amount of stored fat). Attainment of this mass or composition apparently also initiates a series of locomotory and investigative behaviours among males.
3. Effects on fitness	Juvenile males probably disperse to reduce chances of nuclear family incest.

4. Evolutionary origins	Strong males biases in natal dispersal characterize all ground squirrel species, other ground dwelling sciurid rodents, and mammals in general. The consistency and ubiquity of the behaviour suggest that it has been selected for directly across mammalian lineages.

6. Inbreeding versus Outbreeding

If inbreeding depression were the only factor involved in dispersal, we might expect individuals to disperse as far as possible from relatives. However, such is not normally the case. **Shields** (1982) found that most species that have been adequately studied are philopatric, remaining close to the place of birth. He cited apparent cases of **outbreeding depression,** in which matings between members of different populations within a species yield less fit offspring. Members of a population may possess adaptations to local conditions that are lost with outbreeding, for example, two areas might differ slightly in temperature, humidity, or in the types of food available. If each population is genetically adapted to these conditions, they would be better off mating with individuals with those same adaptations. Perhaps that is why white-crowned sparrow (*Zonotrichia leucophrys*) females respond sexually to the songs of males from their natal areas but not to males from other areas (**Baker,** 1983).

4.3. HABITAT CHOICE AND REPRODUCTIVE SUCCESS

In 1980, **Thomas Witham** reported the life history of the aphid *Phemphigus betae,* a plant parasite insect, about 0.6 mm long that feeds on leaves of the cottonwood poplar tree. In the spring, after hatching from eggs laid the previous fall in the bark of tree, females (called **stem mothers**) move up the trunk and select a leaf, settles by its midrib, almost always near the base, and in some way induce the formation of a hollow ball of tissue, called *gall* (Fig. 4.10). In

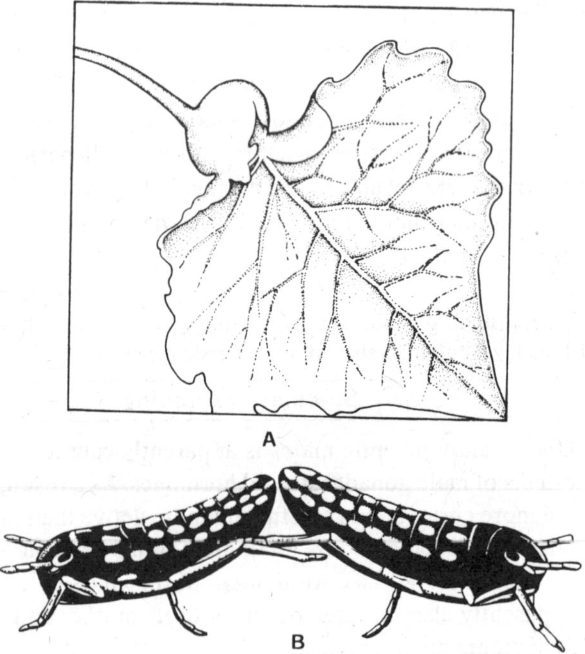

Fig. 4.10. A-Gall formed by a poplar aphid at the base of a large leaf. In this gall she produces her young. B-Territorial dispute between two poplar aphids. B Females may spend hours kicking one another (after Alcock, 1993).

this gall, she will live with the offsprings she bears parthenogenetically. When her daughters are mature, the gall splits, and the aphids within disperse to new plants (see **Alcock,** 1993).

Witham found that females settling on large leaves have higher reproductive success than females on small leaves. Not surprisingly, aphids select the largest leaves, leaving small ones vacant. However, latecomers may have to choose whether to take an already occupied large leaf or an unoccupied small one. If she takes on an occupied leaf, she will have to settle farther from the base (*i.e.,* farther out on the mid rib), where there is less food. Stem mothers farther from the

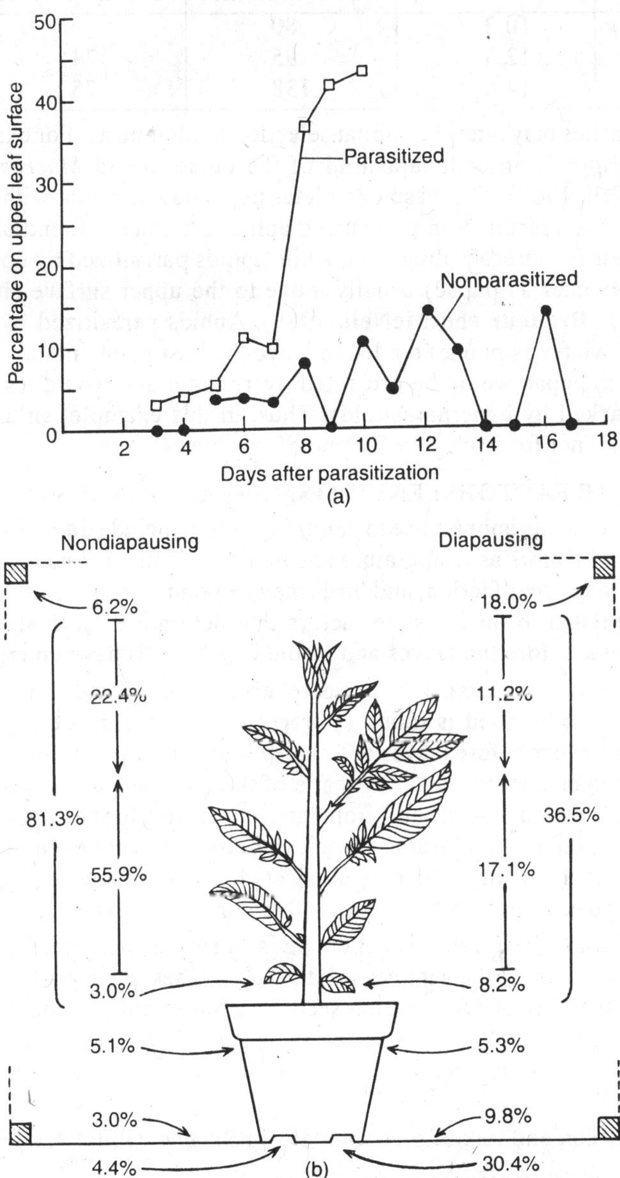

Fig. 4.11. The effects of a parasitic wasp on the behaviour of the potato aphid (host). A–The effect of parasitism by wasp larvae on the distribution of aphid adults. Squares denote parasitized aphids, circles denote nonparasitized aphids. B–The distribution of aphid mummies containing diapausing and non-diapausing individuals of the parasitoid wasp (after Drickamer *et al.,* 2002).

base were found to be smaller in size, and they produced fewer young than those closer to the base. Stem mothers may engage in **pushing and kicking contests** that lasts for days with the largest aphid usually getting the basal position (Fig. 4. 10 B). Choice of habitat in these insects is non-random and results in higher average fitness than would random leaf selection.

Table 4.3. Effects of leaf size and position of gall on the reproductive success of females poplar aphids (Source : Alcock, 1993).

Number of galls per leaf	Mean leaf size (cm)	Mean number of progeny produced		
		Basal female	Second female	Third female
1	10.2	80	–	–
2	12.3	95	74	–
3	14.6	138	75	29

However, parasites may alter the habitat selection of their hosts. For example, the parasitic wasp *Aphidius nigripes* is an endoparasitoid of the potato aphid *Macrosiphum euphorbiae* (**Brodeur** and **McNeil,** 1989). The wasp completes pupal development within the aphid, which becomes mummified as a result. Non-parasitized aphids are usually found on the under surface of potato leaves, their favoured feeding area, while aphids parasitized by nondiapausing larvae (those that not overwinter as pupae) usually move to the upper surface shortly before being mumified (Fig. 4.11; **Brodeur** and **McNeil,** 1989). Aphids parasitized by diapausing larvae (those that will overwinter as pupae) tended to leave the host plant and mummify in concealed sites. Therefore, such pupae would be protected over winter and would less likely to be eaten by predators or attacked by hyperparasitoids. Thus, in this example, an animal may select a habitat that is optimal not for itself, but for one of its parasites.

4.4. PROXIMATE FACTORS: ENVIRONMENTAL SIGNALS FOR DISPERSAL

The proximate cues the animals use in habitat selection include direct locomotive responses to such environmental factors as temperature and humidity. Little is known about the cues used by vertebrates; fish may use olfaction, and birds may respond to vegetation types. The cues may be only indirectly related to the ultimate factors that determine survival; for example, birds select breeding habitat before the leaves and staple insect foods have emerged.

The blue tit (*Parus caeruleus*), a European relative of the chickadee, lives in oak woodlands where most of its preferred food is found (**Partridge,** 1978). But the blue tit establishes its breeding territory each year before leaves and caterpillars (the staple food) have even appeared, so it must be using other features, such as shape of the the trees, or cues to the habitat. Birds probably responds to cues such as stimuli from landscape; sites for nesting, singing, or feeding; food itself; or other animals. In migratory species of birds, it is not even clear when in the life cycle a choice is made. Breeding sites may by slected in late summer or fall before migrating, rather than in the spring, as it usually assumed (**Brewer** and **Harrison,** 1975).

Wood frogs (*Rana sylvatica*), lay their eggs during a brief period of each spring in temporary ponds that dry up in the summer. All the egg masses are deposited in one place in the pond. The physical features of this location seem less important than does the presence of an egg mass, which triggers other females to lay their eggs there (**Howard,** 1980). Eggs in the center of such a mass may be protected from predators and from temperature fluctuations (**Berven,** 1981).

Lastly, competition and risk of predation also influence habitat selection.

4.5. DETERMINANTS OF HABITAT SELECTION

1. Roles of Genes in Habitat Selection

If two animals reared from birth in identical environments are found to differ in habitat preference when they are tested as adults, one can conclude that these differences must be due to hereditary factors or genes. For example, when coal tits (*Parus ater*) and blue tits (*Parus caeruleus*) were reared in aviaries with no vegetation and then presented with a choice between oak and pine branches, coal tits preferred pine and blue tits preferred oak (Fig. 4.12). The differences correspond to the distribution of these birds in nature and to the response of wild titmice in aviaries (**Partridge,** 1974, 1976, 1978). In fact, blue tits feed more efficiently in oak trees than do coal tits (**Partridge,** 1976).

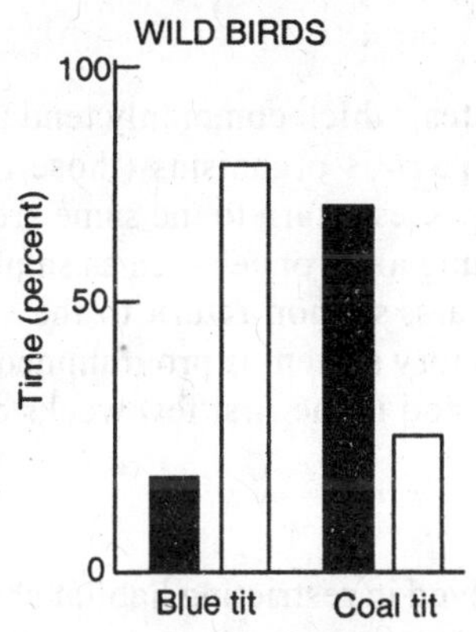

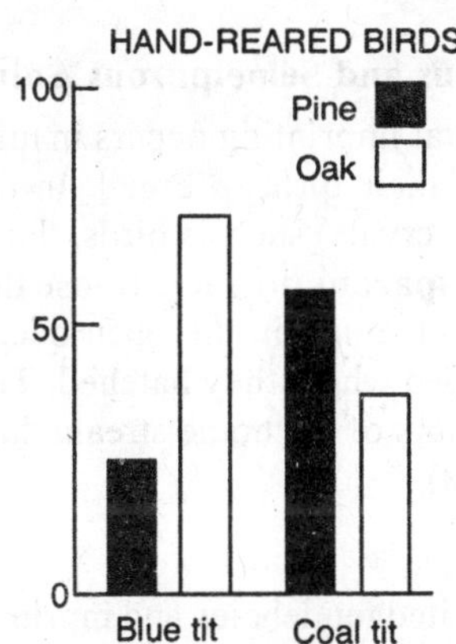

Fig.4.12. Time tits spent in oak and pine habitats in the laboratory. Both wild and hand-reared blue tits preferred the oak habitat to the pine habitat, wild and hand-reared coal-tits, however, preferred the pine habitat. These preferences reflect those of each species in the wild (after Drickamer *et al.*, 2002).

A genetic basis for habitat selection has been shown for a wide variety of invertebrates including insects, molluscs and crustaceans. Included are preferences of adults for general features of the environment (*e.g.*, meadow versus dense woods in different species of *Drosophila*); background matching in moths and butterflies; where to lay eggs (*e.g.*, different species of plants in butterflies) and preferences for different habitats by larvae (*e.g.*, *Drosophila*).

2. Habitat Imprintining

Some experimenters have modified the environments of young birds to test whether their genetic predisposition to respond to certain stimuli can be altered by early experience, or **habitat imprinting.** For example, when **Klopfer** (1963) placed wild-caught chipping sparrows (*Spizella passerina*) in a room containing both pine branches and oak branches, these birds preferred the pine, as they usually do in nature.

The classic studies of **Wecker** (1963) on habitat selection in mammals was on deer mice (*Peromyscus maniculatus*). This species includes many subspecies, but all are two main types: 1. the long-eared, long-tailed forest form, and 2. the smaller, short-eared, short-tailed grassland form. In the laboratory, the grassland form (*Peromyscus mainiculatus bairdii*) does well in forest conditions, where its food preference and temperature tolerance are similar to those of the forest subspecies; thus, experimenters assumed that the avoidance of forests in the grassland deermouse is a behavioural response (**Harris,** 1952).

The objective of Wecker was to assess the genetic basis of this behaviour and to test the idea that habit imprinting is important (**Thorpe,** 1945). He constructed an enclosure halfway in a forest and halfway in grassland, released the mice in the middle, and recorded their locations. The animals he tested were of the grassland subspecies (*bairdii*), and were of three basic types : 1. Wild-caught in grassland; 2. offspring, reared in laboratory, of wild-caught; 3. reared in laboratory for 20 generations. Both wild-caught mice and their offspring selected the grassland half of the enclosure, regardless of previous experience. Laboratory stock and their offspring showed no preference, whether or not they had been raised in forest conditions. However, laboratory stock reared in a grassland enclosure until after weaning showed a strong preference for the grassland when tested later.

Later on, in a series of laboratory experiments designed to explore the importance of early experience on bedding preference in inbred mice (*Mus domesticus*). **Anderson** (1973) raised animals either on cedar shavings or on a commercial cellulose material. When he tested them later, he found that the mice preferred the bedding on which they had been raised, although females raised on cellulose drifted toward cedar shaving in subsequent tests. Naive mice preferred cedar shavings.

Introparous and Semelparous Animals

Habitat imprinting occurs in migratory vertebrates, which commonly tend to return to the vicinity of their birth to breed. In the case of **introparous** organisms (those that have their young at intervals) such as birds, the adults of most species return to the same area to nest each year. **Semelparous** breeders (those that have their young all at once) such as salmon breed only once. After feeding in the open ocean for several years, salmon return to the same upstream spawning bed where they hatched. The salmons' olfactory system is programmed to respond to unique odours of the home stream during a critical period in the first few weeks of life (**Hasler** *et al.*, 1978).

3. Tradition

Inherited tendencies and inprinting may be involved in restricting habitat choice to a small part of the potential range, but **tradition** (Behaviour passed from one generation to the next through the process of learning) may also be an important factor. For instance, many species of waterfowl having **staging areas,** where they rest and feed during migration from breeding to winter grounds. The same areas tend to be used year after year.

Mountain sheep (*Ovis canadensis*) live in unisexual groups, females are likely to stay in the natal group but may switch to another female group when they are between one and two years of age (**Geist,** 1971). Young ram desert the natal group after the second year of life and join all-ram bands. Mothers do not tend to chase their young away at weaning, as most other mammals do. Females follow an older, lamb-leading female; male follows the largest-horned ram in the band. When ram matures, they are followed by younger rams and pass on their habits preferences to them.

In comparison to present time, until the last century, mountain sheep occupied a much larger range in North America and Asia. In contrast, deer (*Odocoileus sp.*) and moose (*Alces alces*) have recolonized areas rapidly and have reached population densities higher than ever (**Geist**, 1971). *Why have sheep failed to extend their range, while moose and deer have done so?* Giest pointed out that deer and moose, which are relatively solitary beasts, establish ranges by individual exploration after being driven out of the mother's range; sheep, in contrast, transmit home-range knowledge from generation to generation and ofter associate with group members for life.

Moose habitats are subject to rapid expansion after fires, but are relatively short-lived, and moose must continually colonize new habitats. Each spring when her new calf is born, the cow drives away her yearling (*i.e.,* a young animal past its first year and not yet two year old) which may wander some distance before establishing its new range.

Thus, mountains sheep habitats are formed by stable, long-lasting climax grass communities which exist in small patches. Geist argued, that given the distance between patches and the ease with which wolves can pick off sheep, the best strategy for the sheep is to stay on familiar ground. Sheep also have at least two and as many as seven seasonal home ranges that may be separated by 20 miles or more. These areas are visited regularly by the same sheep year after year at the same time; knowledge of the location of these ranges and the best time to visit them is transmitted from one generation to the next. Because new habitats rarely become available, there is no advantage to an individual's dispersing and attempting to colonize other areas.

4.6. THEORY OF HABITAT SELECTION

Following two theories have been forwarded by the ethologists regarding the habitat selection in the animals :

1. Optimal Foraging Model

One model think of habitats as patches, or areas of suitable habitat interspersed among areas of unsuitable habitat, and to apply optimal foraging theory, first developed by **MacArthur** and **Pianka** (1966). This theory enables us to predict which habitat patches an animal should select and when it should leave one habitat and move to another so as to get the greatest benefit for the least cost. This economic model incorporates such factors as the availability of resources in various patches and the costs of getting from one patch to another. Although the resource is usually assumed to be food (*i.e.*, energy), nest sites or mates are other possibilities.

2. Ideal Free Distribution.

This model predicts how individuals distribute themselves so as to have the highest possible fitness (**Fretwell** and **Lucas**, 1970). It assumes that animals have complete and accurate knowledge about the distribution of resources (ideal) and that they are passive toward one another and can go to the best possible site (free). Individuals settle in habitats so that the first arrival gets the best resources. As density increases, less-desirable areas are occupied, and animals spread themselves out so that all have the same fitness in the absence of intraspecific competition. One obvious result of such distribution is that rich habitats will have more individuals than poor ones.

If intraspecific competition occurs via **dominance** or **territory**, a **despotic distribution** develops, with some indiduals monopolizing the best resources (**Fretwell,** 1972). Another variation on the ideal free distribution is the **ideal preemptive distribution** (**Pulliam,** and **Danielson**, 1991). Potential breeding sites differ in quality, that is, the expected reproductive success of their occupants, and individuals choose the best unoccupied sites. These best sites are thus acquired and are no longer available to others, but their occupancy does not influence the expected reproductive success of occupants of other sites. Several studies have shown that individuals in the preferred habitat have highest fitness, as measured by reproductive success and survival, than those in less preferred habitat. For instance, **Grant** (1975) found that meadow voles (*Microtus pennsylvanicus*) in the preferred grassland habitat had higher survival and reproductive success than in the less preferred woodland.

QUESTIONS

Long Answer Questions

1. Describe the factors which restrict the habitat use by the animals.
2. Describe the mode of dispersal of a species from its place of birth.

Short Answer Questions

1. Write short notes on the following :
   (*i*) Inbreeding avoidance hypothesis
   (*ii*) Dispersal of lions
   (*iii*) Determinants of habitat selection
   (*iv*) Optimal foraging model.

Very Short Answer Questions

1. What is habitat selection ?
2. Define the following terms :
   (*i*) Philopatric sex; (*ii*) Stem mothers; (*iii*) Habitat imprinting; (*iv*) Semelparous animals.

Ecological Aspects of Behaviour : Food Selection, Anti-Predator Behaviour and Host-Parasite Relation

When an animal forages for food, the forager has to find a food item, which may be a living prey that is skillfully hiding from its predator. Having found an item, the hunter may then have to decide whether an attack is likely to yield benefits in excess of the costs of the attempt. Having launched an attack the next trick is to capture the meal, which may very well have another outcome in mind. Finally, even after a prey is in hand, there may still be difficulties to overcome with respect to consuming the food, whose useful calories and nutrients may be extracted. This chapter, examines each of these obstacles in turn, showing how behavioural ecologists have treated these four kinds of decisions as cost-benefit problem that foraging animal solve.

5.1. OPTIMALITY THEORY

Foraging is very closely related to animal's fitness. It is natural to conclude that animals have subjected to natural selection to be effective foragers. A tool of behavioural ecologists, called **optimality modeling** has been frequently used in the study of foraging behaviour. Optimality model projects which decisions an animal should make in order to maximize its inclusive fitness under a given set of conditions hypothesized to drive the behaviour. Comparisons of actual and predicted behaviour shape our understanding of the behaviour's function. According to **Stephens** and **Krebs** (1986), the optimality model have the following three parts :

1. Decisions;
2. Currency; and
3. Constraints.

1. Decisions. A set of decisions or strategies are available to the animal. For example, a foraging bird may choose to eat a particular piece of food, or search for another one instead; a spider may stay where it is, or move its web to a new place. By the words 'decision' or 'strategy' ethologists do not imply that animals are consciously aware, or that as human beings do. All that is meant is that an animal performs one action out of a variety of alternatives available to it.

2. Currency. The currency or criterion used to compare the value of different decisions. In order to decide among the decisions available to the animal, we must be able to compare them with a common measure. For example, many foraging models use the rate of energy intake of the currency. The best strategy would be one that maximizes this rate. In other models, animals might maximize the time they spend foraging. A choice of currency tends to influence the outcomes of a model.

3. Constraints. The constraints means the limits on the animal. Constraints can be **internal** or **intrinsic** to the animal (such as particular nutritional needs, or the ability to see only certain colours) or **external** (such as levels of temperature or light that affects an animal's ability to forage effectively). An animal can optimize its behaviour only within the range of its capabilities and needs, and constraints define this range.

The optimality models may also include other variables besides these three basic parts. In addition, optimality models can be used to describe other aspects of behaviour, such as mate acquisition and habitat selection.

Advantages of Optimality Model

One cannot expect animal behaviour to be perfectly optimized. Many forces, such as genetic drift or rapid environmental change coupled with evolutionary lag (delay), may prevent animals from being optimal. Despite these forces, optimality model approach has the following advantages :

1. Optimality model helps one's clarify our thinking. When constructing a model one must clearly lay out all the elements that one thinks might be important in determining an animal's behaviour.

2. The results of a model can provide a quantitative prediction that can then be tested. These predictions are often more precise than those one can make without the aid of a model. For example, it might be logical to predict that a bird eating berries on a bush should not necessarily search for every last berry, no matter how long it takes, before leaving for another bush. This still leaves many questions open, however : when should it leave! How does the availability of other bushes affect its decision ? By modeling this problem one can generate more exact predictions. Finally, a model that successfully predicts decisions of animals gives the ethologists more confidence, that they really understand the factors that affect behaviour.

5.2. FORAGING MODELS

For understanding the nature of foraging of animals, **Drickamer** *et al.*, (2002) asked the following questions : A lion encounters a wildebeest. Should it eat it or keep searching? Very different species of animals face very similar questions. Modelers have paid particular attention to two types of models : diet selection models or **prey models** and **patch models.** The prey models deal with the types of prey a forager should eat, while the patch model deals with how long a forager should stay in a food-containing patch.

1. The Prey Model (Locating Food)

Almost everything of biological origin serves as food for some animal or other. Even the most extraordinarily deceptive prey, or poisonous plant, or repellent organic substance (from a human perspective) contains calories and nutrients that some animal exploits. All of these things provide stimuli as by-products of their existence that can potentially be detected if a foraging animal is sensitive to the appropriate cues. For example, dung contains volatile substances, and dung-eating beetle (and other animals) are extremely sensitive to these airborne chemicals. **Heinrich** and **Bartholomew** (1979) reported that dung beetle quickly take to the air when they smell far-off faeces (cow dung) and fly zigzagging up the odour trail to them.

Living prey also generate a spectrum of cues, of factory, accoustical, visual and otherwise, that predatory receivers can exploit as a means to detect victims. Calling tungara frogs, for example, provide information to fringe-lipped bats that may result in their being snatched from the water. Male frogs call loudly at night to attract mates; in this way they sometimes succeed, demonstrating that their cells produce reproductive benefits. But the calls also attract bats that find calling males, which they sweep from the water and devour.

(*i*) Choice of Food Items

Most species of animals are surrounded by all manner of things that they might consider eating. The barn owl (*Tyto alba*) is a nocturnal predator of small mammals. In southwestern New Jersey, these owls roost in three cavities or silos and forage in fields over a radius of

Fig. 5.1. Predation pressure can affect the evolution of communication signals. The fringe-lipped bat is an illegimate receiver that tracks its prey, male tungre frogs, by listening their calls (after Alcock 1993).

several kilometers. **Colvin** (1984) investigated the prey population of this predatory bird and reported that more than 90 per cent of the available small mammals are white footed mice (*Peromyscus leucopus*) and house mice (*Mus domesticus*), and less than 5 percent are meadow voles (*Microtus pennsylvanicus*). However, 70 percent of the owl's diet were meadow voles (Fig. 5.2.). Obviously, owls are not simply taking prey species in proportion to their abundance in the habitat. Let's see how ethologists model a problem of prey choice such as this.

The situation is that a forager is searching for food, and it finds one prey at a time. This optimality model has three parts: The **decision variable** is whether the forager should eat the prey it finds, or whether it should continue searching for another type of prey (For simplicity's sake modlers refer to all types of food as **"prey"** whether it is another animal or a piece of plant material). The **currency** is the rate of energy (caloric) intake : We assume that the animal benefits by maximizing this rate, and we will measure the relative value of the available decisions in this currency. Finally, there are various **constraints** that we can include. Prey, once found, need to be cracked open, live prey need to be subdued. This takes a certain amount of time, called **handling time.** Different types of prey may have different handling times. It is also clear that foragers cannot handle prey and search for it at the same time.

The easiest way to put all these variables together is to use a mathematical expression. Let's define some variables to represent different numbers in the model. First, let's simplify matter by considering only two different types of available prey, type 1 and 2. Each of these prey might provide a different amount of energy for the predator animal, which we define as E_i, where the subscript i represent different prey types. So, for example, the number of calories in an item of prey type1 would be represented as E_1, and that in type2 as E_2. Similarly, each prey type can have its own handling time, usually measured in seconds, which we will represent as *hi*.

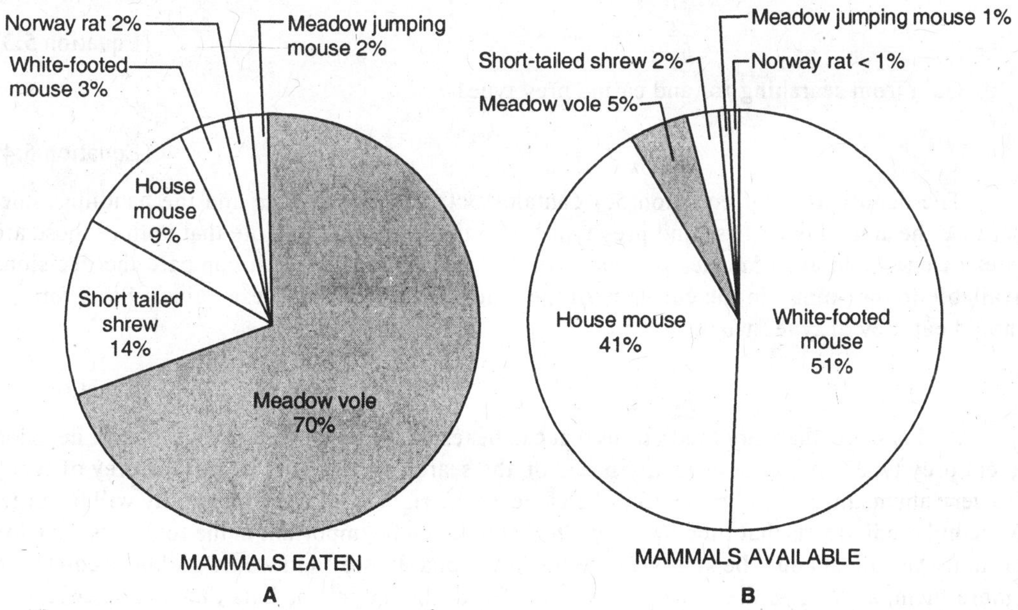

Fig. 5.2. The percentage of small mammals eaten by barn owls (A), compared to the percentage available based on trapping (B). In this study of Colvin (1984) from southwestern New Jersey, barn owls are meadow vole specialists (after Drickamer *et al.*, 2002).

Because ornithologists remain interested in maximizing the rate of energy intake, they need to figure out what this rate would be for each prey item. This is called **profitability,** or the ratio of energy gained to the handling time of each type of prey.

$$\text{Profitability of prey } i = \frac{E_1}{h_1} \qquad \text{(Equation 5.1)}$$

Suppose, prey type1 as the more profitable prey type, so that

$$\frac{E_1}{h_1} > \frac{E_2}{h_2} \qquad \text{(Equation 5.2)}$$

Finally, each prey type can have its own search time, which we will represent as *si.* **Search time** is the amount of time it takes for an animal to locate an item of a particular type. As the density of a prey item increases in the environment, search time for that prey item goes down. There is one more vital assumption : the foragers know the values of the variables we have defined. They know, for example, the probability of finding a prey of a particular type, much as a good gambler knows the odds of drawing a particular set of cards.

Suppose a searching animal has found a prey item. At this stage one can apply the variables which have been defined to ask the question: should the animal eat the prey item or continue searching to find a new type ? If the prey item is type1, the more profitable prey, the choice is clear : the animal should eat it, because it will never find anything better. According to **Drickamer** *et. al.,* 2002), so one can conclude the first prediction of the model : *always eat the most profitable prey.*

The question becomes more challenging if the animal finds prey of type2; then, should the animal eat the less profitable prey or continue searching until the better prey is found ? We must compare the rate of energetic intake for these two choices :

Gain from eating prey type2, once it is found :

$$\frac{E_2}{h_2} \qquad \text{(Equation 5.3)}$$

Gain from searching for and eating prey type1 :

$$\frac{E_2}{S_1 + h_1} \qquad \text{(Equation 5.4)}$$

The denominator of equation 5.4 contains both the search time and the handling time, because the animal must first find prey type1, and then consume it. Note that both of these are rates of energy intake (calories per second). This means that one can compare the decisions available to the animal in the currency of the model by comparing these values. So, a forager should eat prey of type2 when

$$\frac{E_2}{h_2} > \frac{E_1}{S_1 + h_1} \qquad \text{(Equation 5.5)}$$

At this stage, there are predictions that can be tested. First, the model says that the decision to eat prey type2 should be partially based on the search time for prey type1. If prey of type1 are very abundant, S_1 will be low, and the value on the right side of the inequality will be large. A second prediction is that time for prey type2 should not be important in the forager's decision to eat it. An animal could be knee-deep in the lower-quality prey and would still be predicted to ignore them, as long as the inequality is met. Third, the model predicts that a forager should instantly switch back and forth between both kinds of prey to eating only the higher-quality prey, depending on whether or not this inequality is met. This is called **"zero-one" rule**. This rule states that an animal should eat the less profitable prey either none of the time (*i.e.*, with a probability of zero) or all the time (*i.e.*, with a probability of one).

Stephens and **Krebs** (1986) expanded this optimality model by including in it a series of different prey items, not just two. This expanded model predicts that prey type should be added to the diet in order of their profitability, and as in the two-prey case, the inclusion of a particular type does not depend on its own encounter rate.

Validity of the two-prey type model was tested by **Krebs** *et al.*, (1977). They conducted an elegant test of this model with birds called great tits (*Parus major*). They set up a little conveyor belt that carried food past the birds. The prey were either large (eight segment) or small (four segment) chunks of mealworms. By varying the number of worms placed on the belt, the researchers could regulate the bird's search time for different prey items. They directly measured the handling times of both prey types. Several predictions of the model were found correct. Indeed birds selected prey on the basis of profitability. Selectively depended on the encounter rate with profitable prey, rather than the encounter rate with unprofitable prey. However, the zero-one rule did not hold well. As the researchers changed the frequency at which the birds encountered the different prey, they failed to switch cleanly from one to two prey types, but instead showed partial preferences, *i.e.*, they ate prey with a probability somewhere between zero and one. Partial preferences have been reported in many other species as well (see **Drickamer** *et al.*, 2002).

The Effects of Other Currencies and Constraints on the Model

1. Effect of environmental factors on the rate of foraging. One assumption is that animals are equipped with perfect knowledge of the variables in the model. Animals may have to sample their environments and learn about the new conditions. This is one possible explanation for the failure of most tested animals to follow the zero-one rule : animals may have to sample their environment and gather information before making a good decision. This can be certainly incorporated into foraging models (*e.g.*, **Hirvonen** *et al.*, 1999).

2. Effect of breeding season on the rate of foraging. Ethologists have assumed that animals are attempting to maximize the rate of energy intake, but instead they may try to minimize time spent searching (*e.g.*, **Schoener,** 1971) or perhaps may pay attention to both calories and some other aspects of food. Sometimes ethologists can experimentally test which variable is being choosen by foragers. For example, overwintering wild white-tailed deer (*Odocoiles virginianus*) were given a choice of four artificial foods that were either high or low in protein, and high or low in energy. They chose, diets high in calories but low in protein (**Berteaux** *et al.*, 1998). In other example, the golden-winged sunbird (*Nectarinia reichenowi*), an African nectar-feeder that resembles a hummingbird, defends a feeding territory even in nonbreeding season. **Pyke** (1979), using data collected by **Gill** and **Wolf** (1975), tested the notion that these birds were maximizing the rate of energy intake. The birds spent far less time foraging and defending and more time just sitting than was predicted. Thus, instead of maximizing energy intake, what they seemed to be doing was maintaining costs — eating and defending just enough to maintain themselves. However, these data were obtained during nonbreeding season. Most probably, foraging rates and territory defense would increase to meet the demands of reproduction inbreeding season. In this way, *the season during which an animal is observed may influence which variation of the model is most appropriate.*

3. Effect of toxin avoidance on the rate of foraging. Some animals may need to reduce their intake of toxins in certain foods. Mammalian herbivores, for example, eat only a limited quantity of plants that produce high concentrations of toxins (**McArthur** *et al.*, 1991). Some animals, such as pikas (*Ochotona princeps*) get around this constraint by gathering food and storing it until the toxins degrade (**Dearing,** 1997). (Pika is a small mammal, a tailless hare, of family *Ochotonidae*).

Concept of Search Image

These optimality foraging models also assume that the probability that an animal finds a particular prey item depends only on its abundance in the environment. However, not all prey types may be equally easy for foragers to find. In fact, animals some times get faster and faster at finding a particular prey type when they have had experience with it. This has been called a **search image. Pietrewicz** and **Kamil** (1979, 1981) tested whether blue jays (*Cyanocitta cristata*), after previous encounters with cryptic moths (*Catocala* sp.), improve their ability to detect that species and thus show a specific search image.

Even animals that depend on senses other than vision can form search images. Striped skunks can form **olfactory search images (Nams,** 1997). When naive skunks were given repeated experiences with a certain type of food, over time they could detect it both faster and from distance.

2. Patch Model (*How long to stay in a patch ?*)

Many types of food are not distributed randomly (haphazardly) in the environment, but instead occur in patches; imagine scattered trees that are bearing fruit, or tide pools holding tasty snails. Foragers hunting these foods have to decide how long to stay in a particular patch, or when to leave and find a new patch. For example, if one has picked berries, he has faced the same problem : should he search a bush so well so he may get every last berry, or should he move on to another bush?

Several factors tend to play into that person's decision. The average richness of the patches should be important : if berries are in short supply, he might do better to stay and search for every last one. If other bushes have many berries, he will probably do better to move on. In addition, the distance between the patches should be important : if one can reach another bush by taking a few steps, he will probably make a different decision about leaving his patch than if he has to walk half a kilometer.

1. Marginal value theorem. Ethologists looked at these factors together with a graphical model, called the **marginal value theorem (Charnov, 1976)**. In this model, they assume that the forager maximizes the rate of energy intake; this is *currency.* The *decision* variable is to stay in a patch, or to leave to seek a new patch. *Constraints* include : searching for patches and foraging within them are mutually exclusive, patches are found one at a time, and the forager knows the parameters in the model and does not have to learn them. In addition, ethologists assume that the cumulative rate of food intake within each patch can be characterized by a energy gain curve (Fig. 5.3A). When the forager first enters a patch, it has gained no energy. As it forages within a patch, its net energy gain increases as it finds food. However, as the food within the patch depleted, the gain curve flattens out : it becomes harder and harder to find each food item. It is important to remember that the plot describes the net, or cumulative, energy gain. In this model, all patches have the same gain curve.

Ethologists have added to this graph the average travel time to a new patch (Fig. 5.3 B). Food is found only in patches, so before a forager finds a new patch, its net energy gain is zero. At this stage, they had all the information which was needed to calculate when a forager should leave a patch. It is maximizing its net energy gain, it should leave when its expected net gain from traveling to and foraging in a new patch (**Charnov, 1976**). In simple words, it should stay until it can do better elsewhere, travel cost, included. This can be determined by drawing a line,

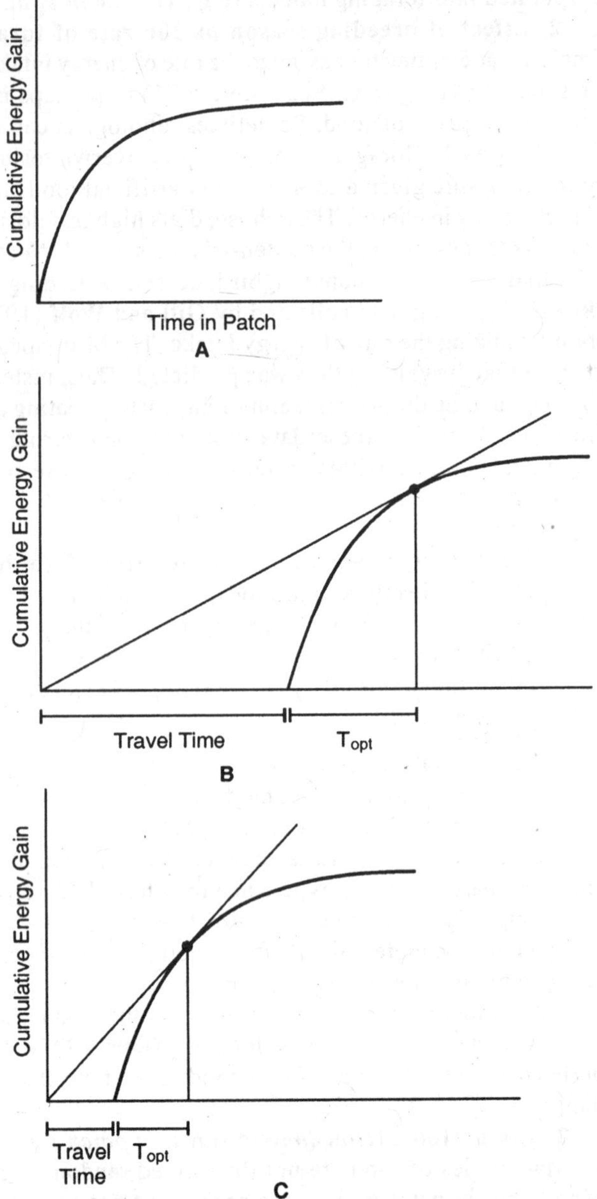

Fig. 5.3. Predicting when to leave a patch : The marginal value theorem. A–A forager's cumulative energy gain in a patch illustrated by the gain curve. The gain curve rises rapidly as the forager first enters a patch, but then levels out as food resources are depleted. B–In next calculation travel time between patches is added. The animal does not gain energy while it is traveling. The optimal time to stay in a patch (T_{opt}) is determined by drawing a tangent from the origin to the gain curve. C–In final calculation, travel time is shortened compared to (B), and the optimal time in a patch is also shorter (after Drickamer *et al.*, 2002).

tangent to the gain curve, to the point on the x axis that represents the travel time. The slope of this line is in the units of the currency (energy gain per unit time) of all the lines that could be drawn between the origin and the gain curve, the tangent is the one with the highest possible slope, or the greatest energy gain per unit time.

When ethologists modify some of the variables, they observe the change in predictions. For example, they could manipulate the travel time between patches exceptionally short (Fig. 5.3C). The optimal time spent in the patch by the forager is now shortened.

Despite the simplicity of this model (*i.e.*, marginal value theorem), studies of a variety of insects, birds, and mammals generally support its predictions (**McNair,** 1982). One test of the marginal value theorem was done in an aviary using great tits (*Parus major*) (**Cowie,** 1977). Each "patch" consisted of a plastic cup filled with sawdust in which six pieces of worm were hidden. Six cups were on each of five artificial trees. Card-board lids were placed on the cups in two ways: hard-to-remove lids fit tightly inside the cups and had to be pried off by the birds, whereas easy-to-remove simply lay on top of the cup and could be flipped off. Birds were tested in two situations : either all hard or all easy. Hard-to-remove lids corresponds to long travel times between patches, easy lids to short times. As can be seen in Figure 5.4, time spent in each patch increased as travel time between patches increased, and the fit to the predicted curve was quite close. However, in real life, all the assumptions of the model are not necessarily met. For example, animals may not know in advance what the rate of energetic return will be in each patch. Foragers faced with two patches of unknown prey density might be expected to sample both (**Shettleworth,** 1984). In a laboratory study, birds (great tits) were given a choice of two feeders, one at each end of an aviary. Each feeder required more hops than the other. Initially, the birds fed on at both feeders, but they gradually shifted to the one that required fewer hops (**Krebs** *et al.*, 1978). Similarly, foraging blue gill sunfish feed on prey that vary both spatially and temporally, so the assumption that they have complete knowledge of all variables in their environment is wrong. These fishes seem to be able to incorporate information into their assessment of patches as they forage (**Wildhaber** *et al.*, 1994).

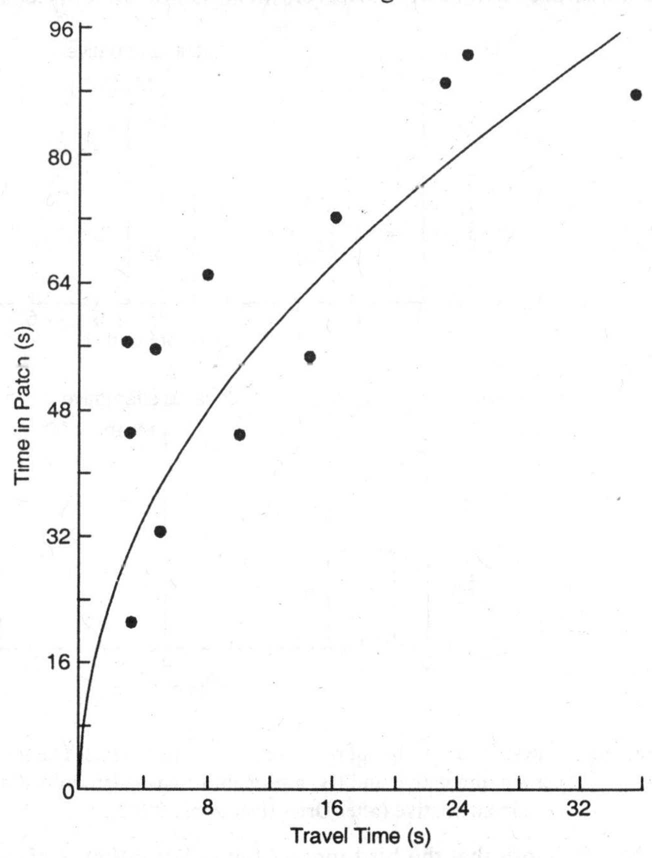

Fig. 5.4. Test of the marginal value theorem in great tits. Time in patch as a function of travel time between patches. Line is the predicted curve based on the theory, adjusted for energy expenditure (after Drickamer *et al.*, 2002).

2. Models for Central-Place Foraging

Behaviour ecologists have studied various central-place forager (*e.g.*, nesting birds, bumble bee, etc.,) in order to understand currencies, decisions and evolutionary constraints in tests of optimality models. The **central place foragers** are animals that carry food back to a central location for storage or for feeding to offspring (**Orians** and **Pearson,** 1979; **Schoener,** 1979). Here, the problem is not only when to leave the patch, but also which and how much food to collect before returning to home base. Time in the patch, load size and selectivity are predicted to increase as the travel distance increases. This prediction has been supported in a number of studies. For example, as travel distance increases, pine squirrels gather cones with a larger number of seeds per cone (**Elliot,** 1988). We can consider an example that combines central place foraging and diet selection is that of a marine diving bird, the rhinoceros auklet (*Cerorhina monocerata*). [**Note.** Auklet means a smaller auk; it is short-winged, web-footed diving bird (family *Alcidae*) of Northern seas]. **Davoren** and **Burger** (1999) found that adults collected different prey items when feeding young than when feeding themselves. Most meals brought back to the nest for chicks comprised one or two larger fish, which are more efficient to deliver, whereas adults generally selected smaller prey when foraging for themselves.

3. Utility Function and Risk Sensitive Foraging

The models which are discussed so far concern a forager's response to the mean (average) amount of prey available. However, mean is not the only characteristic that may vary across

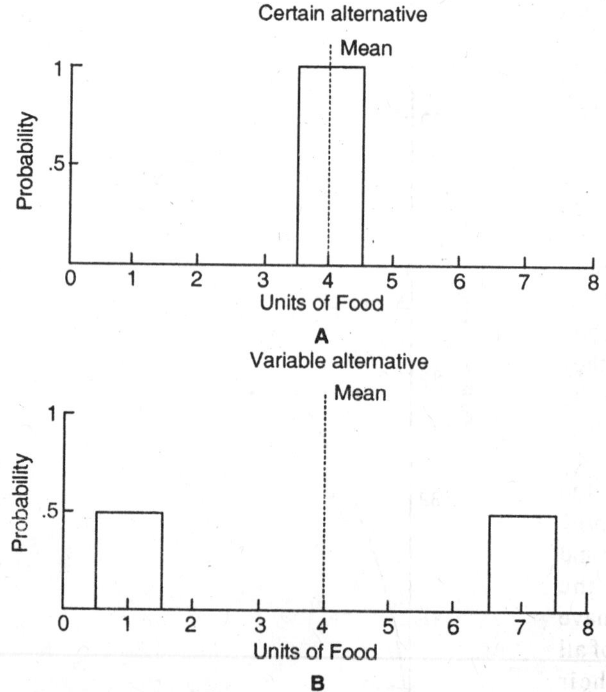

Fig. 5.5. The standard design of risk sensitivity experiments. The forager is offered a choice between (A) a certain alternative and (B) a probabilistically determined alternative with the same mean as the certain alternative (after Drickamer *et al.*, 2002).

patches. Suppose that the bird junco (*Junco hyemalis*) is given a choice between two trays of food. When it selects one tray, the experimenter removes the other. One tray always have four seeds in it. The other tray varies: one half the trials, it has one seed, and on the other half of the trials, it has seven seeds (Fig. 5.5 ; **Stephens** and **Krebs,** 1986). Thus, these two patches have the same mean amount of seeds, but the second patch has a higher variance around that mean. According to ethologists the second patch has higher risk. (Note that "risk" in this context does not refer to danger from predators; it is analogous to the risk a gambler takes). If an animal can distinguish between these two patches, ethologists say they are **risk-sensitive.** If the forger prefers the variable patch, it is **risk-prone;** if it prefers the stable patch, it is **risk-averse.**

Ethologists pose the question : why should animals (*e.g.,* junco) ever be risk sensitive ? One method involves consideration of **utility;** or the value that a resource has to an animals. Often this value is measured in units of fitness. The relationship between each unit of a resources and its utility is called the **utility function,** and it may have different shapes. For example, in Figure 5.6 is a straight - line utility function ; for each seed that a bird finds, its fitness increases by the same amount. This is likely be rare. More common is the utility function pictured in Figure 5.6B, where the utility function flattens out with increasing number of seeds. For example, this might occur if an animal can consume only a certain number of seeds, and seed found after that are useless. The shape of this function is described as concave-down. In other cases, the utility function is concave up (Fig. 5.6 C) : each unit brings more fitness than the previous one. Suppose, for example, that an animal is 20 steps from a water hole : taking 20 steps instead of 19 steps will make it bigger difference in its fitness than taking 19 steps instead of 18 steps (**Stephens** and **Krebs,** 1986).

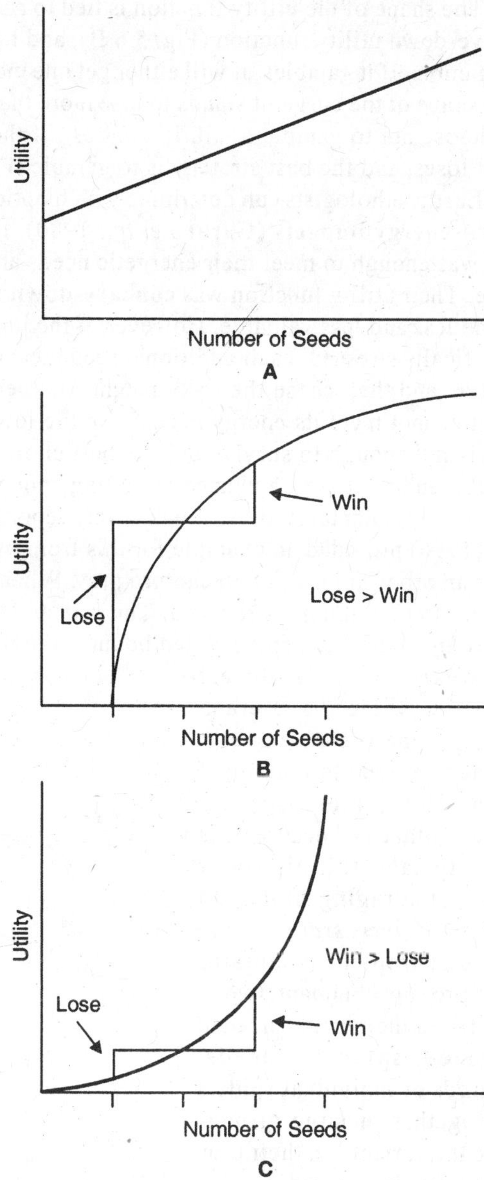

Fig. 5.6. Utility functions and an animal's willingness to accept risk. Utility is the value that a resource has to an animal. A – In a linear utility function, every unit of resource has the same value to the animal. B – In a concave-down utility function, the resource becomes less valuable to the animal as it gains more of it. Imagine an animal can gamble and either win or lose one unit of resource. It stands to lose more utility than it can win, and it should avoid gambling and be risk-averse. C–In a concave-up utility function, the resource becomes more valuable to animal as it gains more of it. Here, a gamble may pay off in a large win, or a relatively small loss. This animal should be risk-prone, and take the gamble (after Dirckamer *et al.,* 2002).

The shape of the utility function is tied to risk. Suppose that an animal is at a point on the concave-down utility function (Fig. 5.6 B), and has a choice about whether to stay where it is, or to gamble. If it gambles, it will either get one more unit of resource, or lose one unit. Because of the shape of the curve, it stands to lose more than it wins. Therefore, it should be risk-averse, and choose not to gamble at all. In contrast, if the curve is concave-up, it stands to win more than it loses, and the best strategy is to gamble (Fig. 5.6 C).

Lastly, ethologists can determine which option did the foraging juncos choose? It depends on their **energy budgets** (**Caraco** *et al.*, 1980). If the amount of food provided in the no-risk patch was enough to meet their energetic needs and sustain their body weights, they were risk-averse. Their utility function was **concave-down :** as the number of seeds increased, each seed became less and less valuable. However, if the juncos were kept at cold temperatures and were energetically stressed, each additional seed meant a lot. These juncos had a *concave-up utility function,* and they chose the risky patch. Another instinctive way to think of this is that if an animal cannot meet its energy needs with the low-risk option (*e.g.*, the known reward of four seeds is not enough to survive on), its best choice is to gamble and take the risky option. The animal then has at least a chance of getting enough resources to survive.

In experimental studies, animals were reported to respond to *variance* in food availability. **Uetz** (1996) provided an example for this from two closely related spider species. Each spider builds an orb web in which to capture prey. When orb webs are placed near to one another, the variance in prey intake is reduced. The primary reason for this is the **"ricochet effect" :** prey that are missed by one spider often bounce into the web of a neighbour. The spider, *Metapeira atascadero* lives in desert grassland, where the average biomass of the available prey is less than or equal to individual needs. Here, most spiders forages solitarily because it is better to take the high-risk option of foraging alone. In contrast, *M. incrassata* lives in the moist tropical mountains where prey are abundant, two to three times the number needed to meet spider's needs. Hundreds of individual spiders join together in large groups, where the variance in their prey capture rate is lowered.

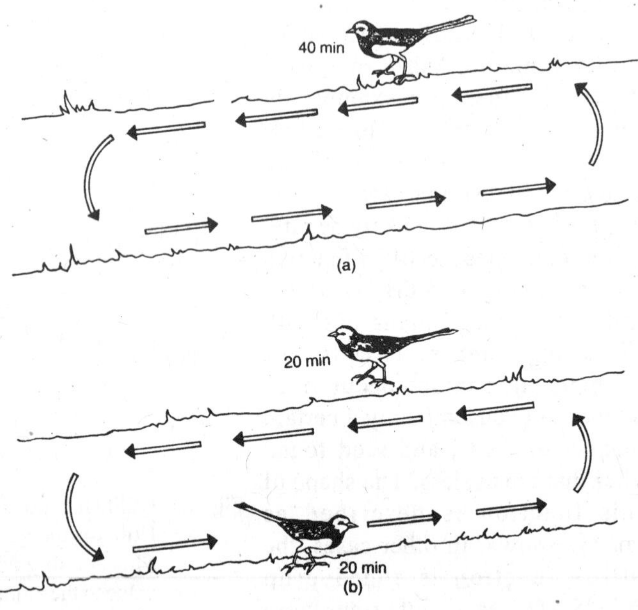

Fig. 5.7. Pied wagtails feeding along a riverbank. A–Pied wagtail exploit their territories systematically. The circuit of the riverbank takes, on average 40 minutes to complete. B–When a territory is shared between two birds, each walks, on average, half a circuit behind the other and to crops only 20 minutes worth of food renewed (after Drickamer *et al.*, 2002).

4. Effects of Competitors on Foraging Behaviour

Foraging behaviour of an animal is also found to be affected by the presence of competitors and predators. For example, competition by members of the same or different species may force them to forage in sub-optimal habitats or include food items they would not otherwise consume.

For example, stickleback fish that had parasite infestations fed on smaller-sized daphnia than compete with uninfested sticklebacks for larger, more profitable prey (**Milinski,** 1984). American crows (*Corvus branchyrhynchos*) change their behaviour in the presence of conspecific competitors (**Cristol** and **Switzer** 1999). Crows fly up and drop walnuts onto the ground in order to break them open. When other crows are present, they do not drop them from as great height, presumably so they have a better chance of swooping down to pick up the nut before competitors snatch it.

According to **Drickamer** *et al.,* (2002), one method to increase net energy gain is by defending a territory against potential competitors. Since a territorial animal have to spend energy advertising its presence and chasing out intruders, defending a territory will be economical only under certain circumstances. Pied wagtails (*Motacilla alba*) are insectivorous birds. Some of them defend winter feeding territories along riverbanks. The others feed in flocks in nearby pools. Territory holders, usually males, follow a circuit up one bank and down the other, a pattern that maximized food intakes as new insects are washed ashore (**Davies** and **Houston,** 1983, 1984); intruding wagtails are chased away. When food in the territories was very scarce, the owners fed elsewhere in flocks but kept returning to the territory to evict intruders. When food was abundant, owners often shared the territory with a **satellite** (usually a juvenile or a female) that walked about one-half a circuit behind the owner (Fig. 5.7). This meant that the food available to the owner was reduced by one-half, but the owner sometimes gained because the satellite helped chase away intruders. If food declined, the owner chased the satellite away. If food became extremely abundant, owners made no effort to defend their territories. Sometimes territory holders fed on the territory even when they could have done better in the flock. Thus, they were not maximizing energy intake in the short run. This system differs from that of the nectar feeders such as the golden-winged sunbirds (**Gill** and **Wolf,** 1975) whose territory sizes vary as a function of food availability. The wagtails keep the same size area but vary their responses to intruders as food supply fluctuates. **Davies** and **Houston** (1984) suggest that territory maintenance is a long-term investment to protect a food source that is more reliable than the pools where the flocks feed.

5. Effects of Predators on Foraging Behaviour

Many foraging animals are also potential prey themselves, and must balance the risk of predation with the benefits of foraging. Often, an animal is less vigilant to the presence of danger when it is attacking or handling prey. In addition, animal that rely on **Crypsis** (blending in with the background) to avoid detection by predators may "blow their cover" when attacking prey, and thus become more vulnerable to attack. The effect of the presence of a predator on diet selectivity has been directly measured in a few of studies and it varies across situations. The rule appears to be that as it becomes more dangerous to capture more profitable prey, selectivity decreases, when it is more dangerous to capture less profitable prey, selectivity increases (**Houtman** and **Dill,** 1998).

Further, risk of predation may also affect the choice of patch. Often there is a trade-off ; high-quality patches may sometimes pose the greatest risk of predation. There are many examples across a variety of taxa of foragers that choose less-valuable food resources if they are closer to cover. For example, desert baboons (*Papio cynocephalus ursinus*) spent more time foraging in a low-risk, but food-poor habitat (**Cowlishaw,** 1997). Similarly, juvenile hoary marmots foraged close to their burrows rather than risk predation by eagles or coyotes in more distant but better quality patches (**Holmes,** 1984). Namib Desert (Arid region, along entire length of coast of Namibia in south-west Africa) gerbils foraging for seeds left patches sooner under the full moon, when the risk of predation was high, than under the darker skies of the new moon (**Hughes** *et al.,* 1994).

6. State-Sensitive Models of Foraging

All members of a population may not make the same foraging decision (*e.g.,* risk-sensitive juncos). Indeed, the same individual may make different foraging decisions at different times.

For example, a very hungry animal may choose to go to a risky patch of food if the quantity of food is higher, whereas a well fed animal may give up food in favour of avoiding predators. Hungry Scorpios (*Buthus occitanus israelis*), for example, are willing to forage under bright moonlight when they are in danger from predators, but well-fed scorpions hide when the moon is bright (**Skutelsky,** 1996). Ethologists have described changeable characteristics of individual, such as hunger level, age or body size, or **states.** They included measures of these states, or **state variables,** into models in order to predict correct animal behaviour. This has been done by computers and by a form of programming, called **dynamic programming.** This tool has enabled researchers to calculate the optimal decisions of animals with different states and of the same animals at different times in their lives (**Clark** and **Mangel,** 2000).

5.3. TECHNIQUES FOR ACQUIRING FOOD

Animals adopt the following techniques for acquiring the food.

1. Modifying Food Supply

***(i)* Growing grasses or algae.** Some animals modify their food supply so that it increases. For example, grazing animals in the grasslands stimulate the growth of some species of grass and prevent succession of grassland communities to other community types, such as forests. Likewise, some species of intertidal limpets increase their food supply with the mucous trail they secrete during locomotion (**Connor** and **Quinn,** 1984). Two species of solitary limpets (*Lottia gigantea* and *Collistella scabra*) routinely return home to a depression on a rock, depositing a mucous trail that acts as an adhesive trap for algae. The mucus also stimulate algae growth along the trail, and the limpets feed on this algae as they retrace their path house.

***(ii)* Growing fungal gardens.** The ultimate in nonhuman manipulation of producers in an ecosystem may be the "agriculture" practiced by certain fungus-growing ants. Leaf-cutter ants (*Atta cephatotes*) are easy to spot in the New World tropics (*i.e.,* Costa Rica) as they walk in long queues, carrying leaf sections back to the nest. Once inside the nest, ants cut the leaf sections into smaller pieces, and work the edges with their mandibles to make the pieces wet and pulpy. They may deposit an anal droplet on the leaf before inserting the leaf into the garden. These processed leaves form the basis for their garden. The ants then plant their crop: they place bits of fungal mycelium on the leaves, and fungus rapidly grows. The ants feed on the tips of the fungus. This arrangement is mutualistic. By converting the indigestible cellulose of the leaves into digestible sugars, the fungus makes the vast energy supply contained in the forest leaves available to the ants. The ants spread the fungus and care for it by weeding out alien fungi, fertilizing it, and producing antibiotics that act against competing fungi and microorganisms.

***(iii)* Livestock farming.** Some ant species act as **livestock farmers,** and raise aphids or other plant-feeding insects in or near nests, protect them, and feed on their "honeydew' excretions from their digestive tracts (**Holldobler** and **Wilson,** 1990). The ants and aphids appear to communicate with each other. Whereas unattached aphids merely eject the honeydew excretions from their anus, aphids that are caressed by an ant's antennae and forelegs will excrete the droplet slowly and hold it on the tip of the abdomen while the ant consumes it. Again, as with the leaf-cutter ants, both partners benefit from this relationship.

2. Construction of Traps

Some insects, such as ant lion (genus *Myrmeleon,* in the order Neuroptera) larvae, make traps to capture food. They make funnel-shaped pits in the sand and bury themselves just below the pit. When ants tumble in, the ant lions knock them down to the bottom of the pit with grains of sand that they hurl at them with tosses of their head. Other traps include the underwater nets of caddish fly larvae (order Trichoptera).

Foelix (1996) has described in detail the traps of spiders. **Orb webs** are the most well studied, and have several different elements. Radial threads are similar to the spokes of a wheel, radiating out from the hub of web. A fine thread, covered with glue, spirals outward from the hub; it is this sticky spiral that traps insects. Most orb webs are flat, and the spider sits either at the hub or on a connecting thread and monitors the vibration of prey hitting the web. Some species, however, modify the orb web further, for example, *Theridiosoma* attaches a lateral tension thread to the hub and tightens it, turning the web into an inside out umbrella. The spider pulls on the radii near the hub with its hind legs, and pulls on the tension thread with its front legs, and sits motionless waiting for prey. When an insect flies near the web, the spider let go with its front legs so that the web snaps back into a two-dimensional shape, and traps the insect.

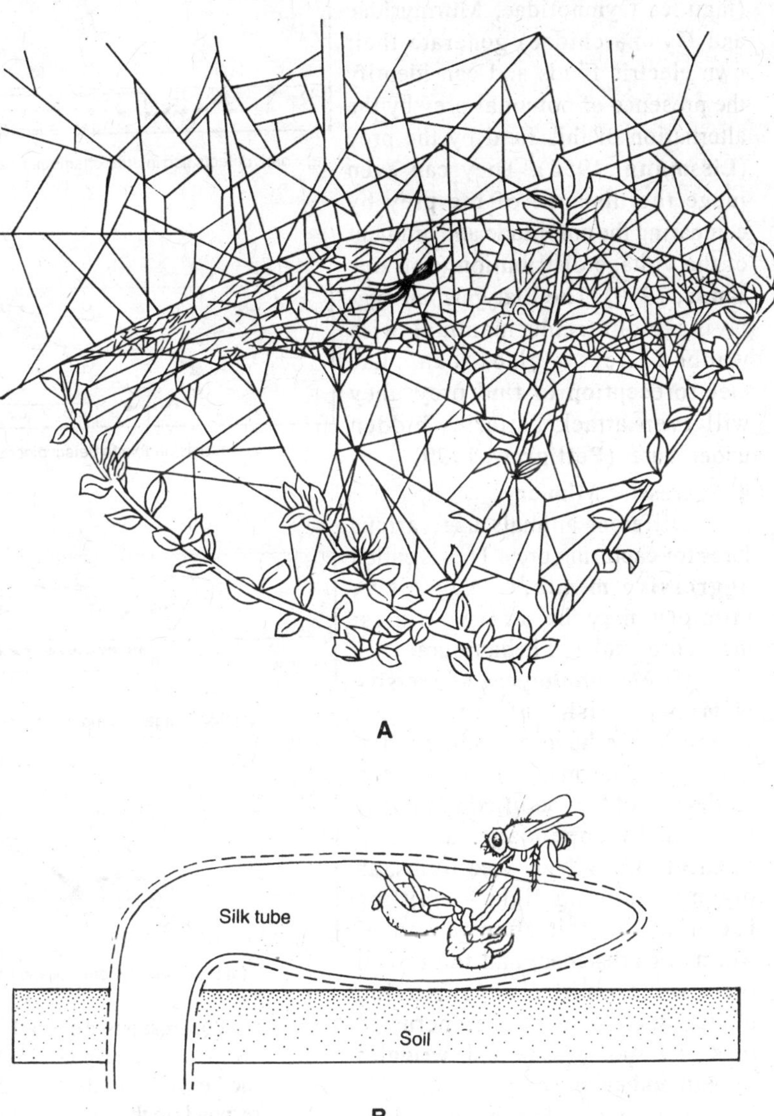

Fig. 5.8. Spider webs : A - Web of spider *Linyphia* ; B - Web of purse web spider (after Drickamer *et al.*, 2002).

Some other types of spiders build **sheet webs,** some with funnels attached in which to hide. These webs are not glue covered, and the spider must depend on speed to snatch up the prey before it disentangles itself. The bolas spider uses a bizarre method of prey capture : it uses a short thread with a large drop of glue on the end and throws it after insects that fly nearby. The spider *Linyphia triangularis* hangs upside down from a dome-shaped web (Fig. 5.8A). The **purse-web** spider *Atypus* lives in a silken tube, and bites prey through the walls of the tube and pulls it inside.

3. Electromagnetic Fields

Certain animals use uncommon method to capture the food. For example, **Kalmijn** (1971) demonstrated that sharks can detect the weak electric fields of fish even when the latter buried in sand (Fig. 5.9). Electric fish (families Gymnotidae, Mormyridae and Gymnarchidae) generate their own electric fields and can identify the presence of potential prey by the alternation of this field by the prey (**Lissmann,** 1958). They can even judge the distance of the prey by comparing the amplitude and gradient of the voltage distribution (**von der Emde,** 1999). The Platypus, an egg-laying monotreme mammal with webbed feet, also can use electroreception to find prey: they will even attack batteries hidden under water (**Pettigrew,** 1999).

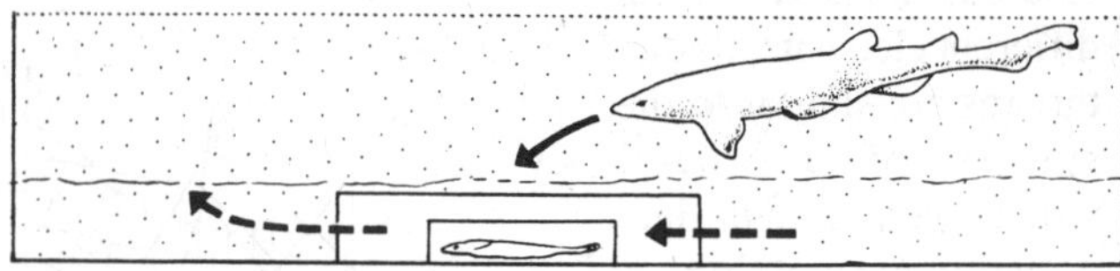

(a) Shark detects fish in agar chamber. Chamber doesn't block electric field.

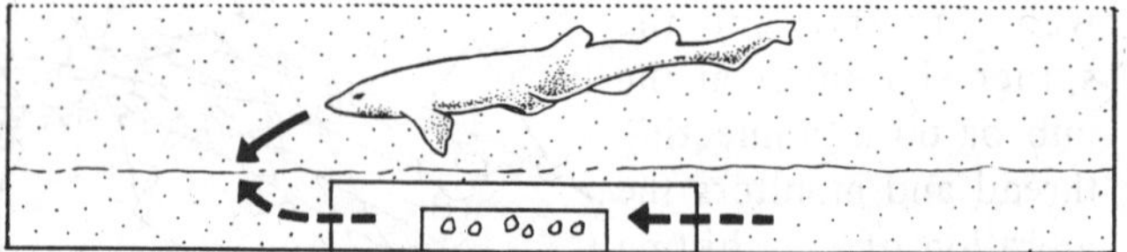

(b) Shark is unable to detect pieces of fish in agar chamber.

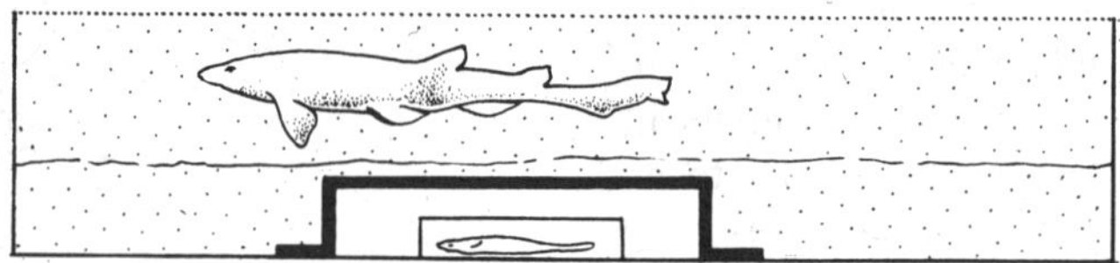

(c) Fish in agar chamber is covered with metallic film.

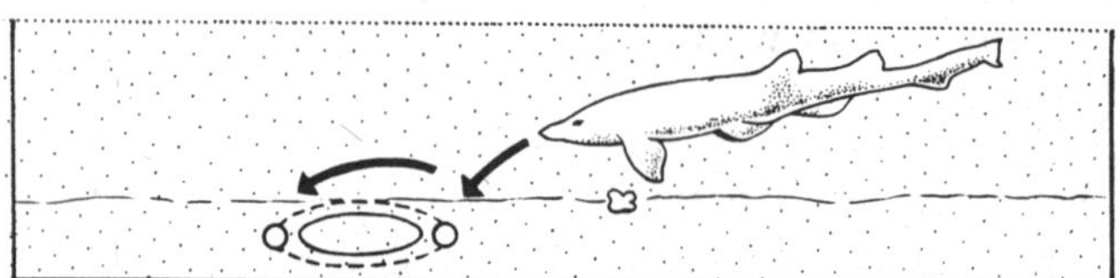

(d) Electrodes produce dipole field that shark detects.

Fig. 5.9. Feeding responses of shark to objects buried in sand. Solid arrow denote responses of shark; dashed arrows denote the flow of seawater through the agar chamber. The shark responds to the magnetic field of intact fish A and B. Pieces of fish in C produce no magnetic field, and the shark responds to odour carried outside the agar chamber by the current. Metallic film in D blocks the magnetic field of the intact fish, and the shark fails to detect the fish. In E, the shark responds to an artificially produced magnetic field, ignoring a piece of fish (after Drickamer *et al.*, 2002).

4. Aggressive Mimicry

Different animals use various lures for capturing prey. This is called **aggressive mimicry.** Aggressive mimicry may be of two types: morphological and behavioural.

(*i*) Morphological aggressive mimicry. Fish of the order Lophiiformes have a modified first dorsal fin spine on the tip of the snout. At the end of the modified spine may be a fleshy appendage, a tuft of filament, or, in deep - sea forms, an organ containing light - emitting bacteria. The bait often resembles worms or crustacea, and the rest of the fish resembles an unmoving object, such as an algae-encrusted rock or a sponge. The fish wriggles the bait and keeps the rest of its body still. **Pietsch** and **Grobecker** (1978) found that in case of angler fish (*Antennarius* sp.), the lure is nearly exact replica of a small fish.

Siphonophores of the phylum Cnidaria capture prey using tentacles armed with stinging cells called **nematocysts. Purcell** (1980) reported that some species have tentacles with branches and clustered nematocysts that look like copepods or fish larvae. When the siphonophore moves its tentacles, it lures zooplankton predators into the web of nematocysts, and the predators become food for an siphonophore instead.

(*ii*) Behavioural aggressive mimicry. In some species of fireflies, a female mimics the mating signal of another species and eats the male that responds. **Jackson and Wilcox** (1993) reported that jumping spider *Portia fumbriata* displays a very impressive repertoire of behaviours. This spider enters the webs of other spiders and vibrates the web in the same way that a struggling prey would. Web-building spiders do not have good vision, so they are attracted to the "prey", but are instead attacked by *Portia. Portia* seems to adjust the signals it makes according to the response of its target, it begins by giving a variety of vibrational signals in different pattern, but then focuses on producing the signals that provoke a reaction from the target spider.

5. Tools

Once, the use of tools was considered an exclusive human trait, but it has evolved independently in several different lineages. In most cases, the tool is an unmodified inanimate object. The sea otter (*Enhydra lutris*), for example, holds a rock on its chest and cracks shell fish, such as mussels, against it. Similarly, the Egyption vulture (*Neophron percnopterus*) picks up rocks and drops them on ostrich eggs. The chimpanzee (*Pan troglodytes*) , however, sometimes makes a modified tool by stripping leaves from a twig, which it then inserts into a nest of ant or termite. The insects cling to the stick, and the chimp eats those that hang on after it removes the stick. The woodpecker finch (*Cactospiza pallida*) of the Galapagos Island uses sticks in a similar way to extract larvae from dead wood and may modified the stick by shortening it (Fig. 7.6).

5.4. FORAGING AND SOCIAL BEHAVIOUR

The food of animals is often spread widely and irregularly over the environment and cannot be defended, so individuals of a species may group in to flocks or herds and may directly benefit from neighbours in several ways. This is done by the following methods:

1. Sharing Information

Foragers may get information concerning food sources from one another, especially if food occurs in dense, rare patches which are unpredictable in time or space. Animals tend to monitor the area around them for other foragers that have found food (**Thorpe,** 1956). Some species produce distinctive sounds when they discover food that give information to conspecifics. For example, when chickens (*Gallus gallus*) hear **food calls** given by conspecifics, they look at the ground as if searching for food (**Evans** and **Evans,** 1999). Groups may also act as " **information centres"** (**Ward** and **Zahavi,** 1973). In this case, an unsuccessful foragers follows previously successful group members back to a food source. An example is ravens who depend on finding large carcasses to survive cold winters. Ravens share roost sites at night and often the entire group is seen departing in one direction in the morning (**Marzluff** *et al.,* 1996).

Animals feeding in groups may forage more efficiently because each individual spend less time scanning the environment for predators. For example, downy woodpecker's (*Picoides pubescens*) wintering in mixed-species flocks scanned less and fed more than did solitary individuals (**Sullivan,** 1984).

Some individuals in social groups may specialize in **stealing food** from others, or in joining others that have already located food. These asymmetrical roles have been called "producing" (generating food) and "scrounging" (stealing food) (**Barnard** and **Sibly,** 1981). This creates a problem for the producers, which have to decide whether to stay with the scroungers or leave for another foraging group.

2. Cooperative Hunting

(*i*) Cooperative hunting in spiders. Some animals can take down larger or more dangerous prey when they hunt as a group. For example, some tropical spider species (*e.g., Anelosimus eximius*) are social, and cooperatively construct large communal webs, which may be several meters across. Together , the spiders attack trapped prey items, drag them back to a central retreat, and feed on them communally (**Buskirk,** 1981).

(*ii*) Cooperative hunting in ants. Perhaps the most spectacular food-catching enterprise is the march of an army ant (*Eciton burchelli*) colony (**Franks** 1989). **T.C. Schneirla** (1948) spent most of his life trying to understand the complex life cycle of several New World ant species. Each night the colony, which consists of a queen, workers, larvae, pupae, and eggs, forms a **bivouac** (*i.e.*, temporary camp without tents). Workers, upward of 50,000 strong, form a protective net (The bivouac) by hooking their legs and bodies together. Each morning the bivouac dissolves and the ants begin moving outward. A column emerges along the path of least resistance and heads away from the bivouac site. There are no true leaders in army ant colony; workers in the lead turn back into the swarm behind them every few centimeters. The ants lay

BOX 5.1.

One can observe army (*Eciton burchelli*) moving along a woody vine carrying white larvae. Along both sides of the column white are rows of "guard workers" including a few soldiers with large white heads. These protect the emigration column from being disturbed (see **Drickamer** *et al.*, 2002)

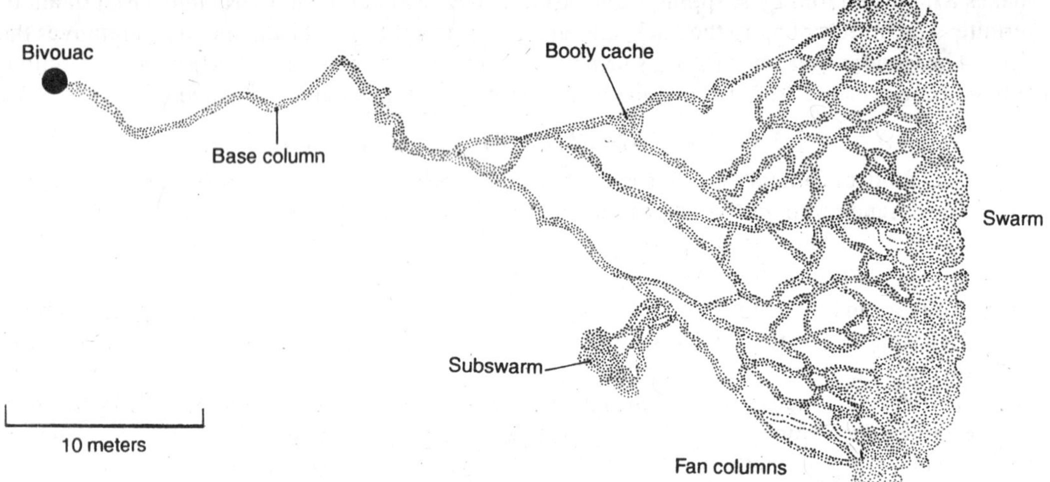

Fig. 5.10. Pattern of raiding employed by army ants. In this swarm raid of army ant, which can be found on Barro Colorado Island in the Panama Canal Zone, the advancing front is made up of large mass of workers. The swarm flushes a variety of prey, mostly invertebrates, but also snakes, lizards and birds. The queen and immature forms remain at the bivouac site (after Drickamer *et al.*, 2002).

down pheromone trails that guide those that follow, and the column may branch into a fan-shaped organisation (Fig.5.10.). Virtually all animals in the path are stung, cut into pieces, and transported to the rear. Army ants are able to attack and consume prey item as large as snake, lizards, and small birds that they would not be able to handle as individuals. Arthropod life in these areas is temporarily depleted. When larvae are developing, the bivouac location changes each night (**nomadic phase**); during the egg-laying and egg development phases of the reproductive cycle, the bivouac remains in the same place each night (**stationary phase**).

***(iii)* Cooperative hunting in carnivores.** Perhaps the most familiar and most intensively studied cooperative hunting have been reported in the mammalian carnivores.

***(a)* Lions.** The lion (*Panthera leo*) lives in closed social units and is most abundant in the grasslands and open woodlands of Africa. They use a **stalk-and-rush** method of hunting. They depend on getting very close to the prey, and then surprising it with a sudden burst of speed. By hunting in groups, lions add to their diet two species of prey that an individual lion could never attack alone: buffalo (*Syncerus caffer*) and giraffe (*Giraffia camelopardalis*). They are also at least twice as successful as lions that hunt alone. Groups can also drive other predators and scavengers from the food. **Schaller** (1972) reported that plains-dwelling prides kill less than half their food, relying on other predators, such as hyenas (*Crocuta crocuta*), to make kills for them.

(*b*) **African wild dogs.** The African wild dogs (*Lycaon pictus;* family Canidae) may seem an unlikely big game predator, as it weighs only about 18 kg. However, these dogs typically catch prey weighing as much as 250 kg (**Schaller,** 1972). Rather than using the stalk- and -rush tactic of lions, these canids are **coursers** , *i.e.,* pursuing prey for many kilometers. The dogs live in mixed -sex packs that average ten adults of particular interest in the prehunt "ceremony", in which dogs draw back their lips to expose their teeth, nibble and lick each other's mouths and run whining (whine means a long, high pitched complaining cry) from pack member to pack member. The greeting seems to represent **ritualized food begging.** Setting out on a hunt, the dogs travel at a trot/jog and fan out loosely over the terrain. When a prey is encountered (such as Thomson's gazelles, *Gazella thomsoni*, or wildebeest *Connochaetes taurinus;* **Carbone** *et al.*, 1997) they cooperatively work to cut off its escape route and bring it down. Wild dogs face competition from the spotted hyena, *Crocuta crocuta* (**Gorman** *et al.,* 1998). Hyenas act as **kleptoparasites** and steal food from dogs (Box. 5.2).

BOX 5.2.

Wild dogs are Vulnerable

Hunting is very energetically costly for wild dogs, so even a small amount of kleptoparasitism can have grave effects. In fact, hyenas may be contributory factor to the decline of wild dogs (**Gorman** *et al.,* 1998), Lions are another problem for wild dogs; they are responsible for 39 percent of natural pup deaths and 43 percent of natural adult deaths. In fact, dogs are at their lowest density where food is most abundant because they seems to be avoiding areas used by lions (**Mills** and **Gorman** 1997). The conservation of wild dogs presents a challenging problem.

(*c*) **Hyenas.** Although hyenas do scavenge food from other carnivores, they get more of their food from hunting (**Kruuk,** 1972). The basic social unit is the **clan,** which consists of 10 to 60 hyenas of both sexes and all ages. There is little exchange between clans; each defends a territory with a centrally located den. Because the prey, mainly wildebeests and zebras, migrate to the woodlands in the dry season, most hyena clans breaks up at that time and some individuals become nomadic and follow the prey. Thus, food availability has a profound effect on social organisation in hyenas. In the Masa Marai reserve in Kenya, three quarters of the hunts observed over 7 years were made by solitary individuals (**Holekamp** *et al.,* 1997). Holekamp and coworkers (1997) have described the chase as follows: Hyenas typically rushed a group of animals, paused briefly to watch, selected an individual, and chased it down over distances ranging from 75m to 4km. Prey then grabbed and disembowelled (Disembowel means to take out the bowels of ; evisecrate or to remove the contents of), taking 0.5 to 13 minutes to die. Other hyenas arrived at the kill site and competed for the prey.

(*d*) **Wolf.** The social carnivore of northern temperate and arctic regions is the wolf (*Canis lupus*), which typically preys on deer (*Odocoileus sp.*), moose (*Alces americana*), buffalo (*Bison bison*), sheep (*Ovis sp.*), Caribou (*Rangifer tarandus*), and elk (*Cervus canadensis*). Pack size varies greatly, but most packs have fewer than eight members (**Mech,** 1970). Most often they use scent to detect prey; the lead animals stop when they detect the odour of prey. They may also use chance encounter and tracking. If they detect the prey at some distance, they proceed slowly and stalk sometimes to within 10 meters of the prey. Once prey is in flight, the wolves rush. Wolves must get close to their prey during the stalk - and - rush or they will quickly give up. If they cannot make an attack, they may chase the prey, usually for less than half a kilometer. Some earlier studies had described wolves as coursers, but **Mech** (1970) concluded that they are mainly stalkers and rushers.

5.5. ANTI - PREDATOR BEHAVIOUR (or Defense Against Predators)

Most animals, except for a few upper trophic - level carnivores, are potential prey. Many animals have behavioural adaptations to reduce the chances of being eaten.

A. Individual Strategies

Anti - predator strategies undertaken by individuals are of the following types:

1. Escaping and Freezing

Many animals keep close to a nest, burrow, or other refuge near the centre of their home range. For example, white footed mice (*Peromyscus leucopus*) usually nest in hollow trees or logs, and the probability of finding the mouse at a particular spot decline rapidly with distance from the nest, so they are rarely in unfamiliar space or far from shelter (**Vessey,** 1987). Familiarity with escape routes makes it more likely that animals can avoid potential predators (**Stamp,** 1995).

Another anti-predator behaviour is to **freeze**. The presence of protective cryptic colouration is often associated with this behaviour, as in the spotted white tailed deer fawn (*Odocoileus virginianus*). Some animals carry this strategy further and pretend death. For example, the opposum (*Didelphis virginiana*), if harassed by a predator such as a dog, remains motionless on its back. Hog-nosed snake (*Heterodon platyrhinos*) shows similar behaviour, although they may precede it by threat behaviour in which the hog-nosed snake resembles a cobra. Freezing and pretending death may work because many predators seem to respond only to moving prey. For example, wolves generally will not attack prey that stand motionless (**Mech,** 1970).

2. Deception

Another tactics used by the vulnerable animals is deception. When threatened, larvae of the moth *Hemeroplanes* sp., inflates its anterior end to form an excellent representation of a snake's head, which it then waves. Likewise, green caterpillar of hawkmoth (*Leucorampha omatus)* to show snake display (Fig. 5.11).

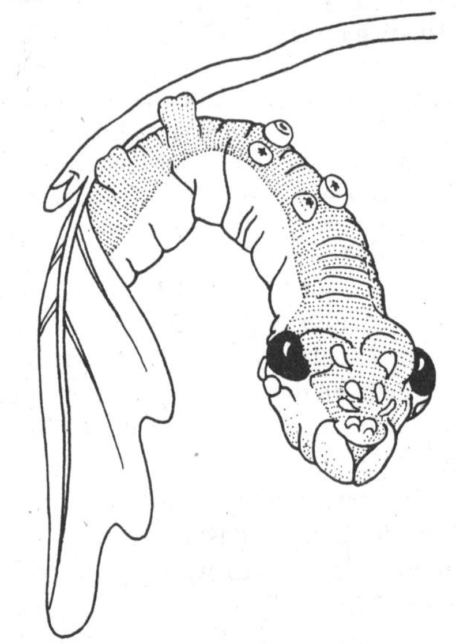

Fig. 5.11. Insect defense display. The anterior end of an alarmed caterpillar resembles the head of a snake and frightens predators. It has a false pair of large shiny eyes and wave back and forth like a serpent.

Other lepidopteran larvae resemble inanimate objects in the environment such as twigs, leaves, bark or even bird droppings.

Many animals have behaviours or markings that may serve to misdirect or surprise predators. Some reef fish have conspicuous eyespots on their tail end that are much larger than the real eyes: these could frighten away potential predators or misdirect their attack to less vulnerable parts of body. Some moths (*Antharea polyphemus* and *Automeris coresus*) are cryptically coloured when at rest on vegetation, but if startled, they spread their forewings and reveal large brightly coloured eyespots on the hindwings (Fig. 5.12).

3. Toxicity

Many organisms are harmful to predators. Plants may contain toxins such as alkaloids; animals may have armor plates (as in armadillo) or spines (as in porcupine, hedgehog and three-spined stickleback). Many invertebrates inject venom into attackers (*e.g.*, scorpions, worker bees, wasps, *Pollistis,* etc.). **Eisner** (1966, 1970) discovered a large number of arthropod defense mechanisms. For example, the bombardier beetle (*Brachinus* sp.) has a plumbing system that resembles a liquid-fueled rocket. It stores quinones and hydrogen peroxide in separate reservoirs.

When the beetle is disturbed, these two liquids are mixed in outer vestibule, and in the presence of enzymes, the solution heats to the boiling points and is discharged as a poisonous spray.

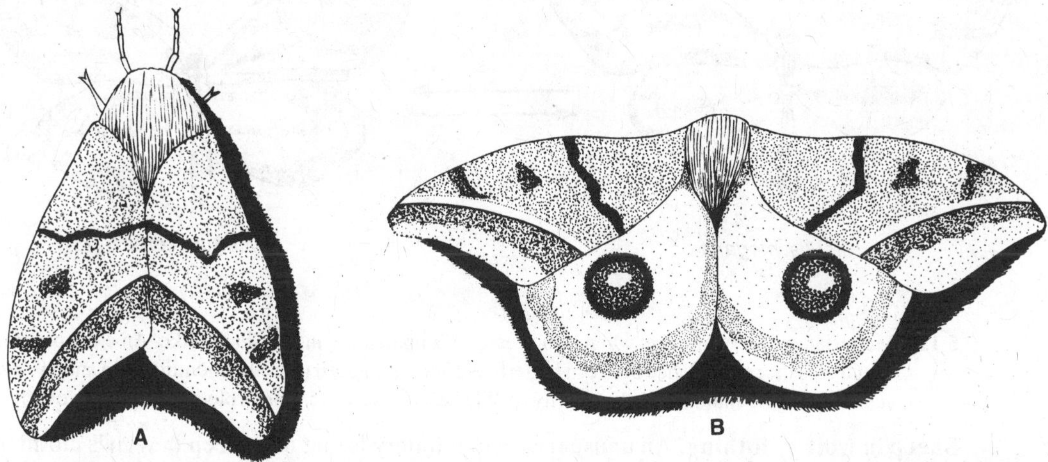

Fig. 5.12. The moth *Automeris coresus* A - At rest; B - Displaying the vivid eye- spots on its hind wings in response to light touch.

Poisonous animals tend to be conspicuously colured, presumably to that predators can easily recognize and thus avoid them. Such **warning coluration,** or **aposematism,** occurs, for example, in skunks (*Mephitis mephitis*; it is a black-and-white striped American mammal) and coral snakes (*Micrurus fulvius;* it has yellow-red coloured rings throughout the body) (**Wickler,** 1968). Many animals sequester poisons for defense. Orange and black monarch butterflies (*Danaus plexippus*) are toxic because of poisons (cardiac glycosides) they obtain from the milkweed plants they feed on as larvae. Adult females lay eggs on the toxic species of milkweed, the larvae have evolved a resistance to the poisons and are able to store them. Milkweed species vary in the amount of toxins they have, so monarchs also vary in their toxicity.

Avoidance of dangerous or unpalatable prey may occur also in the absence of opportunities for learning. For example, laboratory-reared motmots (*Eumomota superciliosa*) – lizard-and snake-eating birds–avoid models painted with the colour patterns of the coral snakes (**Smith,** 1975). Naive domestic chicks were less likely to eat meal worms that are painted black and yellow than those painted green or olive (**Schuler** and **Hesse,** 1985).

4. Mimicry

Among arthropods such as butterflies, unpalatable species tend to resemble each other and are referred to as **Mullerian mimics,** for example, the monarch and queen (*Danaus gilippus*) butterflies are usually both toxic because of the plants on which they feed. Such Mullerian mimics are also found in Trinidad butterflies (Fig 5.13). Mullerian mimics seem to have evolved because by looking alike and acting similarly, they provide fewer different prey types for predators to learn to avoid and thus are less likely to be eaten by mistake.

Some palatable species, called **Batesian mimics,** may evolve morphologies and behaviours similar to those of unpalatable species. For example, viceroy butterflies (*Basilarchus* (*Limenitis*) *archippus*), a favourable meal of many birds, have colour pattern closely resembling those of monarch butterflies [*Anosia* (*Danaus*) *plexippus*] which bird find inedible (Fig. 5.14). **Brower** *et al.* (1968) demonstrated "one-trial conditioning" of blue jays (*Cyanocitta cristata*) fed monarch butterfly often led to vomiting by the jay and avoidance of the monarchs or their mimics.

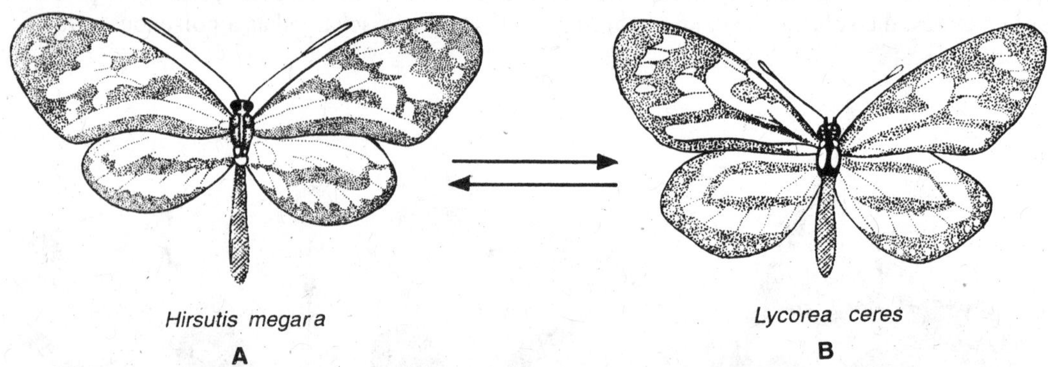

5.13. Mullerian mimicry appears when two species of unpalatable insects look alike. An example from Trinidad is the resemblance between butterflies *Hirsutis megara* (Family Ithomiidae, A) and *Lycorea ceres* (Family Danaidae, B). Similarity enables each species to gain protection from the other.

Sheep in wolf's clothing. An unusual case of mimicry is that of a sheep in wolf's clothing: a fly mimics its predator, a spider (**Mather** and **Roitberg,** 1987). Snowberry flies (*Rhagoletis zephria*) are among the prey of solitary zebra jumping spider (*Salticus scenius*). The flies have a wing banding pattern that resembles the legs of the jumping spider. Flies display with their wings, mimicking a walking spider. Other spiders flee from displaying flies, apparently mistaking them for an aggressive conspecifics. Nondisplaying flies, or those with their bands covered, are more likely to be eaten by the spiders.

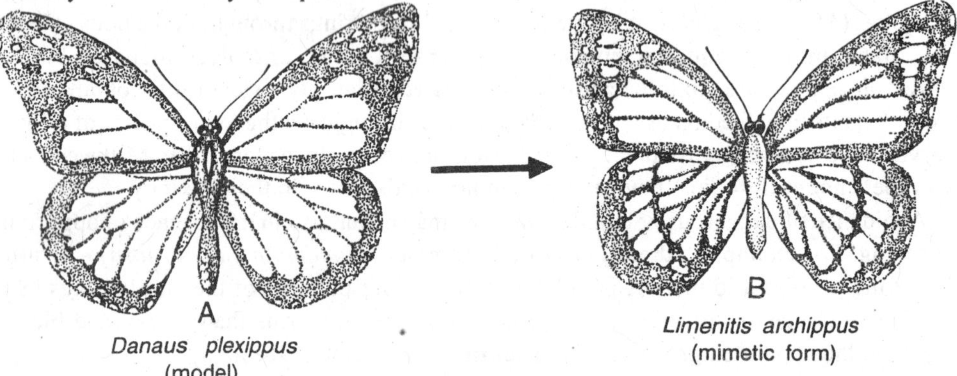

5.14. Batesian mimicry shown by North American butterfly *Limenitis archippus* (B) that mimics the monarch butterfly *Danaus plexippus* (A).

5. Distraction Displays

Individuals may use distraction displays, for example, the broken-wing displays of avocets (*Recurvirostra americana*) to attract the attention of a predator and draw it away from the nest (**Sordahl,** 1981). Broken-wing display by American avocets presumably function to distract predators from the nest or young. Although considered by some to be altruistic behaviour, it really is most easily understood as a form of parental investment (see **Drickamer** *et al.,* 2002).

Further, many mammals have white rump or tail patches; in the presence of predators the hairs are erected and the tail waved, making a conspicuous display. Observers have suggested several functions to this **flagging behaviour** such as distracting the predator from other members of the group, warning other many group members, confusing the predator when many group members are displaying, signaling the predator that it has been detected and eliciting premature pursuit (**Smythe,** 1977). These hypotheses are difficult to tease apart, as both the predator and conspecifics are likely to be in a position to receive a signal from a flagging animal. In white-tailed deer, deer that flag generally run faster than deer that are not flagging, suggesting that flagging is meant to deter the predator from pursuing, the hypothesis that flagging warns conspecifics was not supported (**Caro** *et al.,* 1995).

B. Social Strategies

Experimental evidence from a wide range of taxa demonstrated that animals can reduce their risk of predation by associating with conspecifics in groups (**Lima** and **Dill,** 1990).

1. The encounter effect. Groups of prey may be less likely to be encountered by searching predators than are solitary prey. The detection of three-dimensional groups does not increase proportionally with group size (*i.e.,* a predator is not five times as likely to find a group with ten members as with two members (**Inman** and **Krebs,** 1987).

2. The dilution effect. Whereas the encounter effect acts when predators are searching for prey, **the dilution effect** acts after a predator has found a group. Simply stated, this is "safety in number". If a predator that attacks a group can eat only one prey, it is better to be in a larger group than a smaller group from a purely probabilistic perspective : the chances of being the individual that is selected by the predator is lower in a large group (**Foster** and **Terheme,** 1981).

3. The selfish herd. Some, early ethologists felt that natural selection at the population level was necessary to explain why animals form groups; that is, the behaviour had to evolve "for the good of the group" rather than for the good of the individual. Several authors, notably **W.D. Hamilton** (1971) described how selfish behaviour can lead to aggregation. For example, frogs surrounding the rim of a pond that are trying to evade a water snake. The snake will snatch the nearest frog. Thus, it is in the best interest of each frog to attempt to get between two other frogs. This rule alone leads to the aggregation of frog into heaps.

There is much experimental evidence in support of the idea that the edges of groups are more susceptible to predation. Many flocking or schooling animals cluster in the presence of a predator, as Hamilton points out. Both tadpole groups and fish schools (**Krause,** 1993) become more compact in the presence of chemical cues indicating that a predator is near. Direct evidence of differences in predation risk in different areas of a group comes from studies of a Mexican spider, *Metapeira incrassata.* Each spider has its own orb web, but hundreds of individuals are clustered together in a colony supported by shared frame lines. Predation attempts by wasps, hummingbirds, and other predators were directly observed by **Rayor** and **Uetz** (1993), and risk of predation was highest on the edge especially for large spiders, which are favoured by predators. Animals may also be protected from parasites in the same way : ungulates in the center of groups are bitten less often than those on the periphery (**Mooring** and **Hart,** 1992). However, not every predator preferentially attacks the edge of a group : black-crowned night herons (*Nicticorax nycticorax*) are more likely to attack nests in the center of a least-tern colony (*Sterna antillarum*) (**Brunton,** 1997).

4. Increased detection of predators. The more animals in a group, the more likely it is that one of them will detect an approaching predator. This has been called the **"many eyes" hypothesis.** One of the most obvious advantages of a cohesive group whose members respond to each other's behaviour is protection against predators. With a number of animals on the alert, the approach of a predator is less likely to go undetected and one alarm signal will suffice for all. Furthermore, each individual animal will be able to spend more time feeding, protected by the many pairs of eyes around it. **Lazarus** (1979) showed this effect in action by watching how effectively red-billed weaverbird, spotted goshawk flying overhead when they were on their own or in groups with others birds. He found that solitary birds often failed to respond at all, but where there were two or more birds, the hawk was much more likely to be seen and responded to. **Elgar** (1989) reviewed over 50 others studies that show that birds and mammals spend less time in vigilance and more time in feeding, the bigger the groups they are in **Saino** (1994) showed that in flocks of carrion crows (*Corvus corone*), feeding rates were very strongly correlated with food density, independently of the amount of food present, there was also a strong effect of group size on vigilance rate. Birds in larger groups looked up less and fed at a higher rate than birds in smaller groups.

It is also reported that there is great advantages in spotting the predator directly rather than relying on the response of other birds. **Elgar** *et al.*, (1986) filmed flocks of house sparrows as they were feeding and then suddenly introduced a predator stimulus. He showed that the birds which happened to have their heads raised at the moment when danger appeared took off faster than those happened to be feeding.

In meerkats, which are socially living mongooses, **vigilance** is undertaken by particular individuals (adult) which take turns to go to a high look-out point such as a tree and keep watch for predators while the other feeds (**Macdonald,** 1986). Such a sentinel adult remains near the nest hole "baby sitting". Such individuals look as though they are being altruistic since they do not feed while on sentinel duty, but meerkats live in close-knit groups where the animals know each other as individuals and where therefore potentially long-term benefits from repeated social interactions.

Some animals give **alarm calls** when they detect a predator. When alarm calling is not particularly dangerous to the caller, it is easy to see how it might evolve. In other species, such as Belding's ground squirrels, callers put themselves at risk of predation. **Sherman** (1977, 1981) showed that close relatives of the caller were likely be close by and thus could benefit from seemingly altruistic behaviour.

Alarm calls may be specific to the type of predator. Vervet monkeys give different calls to leopards, eagles and snakes. Monkeys that hear playbacks of these calls responds appropriately: they run into trees when they hear

Fig.5.15. The response of vervet monkeys (*Cercopithecus aethiops*) to three different calls. A – That produced in response to leopard call; B – That produced in response to eagle call; C – That produced in response to snake call.

leopard alarm, look up or run into bushes with **eagle alarm,** and stand bipedally and look down when they hear the **snake alarm** (Fig. 5.15). Interestingly, infants gave alarm calls to animals that were not dangerous, but were likely to give leopard calls to other terrestrial animals, and snake calls to long, snakelike objects (**Cheney** and **Seyfarth,** 1990). Chickens also produce qualitatively different calls in response to terrestrial and aerial predators. Hens crouch when they hear an **aerial call,** and stand erect and vigilant when they hear a terrestrial call (**Evans** *et al.,* 1993).

5. Confusion effect. A variety of predators exhibit a lower prey capture rate and a longer period of hesitation when they attack grouped prey than when they attack solitary prey (**Miller,** 1922). Factors that increase this *confusion effect* are the number and density of prey swarms and the uniformity of appearance of swarm members (**Milinski,** 1979). The predator tries to isolate an individual from the group, and any individual that strays from the group or is in any way different from the rest is likely to be attacked. For example, when experimenters presented hawks with sets of ten mice, the hawks were more likely to catch the oddly coloured mouse (**Mueller,** 1971). Predators can improve with experience : in laboratory experiments with three-spined stickleback fish (*Gasterosteus aculeatus*) feeding on water fleas (*Daphnia*), fish with experience in feeding on dense swarms fed more efficiently on them than did fish with experience only with low-density population (**Milinski,** 1979).

6. Mobbing. Some prey species are able to turn the tables and attack the predator. Group defense by **mobbing** of predators by many members of the prey species is common. Many species from a variety of taxa mob snakes, including Fromosan squirrels (*Callosciurus erythraeus thaiwanensis*) (**Tamura,** 1989), scrub jays (*Aphelocoma c. coerulescens*) (**Francis** *et al.,* 1989), and cotton-top tamarins (*Saguinus oedipus*) (**Hayes** and **Snowden,** 1990). Raptors (*i.e.,* raptorial birds such as hawks, vultures, eagles, owls, etc, which seize and devour living prey) are also often mobbed by mixed groups of birds from several species. For example, the mobbing call of the blackcapped chickadee (*Parus atricapillus*) is recognised by at least ten other bird species that then join in mobbing (**Hurd,** 1996). Mobbing can be very effective. Captive owls (*Athena noctua* and *Strix aluco*) showed great distress when mobbed by blackbirds and moved their perch sites.

5.6. HOST-PARASITE RELATION

Many animals (mostly invertebrates) live in intimate association with other animals or plants. This kind of association is termed as a symbiotic relationship, or simply **symbiosis.** Symbiosis was first defined in 1879 by German mycologist **H.A. DeBary** as "unlike organisms living together". In most symbiotic relationship, a larger organism (called the **host**) provides an environment (its body, burrow, nest, etc.) on or within a smaller organism (the **symbiont**) lives (see **Brusca** and **Brusca,** 2003). Some symbiotic relationships are rather transient – for example, the relationships between ticks or lice and their vertebrate **host** – whereas others are more or less permanent. Some symbionts are opportunistic (**facultative**) whereas others cannot survive without their host (**obligatory**).

Symbiotic relationships can be subdivided into several categories based on the nature of the interaction between the symbiont and its host. Perhaps the most familiar type of symbiotic relationship is **parasitism,** in which the symbiont (a **parasite**) receives benefits at the host' expense (Box 5.3). Parasites may be external (**ectoparasites**), such as lice, ticks and leeches; or internal (**endoparasites**) such as liver flukes, some roundworms and tapeworms. Other parasites

BOX 5.3.

A parasite is an organism that obtain its nutrients from one or a very few host individuals, normally causing harm but not causing death immediately (**Begon** *et al.,* 1996). The host can live without the parasite, but the parasite cannot survive without the host.

may be neither strictly internal nor strictly external; rather, they may live in a body cavity or area of the host that communicates with the environment, such as the gill chamber of a fish or the moth or anus of a host animal (**mesoparasites**). Some parasites live their entire adult lives in association with their hosts and are **permanent parasites,** whereas **temporary** or **intermittent parasites,** such as bedbugs, only feed on the host then leave it. Parasites that parasitize other parasites are **hyperparasites.** Temporary parasites, such as mosquitoes and aegiid isopods, are often referred to as **micropredators,** in recognition of the fact that they usually "prey" on several different individuals. **Parasitoids** are insects, usually flies or wasps, whose immature stages feed on their "hosts' bodies", usually other insects, and ultimately kill the host. A **definitive host** is one in which the parasite reaches reproductive maturity. An **intermediate host** is one that is required for parasite's development is in which the parasite does not reach reproductive maturity.

A few groups of invertebrates are predominantly or exclusively parasitic, and almost all invertebrate phyla have at least some species that have adopted parasitic life-styles. Parasites exploit at least two different habitats in their life cycles. This practice is essential because their hosts eventually die. Thus, the development period from zygote to adult parasite involves either the invasion of another host species or a free-living period. When more than one host species is utilised for the completion of the life cycle, the organism harboring the adult parasite is the primary or definitive host, and those hosts in which any developmental or larval forms reside are the intermediate hosts. The completion of such a complex life often requires elaborate methods of transfer from one host to the other, and of surviving the changes from one habitat to another. Losses are high, and it is common to find life cycle stages that compensate by engaging in periods of rapid asexual reproduction in addition to the sexual activities of the adult. Thus, we find that many parasites enjoy some of the benefits on indirect development (*e.g.*, dispersal and exploitation of multiple resources) while being subjected to accompanying high mortalities and the dangers of very specialized lifestyles.

Problems of Parasites

Parasitic animals face certain problems in the establishment and maintenance of a parasitic relationship with the host. A primary problem for the parasite is that of locating and infecting new hosts. Hosts often have defenses and respond to the invading parasites (*e.g.*, helminths, protozoans, bacterial and viral parasites). In host, any tissue damage causes inflammation and phagocytic reactions of leucocytes. The host's immune system is also stimulated by the protein antigens on the parasite's surface. The parasites which overcome these host defenses may be walled off in the host's body. Hydrochloric acid and proteases (enzymes) found in the vertebrate stomach are effective barriers which few parasites can cross. Low levels of oxygen in the gut contents and high body temperature of hosts (birds and mammals) are also barriers for the parasites. These barriers indicate that the parasites have adaptations to overcome them.

For example, endoparasites face a problem of migration from one host to another. Many are likely to perish in this adventure. Numerous capsules (containing egg cell) are therefore produced by liver fluke, so that enough survive to continue the race. One liverfluke (*Fasciola hepatica*) may lay about 50,000 capsules. A single tapeworm (*Taenia solium*) can live for about 30 years, and every year, it sheds about 2,500 gravid proglottids, each containing 30,000 to 40,000 onchospheres. Formation of several larvae from one zygote in flukes and budding in the hydated cysts in *Echinococcus* further add to multiplication. Bisexuality also increases the reproductive rate, as all the individuals lay eggs.

Host-Parasite Relationship

A "successful" parasite usually does not kill its host because the death of the host means the end of parasite also. The parasite or its progeny (egg or larvae) must escape before the host dies. In many cases, a host can tolerate a small population of parasites without any serious consequences. If the effect of a parasitic infection is manifested in the host, the condition is called a **disease.** The parasitic disease may damage host cells and tissues, and may block gut, blood vessels or ducts. The parasite may produce wastes or substances which have a toxic or allergic effect on the host. The disease-causing parasites are called **pathogenic parasites** or **pathogens.**

QUESTIONS

Long Answer Questions

1. What is optimality theory? Give its advantages.
2. Describe in detail the prey model of foraging.
3. Give a detailed account of patch model of foraging.
4. Describe the effects of predators on the foraging.
5. Describe how sharks detect their prey which are buried in sand.
6. Give an account of cooperative hunting by various predators.
7. Describe the anti-predator behaviour of prey animals.
8. Give an account of host-parasite relationships.

Short Answer Questions

Write short notes on the following :

(*i*) Concept of search image
(*ii*) Central-place foraging
(*iii*) Risk-sensitive foraging
(*iv*) Fungal gardens
(*v*) Aggressive mimicry
(*vi*) Confusion effect
(*vii*) Mobbing.

Multiple Choice Questions

Choose the correct answer from the four alternatives given

1. Cause of mimicry is
   (*a*) concealment (*b*) attack (offence)
   (*c*) protection (defence) (*d*) parasite
2. The individual that shows mimicry is called
   (*a*) mimic (*b*) predator
   (*c*) prey (*d*) parasite
3. The snake *Heterodon* flattens its head, produces frequent hissing and strikes to advertise as if it is dangerous. This is an example of :
   (*a*) alluring mimicry (*b*) warning mimicry
   (*c*) concealing (*d*) batesion mimicry

Answers

1. (*a*); **2.** (*a*); **3.** (*b*).

Genetic Basis of Behaviour

There are ample examples, to support that behaviour has the genetic basis. Like any other morphological or physiological character, behaviour patterns are also heritable. This is most apparent in case of stereotyped (*i.e.,* FAPs) behaviours which are performed in the same way by all members of the same species, generation after generation with near perfect execution. But in learned behaviour also the ability to learn various skills is inherited. Determining the role of heredity in such cases is not easy.

6.1. ROLE OF GENES IN SHAPING BEHAVIOUR OF AN ANIMAL

The study of behavioural genetics includes both proximate and ultimate questions. The first set of questions are proximate in nature. They concern how genes can determine behaviours.

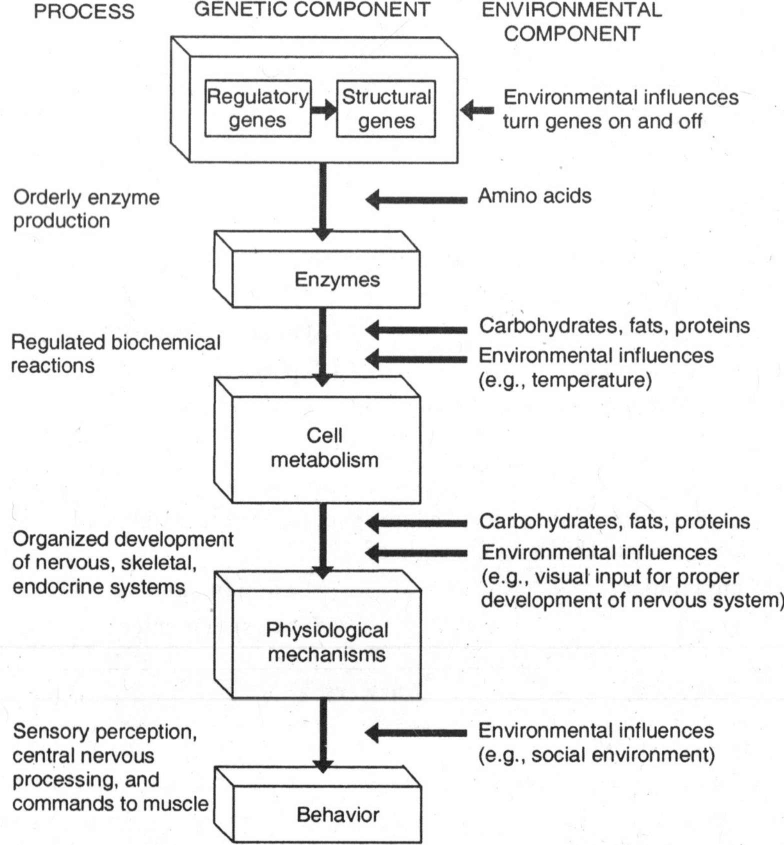

Fig. 6.1. Model showing relationship between genes and environment in the control of behaviour (after Drickamer *et al.,* 2002).

Genes, after all, are simple strings of nucleotides, and it is long path between that and, for example, a behaviour pattern such as predatory behaviour :

Gene (Nucleotides of DNA) $\xrightarrow{\text{transcription}}$ RNA $\xrightarrow{\text{translation}}$ Polypeptide $\xrightarrow{\text{folding, etc.,}}$ Structural/Functional (enzymatic and motility) or Regulatory protein $\xrightarrow{\text{+ environment}}$ Phenotype (*e.g.*, predatory behaviour)

Some of the influences on the pathway connecting genes and behaviour have been depicted in Figure 6.1. This is a complex but exciting area of research that draws from many fields, including molecular biology, development biology and neurobiology. This field is in a rapid phase of growth.

The second set of questions are ultimate in nature. In order for behaviour to evolve, it must be at least partly under genetic control, and behavioural geneticists often study the extent to which this is true. Sometimes, this is phrased as nature versus nurture, a phrase that originated with Shakespeare's the Tempest : a trait that is influenced by the genes is determined by "nature", while a trait under environmental control is determined by "nurture". However, it is simplistic to think that there is a black and white dichotomy. Nearly all traits are influenced by both genes and the environment; thus, one goal of behavioural genetists is to explain their relative effects. Such questions are important in the understanding of evolution of traits. This area of behavioural genetics is also in a phase of explosive growth (see **Drickamer** *et al.*, 2002).

6.2. GENETIC BASIS OF BEHAVIOUR

A. Studying Single Gene Effects

I. Mendelian Crosses

1. Genetic control of mating behaviour in *Drosophila*. In *Drosophila*, a mutant gene which affects some morphological or physiological characters also affects its behaviour. For example, the gene called *bar* reduces pigmentation in eyes; *forked* and *hairy* genes affect the number and structure of the bristles; gene *dumpy* alters the shape of wings and gene *yellow* or *black* change the normal *grey* body colour. These mutant genes also have other effects and some of them affect mating behaviour of *Drosophila*. Males with any of these mutants are less successful in stimulating females to mate with them. Following reasons have been attributed to it:

1. The male fruit flies with bar eyes or white eyes can not see and have difficulty in finding females to mate.

2. Male flies with forked or hairy genes have defective bristles and lack tactile sense causing failure of mating.

3. Flies with vestigial and dumpy mutant genes have malformed wings that can not vibrate properly and fail to sexually arouse the females for copulation.

4. Yellow or black (mutant) males too take longer to mate than the normal grey males (**Manning**, 1979).

2. Genetic control of hygienic behaviour in honey bees. Though it is very difficult to isolate any one gene as an determiner for particular behaviour pattern, **W.C. Rothenbuhler** (1964) demonstrated that **hygienic behaviour** (*i.e.*, cleaning of nest or beehive by worker bees of species *Apis mellifera*) is controlled by two pairs of genes (*U* and *R*). There are two strains of honeybees :

1. Normal hygienic strain (brown strain). The bees of this strain clean those waxen cells of the hive which contain dead larvae.

2. Unhygienic strain (van Scoy strain). The members of this strain do not clean the waxen cells. They suffered from serious epidemic disease known as **American Foul brood** (Caused by the bacterium *Bacillus larvae*).

The hygienic behaviour has two components :

1. The removal of wax cap of waxen cell.
2. The removal of dead larvae lying in the cells of hive.

Rothenbuhler crossed the above hygienic and unhygienic strains. The hybrids of F_1 generation were all unhygienic, showing that unhygienic trait is **dominant.** A test cross between F_1 hybrid and hygienic strain produced four behavioural phenotypes in approximately equal proportions.

1. Worker bees that neither uncap the cells of dead larvae nor remove the corpses of dead larvae even though the cap was removed by the experimenter.
2. Worker bees that both uncapped the cells and also removed the corpses of larvae.
3. Worker bees that opened the caps of waxen cells but left the dead larvae untouched.
4. Worker bees that did not uncap the cells but when cells were uncapped by the experimenter, the corpses were removed.

From the above results it was concluded that unhygienic behaviour is controlled by two pairs of genes. 1. The ability to uncap is by one pair of genes; and 2. The ability to remove dead larvae from cells by another gene. It is likely that most behaviour patterns are controlled by more than two gene pairs. In complicated behaviour patterns large number of genes are associated. Along with these genes, gene complexes have a switching mechanisms also. **Switch-genes** control the operation of a group of genes which inherits as a block.

3. Genetic control of courtship display in ducks. The courtship display in ducks consists of a series of patterns, most of which can be observed with slight modifications throughout the closely related families of species. **Konrad Lorenz** (1941) studied, **down up** patterns in ducks. In this behaviour pattern the drake (male duck) dips bill into the water and then suddenly rises his head and with it a plume (a long spreading cloud of vapours). This behaviour is lost in yellow-billed teal (*Nettion flavirostre*) and pintail ducks (*Dafila acuta*). When these two species crossed, the hybrid showed "down-up" behaviour pattern.

4. Onset of receptivity to insemination in mosquitoes. Gwadz (1970) worked on wild populations of mosquitoes. The females of different populations of these mosquitoes become sexually receptive at different times after emergence. The females of one strain, called **GP,** because it was originally collected at Gunpowder Falls, Maryland, are receptive to insemination quite early—a mean of 38 hours after emergence. Females of another strain, called **TEX** because it was originally collected at Austin, Texas, take much longer to accept males—with a mean of 120 hours (Fig. 6.2). Hybrids between the two strains (either GP/TEX or TEX/GP) have an intermediate time of insemination, with a mean of about 54 hours, but with slope more similar to the GP strain. When the F_1 hybrids are back-crossed to the GP and TEX parental strains, the results were accordant with the idea that early receptivity was due to a single, autosomal, semi-dominant gene.

Gwadz also showed that late-emerging females of the TEX strain could be made receptive much earlier by application of the juvenile hormone normally produced by one of insects endocrine organs (corpora allata) (Box 6.1). Given early hormone artificially, they behaved like GP females. Thus the genetically controlled difference between the two strains may well be due to changed timing of hormone production and this results in changes to the rate at which female receptive behaviour develops.

Box 6.1.

Juvenile Hormone (JH)

JH is an insect hormone that maintains the presence of juvenile features of the larva by suppressing the development of the imaginal buds, precursors of the adult organs. In its absence, adult features appear on molting or following a pupal stage. It is also involved in normal egg production by female insects.

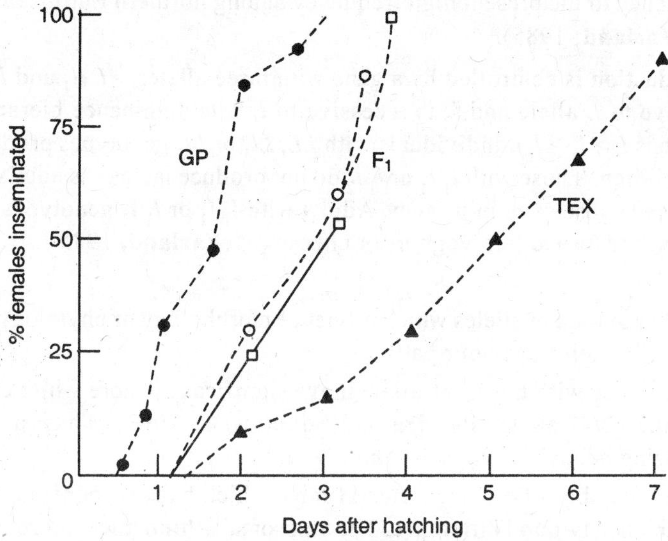

Fig. 6.2. Onset of receptivity to insemination for two parental (GP and TEX) strains of mosquito, and the F_1 hybrids (open circles show GP/TEX) hybrids and squares show the reverse TEX/GP hybrids (after Manning and Dawkins, 1998).

5. In *Drosophila,* 80% of females with copies of the wild-type form of *apterous* gene are sexually receptive 3 days after becoming an adult. In contrast, less then 15% of individuals with the ap^4/ap^4 genotype will permit courting males to copulate with them at this stage. In 1991, **John Ringo** and his associates have found that these females have a great deficiency in the production of a substance called **juvenile hormone (JH)** compared to individual with the ap^+/ap^+ genotype. As a result of the absence of sufficient juvenile hormone, egg development is greatly retarded and the typical pattern of sexual receptivity is disrupted. (Note that this is a case of **pleiotropy,** with one gene's information affecting the development of multiple traits). A female with the ap^4/ap^4 (mutant phenotype) genotype will become sexually receptive if juvenile hormone is applied to her cuticle, a finding that supports the hypothesis that the gene regulates hormone production, which in turn affects sexual behaviour (see **Alcock,** 1993).

6. Lactase production in human beings. In human beings single-gene effect and multiple allelism can be seen in the production of lactase enzyme to digest milk-sugar lactose. Lactase is not universally found in all human beings, few of them lack it. We can trace the following reason of lactase deficiency that is prevalent in certain racial groups of humans.

Before the domestication of animals (10,000 years ago), the only source of milk for human babies was the human mother's milk given during the early years of life. The child's stomach produces the enzyme lactase, which is necessary for the digestion of lactose sugar present in milk. In later life the enzyme disappears, and among many present day human populations it is completely absent in adults. People who lack lactase are intolerant to lactose and experience nausea, vomiting and abdominal pains if they drink milk. The retention of lactase in late life is an inherited trait, which is thought to be the result of a mutation that became advantageous in populations for which milk became a regular part of diet. Presumably, the mutant was rare before the domestication of animals, when humans lived entirely by hunting animals and gathering plants. Thus, in some populations the lactose-tolerance genotype must have increased from initially low frequency to the present high frequency among northern Europeans and other milk drinkers (see **McFarland,** 1985).

Lactase production is controlled by a gene with three alleles – L, l_1 and l_2. Both l_1 and l_2 alleles are recessive to L allele and l_2 is recessive to l_1 (the dominance hierarchy of multiple alleles of this gene is $L > l_1 > l_2$). Individuals with LL, Ll_1 or Ll_2 genotypes produce lactase both as adults and as children. Those with $l_1\,l_1$ or $l_1\,l_2$ do not produce lactase as adults, and those with $l_2\,l_2$ cannot produce lactase even in infancy. Adults with $l_1\,l_1$ or $l_1\,l_2$ genotypes can digest milk that has been soured or turned into yoghurt or cheese (**McFarland,** 1985).

II. Mutations

A large number of mutant alleles which influence morphology or phyiology of the organism are found to influence their behaviour patterns.

1. Larval fruit flies with the *rover* allele move significantly more while eating than those homozygous for the *sitter* allele (**Pereira** and **Sokolowski,** 1993). Thus, a single gene can influence the foraging behaviour in *Drosophila.*

2. There are a number of mutant forms of mice which have recognizable morphological character and exhibit behavioural difference. The autosomal **albino** mice (morphological variant) are less active, are unable to nibble but more successful in mating. Gene of albinism is found to be pleiotropic (see **Dewsbury,** 1978).

3. In *Drosophila, yellow* mutation is sex-linked recessive. It alters the colour of cuticle (morphological change). The flies with yellow allele are found to be less successful in mating and exhibit abnormal courtship with more orientation but less vibration and licking.

III. Knockout Genes

A new and potentially powerful technique involves the introduction of a genetically engineered mutant or **knockout gene** into an embryo. A knockout gene is one that has been targeted for disruption so that it no longer functions normally. In this case, the mutation affects a specific gene of interest, rather than randomly affecting the genome as has been the case with older mutagenizing techniques. The embryos are grown and animals are screened to find those that have the inactivated knockout gene in their DNA. Those individuals are bred to create a line that is homozygous for the inactivated gene. This allows investigators to study the animal without that particular gene being expressed.

These techniques have yielded a number of examples of mutations at single loci that result in a some major dysfunction of some aspects of behaviour. For example, mutant strains of paramecium vary in their movement patterns (**Kung** *et al.,* 1975); these have the evocative names of *sluggish, spinner* and *paranoiac* (which moves backwards). Mutations that have been identified in *Drosophila* include *spinster* (unreceptive females), *dunce, rutabaga* and *amnesiac* (poor performance on learning and memory tasks) and *fruitless* (males court but never attempt to copulate).

Single gene effects have been found for some surprisingly complex behaviours, including aggression, sexual behaviour, and anxiety and fear (**Nelson** and **Young,** 1998). For example, the *fosB* mutant in mice affects nurturing behaviour. Mutant mice mothers do not retrieve their offspring or keep them warm (**Brown** *et al.,* 1996). *Nurturing behaviour* in normal mice results both from changes in the hormone, liters of the mothers at birth, as well as from responses that are induced over a period of several days by exposure to the pups. The hormonal profiles of *fosB* mutant mothers are normal; the effect of the gene appears to be on the induced response (**Brown** *et al.,* 1996).

IV. Genetic Mosaic Fruitflies

For understanding the role of genes in nervous system, **Benzer** and his colleagues (**Hotta** and **Benzer,** 1976), have made use of some extraordinary genetic manipulations that are possible in *Drosophila* to produce **"mosaic flies"**, *i.e.,* flies that are genetically different in different cells of their bodies. The most interesting mosaics are **gynandromorphs,** individuals in which some cells are male, some female. About one third of *Drosophila's* genetic materials carried on the X(sex) chromosome, of which males carry one copy and females two. If, during cell division in a female embryo, an X chromosome is lost from one of the two daughter cells, this cell and all its descendants from then on will carry only a single X chromosome and hence the male. Loss of an X chromosome is quite common in certain genetic stocks and can occur at any point in development. Thus, the proportion of male tissue can vary from 50% (the X is lost at the first cell division of the zygote) to a minute patch of male cells only (Fig. 6.3). So, if there are genes on the X Chromosome which affect male sexual behaviour and these mutate, and, as is usual, the mutants are recessive, it may nevertheless be possible with gynandromorphs to identify which cells in the fly's nervous system must be male if it is to show masculine behaviour.

Fig. 6.3. Diagrammatic representation of mosaic or gynandromorph fruit flies (*Drosophila*). Female cuticle is represented dark, male cuticle light. The varying proportions result from the different stages of development at which an X chromosome is lost from one daughter cell following cell division in a female embryo. This cell and its descendants will become male. The top left hand example is exactly divided into male and female halves: the X chromosome must have been lost from one daughter cell at the very first division of the fertilized egg (after Manning and Dawkins, 1998).

Male fruit flies vibrate their wings as they face females during courtship. They produce two types of vibrating "song", so-called **pulse song** and **sine song.** Only if the mesothoracic ganglion has genetically male cells carrying the right genes will the insect produce pulse song. Though all its other central nervous system be male, if just these neurons are females in genotype, no pulse song is produced. This shows that the presence of genes in *particular cells* can determine how the fly behaves. Such a remarkable result is a consequence of the way insects are organised: *each cell determines its own sex genetically.* We could have such results in vertebrates where sexual behaviour develops as a result of hormones secreted by the gonads which affect all the cells of the body.

B. Studying Multiple Genes Effects

Most behavioural traits are caused by multiple genes rather than single genes. In this case, it is more difficult to examine the mechanisms by which gene action occurs. A great deal of research has focussed on developing techniques that permit ethologists to determine the extent to which a trait is influenced by the genetics, the environment, or both.

(*a*) Cross-fostering experiments. One can transfer neonatal animals from the parent female to another female of the same species or strain, or to a female of a different species or strain (if she will successfully rear the young). We can then compare genetically similar animals with different rearing environments and assess the relative importance of the effects of genotype versus the effects of maternal-care environment on certain behavioural traits. This technique of cross-fostering helps the ethologists to differentiate species-specific behaviour from environmentally influenced behaviour. If genetically similar animals reared under different conditions exhibit similar behaviour, then one can conclude that genetic control of the behaviour is fairly rigid. However, if the behaviour differs significantly, we can conclude that it is strongly influenced by environment. This technique has been widely used in bird studies, as many species will accept foreign eggs as their own. For example, great egret (*Casmerodius albus*) chicks regularly attack and kill younger nestmates (**siblicide**). Great blue heron (*Ardea herodias*) chicks, in contrast, rarely do. **Mock** (1984) tested the hypothesis that the difference in behaviour has to do with prey size : egret parents provide small fish that are easy for aggressive chicks to monopolize, whereas herons feed their young larger prey. Cross-fostered heron chicks raised on small prey by egret parents became siblicidal, showing that environmental conditions can induce siblicide in this species. In contrast, cross-fostered egret chicks were still very aggressive (see **Drickamer** *et al.*, 2002).

(*b*) Twin and adoption studies : The study of the behavioural genetics of humans is difficult, as many of the experimental techniques open to researchers on other animals are not available. Two methods have been commonly used to separate environmental and genetic effects on human behaviour : twin studies and adoption studies.

1. **In twin studies,** researchers take advantage of the existence of following two types of twins: **Identical** or **monozygotic twins** develop from the splitting of a single fertilized egg and have identical genotypes. **Fraternal** or **dizygotic twins** develop from two separate fertilized eggs and share half of their genes. We can compare the resemblance between identical twins to that between fraternal twins. If genetics is important in determining a particular trait, we expect the resemblance to be stronger between identical twins.

2. **In adoption studies,** adopted children are compared to their adopted parents, with whom they share an environment, and birth parents, with whom they share genes (**Drickamer** *et al.,* 2002).

These studies have provided evidence for genetic effects on a number of behavioural traits. For example, both *verbal* (spoken rather than written) and *spatial* (having to do with space) performance appear to be under partial genetic control. Both are strongly correlated between identical twins than between fraternal twins (**Plomin** *et al.,* 1997). Adopted children do not significantly resemble their adoptive parents in these traits, but do resemble their birth parent (**Plomin** and **Craig,** 1997).

Most behaviour biologists believe that any study on humans must be treated with great caution because it is very difficult to control environmental factors as one might like to do. For example, identical twins are often treated more similarly than fraternal twins, and therefore would experience more similar environments (**Ehrman** and **Parsons,** 1981). Adoption agencies often try to place children in homes that are similar to those they came from, again making it difficult for researchers to distinguish between genetic and environmental effects. There is likely to be a strong interaction between genotype and the environment as well : genetically different traits in children elicit different responses from family and peers (persons of same age, status or ability as another specified person), thus creating differences in the children's environments (**Plomin,** 1994). Interactions such as these make it very difficult to tease apart the effects of genes on traits. Additional problems arise from the fact that twins share the same environment in the womb : it is possible that we might attribute similarities between twins to genetics, when it is in fact due to shared intrauterine environment (**Devlin** *et al.,* 1997). It is likely, however, that there are significant genetic influences on at least some aspects of human behaviour. This does not mean enviornment influences can be ignored. In fact, environmental influences are extremely important in determining human behaviour, and interact with the genotype to have broad effects. Therefore, educational and other social opportunities play a critical role in human behavioural development despite a genetic influence on many psychological traits (**Bouchard,** 1994).

(*c*) Inbred lines. One method to study the effect of environmental factors on behaviour is to hold the genetic component constant by using homogenous strains of animals. One technique of achieving genetic homozygosity (a population of animals with the same homozygous genotype) is by inbreeding, or allowing only brother-sister matings for many generations. In mice, after about thirty generations of inbreeding, virtually 100 per cent of the allele pairs are homozygous. During the inbreeding process, many recessive alleles that are lethal or otherwise deterimental to successful reproduction may also become homozygous. Thus, to obtain available inbred strain of mice, we must begin with numerous brother-sister pairs because many lines will die out before a high degree of homozygosity is achieved.

In obtaining a homozygous strain, we have in our hands an important genetic tool. Because we know that all the animals are genetically identical, we know that any phenotypic differences among them are a result of environmental differences. Conversely, we can expose two different inbred strains to the same environment, and know that differences we see, are due to genetic difference rather than environmental effects.

Examples :

(*i*) Maze-learning ability of mice. A classical experiment on maze learning in rats (**Cooper** and **Zubek,** 1958) illustrates how both the environment viornment and genetic makeup can affect behaviour in an inbred strain. Two strains of rats were established by **selective breeding.** Animals in the **"maze-bright"** strain were adept at learning mazes and making relatively few errors, while **"maze-dull"** animals did not learn as fast. Young animals from each strain were raised under identical conditions until weaning, at which time some animals from each strain were reared in an enriched environment and some in a restricted environment. The **enriched environment** was

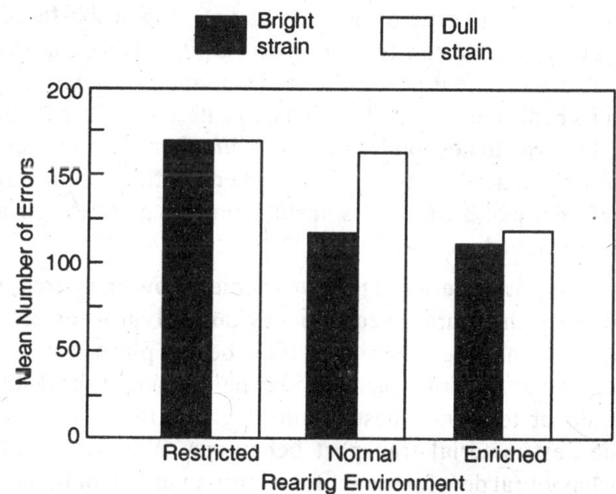

Fig. 6.4. Results of Cooper and Zubek's (1958) classical experiment showing the effects of genes and environment on maze learning in rats. The "maze-bright" strain of rats made fewer errors than the "maze-dull" strain in a normal environment. An enriched environment reduced the difference between the two strains significantly. In addition, a restricted rearing environment caused animals in the two strains to perform almost identically (after Drickamer *et al.,* 2002).

brightly coloured and many toys that the rats could manipulate. The *restricted environment* was grey and had no toys. The enriched animals from the maze-dull animals (Fig. 6.4), demonstrating that the environment can have a strong influence in a genetically homogenous strain. However, it was also clear from the results that the genetic differences between the two strains also affected their maze-learning abilities.

(*ii*) Escape behaviour of paradise fish. Paradise fish are small insectivores that live in shallow marshes and rice fields in east Asia. Several other fish species prey upon paradise fish, so being able to recognise predators and respond appropriately is important. When paradise fish see fish of other species, especially fish with large eye spots (markings that appears similar to eyes), they investigate them, and if the fish are dangerous predators, the paradise fish flee (**Gerlai,** 1993; **Csanyi,** 1986). Two inbred strains of paradise fish, raised under identical conditions, were presented with model predators. One strain was more likely to flee or back away and less likely to approach the predator than was the other strain (**Miklosi** *et al.*, 1997). The difference is attributable to genetic differences between the strains of fish.

C. Studying Polygenic Effects on Behaviour

Many of the behavioural traits are expressed in degrees, *i.e.*, they are quantitative traits and controlled by many genes or polygenes. For example, the sprint (race) speed of scorpions varies across individuals and encompasses a wide range of speeds. Variations without natural discontinuties is called **continuous variations.** Most traits like this depend on a large number of genes, each with a small effect, and are said to be under **polygenic control.** In order to understand the evolution of continuously varying traits, we can use the tools of **quantitative genetics.**

Examples :

1. Nest-building in lovebirds. In a classic hybrid study, **Dilger** (1962) crossed two species of love bird (a type of delicate parrot), the peach-faced love bird (*Agapornis roseicollis*) and Fischer's love bird (*Agapornis personata*). The females of these two species have characteristic but very different nest-building behaviour. *A. personata* female picks up bits of bark or leaves in its beak and carries them back to a nest site. In the laboratory paper strips may be transported in this way to nest boxes. *A. roseicollis* female, however, picks and painstakingly tucks strips of nesting material into the feathers on its flank. Many such strips are tucked into the feathers before the bird flies to its nesting site. If any of the strips drop out on the way to the nest, they are retrieved.

Hybrids between the two species show an interesting but highly unsuccessful compromise between the 'pure' species behaviours. Hybrid females when in a nest-building mood, would picks up nesting material in their beaks, place it in their flank feathers, remove it again and repeat the pattern over and over before finally back to the nest with the paper in either their beaks or feathers. These results suggests that many genes are involved in the development of the nest-material transport behaviour. When these genes scramble together in hybrid, the behavioural development of the individual is scrambled and a hybrid behaviour with continuous pattern results.

2. Male crickets attract females over long distances by a calling song. The sound is produced by rhythmic opening and closing of specialised forewings that carry friction mechanisms. Each closing stroke of the wings produces a sound pulse, while the opening stroke is silent. The songs are remarkably stereotyped among members of a local population, but they differ considerably from one species to another. The differences occur primarily in the temporal patterning of the pulses, as illustrated in Figure 6.5.

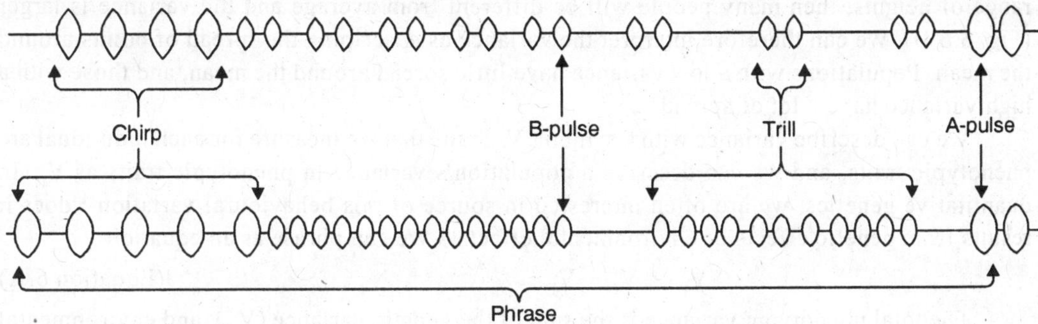

Fig. 6.5. Phrase or expression structure of the calling song of the cricket *Teleogryllus.* Each phrase is composed of two types of pulse : A–**pulses** contained in the chirp portion of the phrase, and B–pulses contained in the trill. The song of *T.oceanicus* is shown in A and that of *T. commodus* below (after McFarland 1985).

Interspecific hybrids are not common in nature, they can be produced in a laboratory. For example, **Leroy** (1964) obtained hybrids between the Australian crickets *T. commodus* and *T. oceanicus.* **David Bentley** and **Ronald Hoy** (1972) found that the songs of the F_1 hybrids are distinctly different from either parental song. Particularly, the intra-chirp and intra-trill intervals of the hybrids are intermediate between those of the parents. **Bentley** and **Hoy** found that reciprocal hybrids differed from each other (indication of sex linkage). The hybrid from female *T.oceanicus* by male *T. commodus* was similar to that of *T. oceanicus* in having well-defined inter-trill interval. On the other hand, the hybrid from female *T. commodus* by male *T. oceanicus* lacked a well defined inter-trill interval, and its song was similar to that of the *T. commodus* parent. These results suggest that the inheritance of the song pattern is **polygenic.**

Some Components of Quantitative Genetics

(*i*) Variance. Variance is a statistical measure of variability. It takes into account how different each individual measurement is from the population average. Imagine for a moment a frequency distribution of the heights of a population of 500. different people. Curves such as this are often normal (bell-shaped) curves, with a lot of people of medium height, and fewer and fewer people out toward the very short and very tall ends. We can assume that bell-shaped curves might take several shapes (Fig. 6.6). If most of the people are similar in height, then each person is close to the average and the variance is small (Fig. 5.6. A). In contrast, if there is a broad

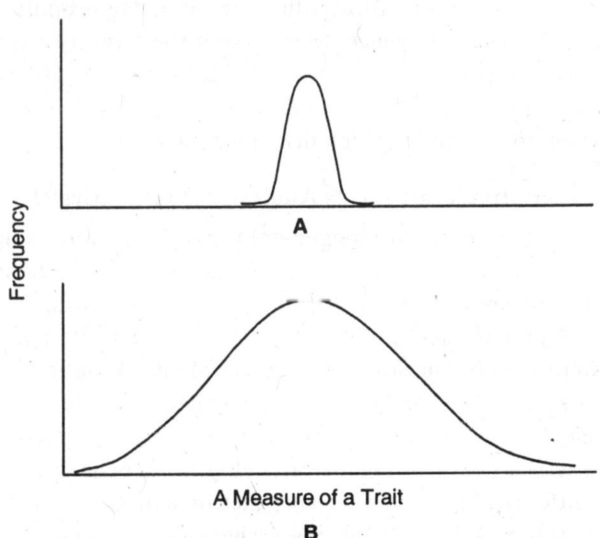

Fig. 6.6. Variance in quantitative traits. Different frequency distributions illustrating populations with (**A**) low variance or (**B**) high variance in a particular phenotypic trait.

range of heights, then many people will be different from average and the variance is larger (Fig. 6.6 B). We can therefore interpret the variance as describing the spread of points around the mean. Populations with a low variance have little spread around the mean, and those with a high variance have a lot of spread.

We can describe variance with a symbol, V. Traits that we measure for each individual are phenotypic traits, and we can describe a population's variance in phenotypic traits as V_P. In quantitative genetics, we are often interested in source of this behavioural variation : does it results from genetic effects or environmental effects ? We can phrase as an equation :

$$V_P = V_G + V_E \qquad \text{(Equation 6.1.)}$$

The total phenotypic variance is the sum of the genetic variance (V_G) and environmental variance (V_E). In other words, some of the phenotypic variation we see in a population results from differences in the genes, and some results from differences in the environment. Brain development, for example, is determined partly by genes and partly by environmental conditions during development, such as nutrition and the amount of stimulation in the environment. Each individual may differ from the population mean partly because of the genes they carry, and partly because of the environment they experience.

Occassionally there is an interaction between the genotype and the environment which is also a part of the total phenotypic V. For example, a genotype that increases territorial aggression may be expressed only in habitats with few predators, whereas another genotype might do the reverse. In this case, we can add another component, $V_{G \times E}$ or gene by environment interaction.

$$V_P = V_G + V_E + V_{G \times E} \qquad \text{(Equation 6.2)}$$

This equation says that the phenotypic variance in a population is due to variance due to genetics, to the environment, and to the interaction between genetics and environment. We can use this equation to estimate **broad-sense heritabilty,** or the proportion of the total variance in the phenotype that is result of genetic variance. We can also calculate a more meaningful measure, **narrow-sense heritability,** that measures the proportion of total phenotypic variance due to additive genetic variance. Narrow-sense heritability can be measured with selection experiments or with measures of the resemblances among relatives, such as parent-offspring regression. Heritability estimates must be used with caution, as they are good only for the populations and the environments in which they are measured.

Quantitative Trait Locus Analysis (QTL Analysis)

Behaviour traits are generally governed by more than one gene, which makes behavioural genetic analyses quite difficult. In particular it is difficult to determine where the particular genes responsible for a behaviour are located on the chromosome(s). **Quantitative trait loci (QTL)** are the set of gene loci that govern a trait (*e.g.*, aggressiveness) that is not completely determined by any one gene acting alone (**Takahasi** *et al.*, 1994). A QTL analysis involves cross-breeding animals that show different levels of a behaviour (*e.g.*, aggressive versus nonaggressive mice). Through a series of genetic crosses and analysis for specific DNA markers possessed by the original strains (Fig. 6.7), behavioural geneticists can determine which parts of different chromosomes have genes for a particular polygenic trait (**Barinaga,** 1994). Similarly, by applying QTL analysis, researchers have found two genomic regions in honeybees that affect foraging behaviour (**Hunt** *et al.*, 1995; **Page** *et al.*, 1995). These regions affect both the amount of pollen that honeybees store and whether foragers will collect pollen or nectar.

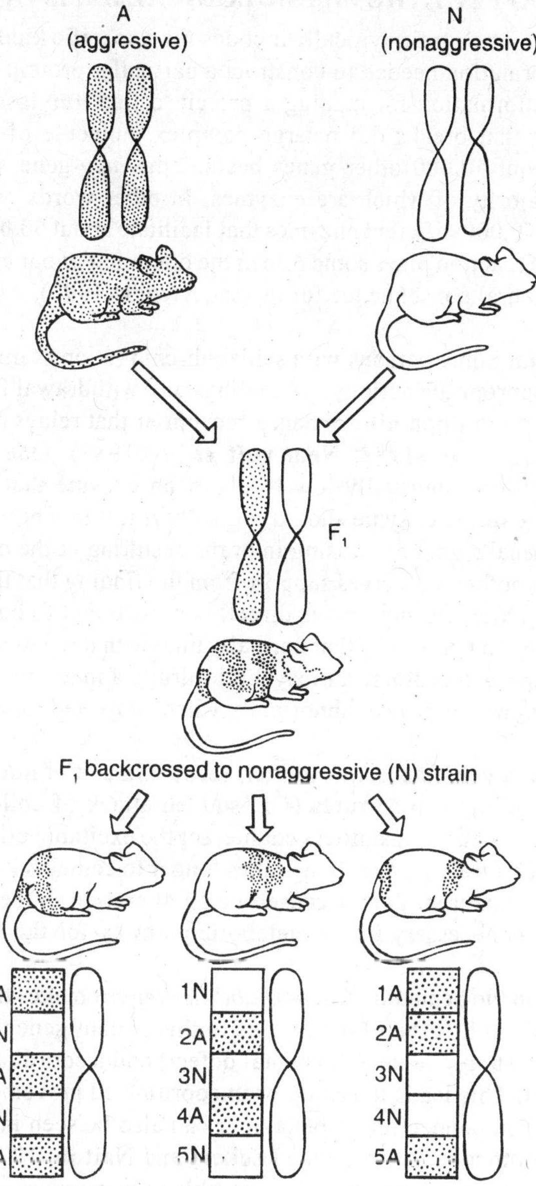

Fig. 6.7. An example of the procedure (**Barinaga**, 1994) used for quantitative trait locus (QTL) analysis. Two inbred strains (*e.g.*, aggressive and non-aggressive) are crossbred, which gives them one homologue from each pair of homologous chromosomes. The progeny from the crossbreeding (F_1) are then mated to individuals from one of the original strains (in this case, strain N). These back-crossed progeny have one homologue that is combination of parts of the chromosomes from each strain (formed when the F_1 mice formed gametes) and one homologue that is from one of the inbred strains. The backcrossed mice are ranked for aggressiveness. The recombinant chromosomes of these mice are then analysed for genetic markers that are unique for each of the inbred strains. For each marker, the mice are sorted into those that have the A-type DNA and those that have the N-type DNA. If the A-type animals possessing a particular marker are more aggressive than the N-type, that particular section of the chromosome represent a QTL that may contain a gene(s) contributing to aggressive behaviour (after Drickamer *et al.*, 2002).

6.3. GENES AFFECT THE PHYSIOLOGICAL BASIS OF BEHAVIOUR

A gene does not make a trait; typically it codes for a specific kind of ribonucleic acid that in turn carries the information needed to construct a particular protein. For example, *amy* gene of humans contains information for making a protein called **amylase.** This salivary enzyme catalyses the reaction that breaks down large complex molecule of glycogen into its sugar subunits. We have about 50,000 other genes besides the *amy* gene, most of which code for proteins, the large majority of which are enzymes. In other words, we have the information needed to make about 50,000 different enzymes that facilitate about 50,000 different biochemical reactions. Each type of reaction plays some role in the operation of our bodies; some biochemical reactions have behavioural consequence for us (see **Alcock,** 1993).

Examples :

1. Schizophrenia. Some persons with schizophrenia (a long term mental disorder whose symptoms include inappropriate actions and feelings and withdrawal from reality into fantasy) appear to have an excess of **dopamine,** a neurotransmitter that relays messages in certain parts of the brain (**Stenberg** *et al.*, 1982; **Nemeroff** *et al.*, 1983). One study found that some schizophrenic patients have unusually low levels of an enzyme that breaks down dopamine molecule. The shortage of the enzyme should logically result in a buildup of dopamine, which could then disrupt normal signal processing in brain, resulting in the behavioural symptoms of schizophrenia. This hypothesis receives support from the finding that the effectiveness of drugs used to control schizophrenic symptoms is directly proportional to how well those drugs bind with receptor molecules on nerve cells that normally bind with dopamine (**Nicol** and **Gottesman,** 1983). By blocking these receptors, antipsychotic drugs apparently prevent dopamine from reaching the receptors, which behave abnormally when subjected to excessive amounts of this neurotransmitter.

2. Seizures. Low level of neurotransmitter, **serotonin** and/or **noradrenalin** is found to be associated with susceptibility to seizures (*i.e.*, sudden attack of epileptic or diseases), when exposed to sound. These neurotransmitters counteract the excitable effect of adrenaline. Due to lower level of these neurotransmitters, body takes longer to comeback to normal and the animal become susceptible to seizures. This is caused by an alteration in the genetic material coding for an enzyme which is necessary in the metabolic pathway for the synthesis of serotonin or noradrenalin.

3. Mutant **unc5** in the nematode *Caenorhabditis elegans* makes the worms unable to move effectively but are able to flex their bodies. Due to this mutant gene the dorsal muscles do not get an organised nerve supply (a developmental defect) and a coordination between dorsal and ventral muscles is lost. This leads to erratic or uncoordinated movement of abdominal parts.

4. The biochemical connection to behaviour can also be seen in studies of the behaviour of the single-celled protozoan *Paramecium* (**Eckert** and **Naitoh,** 1972; **Kung** *et al.*, 1975). A paramecium is a thoroughly competent swimmer, able to manoeuvre rapidly through its watery universe. When the anterior end of the organism collides with an obstacle, the cilia that covers its body reverse their beating stroke for a few seconds and the paramecium backs up. The cilia then resume their forward drive stroke. By the time this happens, the orientation of the paramecium has usually changed somewhat, so that as it moves forward it moves past the obstacle. If not, it switches into reverse and repeats its avoidance response (Fig. 6.8).

By screening large number of paramecia, observers have found mutant individuals of various sorts, among them the slow-swimming **sluggish,** the extremely rapid-swimming **fast,** and various avoidance mutants, including **paranoiac** (which swims backward for a much longer time than is normal) and **pawn** (which is named for the chess piece that is forbidden to back up at all). The appropriate set of genetic crosses revealed that each of these behavioural mutants was linked to its own single-gene mutation.

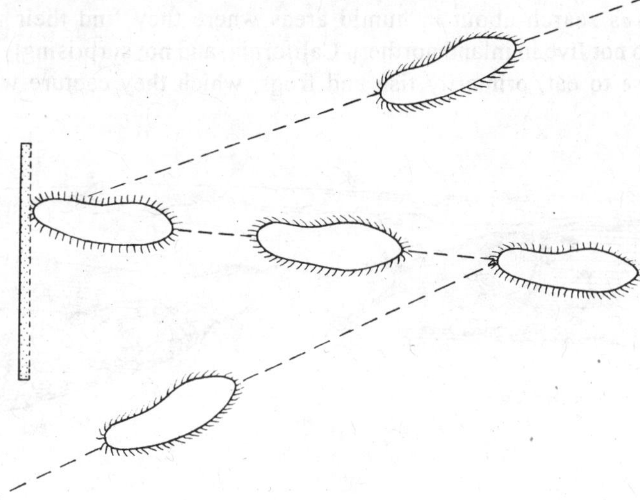

Fig. 6.8. Avoidance response. When a paramecium runs into an obstacle, its cilia reverse their beating stroke for a short time, then resume the forward stroke pattern (after Alcock, 1993).

The normal avoidance response is mediated by the effect of a tactile stimulus on an electrical charge gradient across the membrane that surrounds the paramecium. It is possible to place minute recording wires within and outside an intact paramecium and measures the charge differential across the membrane (called the **membrane potential**). Thin tiny, but significant, charge difference is caused by the way charged particles (ions) are distributed inside and outside the membrane. When the paramecium touches something with its anterior end, the permeability of the membrane changes so that calcium ions from the water enter the cell. The change is surely mediated by one or more enzymes. As the positively charged calcium ions enter, they change the membrane potential of the protozoan. This causes the cilia to reverse beating stroke, and the animal backs up. Within the paramecium, physiological systems begin expelling the calcium ions almost at once, and when the original membrane potential is restored (after a second or two), the cilia start beating in the forward position and animal moves ahead once more.

The behavioural effect a *pawn* mutation arises because the membrane does not respond to a tactile stimulus in the normal fashion. There is no influx of calcium ions and thus no membrane signal to the cilia. The cilia do not reverse their beating stroke, and the creature does not back up (see **Alcock,** 1993).

6.4. PROXIMATE AND ULTIMATE LEVELS OF GENETIC ANALYSIS OF FEEDING BEHAVIOUR OF GARTER SNAKES

At the proximate level, an individual's genetic make up affects the enzymes its cells can produce. The enzymes present in an individual influence the rats at which certain biochemical reactions occur, regulating how cells develop and function and thereby affecting the cellular basis for particular behavioural ability.

At the ultimate level, we expect alleles with reproduction-enhancing information to spread through population. But is there evidence that evolution of this sort has occurred in nature ? A particularly satisfying investigation by **Arnold** (1980) shows how it is possible to integrate proximate and ultimate levels of analysis. Arnold worked with a garter snake, *Thamnophis elegans,* that is found over much of western North America in a wide variety of habitats, including both foggy, west coastal California and the drier, elevated, inland areas of that state. There are marked differences in the diets of snakes living in two areas (referred to here after as "coastal" and "inland").

Coastal snakes search about in humid areas where they find their major prey, slugs (Fig. 6.9). Slugs do not live in inland northern California, and not surprisingly, the inland snakes find something else to eat, primarily fish and frogs, which they capture while swimming in lakes and streams.

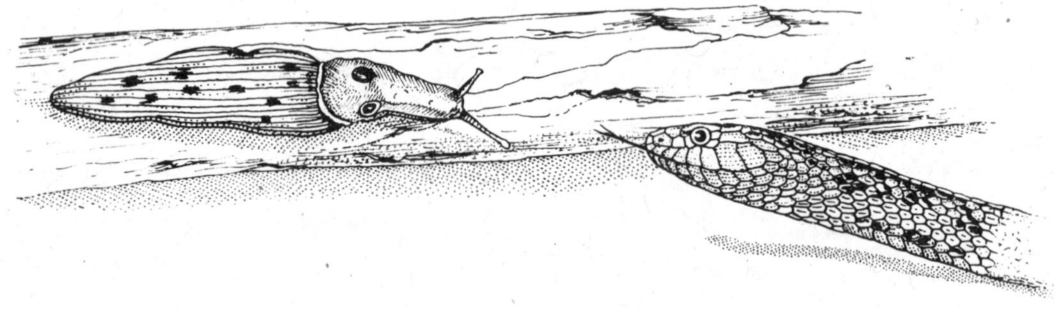

Fig. 6.9. Snail the food of "coastal" garter sanke (after Alcock 1993).

Having established that there is a difference in the feeding behaviour of coastal and inland garter snakes, Arnold wished to determine whether genetic differences were involved. He took pregnant female snakes from the two populations into the laboratory, where they were held under identical conditions. When the females gave birth to a litter of babies (garter snakes produce live young rather than laying eggs), each baby was placed in a separate cage away from its littermates and its mother to remove these possible enviornmental influences on its behaviour. Some days later **Arnold** offered each snake a chance to eat a small chunk of freshly thawed banana slug by placing it on the foor of the young snake's cage. Naive young coastal snakes usually ate all the slug pieces they received; the inland snakes usually did not (Fig. 6.10). In both populations, slug-refusing snakes did not even make contact with the slug food, but ignored it completely.

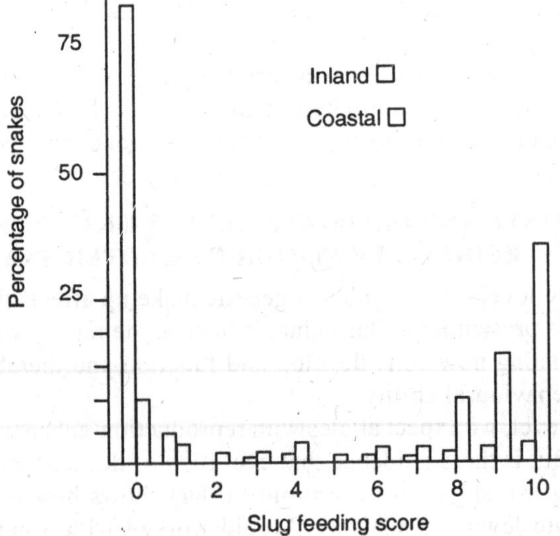

Fig. 6.10. Response of newborn, naive snakes to slug chunks. Garter snakes from coastal populations tend to have high slug feeding scores (*e.g.*, a score of 10 indicates that the snake ate a slug cube on each of the 10 days of the experiment). Inland garter snakes rarely eat even one slug chunk (which would yield a score of 1) (after Alcock, 1993).

Arnold took another group of isolated newborn snakes that had never fed on anything and offered them a chance to respond to the **odours** of different prey items. He took advantage of the readiness of newborn snakes to flick their tongues at and even attack cotton swabs that have been dipped in fluids from some species of prey. By counting the numbers of tongue flicks that hit the swab during 1 minute trial, he measured the relative responsiveness of inexperienced baby snakes to different odours. (Snakes have an affactory organ, called **Jacobson's organ**) in the roof of the mouth that analyzes odours carried to it by the tongue. When the animal's tongue touches a source of chemicals, it carries some molecules to this organ, which play a role in prey recognition).

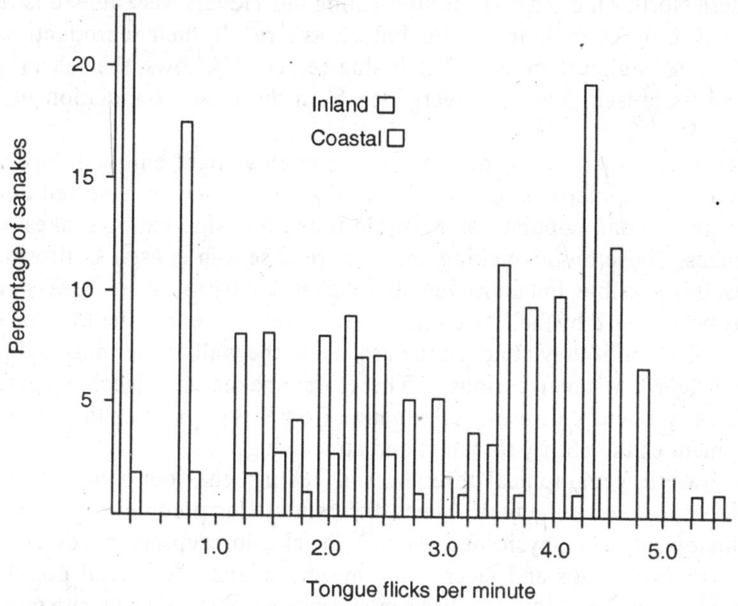

Fig. 6.11. Odour preferences of inland and coastal garter snakes as measured by the frequency of tongue-flicking in response to cotton swabs dipped in slug extract. Most-coasted snakes tongue-flicked much more than inland snakes did (after Alcock, 1993).

Population of inland and coastal snakes reacted about the same to swabs dipped in toad tadpole solution (a prey of both groups), but behaved very differently toward swabs daubed with slug scent (Fig. 6.11). Within each group there was some variation in response, but most inland snakes ignored the slug odour, whereas most coastal snakes responded strongly to it. Because all the young snakes had experienced the same environments, the differences in their willingness to eat slugs and to tongue-flick in reaction to slug odour had to have been caused by genetic difference among them.

Arnold then did a **heritability study.** By comparing the tongue-flick scores of siblings, Arnold determined that *within* each population only about 17 percent of the *differences* in chemoreceptive responsiveness to slug odour stemmed from genetic differences among individuals. The *low heritability* of tongue-flicking within the coastal or the inland population means simply that most of what little phenotypic variation exists with each population is caused by subtle environmental variables. This makes sense if past selection has eliminated much of the genetic variation that affected the development of feeding behaviour in coastal snakes and in inland snakes as well.

If the feeding differences *between* the two populations of garter snake arise because most coastal snakes have a different allele or alleles than most inland snakes, then crossing adults from the two areas should generate a great deal of variation in the resulting group of "hybrid" offspring, more variation than is found in either parental population. Arnold conducted the appropriate experiment and found evidence in support of the expected result, confirming again that the differences between populations have a strong genetic component.

After recognising the genetic and physiological basis for differences in food preference between coastal and inland garter snakes, Arnold turned his attention to the **evolutionary basis** for these differences. He suggested that among the original colonizers of the coastal habitat were a very few individuals that carried the then rare allele(*s*) for slug acceptance. (There are reasons for thinking that coastal California was colonized by *Thamnophis* more recently than inland western North America). These slug-eating individuals were able to take advantage of an abundant food resource in their new habitat. If, as a result, their reproductive successes was a little as 1 per cent higher than that of their slug-rejecting fellows, the coastal population could have reached its present state of divergence from the inland population in less than 10,000 years.

It is easy to visualise why slug-accepting alleles might enjoy an advantage and spread rapidly in coastal populations. But why have they been actively selected against (and nearly eliminated) from inland populations ? Arnold found that slug-eating snakes will also consume aquatic leeches. These blood-sucking animals are absent in coastal California but plentiful in inland lakes. It is possible, but unproven, that snakes that try to eat leeches (which slug-acceptors will do) may be damaged by their "victims". Leeches can live even after they have been swallowed by a garter snake, and if they attached themselves to the wall of the snake's digestive tract, they might injure their consumer seriously. Therefore, snakes with leech-accepting alleles might reproduce less in inland California, eliminating from this population the alleles that also lead to the development of the ability to detect and attack slugs.

In this way, the geographical differences in feeding behaviour of the snakes can be explained in terms of their proximate genetic basis (different alleles predominate in the two populations of snake), their proximate physiological basis (the chemoreceptors that detect certain molecules associated with both slugs and leeches are more prevalent in coastal populations), and their ultimate ecological and evolutionary basis inland snakes must compete with potentially dangerous leeches whereas coastal snakes are exposed only to edible slugs (**Arnold,** 1980). Arnold's study of snake behaviour genetics provides a model of how to integrate proximate and ultimate levels of analysis in behavioural research (see **Alcock,** 1993).

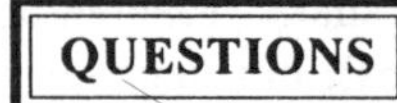

QUESTIONS

Long Answer Questions

1. Describe the genetics of nest-building by the love birds.
2. Give an account of genes affecting the physiological basis of behaviour.
3. Explain in detail the Arnold's genetical and evolutionary studies on the feeding behaviour of garter snake.
4. What are the advantages and disadvantages of following techinques for studying genetics of animal behaviour: inbred strains, artificial selection and cross-fostering ? For each technique give an example.

Short Answer Questions

Write the short notes on the following :

(*i*) Genetic control of mating behaviour in *Drosophila*

(*ii*) Genetic control of hygienic behaviour of honey bees

(*iii*) QTL of aggressive behaviour.

7

Evolution of Behaviour

7.1. GENES AND BEHAVIOUR

The total set of genes an individual possesses is its **genotype.** The genotype and the environment determine the **phenotype,** or the observable traits of an organism. Genes are located on homologous pairs of chromosomes, and may have different forms, or **alleles.** Genes can have either structural or regulatory functions.

Genetic variability in populations came from several sources. When gametes (egg and sperm) are formed, homologous chromosomes exchange parts in the process of **crossing over.** After crossing over, during gamete formation the homologues are separated independently. Hence, **sexual reproduction** shuffles the genetic make up of each offspring relative to its parents. The ultimate source of new variability, however, is **mutation,** which is a heritable change in the genetic material.

Evolution is a change in the **frequency of alleles** in a population over generations. **Darwin's theory** of evolution by **natural selection** proposed that some organisms had heritable traits that made them better suited for survival and reproduction than organism with different traits. Therefore, more offspring are born with those traits, and the traits increase in frequency in the population. Such traits are said to provide an animal its **fitness** or probability of surviving and reproducing. Traits that have been shaped by the process of natural selection are called **adaptations.**

Evolution may result not only from natural selection, but also from **genetic drift,** the random changes in frequencies of alleles, particularly in small populations. **Mutations** and **gene flow** also contribute to evolutionary change. If evolutionary forces are at work in a population, the genotype frequencies will usually not be in Hardy-Weinberg equilibrium.

Selection can work at the level of genes, individuals, kin groups, or unrelated groups. We should be careful about uncritically referring to adaptive explanation for the behavioural patterns that we see. Traits may be environmentally induced or result from other evolutionary forces besides natural selection. Selection pressures change over time. Traits are not always independent of one another. Finally, natural selection can act only on the genetic variation available. Inspite of this, natural selection has provided a very useful framework for the study of behaviour.

7.2. MICROEVOLUTION AND MACROEVOLUTION OF BEHAVIOUR

Traits with a genetic basis are capable of evolving over time. Evidence of behavioural evolution can be obtained both at the level of **microevolution** or genetic changes within populations or species and at the level of **macroevolution,** or the evolutionary patterns of behaviour recognisable above the species level.

A common error about evolution, particularly in nonbiologists, is that it is regarded to be **progressive :** that it proceeds in a predetermined, linear direction toward an optimum phenotype (see **Drickamer** *et al.,* 2002). Even the language that is sometimes used to describe evolution suggests this idea (for example, when traits or species are referred to as "lower" and "higher", or "primitive" and "advanced"). In fact, *evolution is not directional or progressive.* For example, environments may vary overtime, and once optimal phenotype may no longer favoured. In some cases, ancient lineages, such as horseshoe crabs (*i.e., Limulus*), have remained essentially unchanged over vast spans of time, presumably because they are in relatively constant environments where they are successful. Some have even become less complex : **internal parasites** (endoparasites) that have evolved from free-living forms have adapted to a new environment where complex sensory structures and complicated nervous systems are not required. Organisms may also be unable to reach an optimum phenotype : genetic variation is necessary for evolution by natural selection to occur, and variation is not always present. It may also not be possible to optimize every part of an animal's phenotype, as an adaptation in one trait may mean a trade-off or compromise in some other area. Nonetheless, many of the examples in this chapter are of cases where we clearly see adaptation in response to a particular known selective force. The examples are conclusive evidences of natural selection.

7.3. MICROEVOLUTIONARY CHANGES IN BEHAVIOUR

Microevolutionary changes occur within species. Evidence of microevolutionary changes in behaviour have been obtained from two sources : 1. domestication and 2. observations of natural selection in the field.

1. Domestication and Behavioural Changes

Human beings have had a long association with many species of animals such as dogs, cats, fowls, sheep, goats, etc. They control all aspects of the lives of many domestic species, including housing, feeding, socialization and breeding. Because we prefer domestic animals with particular traits such as hogs (A hog is a castrated male pig reared for slaughter) with more meat (park), or horses that obey their riders, so in many animals we have selectively bred animals with particular traits. This process is called **artificial selection** and it has led to some remarkable changes in the behaviour and morphology of domestic animals. These changes provided a big piece of the evolutionary puzzle for **Charles Darwin :** he understood that natural selection could produce similar changes in animal populations in the wild.

Dog behaviour provides a wonderful example of evolution under domestication. We can consider the morphological differences between, say pugs and Great Danes. At first glance, these two dogs may appear to be an adult and juvenile of the same breed, but they are actually adults of two different breeds. This demonstrates the evolution of the remarkable diversity in domestic dogs in a short span of time. We can also consider the breed-specific behavioural characteristics in dogs, such as **sheepherding dogs** that can direct their charges through complicated paths, and **hunting dogs** that will carry prey back to their masters instead of eating it themselves. Effective artificial selection requires both the presence of **genetic diversity** as well as **selective breeding**. Let's examine these two factors for the case of artificial selection in dogs.

1. Genetic diversity of dogs. It has long been suspected by biologists that dogs arose from wolves, which were the major carnivore in the northern hemisphere (**Scott** and **Fuller,** 1965; **Morey,** 1994), but the fossil record is inadequate to give much detail about when and how this happened. **Molecular techniques** have been used to reconstruct the history of the group through genetic analysis. By comparing the strings of nucleotide sequences (of DNA) across a variety of groups, biologists have deduced the branching order in which mutations arose, and determine which groups are likely to be closely related. These studies suggest that either wolves were domesticated in several places and at different times, or that dogs mixed with wolves repeatedly after their first domestication (**Vila** *et al.*, 1999). This accounts for the great genetic diversity in the domestic dog, which provide the raw material for the evolution of different dog breeds.

2. When did selective breeding lead to diversity of dogs? Fossilised bones of wolves have been found in association with hominids from up to 400,000 years ago (**Clutton-Brock,** 1995). However, it probably was not until 10,000 to 15,000 years ago, when hominids shifted from nomadic hunter-gatherer societies to more sedentary agricultural societies, that dogs began to diverge from wild wolves (**Vila** *et al.*, 1997). Even after divergence from wolves, dogs did not immediately diversify.

For several thousand years, dogs were likely to interbreed with each other; they probably accompanied travelling humans as companions or for trade, and thus, gene flow among dogs was very high and differentiation among different groups was low. It was not until recently that modern breeding methods led to the breeds we see today (**Vila** *et. al.*, 1997). Evidence of this evolutionary history is disclosed by the high diversity of **mitochondrial DNA** within dog breeds, which would have been lost if dog breeds had been under selection for a long time.

Much of the differences in behaviour between dogs and wolves, and among dog breeds, had been the result of selection on agonistic (aggressive) and investigatory behaviour (**Scott** and **Fuller,** 1965). For example, **bulldog breed** (Box 7.1), in the interest of an English sport, was selected for its tendency to attack the nose of a bull and hang on, in contrast with the more typical slashing attack from the rear used by wolves. **Terrier breeds** have been selected.

Box 7.1.

Bulldog is a medium-sized, short-haired, powerful dog, originally bred in England for use in bull-baiting (a sport). This breed of dog has a powerful jaw and a flat wrinkled face.

for their tendency to attack prey ruthlessly, regardness of any injury suffered; the normal wolf pattern is to snap and then withdraw to avoid injury. In other breeds of dogs, selection has been to reduce agonistic behaviours : scent hounds and bird dogs are examples of peaceable animals that can be kept in groups in a kennel.

Wolves and dogs search for prey rather than lying in wait and therefore have a well-developed investigatory range. Some dogs, such as **scent hounds,** are bred for the ability to follows a trail. Others, such as **bird dog,** use their visual and olfactory senses much more equally : after visually searching the ground, they locate prey by scent only when they are a few paces away, and then they freeze. Dogs must be trained for all of these tasks, but in each case, artificial selection has produced a phenotype that is favourable for specific behavioural traits (**Scott** and **Fuller,** 1965).

In Russia, an experimental approach to the study of the effects of domestication has been continuing for over 40 years (**Trut,** 1999). A Russian ethologist **Dmitry Belyaev,** believed that the patterns of morphological changes that occur in domesticated animals are the result of selection directly on behaviour rather than on morphological traits. He and his colloborators tested this idea by using artificial selection on a species that had never been domesticated, the silver fox (*Vulpes vulpes*). In their experimental design, the only trait that was used in choosing which animal would reproduce was their friendliness toward humans in a standardized test. The results of their experiments were quite interesting. As expected, foxes became more and more friendly over generations of selection, so that, after 30 to 35 generations of selection, 70 to 80 per cent of the foxes whimper to attract attention of humans and they were quite eager to make contact with humans. Morphological changes are also clear in the selected line, many of which are suggestive of domestic dogs : the foxes are more likely to have a white pattern on their forehead, brown mottled fur, floppy ears, shortened legs and tail, shorter and wider snouts and a tail curled up in a circle. In wild foxes, many of these traits are characteristic of very young animals but disappear during ontogeny. It is likely that selection for friendliness has acted on the timing of development. These results suggest that **Belyaev** was right, *i.e.,* selection on behaviour alone can lead to striking morphological differentiation.

Further, obnoxious behavioural traits may also appear as a side-effect of selection for other traits. Hatchery-reared trout, for example, are much more aggressive than wild fish, and may devote so much energy to unnecessary aggression that their survival in the wild is decreased (**Deverill** *et al.*, 1999). These examples explain that morphological and behavioural traits can be linked in both artificial and natural selection.

3. Natural Selection in the Field

A. Observing Change Over Time

It is extremely difficult to observe evolution in action in the field. The best chance of seeing natural selection at work occurs when there is a strong selection pressure acting on a population with a short enough generation time, so that humans can track changes from one generation to the next. The following example shows this fact.

(*i*) Industrial Melanism

Industrial melanism means changes in colour in the peppered moth, *Biston betularia* (**Kettlewell,** 1965). This moth is active at night and rest on trees' bark during the day. Moths that do not match the background up on which they rest are more likely to be picked off by birds (predators). Selection-pressures on the colouration of this moth changed with the advent of the industrial revolution in Britain, when pollution killed the light-coloured lichen that covered the bark of many trees. Thus, initially moths were subject to selection for light colour, but as the lichen disappeared darker moths more likely to survive and reproduce and the frequency of dark moths in the populations increased. As pollution control laws went into effect, the frequency of the dark morph declined again as the air became cleaner (**Cook** *et al.*, 1986, **Cook** *et al.*, 1999).

(*ii*) Galapagos Islands' Finches

Grant and colleagues have been studying the action of natural selection on finches in the Galapagos Islands (**Grant,** 1986). On one of the islands, Daphne Major, regular rainfall throughout the early 1970s produced abundant seeds and seed-eating finches thrived. In 1977, a drought occurred and the ground finches (*Geospiza fortis*) failed to breed, declining in number by 85 percent. The corresponding decline in seed supply was nonrandom, with small seeds becoming scarce and large, hard seeds, remaining relatively common. The result was strong selection for birds with large beaks that could handle the seeds, and within a short period of time, the average beak size increase effectively (**Boag** and **Grant,** 1981). When the El Nino year brought remarkable rains in 1983, selection pressure again shifted to favour small-billed birds that ate small seeds (**Gibbs** and **Grant,** 1987). (**El Nino** is a complex cycle of climatic change affecting the Pacific region).

(*iii*) Change in Behaviour of Guppies (fishes)

An example of behavioural change in short time period is described by **Magurran** (1999). In 1957, a researcher collected 200 guppies from a site on a river in Trinidad that also had predatory pike cichlids, and introduced them to a site that had neither guppies nor predators. Thirty-four years and 100 guppy generation later, fish were collected from these population and their offspring were reared in the laboratory. When tested as adults, these guppies showed reduced

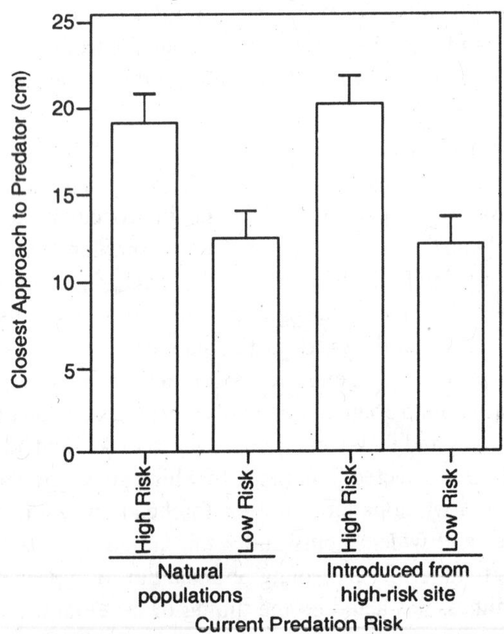

Fig. 7.1. Guppies (fishes) were taken from a site with many predators and introduced into the Turare River, the introduced guppies behaved in an appropriate way : guppies in high-risk areas were cautious about approaching the predators, and those in low-risk areas were not. They were similar to guppies in natural high- and low-risk population (after Drickamer *et al.*, 2002).

schooling behaviour (schooling reduces predation on individuals) and were more inclined to inspect predators (a dangerous behaviour which nontheless provides useful information about the predator) than were guppies from sites where predators occurred (Fig.7.1). This suggests that the descendents of the transplanted guppies had evolved to be less cautious.

B. Understanding Evolution From Geographic Patterns of Behaviour

For inferring evolution of behaviour, one effective approach is to compare population of the same species that are found under different environmental conditions. In this way, one can identify possible ecological causes of adaptation (**Foster** and **Endler,** 1999).

Spiders display a predatory behaviour. The spider, *Agelenopsis aptera* lives in the southwestern United States. This species of spider builds a web shaped like a flat sheet with an attached funnel, where it sits while waiting for prey. **Susan Riechert** (1999) has studied population of this spider in two different habitats. The desert grassland is a harsh environment, very dry and hot with few insect prey. The riparian (streamside) woodland habitat is more tolerable : spiders live alongside a permanent spring-fed stream, with shelter from the sun and plenty of insects. However, birds also do well in this habitat, and the risk of predation by birds is very high.

These spider populations exhibit behaviours that appear to be adaptation to these environments. 1. The grassland spiders attack a greater proportion of prey that hit their webs, fight more vigorously over web sites, and have larger territories than do the riparian spiders. 2. Riparian spiders in an isolated Texas population show stronger antipredator behaviour; they are more likely to retreat into a funnel when started, and are slower to return to foraging after they have retreated.

In the riparian population of spiders in Arizona, not all behaviours of the spiders appear to be adaptive. For example, they increase to high level of fighting more often than would be optimal, and attack even prey that is not very profitable (thus unnecessarily exposing themselves to predation risk). A series of experiments established that gene flow from surrounding populations is probably evolutionary force that keep this population from being completely adapted to its environment. Further, *Agelenopsis* spiders move from populations in one habitat type to those in another and thus prevent behaviour from evolving to become perfectly adapted to the environment. If spiders are prevented from entering a population for just a single generation, a measurable shift in the behaviour toward the optimum occurs (Fig.7.2). This work provides a very clear example of the different evolutionary forces that interact on behavioural traits.

Arid Habitat

Extreme temperatures
Low prey levels
Low predation risk

Spiders are :
Aggressive
Not cautious
Well adapted to environment →

Riparian Habitat

Moderate temperatures
Plenty of prey
High risk of predation
TX riparian : No gene flow

Spiders are :
Not aggressive
Cautious
Well adapted to environment
AZ riparian : incoming gene flow

Spiders are :
More aggressive and less cautious than predicted
Not completely adapted to enviornment

Experiment to Test the Effects of Gene Flow

	Experimental treatment		Offspring of Spiders Exposed to treatment	
AZ riparian ↗	Gene flow cut off Predators excluded	→	Broad range of phenotypes, both aggressive and cautious	Shows importance of predation as a selection force
AZ riparian ↘	Gene flow cut off Exposed to predators	→	Not aggressive Cautious	These are now better adapted to the environment, and are similar to TX population

Fig. 7.2. The spider *Ageleopsis aperta* are found in both arid habitats, where conditions are poor but predators are few, and riparian habitats, where conditions are good but there are many predators. Spiders from arid habitat are aggressive and will fight for the few good web sites and are not cautious (they are quick to return to foraging, and will come out on their web even for low quality prey. Spiders from Texas (*TX*) riparian habitats are not aggressive and are cautious, as predicted. However, spiders from Arizona (*AZ*) riparian habitats appear to not be adapted, because of spiders immigrating from arid habitat. This hypothesis was tested by cutting off gene flow and manipulating exposure to predators (after Drickamer *et al.*, 2002).

7.4. MACROEVOLUTIONARY CHANGES IN BEHAVIOUR

Macroevolutionary change occurs above the level of species. Ethologists have got evidence for the macroevolution of behaviour from the following sources : 1. Fossil evidence; 2. Phylogenetic reconstruction and the comparative method; 3. Detailed studies of adaptive radiation, and 4. The role of behaviour in speciation.

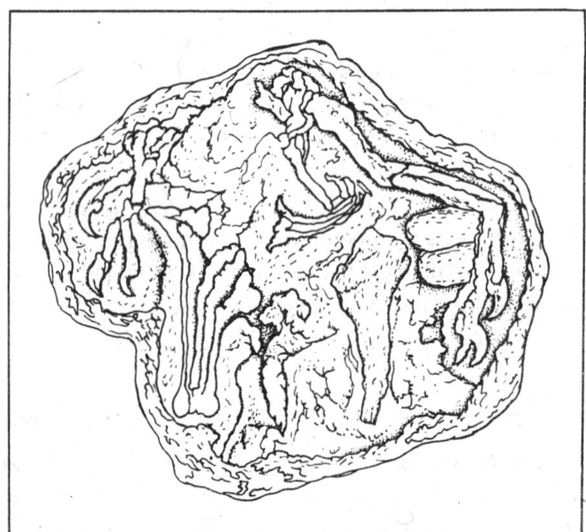

Fig. 7.3. A fossil *Oviraptor* (dinosaur) and its eggs' arrangement in the nest suggests that the dinosaur exhibited the parental care (after Drickamer *et al.*, 2002).

1. Behaviour and the Fossil Record

The fossil record provides a great deal of data about the evolutionary history of organisms. Behaviour itself does not have fossil remains. However, ethologists can infer much about behaviour from bones, teeth, horns and tracks. Some species construct structures, or artifacts (An *artifact* is a useful or decorative, human-made/animal-made object such as nest, hive, burrow, etc.), that can be studied in fossil form (**Hansell,** 1984).

Fossils of dinosaur nests have provided vital information regarding the parental care. Birds are descendants of dinosaurs, and researchers have wondered whether advanced parental care originated in the birds or in dinosaurs. Fossils show that some dinosaurs incubated their eggs much as modern-day birds do. **Norell** *et al.*, (1995) have found that fossil dinosaur *Oviraptor* had arranged its eggs in its nest in a circle in two layers, with the broad end toward the middle (Fig.7.3). This indicates that the parent manipulated the eggs into this configuration, much as is done by living birds.

Fossilized morphological structures have been found to be good sources of information about behaviour. For example, the horse lineage provides a distinctly clear fossil record. Earlier horses were browsers, eating leaves and succulent plants. However, when the climate in North America became drier during the Miocene, grasslands became more wide-spread. Under this new selective regime, fast running and the ability to chew tough grass became favoured, and this is reflected in the foot and tooth structures (**Stanley**, 1993).

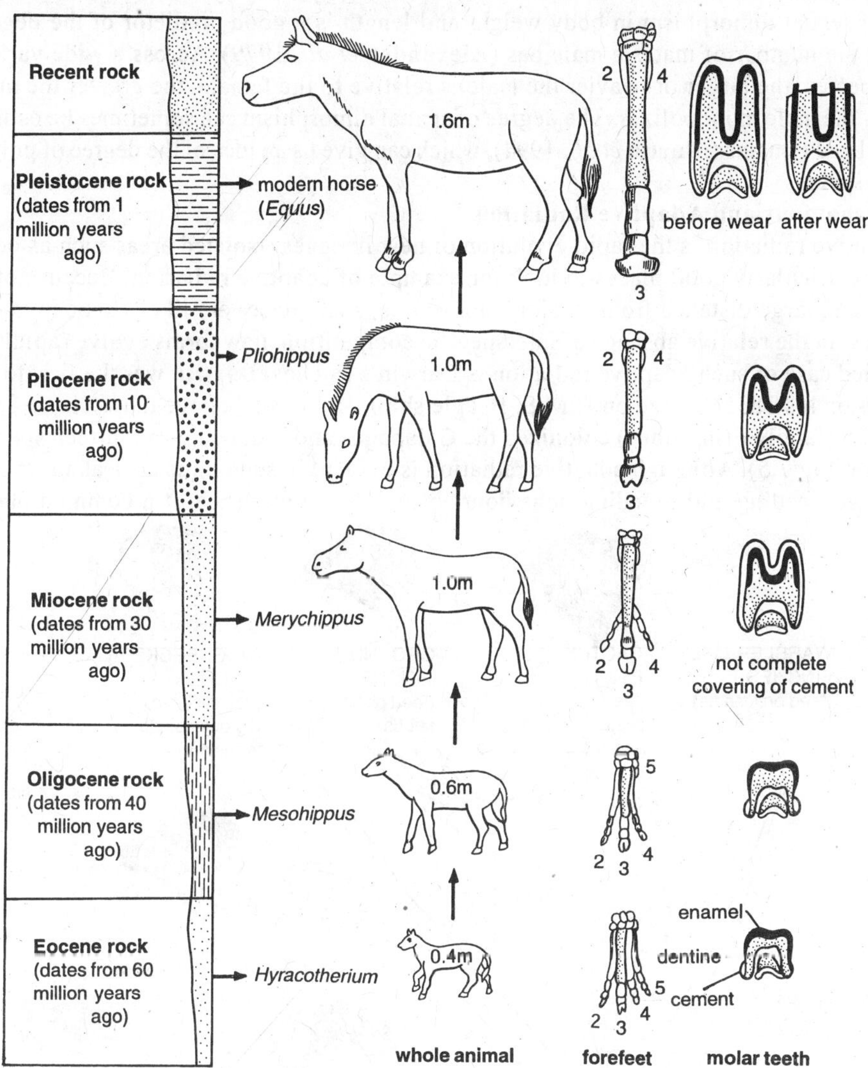

Fig. 7.4. Evolutionary history of horse.

Evidence of predation has been gained from drill holes in the fossilized shells of bivalve molluscs from the late Triassic (about 200 million years ago). These holes were made by carnivorous snails that penetrated the shells of this prey. Oddly, such habit (*i.e.,* of predation) seems to have disappeared, only to reappear some 120 million years later (**Fursich** and **Jablonski,** 1984). Thus, what we might consider a real breakthrough in predatory technique did not persist; the presumed selective advantage of such behaviour pattern may not be a good predictor of its long-term survival. In further studies of snail predation on molluscan prey during the last 100 million years, **Kitchell** (1986) concluded that prey selection was nonrandom and governed by foraging rules similar to those used by present day animals.

Hint about the behaviour of extinct mammalian carnivores was revealed by **Hunt** *et al.*, (1983), who excavated ancient burrow system containing the skeletons of bear dogs from the early Miocene, 20 million years ago. These animals were about the size of a wolf or hyena and apparently used the dens much as do modern carnivores.

Ethologists also find out information about mating systems by fossils. In mammals the amount of sexual dimorphism in body weight and length is a good predictor of the degree of polygyny, the number of mates a male has (**Alexander** *et al.*, 1979). Across a wide variety of modern species, the larger or heavier the male is relative to the female, the greater the number of females the male monopolizes. The degree of sexual dimorphism can sometimes be estimated from fossilized remains (**Martin** *et al.*, 1994), which can give us an idea of the degree of polygyny in extinct species.

2. Behaviour and Adaptive Radiation

Adaptive radiation is the rapid evolution of new lineages. Isolated areas such as oceanic island are particularly good places to look for example of adaptive radiation. Because of their small size and large distance from a colonizing source, such places are likely to be invaded by few species. In the relative absence of interspecific competition, new forms evolve rapidly. The best–studied case of such adaptive radiation is Darwin's finches. **Darwin** was the first to study these birds on his 1835 voyage on H.M.S. Beagle ship. Biologists believe that a single species of bird finch (family Fringilidae) colonized the Galapagos and radiated into fourteen species in four genera (Fig.7.5). Although adaptive radiation is evident in such traits as beak morphology and plumage, feeding and breeding behaviours also provide evidence of a common ancestor

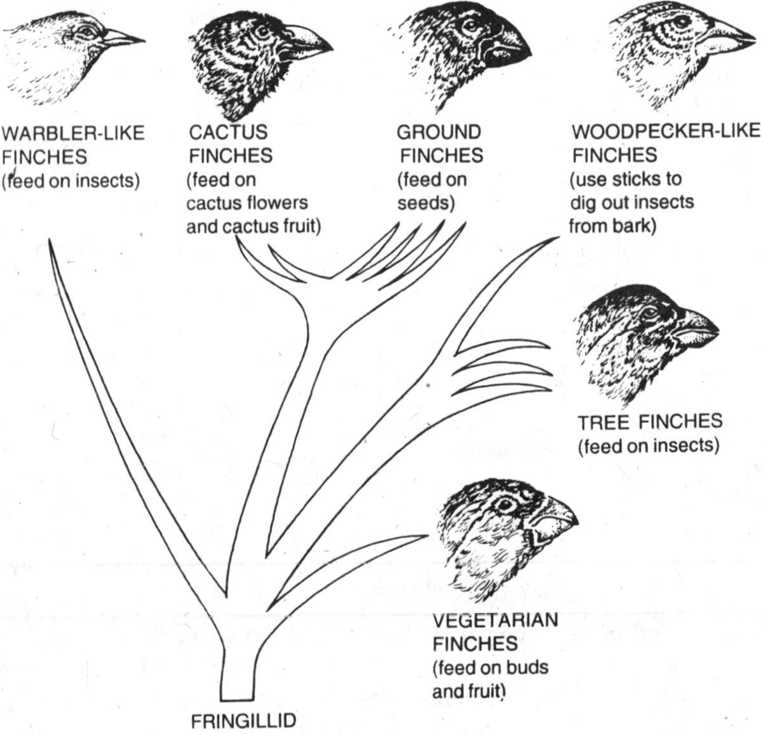

Fig. 7.5. Adaptive radiation of Darwin's finches on the Galapagos Islands. Small, isolated island with low colonization rates serve as useful sites of study evolution and adaptation. It is likely that a single species of finch gave rise to all these types (after Drickamer *et al.*, 2002).

and adaptive radiation. The ancestral form was probably a heavy-beaked ground finch that fed on seeds, from which evolved : modern ground finches; cactus-feeding finches with long, decurved bills, insectivorous forms, including the famous woodpecker finch, which uses a tool in the form of stick in its beak to probe underneath bark for insects (Fig. 7.6) and slender – beaked barbler-like finches that feed on small insects.

Fig. 7.6. The wood pecker-finch. It is shown to carry a stick or cactus spine to dislodge insects from bark crevices.

3. Phylogeny and Comparative Approach

Ethologists have long used the technique of comparing different species in order to understand the evolution of trait.

***(i)* A classic comparative study : Balloon flies.** According to **Kessel** (1955), in 1875, **Baron Osten-Sacken** was visiting the Swiss Alps, and he noticed groups of bright silvery flashes in the shadows of fir forest. Believing that they were silvery insects, he captured some in a net but found instead that they were dull-coloured flies in the family Empididae, or empidids. However, along with the flies, he had netted packets of filmy material (Fig. 7.7). At first, the function of these balloons was unclear : various people hypothesized that they acted as surfboards or perhaps as warning signals

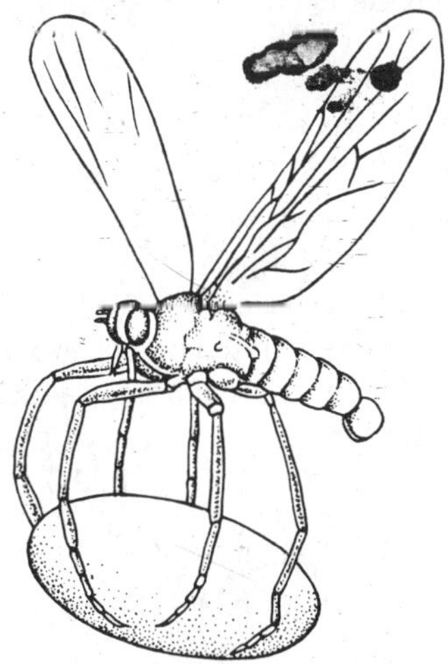

Fig. 7.7. Male balloon fly (*Hilara sartor*) carrying balloon. Males with balloons fly in a swarm from which a female selects a mate. That mate gives the balloon to the female prior to copulation (after Drickamer *et al.* 2002).

to predators. Instead, it is suspected that males use the balloons to attract mates. The balloons apparently are not valuable to the females in any way : they contain no nutrients. How could something like this have evolved ?

According to **Drickamer** *et al.*, (2002), one can trace out the possible evolutionary routes that a behaviour has taken by examining other species that have similar behaviour. For example, **Kessel** (1955) described possible evolutionary stages in the origination of this behaviour.

Stage 1. In the majority of empidid species, both sexes capture insects independently, and no presentation of prey is associated with mating. These flies sometimes prey on conspecifics; when the male attempts to copulate, he may be eaten by the female.

Stage 2. The male captures a prey item and presents it to the female as a nuptial gift. He copulates with her as she eats the prey instead of him. Several empidid species and many other species use nuptial gifts in mating. In an unrelated insect, the scorpionfly *Hylobittacus apicalis* (order Mecoptera), a male with a prey item advertises to female by means of a pheromone, a chemical signal. But, sometimes a male assumes the behavioural posture of a female; he approaches a male that has a prey item, lets the male try to copulate then steals the prey item and either eats it or uses it to attract a female (**Thornhill,** 1979).

Stage 3. Instead of taking the prey and searching for a female, some male empipids join other males, each with a prey item, in an aerial dance. The prey, **Kessel** proposed, is now a stimulus for mating rather than a distraction to avoid **mate canibilism.** The female enters the swarm and selects one of the males. Several empidid species represent this stage.

Stage 4. In many species of the empidid genus *Hilara*, the male wraps the prey loosely with some silken threads, an action that seems to quite the prey.

Stage 5. In several species of the genus *Empis*, from the western united states, the male applies elaborate silken wrappings to the prey, which then resembles a balloon. When male and female meet in midair, the male transfers the balloon to the female and climbs on her back. The pair alights on a plant, and the female rolls the balloon, probes it, and eventually consumes the prey item while the male copulates with her. The balloon itself stimulates mating.

Stage 6. The male catches a small prey and may consume its fluid so it is no longer edible, he then constructs a complex balloon. The female accepts the balloon and plays with it during copulation but gets no meal from it. *Empis aerobatica* may represent this stage.

Stage 7. The prey item is very small, of no food value to either sex, and pieces of it are plastered at the front end of the balloon. The balloon is now the sole stimulus for copulation; *Epimorpha geneatis* illustrates this stage.

Stage 8. The *Hilara sartor* male gives the female a balloon that has no prey at all. **Kessel** hinted that this behaviour was the final stage in the series.

Thus, arranging existing species with different behaviours along a continuum (called a **transformation series**) is a very simple way of doing a comparative study. It offers some understanding as to how a complex trait may have evolved in stages. However, there are problems with this method. For example, we do not know whether the balloon species went through every stage during the course of its evolution. In addition, we may not have the order of the stages right. As discussed at the start of this chapter, *traits do not always evolve from simple to complex*. For example, it may be that the balloons evolved first to make the male appear larger and more attractive, and later evolved into nutritional gifts (see **Drickamer** *et al.*, 2002).

From just described transformation series, ethologists have drawn the following inferences. For example, for stage 2 species, **Kessel** suggested that the nuptial feeding of the female initially functioned to reduce the chances of the male's being consumed by the female during mating. **Thornhill** (1976) suggested an alternative function based on *parental investment*. Male insects may invest in their offspring by providing nutrition to their mates in the form of glandular secretions, prey captured by the male, regurgitating food offering or even the male itself. Females may select males on the basis of the quality of food the males offer. In other example, recent work has shown that a balloon-bearing species, *Empis snoddyi* (stage 8) faces conflicting selection pressure. The males that are most likely to be successful at attracting mates are large males with intermediate-size balloons. Intermediate-size balloons represent a trade-off between long-range attraction of females with larger balloon, and improved flying efficiency with smaller balloons (**Sadowski** *et al.*, 1999).

(*ii*) Use of Phylogeny in Studying Evolution of Behaviour

Phylogeny means the history of descent of a group of taxa. A phylogenetic approach is helpful with several of the problems that confront students of behavioural evolution :

1. It can help us ascertain the order in which traits evolved.
2. It can help us to distinguish **homology** (similarities due to common ancestry) from **analogy** (similarities due to convergent evolution).

(*a*) Methods of reconstruction of phylogeny of behaviour. It has long been recognised that closely related species are more likely to share traits than are distantly related species. **Darwin** introduced the metaphor of **tree** to describe relationships among taxa. The **trunk,** or the earliest ancestor, grows into separate **limbs**, limbs give rise to **branches**, which in turn give rise to **twigs**. Some branches have many twigs or descendents, while others have few. Some branches fall off, representing extension. The tips of the twigs represent species that are alive today. Twigs on the same branch should share characteristics because of common descent. Today, many scientists prefer the term **bush** instead of tree; however, both terms are commonly used (see **Drickamer** *et al.*, 2002). The most common problem faces the modern-day systematics is to reconstruct the shape of the tree or bush from present-day species and fossils.

The primary step in phylogenetic reconstruction is to identify particular **characters** of interest. A character is an inherited trait. The characters of all the specimens of interest are examined and **their character states** are described. For example, in spiders, the plane in which the jaws are moved is a character. There are two possible character states : jaws can either move from side to side or front to back. Morphological traits were historically the most often used characters, but in recent decades, new methods have led to increased use of **genetic characters**. For example, a distinct spot on a gene is a character, and the particular nucleotide in that spot in the character state. The assumption is that character states of more closely related species are likely to be similar, because they have been inherited from a more recent common ancestor.

After characters are recorded, one can reconstruct a phylogeny. One well known method is **cladistics**, which was developed in the 1950s by **Willi Henning**. This method depends on identifying **monophyletic groups**, or groups of taxa that share a more recent ancestor with each other than they do other groups. **Polyphyletic groups**, in contrast, are taxa that arose from several different ancestors.

(*b*) Using behaviour to reconstruct phylogeny. Ethologists have been interested in knowing whether behavioural patterns can be used like morphological or genetic characters in building a phylogeny. It is not clear whether behavioural traits are useful for this purpose : as we have already seen, behaviour is often under a great deal of environmental influence, and might be too variable to shed light on phylogenetic relationships. The ethologists were of the opinion that behaviour could be useful. During early 1990s, one of the fundamental basis of their research, was the idea that behavioural patterns could be treated like the anatomical features. They "dissected" behavioural patterns to learn their "anatomy" (see **Drickamer** *et al.*, 2002). These ethologists also demonstrated homologies in behaviour, which are patterns shared by species through descent from a common ancestor. For example, **Van Tets** (1965) studied birds of the order Pelecaniformes, scoring species for the presence or absence of behavioural traits, such as head wagging (*wag* means to move rapidly to and fro), and comparing the pattern with the phylogeny as inferred from morphology (Fig. 7.8). He concluded that behavioural data could

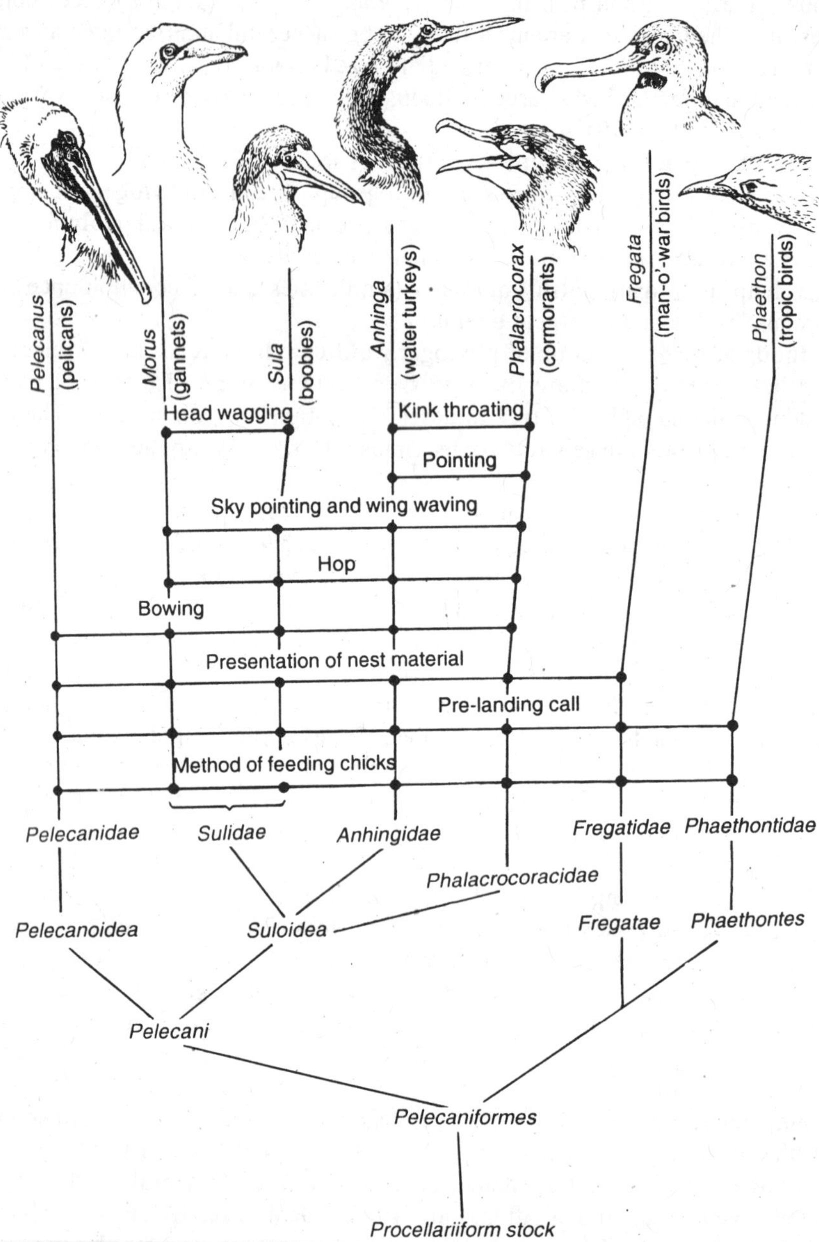

Fig. 7.8. Behavioral characters or traits and phylogeny in the bird order Pelecaniformes. The tophalf of this figure shows the distribution of behaviours among genera. Dots indicate that the behaviour is present in that genus. In the bottom half of the figure is a phylogenetic tree based on morphological traits. Note that genera that are closely related based on morphology also share the most behaviours, where as those distantly related share the fewest (after Drickamer *et al*., 2002).

be used to provide accurate assessments of phylogeny. In recent years, his data were further analyzed by **Kennedy** *et al*., (1996) who constructed trees from the behavioural data and compared them to phylogenetic trees based on morphology and genetics, and found that all produced similar results. Similarly, **Paterson** *et al.*, (1995) used 72 behavioural and life history characters to construct phylogenetic tree, for 18 species of albatrosses, petrels, and penguins, and found that these trees were similar to those based on molecular data.

C. Use of Phylogenies in Testing of Evolutionary Hypotheses about Behaviour

Phylogenetic trees can also be used to test hypotheses about the evolution of behavioural traits. With this approach, a phylogenetic tree is constructed, generally using morphological and/or genetic traits. Then, behavioural characters are mapped on to the tree. (It is important not to construct the tree using the behavioural characters under test, because that would be circular reasoning). By examining such a phylogenetic tree, it is often possible to determine the direction of evolution of a behavioural trait. It is also possible to identify **selection pressures** that are associated with the evolution of a particular trait. In order to do this accurately, we need to know how many times particular traits have been evolved. This is illustrated in figure. 7.9.

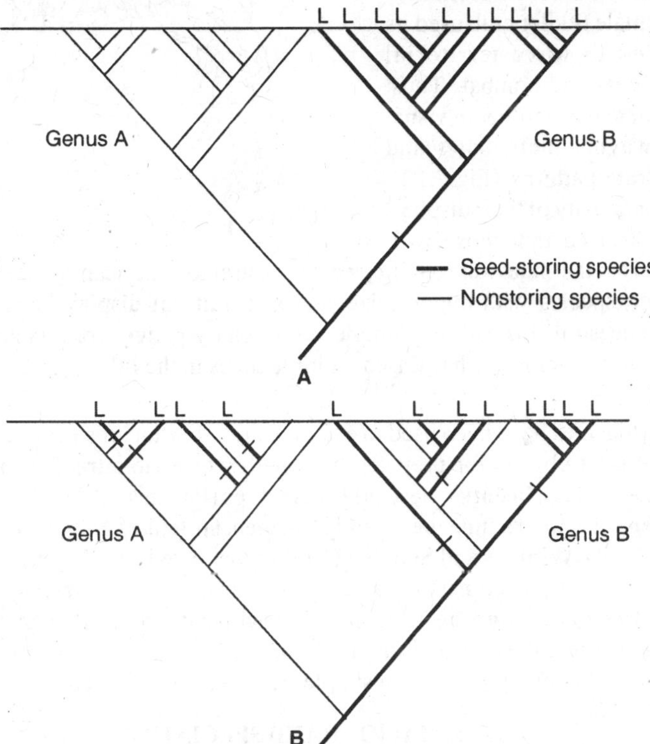

Fig. 7.9. Imagine that one researcher is interested in knowing the relationship between food storing and spatial learning. He hypothesizes that animals that depend on finding seeds that they store will perform better in laboratory tests of spatial learning. If he examines 10 seed-storing species and find that they all are good learners, and 10 nonstorers, and find out they are poor learners, one may conclude that this hypothesis is correct. **Rosenheim** (1993) believed that perhaps these are not independent sample : If all the seed-storers are closely related, and all the non-storers are closely related, there is likely to have been only one evolutionary event (A), and your sample size is one, not 20 (L indicates species able to learn ; crossbars indicate evolutionary events). A different phylogeny would generate another answer ; in (B), there are eight evolutionary events, and that would be researcher's sample size (after Drickamer *et al.*, 2002).

The tiger moth (Family Arctiidae) provides an example of the use of phylogeny in understanding the evolution of a behavioural trait. Some species of this moth use high-frequency sound in their mating displays. High-frequency sound is also used as a defensive display against bats and other predators with high-frequency hearing : these moths are distasteful to predators and the sound detect them. A phylogenetic analysis shows that the use of sound in defense against predators came first, and sound subsequently took on a courtship function in some species (**Weller** *et al.*, 1999).

Fiddler crabs (*Uca* sp.) provide an example where ethologists' initial perception for the direction of evolution was apparently incorrect (see **Drickamer** *et al.*, 2002). The male fiddler crab used its enlarged cheliped to wave territorial displays and in threat and combat. Thus, fiddler crabs signal to one another in visual displays, often waving their claws and dancing in elaborate patterns (Fig.7.10). Species vary in complexity of their displays and it was hypothesized that those with simpler displays, from the Indo-west pacific, were ancestral to American species with more complex displays. However, species with hypothesize "derived trait" of display behaviours appear to be phylogenetically ancestral. Instead of a simple evolutionary pattern from simple to complex, it is likely that behavioural complexity has arisen multiple times in the fiddler crabs (**Sturmbauer** *et al.*, 1996).

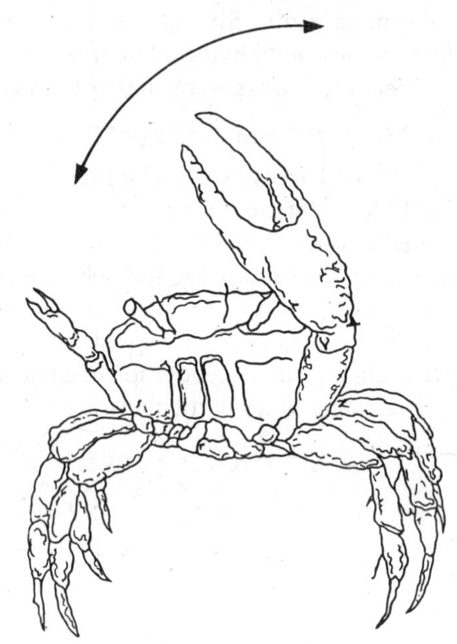

Fig. 7.10. A fiddler crab display.

Further, phylogenies have been used to examine the evolution of structures that animals make. Many species construct objects for prey capture, shelter, or mate attraction, for example, **spider webs, caddisfly cases, nets, beehive, nests of** paper wasps (Polistes) and of bee *Melipona* and birds' nests. It is relatively easy to examine these semipermanent structures and to compare closely related species (**Hansell,** 1984). **Winkler** and **Sheldon** (1993) constructed a phylogeny with DNA data for 17 shallow species, and then placed nest characters on it. The primitive nesting mode appears to be **burrowing**. Construction of **mud nests** originated once, and mud-nesting species have primarily diversified in Africa, where the climate is dry and mud is a feasible construction material. Mud nests seem to have increased in complexity from simple cups to fully enclosed nests.

7.5. BEHAVIOUR AND SPECIATION

When an ancestral species gives rise to new species (**speciation**), the process usually starts with the interruption of gene flow by some type of physical barrier. This is termed as **allopatric speciation** (**Mayr,** 1963). If the populations diverge from one another while they are separated, they may not be able to interbreed again if they are reunited, they have become separate species.

Barriers to interbreeding between groups of organisms are called **isolating mechanisms**, and are divided into two types : **1. Prezygotic isolating mechanisms** are those that occur before the formation of zygotes. **2. Postzygotic isolating mechanisms** are those which occur after the fertilization, and are generally physiological in nature such as the zygote dies or the hybrid has reduced viability or fertility.

Here, prezygotic mechanisms are more interesting, because many of these are behavioural. For example, members of different species may be active at different times of the day or the year or may use different parts of the habitat, and would thus be unlikely to come into contact with one another.

In another type of prezygotic isolating mechanism, animals may contact each other but not mate, either one partner fails to court, or the other partner is not receptive. Which signals or cues do animals use in species recognition ? In many cases, morphological traits transmit information. This is demonstrated by a recent study of cichlid fishes in Lake Victoria (Second largest freshwater body in the world after Lake Superior in North America; it is located in East Central Africa and its southern half is in Tanzania and northern half is in Uganda) (**Seehausen** *et al.*, 1997). Different species live sympatrically in the lake, but are sexually isolated by mate choice. In this fish species, mate choice is determined on the basis of colouration. In recent years, human activity has led to **eutrophication** of the lake, and it is now much more turbid. The fish cannot see colours as well, and now mate choice is relaxed, fish are more likely to mate with members of others species (see **Drickamer** *et al.*, 2002).

Indeed behavioural cues are very important in courtship and species recognition. For example, there are several species of green lacewings (insects with long, delicate wings) that look exactly alike to human eyes. **Wells** and **Henry** (1992) has reported that males and females of green lacewings sing to each other in a duet composed of low-frequency sound. Females respond most strongly to the songs to their own species.

Closely related species that are **sympatric** (exist in same location), are more likely to have divergent courtship signals than those that occurs in separate

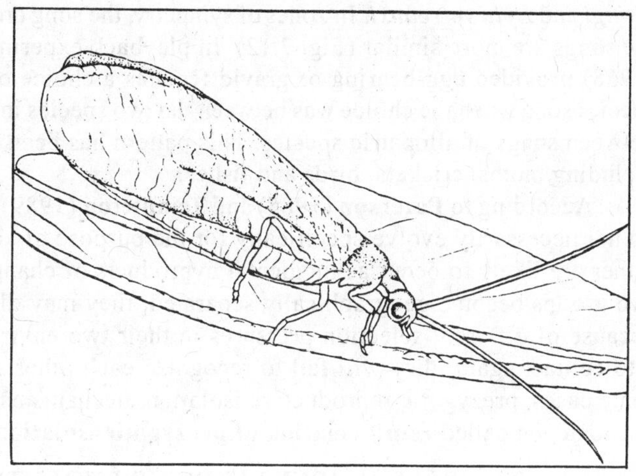

Fig. 7.11. A green lacewing.

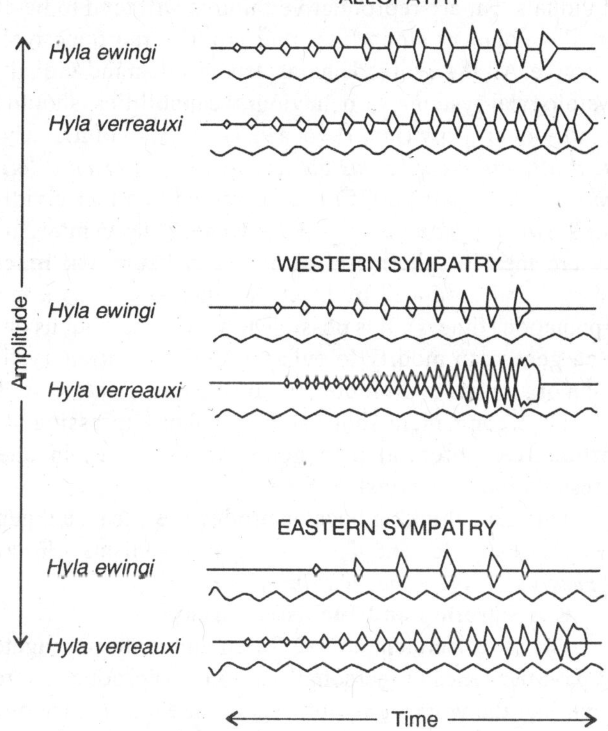

Fig. 7.12. Oscillograms of songs of two Australian frog species. The amplitude of each note of the song is shown on the vertical axis, and time is shown on the horizontal axis. Under each song recording is a 50-cycles-per second reference line. The songs of the two species are similar when comparing individuals caught in areas where the two species do not overlap (allopatry). Where the two species do overlap (sympatry), song differences are noticeable (after Drickamer *et al.*, 2002).

locations. **Littlejohn** (1965) recorded the songs of two species of Australian tree frogs – *Hyla ewingi* and *Hyla verreauxi.* In zones of sympatry, the song are quite distinct; in zones of allopatry, the songs are more similar (Fig. 7.12). In playback experiments, **Litttlejohn** and **Loftus-Hills** (1968) provided egg-bearing or gravid females a choice of two songs. They chose their own species song when the choice was between the two species in sympatry, but showed no preference between songs of allopatric species. This pattern has been seen in numerous other animal taxa including moths, crickets, birds and fishes.

According to **Paterson** (1985) and **Templeton** (1989), reproductive isolating mechanisms do not necessarily evolve specifically for the purpose of keeping groups of animals apart, but rather are likely to occur as incidental byproducts of change between groups. For example, if two groups become geographically separated, they may diverge in their courtship behaviours because of different selection pressures in their two environments. Then, if the groups come into contact again, they will fail to recognize each other as the potential mates. However, in some cases, prezygotic reproductive isolation mechanisms may be under direct selection, in a phenomenon called **reinformation of prezygotic isolation.**

7.6. SELFISH GENES AND THE EVOLUTION OF BEHAVIOUR

The logic of natural selection needs that over time animals will evolve behavioural daptations that promote the survival of genes that underlie advantageous traits. Genes in individuals that are reproductive failures will tend to be eliminated; genes with the capacity to steer development along lines that produce reproductively successful individuals should tend to survive. All the evolved characteristics of an individual, be their developmental mechanisms, physiological systems, or behavioural capabilities, should be related to survival of genes within the individual (**Dawkins,** 1976, 1989). Many evolutionary biologists have concluded that *life is essentially an unconscious contest among different alleles to survive and replace alternative forms.* **E.O. Wilson** (1975) has expressed this fact vividly with his aphorism "*an organism is DNA's way of making more DNA".* Genes (DNA) may survive indefinitely. Individuals do not; they are merely ephermeral throw away **"survival machines"** (to use Dawkin's metaphor) designed and produced by a set of genes simply as a means of enhancing gene survival and perpetuation (fitness). It is possible to view behaviour as one design feature of 'survival machine' which genes can modify to enhance their evolutionary fitness. In this sense, genes appear to design organisms to promote their own 'selfish' interests. The concept of selfish gene, however, is merely a convenient shorthand metaphor expressing an evolutionary truism – the enhanced reproductive potential of a gene (or an allele) in one generation leads to its enhanced representation in the next.

This gene thinking became productive because it generated a basic question that could be asked about any element of the biology of organisms – *How does a trait help individuals maximize the survival of the genes within them ?*

Kin Selection and Inclusive Fitness

The most frequent course for an animal to propagate its genes is to make copies of them and pass the genes on personally, through reproduction, to its offspring. When an animal does something, the working hypothesis of an evolutionary biologist is that the action in some way helps to maximize the individual's reproductive success, or **personal fitness**, as measured by the number of surviving reproducing (fertile) offspring that carry its genes. Thus, in a way, animal behaviour too can be interpreted as an '*evolved adaptation'* that leads individuals to try to produce more surviving progeny than other members of their species.

In contrast, there are many examples in which an animal tends to behave in ways that clearly reduce its own reproductive success but instead elevate the personal fitness of another individual. For these special cases, the evolutionary approach predicts that the behaviour must still benefit the genes of the **helpful individual. Self-sacrificing** or **altruistic behaviour** can lead to gene survival if it is directed towards a relative that share genes in common with the helper. The closer the degree of relatedness, the more likely the two individuals are to share identical genes, and the greater the potential genetic benefit of helpful acts to the aid-giving animal (**Hamilton,** 1963, 64). Selection for activities that lower an individual's personal reproduction but raise a relative's fitness is called **kin selection.** The point is that elevating the fitness of relative is a way to propagate one's genes, transmitting them through a relative's offspring rather than through one's own progeny. This may sometimes be a more effective way to promote the survival of one's genes than by "selfish" attempt to reproduce personally (**Alcock,** 1979).

In this way, an individual's genetic success is measured by the sum total of genes passed on to the next generation and there are two ways to transmit genes : 1. directly through the individual's progeny and 2. indirectly through those offspring that relatives were able to produce, because they were helped by the individual (**Eberhard,** 1975). This total is the measure of an individuals **inclusive fitness.** Animals with traits that give them higher inclusive fitness will spread their genes through a population more effectively than animals whose traits are associated with lower inclusive fitness. Following examples will prove this point :

1. The castes of sterile females among the social hymenopterans are found to sacrifice their own reproductive potential and devote their time and energy to raise the progeny of the colony queen.

2. In some bird species, such as Florida scrub jay (*Aphelocoma coerulescens*), adult birds help parental pairs raise offspring (by contributing food and defending the nest from predators) instead of setting up breeding territories of their own.

Similar helpers are found in jackals (*Canis mesomelas*) and red foxes (*Vulpes vulpes*). Various species of mammalian carnivora hunt cooperatively. In hunting dog (*Lycaon pictus*), the spoils of the hunt are shared with pups and guard adults which did not participate in the hunt. Lions (*Panthera leo*) hunt cooperatively, they cooperate in driving intruders out of the pride and the young are suckled communally. In some cases, as in the guard bees of honey bee hives, individuals may even sacrifice their lives defending the interests of others.

Group Selection

Most biologists tend to ask the following evolutionary question : How do animal's traits contribute to the survival of the species ? This question means that species or populations are competing to survive and that traits will evolve that prevent the extinction of a group. In his classical book, *Adaptation and Natural Selection* **Williams** (1966) pointed out that this question is not logically derived from the theory of natural selection. In principle, natural selection could act at any of a variety of levels (*e.g.,* at the level of individual, the population and the species). The concept of **group selection** was proposed by **Wynn-Edwards** (1962) and it implies that individuals will sacrifice their reproductive output (*i.e.,* fitness) in order to maintain the group as a whole. The genetic self-sacrifice required by group selection cannot possibly be the result of natural selection, which, by definition, does not favour traits that reduce individual inclusive fitness and thereby lower the survival of gene associated with the trait. If within a population, there are two competing alleles, one of which appears in individuals that make as many copies of their genes as possible and one of which appears in individuals that sacrifice themselves (and therefore the genes within them) for the benefit of population, which allele tend to disappears ? Natural selection favours genes that are selfish, that survive and reproduce themselves at the expenses of the other alleles, even if this threaten the long-term welfare of the species (**Dawkins** 1989). Natural selection is not a thing that can plan ahead for the survival of groups (as is indicated by the fact that an estimated 98 per cent of all the species existed are now extinct, see **Alcock,** 1979). Natural selection is a process, the results of individual differences in inclusive fitness. Because individuals differ in their capacity to propagate their genes, some genes are selected for (survive) and other genes selected against (become extinct).

There are other reasons besides the high extinction rate of species for believing that *adaptations have the functions of boosting individual fitness for the present* and do not act to prevent the extinction of species over the long haul. There are many examples, where animals behave in ways that are simply contradictory with the notion that their actions help their group or species survive.

1. Starving infant seals that have become separated from their mothers in the chaos of a rookery make repeated attempts to approach other adult females to nurse. Despite the desperation of the orphans, females always rebuff them and sometimes attack them brutally. As a result, orphan infants of seal die on a beach where hundreds of lactating females are feeding their single pups (**Williams,** 1966). This simply indicates towards the "selfish" behaviour of the seals as a "selfish" adaptation that increases the reproductive success of individual females. A seal that supported an infant other than her own with her milk and time would be promoting some other seal's genes at her genetic expense. Her genes and her generous behaviour, even though they might help the species as a whole, would be replaced in the long run by the genes and behaviour of genetically selfish seals.

2. Another example involves the behaviour of groove-billed anis (*Crotophaga sulcirostris*), a peculiar looking, communally-nesting black bird with a long floppy tail (**Vehrencamp,** 1977). It is an unusual species in that two or three females will lay their eggs in the same nest, but these females compete to lay the largest percentage of the eggs in the **communal clutch.** A female anis will often throw eggs laid by other females out of the nest before laying her first egg in the communal nest. Thus, the behaviour of female anis is completely consistent with the notion that traits evolve because individuals differ in gene-copying success. It is much harder to give a group selection interpretation to the egg-killing behaviour of anis and other birds (*e.g.*, long-billed marsh wren, *Telmatodytes palustris;* **Picman,** 1977).

3. **Sarah Hrdy** (1977) reported a strange case of **infanticide** in old world monkeys, the Hanuman langurs of Abu, India. In langurs, bands often consist of one reproductively active male and a harem of adult females and their offspring. Adult males fight violently among themselves for control of a harem. A male that acquires a harem, after ousting the old male, systematically harasses and when possible bites the infant members of the troop despite the efforts of the mothers to protect their progeny. Because males are much larger than females, they regularly succeed in killing the female's babies.

Hrdy (1977) has pointed out that the behaviour of new harem owner is biologically sensible if male behaviour has evolved to maximize the survival of a male's genes. With the death of each infant sired by the previous harem master, the new male directly removes some of his competitor's genes. Moreover, after the death of her infant, it is to a female's advantage to come into estrus promptly, at which time she is immediately fertilized by the new male, who transmits his genes to the resulting offspring (if the females were to care for the child of the old harem owner, she would not ovulate for 2 to 3 years). **Hrdy's hypothesis** is that male langurs practice infanticide in a way that tends to benefit the infanticidal male's genes (see **Alcock,** 1979).

Evolutionary Stable Strategies (ESS)

The use of cost-benefit arguments in modelling evolution has led to another important theoretical development. For each evolutionary problem facing an organism, there will usually be a range of options or **strategies** (*i.e.*, in biology, a strategy is simply one of a series of alternative courses of actions ; **Parker,** 1984) available for solving it. In evolutionary terms, we can imagine mutant alleles arising which code for each strategy. Selection will then favour the allele coding for the strategy which yields the highest net benefit. What will be the net result of selection between these pre-programmed strategies ? The answer is neatly summed up in a term coined by **Smith** (1974) – an **evolutionarily stable strategy (ESS)** which emerged during his application of the mathematical theory of games to the study of evolution.

An ESS is defined as a strategy which, if adopted by most members of a population, cannot be bettered by an alternative strategy. In other words, the best strategy for an individual to adopt depends on which most of the population adopt. Since each individual will be selected to maximise its own reproductive potential (and hence the inclusive fitness of its genes), the strategy that ultimately persist in the population (*i.e.,* is evolutionary stable) is one that cannot be bettered. If a better alternative evolves or environmental pressures change to favour a different strategy, a new strategy will become the ESS. Once evolved, the ESS is maintained by selection acting against any deviation from it. **Smith** used mathematical models to consider the adoption of various strategies by individuals and implications of such strategies for the population as a whole. The notion of ESS can be applied to questions such as those of sex ratio, pattern of aggressive behaviour and pattern of social organization. **Dawkins** (1976, 1989) has presented the ESS in non-mathematical terms and expanded it in several directions.

Types of ESS. There are two basic types of ESS :

1. Pure ESS;

2. Mixed ESS or Conditional ESS. A conditional ESS is really a set of solutions to a set of conditions. In biological terms, conditional ESS is of following two types :

***(i)* Environment determined strategies.** Individuals play strategies in relation to external cues.

***(ii)* Phenotype-limited strategies.** Individuals play strategies in relation to their own phenotype.

Biological Games and ESS

In game theory, a distinction is often made between two person games and *n*-person games. In biology this distinction has also been found useful by **Parker** (1984). Games are of following types :

1. Contests. Individuals meet in pair, so that the game concerns the interaction between pairs of strategies. **Smith** (1982) has termed these **'pairwise interactions'.** Obvious example of contests are fights between pair of individuals over resources such as territories, females, nest sites, or food items.

2. Scrambles. Most competitive games do not involve pairwise interactions, they involve *n* players and *n* may sometimes be the number of individual in the population. **Smith** (1982) called these sort of interactions, **'playing the field'**. Scrambles are of two types :

1. One-option scramble. A one-option scramble is an *n*-player game in which a given individual can gain more of some fitness-related commodity by increasing its expenditure beyond that of its competitors. It is a form of biological treadmill that is somewhat equivalent to advertising in economics. Models of one-option scrambles have been constructed for **parent offspring conflict, territorialtiy, sexual advertisement** and **arms races.**

2. Alternative-option scrambles. These are also *n*-person games, but instead of there being basically just one way to obtain benefits, there are two or more alternative options. These scrambles are exemplified by **ideal free searching** of habitat, mate or food among competitors. Evidence for ideal free searching (or distribution) have been found both in **mate searching** of dung flies and toads and in fish and ducks foraging for food. Other examples of alternative-option scrambles include the typical **producer-scrounger systems** such as **food stealing, food hoarding, digging** and **entering** in *Sphex* wasps, and alternative **mating strategies** (sneaks/guarders; callers/satellites).

At this stage, it would be quite appropriate, if some important types of biological games may be assessed in terms of ESS :

1. Animal contests. Animal contests probably offer the best demonstration of the need for ESS approach. Animals fight each other to gain some fitness-related commodity. Thus, if two individuals meet by an item of food or a territory, either would gain in terms of fitness if they were able to repel the other individual and exploit the resource alone.

What should an animal do to win ? This clearly depends on how his opponent is likely to respond. If the response is likely to be fight to the death, it will probably pay to withdraw and search for an alternative, unoccupied resource. In contrast, if an opponent is likely to flee after receiving a few threats, it will clearly pay to offer some demonstration of aggressiveness.

There have been essentially three types of approach to the modelling of animal contests, depending on the sort of decision rule for winning.

1. War of attrition models (Smith-Price). Victory is decided by persistence – the winner is the individual that chooses to continue longer than his opponent. The choice of persistence time is made before the contest. (*Note.* Attrition means abrasion, friction, harassment).

2. Hawks-doves models (Smith-Parker). If there is intensified fighting, the winner is the individual who injures his opponents.

3. Information acquired during contests (Smith-Parker). The theory of animal disputes began with the assumption that contests are **symmetric,** *i.e.,* opponents are equal. Later on, there arose a growing need to include inequalities between opponents in the models and the theory of **asymmetric** contests then began (see **Krebs** and **Davies,** 1984).

(Consult Chapter 8 for Game Theory Model)

QUESTIONS

Long Answer Questions

1. Give an account of microevolutionary changes in the behaviour of animals.
2. Describe macroevolution of behaviour.
3. Write down about the eight evolutionary stages of balloon-gifting behaviour of ballon-flies as suggested by **Kessel** (1955).
4. Give an account of methods of reconstructions of phylogeny of behaviour.
5. Describe evolutionary stable strategies (ESS).

Short Answer Questions

6. Write **short notes** on the following :
   (*i*) Animal contest
   (*ii*) Group selection
   (*iii*) Kin selection
   (*iv*) Inclusive fitness and
   (*v*) Diversity of dogs.

Multiple Choice Questions

Choose the correct answer from the four alternatives given

1. The phenomenon of industrial melanism demostrates
   (*a*) natural selection (*b*) reproductive isolation
   (*c*) geographic isolation (*d*) induced mutation
2. Darwin's finches is an example of
   (*a*) adaptive radiation (*b*) geographic isolaton
   (*c*) intraspecific competition (*d*) all
3. Which of the following is a postzygotic mechanism to prevent hybridization?
   (*a*) temporal isolation (*b*) behavioural isolation
   (*c*) mechanical isolation (*d*) hybrid sterility

Answers

1. (*a*); **2.** (*a*); **3.** (*d*).

8

Individual Behaviour (Conflict and Aggression)

There are some behavioural patterns which an animal displays for acquiring food, water, oxygen, energy, homeostasis, reproduction, social status and harmony within its environment. Unlike the stereotyped behaviours, these behaviour patterns are characteristics of the individual. Some of them which are most common to an individual in its life cycle be listed as follows :

1. Exploration;
2. Conflict;
3. Aggression;
4. Fear;
5. Communication;
6. Avoidance;
7. Play;
8. Locomotion;
9. Feeding and drinking;
10. Sleep;
11. Social behaviour;
12. Biological rhythms.

Of these, only conflict and aggression will be considered in this chapter. Some others such as communication, social behaviour and biological rhythms have been dealt separately elsewhere. **Drickamer** *et al.,* (2002) have considered conflict and aggression as same type of behaviour. However, they have differentiated between *aggression* and *agonistic behaviour*.

8.1. CONFLICT

Conflict refers to the conflict between behavioural tendencies within an animal (**Kumar,** 1996). There are two ways in which we come across a conflict situation : 1. Condition when two tendencies, which are mutually incompatible, arouse together and compete one another for dominance in regulation of a behaviour. For example, a hungry animal wants to approach a food source but avoids to do so, because of some fear. Thus, animal is in conflict between approach and avoidance of the food. 2. In other condition, only one tendency is aroused in animal to achieve a goal but the goal itself remain inaccessible, and animal thus be thwarted. For example, a hungry animal wants to reach a food source but finds inaccessible as its runway is blocked by a glass plate.

Conflicting Situations

In animals, following three situations of conflict arise :

1. Avoidançe-avoidance conflict. It occurs when two behavioural tendencies for avoidance are aroused simultaneously. It is unstable and movement of the individual to either direction from equilibrium point will increase avoidance to the object coming close to the animal.

2. Approach-approach conflict. It occurs when two behavioural tendencies aroused simultaneously are directed towards the different goals. This is transitory and unstable. A slight swift to either direction from equilibrium point will enhance approach in opposite direction.

3. Approach-avoidance conflict. It occurs when two behavioural tendencies aroused together to approach and to avoid the same object. It is stable, as animal always moves towards the equilibrium point.

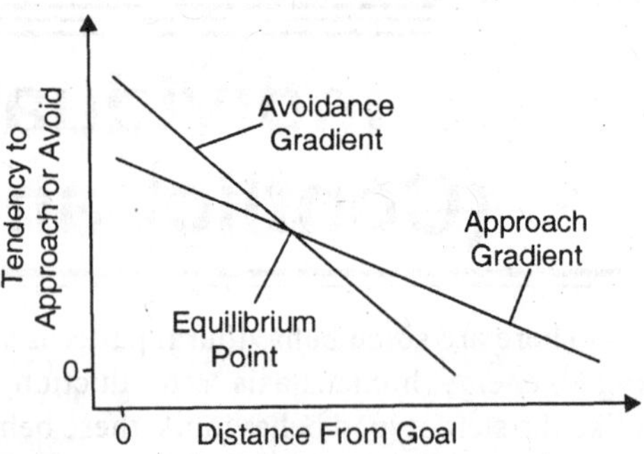

Fig. 8.1. A curve of approach–avoidance conflict.

Behavioural Display as an Evidence for Conflict

There are four lines of evidence in which a behavioural display is interpreted as conflict (**Tinbergen,** 1962; **Hinde,** 1966).

1. The situation. The situation of behavioural display is important for its consideration as a component of the conflict. For example, and animal near its territorial boundary always remain frightend and aggressive. But, then the presence of potential mate may arouse its sexual motivational system.

2. The behaviour accompanying a display. The behaviour accompanying a display is important, as it may or may not be due to conflicting tendencies. For example, an animal may express approach and retreat from its rival. It may also show colour changes as a result of the sexual motivation.

3. The behaviour preceding display. The behaviour preceding a display is assessed to interpret display as a conflict behaviour (**Mogniham,** 1955).

4. The nature of display. The posture of the animal during a behavioural display is one of the important point that distinguishes a behaviour as a conflict behaviour (**McFarland** and **Bahar,** 1968).

Types of Conflict Behaviour

The two behavioural movements conflicting each other cannot be expressed simultaneously. The behaviour observed in an animal in conflict is thus very different from its normal behaviour. Following types of conflicitng behaviours can be seen in animals under conflicting situations.

1. Displacement Activities

In conflicting situations, new behavioural pattern emerges which is apparently unrelated to the normal preceding types. It is irrelevant and defined as a **displacement activity (Kortlandt,** 1940; **Tinbergen,** 1951). For example, a male three-spined stickleback fish while courting an unreceptive female suddenly swims to its nest and perform fanning movements. It is evident even when there are not eggs in its nest to care. Male still comes to nest in between his courtship behaviour. This is regarded as displacement activity as it remains unrelated to the preceding normal (courtship) behaviour.

In another case, it has been observed that chickens preen more they had woken up or are able to go to sleep in the roost (**Duncan** and **Wood-Gush,** 1972).

Fighting gulls may show "grass-pulling" in the middle of a fight (Fig. 8.2) (**Manning** and **Dawkins,** 1998).

Characteristics of Displacement Activities

1. It is apparently an irrelevent behaviour.
2. It emerges in intermediate context of the two conflicting motivational system.

Fig. 8.2. "Grass-pulling" in the lesser black-backed gull. A male faces his rival across the boundary between their territories. Interspersed with bouts of threat or actual fighting, they seize grass or turf and tug at it flamboyantly, deliberately choosing material that will resist their efforts (after Manning and Dawkins, 1998).

3. It occurs in absence of any eliciting stimulus.
4. It remains unrelated to the preceding behaviour.

Hypotheses regarding Causation of Displacement Activities

***(i)* Overflow hypothesis. Tinbergen** (1951) assumed that two motivational systems in a conflict situation result in 'overflow' or 'sparking-over' of energy that leads to the activation of a third motivational system which is responsible for the appearance of an irrelevant (displacement) behaviour.

***(ii)* Disinhibition hypothesis. Von Iersel** and **Bol** (1958) assumed that two behavioural tendencies which are in conflict inhibit one another. While doing so, a third motivational system is activated leading to expression of an irrelevant behaviour.

***(iii)* Rowel's concept. Rowel** (1961) worked out apporoach-avoidance conflict in chaffinches. He generalized his study and proposed that the displacement activity in animals occurred when displacement activity in animals occurred when three following conditions are met together. So, according to this concept :

(*a*) The two conflicting conditions must be arrived at equilibrium.
(*b*) The equilibrium should be long lasting.
(*c*) The normal activity of the animal remains guided by the external stimuli.

***(iv)* McFarland's hypothesis. McFarland** (1965) put forward a hypothesis which is a variant of the disinhibition hypotheis. According to him, a motivational conflict was not necessary condition for the displacement activity. Working with thirty Barbary doves **McFarland** made a distinction between irrelevant activities during the conflict and thwarting. He noted that in a conflict between two behavioural tendencies, a displacement of animal's activity was brought about by disinhibition, as stated earlier. But, displacement in a thwarting situation was a switch of attention from the inaccessible goal to another object.

(v) Neurophysiological approach. Any stimulus perceived by CNS (brain) evokes two types of responses : (*i*) it activates specific loci of the brain, and (*ii*) it also stimulates activation of the reticular formation which comprises a diffuse series of fiber tracts. Via cortex these reticular formations affect the arousal in an animal (**Manning,** 1979) which results in some irrelevant or displacement activities. For example, stimulation of specific loci in the brain results in displacement activities such as preening and sleeping in gulls (**Delius,** 1967). Most probably, the displacement activities are direct consequences of changes in the physiology of animal that may be brought about by some external factor.

2. Redirected Activities

This is another type of performance of animal during conflict situations. Here, a third motivational system is not aroused as a result of two conflicting tendencies. Instead, one of the two motivational system in conflict is redirected towards an object other than which stimulated them. For example, herring gull in conflict during aggressive encounters redirect their pecking on objects in the environment, rather than on enemy.

3. Alternation Movements

Sometimes, an animal in conflict alternates its behavioural pattern between two conflicting tendencies, *e.g.*, an animal desiring to approach a food source, but frightened to do so, alternates its movements between approach and avoidance movements.

4. Intension Movement

The intension repeats a modified form of behavioural movement during the conflict period. Flying of a bird, for example, is preceded by some events indicating of its readiness to fly. The particular events in this case are crouching to bird, withdrawal of head and raising of its tail, etc. During conflict these movements are repeated as such or in a modified form.

5. Ambivalent Behaviour

Sometimes the animal in conflict maintains its posture in a way to contain both types of behavioural tendencies. For example, in ruminats (*e.g.*, cow) on offering an unusual food while position of neck and head express approach towards the food, the legs and weight of it are positioned in a way to facilitate immediate withdrawl from sources.

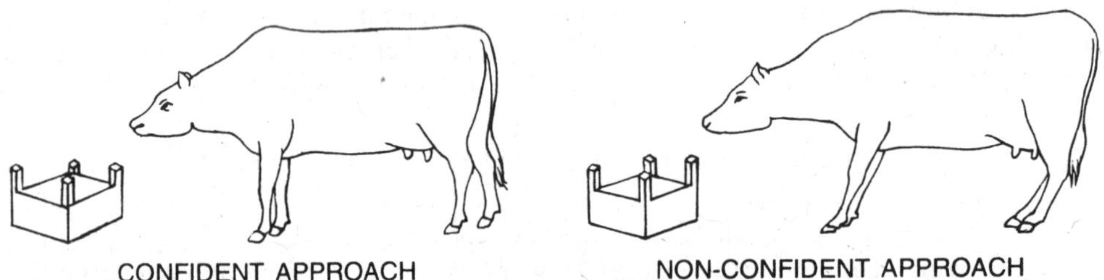

Fig. 8.3. Demonstration of ambivalent behaviour by cow.

6. Compromise Behaviour

This is similar to ambivalent behaviour. However, unlike ambivalent pattern, here only one behavioural pattern is expressed which is for both of the tendencies in conflict.

Prolonged Conflict

Prolonged conflict is the natural state for the wild animals for much of their lives. While sometimes decisions have to be made at once (for example, delaying escape from an approaching predator whould be strong by selected against), there are many other occasions when it is much better to wait and collect information before a decision is made. This is particularly true in social situations where animals may take considerable time to "assess" each other. For example, red deer stags, may take half an hour or more to assess each other's fighting ability before deciding whether to attack or flee (**Clutton–Brock** and **Albon,** 1979). The roaring matches and parallel walking are used by the stags indicators of the likely outcomes of a real fight and mean that animals remain in a state of contiuous motivational conflict while they gather the relevant information. Even escape from predators may be preceded by a period of lengthy assessment. Prey animals do not flee every time a predator appers but weight up the risk that a predator will attack against the loss of feeding time they would gain by moving away (**Ydenberg** and **Dill,** 1986). Zebras watch hyaenas carefully and are much more likely to flee on some days than others, apparently sensitive to small differences in the hyaenas behaviour and the size of the hunting group that indicates what prey the hyaenas are after on any one occasion (**Kruuk,** 1972). Such adaptations are highly adaptive but take time to make. The motivational conflict is resolved only when, eventually, the new information received tips the balance of causal factors in the direction of one behaviour or another (**Manning** and **Dawkins,** 1998).

Conflict Behaviour and Animal Husbandry

Understanding of the conflict behaviour among domestic animals is of great significance for the welfare of cattles, fowls and other domesticated animals. Thus, the study of their behaviour patterns serves as a useful guide to look into their frustration and problems. It is, therefore, necessary to get the best results out of livestock.

8.2. VACUUM ACTIVITIES

When very highly motivated animals sometimes carry out behaviour even when the appropriate stimuli are not present. **Lorenz** (1951) describes the case of a starling which went through all the movements of catching and eating an insect even though no insect was present – so the behaviour was performed 'in a vacuum'. Hence kept in wire–floored cages sometimes go through all the movements of dustbathing, even though there is nothing for them to dustbath in. **Vestergaurd** (1980) calls this "vacuum" dustbathing and suggests that when behaviour is performed in the absence of suitable stimuli, or at least with very minimal stimuli, the animal is very highly motivated to perform the **"frustrated behaviour"** (see **Manning** and **Dawkins,** 1998).

8.3. AGGRESSION

Aggression is a group of behavioural activities including **threat postures, rituals** and occasionally **physical attacks** on other organisms, other than those associated with predation. They are usually directed towards members of the same sex and species and have various functions including the displacement of other animals from an area usually a territory or a source of food, the defence of a mate or offspring or establishment of rank in a social hierarchy.

The term 'aggression' is controversial and suggested an existence of unnecessary violence within animal groups; the alternative term **"agonistic"** is preferable (Box 8.1).

Box 8.1.

Agonistic Behaviour

Behaviours usually lumped as agonistic are among the most frequently studied of all animals and human activities because of their distinctness and evident significance in the lives of social animals. **Hinde** (1970) has defined agonistic behaviour as a complex behaviour including **attack, threat, submissive** and **fleeing behaviours.** He pointed out that such a concept is useful because of the presence of simultaneous tendencies to attack and to flee and the associated assurance of threat behaviour. **Manning** (1979) used the term 'agonistic behaviour' to cover all different types of responses seen in fighting and territorial behaviour. According to **Dewsbury** (1978) agonistic behaviour encompasses a wide range of behavioural patterns related to inter-animal conflict, including fighting, defensive behaviour, fleeing and freezing.

Agonistic behaviour is an alternative term for **aggressive behaviour** (**R. J. Miller,** 1978) which refers simply to fighting and competitive behaviour, usually in animals and includes threats and offensive attacks as well as defensive fighting. Agonistic behaviour tends to occur in situations involving conflict or competition for space, resources, mates or status or for protection of self or young.

Huntingford (1976) used the term *aggression* in a broad sense to cover behaviour ranging from overt fighting to "supposed symptoms of boldness or confidence". She viewed *conflict* between animals in terms of three main situations: between members of the same species (**social aggression**), hunting and stalking as shown by predator toward its prey (**predatory aggression**) and attack on a potential predator by a prey species (**antipredatory aggression**).

Aggression and Agonistic Behaviour

Some definitions of *aggression* include predatory behaviour in which the animal being attacked is eaten in the process. Psychologists have defined aggression as behaviour that appears to be intended to inflict noxious stimulation or destruction on another organism (**Moyer,** 1976). The notion of purpose is necessary to exclude such destructive behaviours as a male elephant seal's trampling a pup while trying to copulate with the mother. Use of the word aggression as a form of *resource competition,* in which an animal activity excludes rivals from some resource such as food, shelter, or mates (**Archer,** 1988).

Agonistic behaviour is a system of behaviour patterns with the common function of adjustment to situation of conflict among conspecifics. The term includes all aspects of conflicts, such as threats, submissions, chases, and physical combat, but it specifically exclude predatory aggression, since, according to **Scot** (1972), ingestive behaviour is part of a separate behavioural system with a different function.

Ethologists (**Moyer,** 1976; **Wilson,** 1975) have listed following forms of aggressive behaviour:

1. Territorial aggressive behaviour in which takes place exclusion of others from some physical space (see chapter 1i for more details).

2. Dominance. In this type of aggressive behaviour control of a conspecific as a result of a previous encounter is done (see Chapter 11 for more details).

3. Sexual aggressive behaviour. In this case, use of threats and physical punishment, usually by males, to obtain and retain mates.

4. Parental aggressive behaviour. In this case, attacks are made on intruders when young are present.

5. Parent-offspring aggressive behaviour. This is disciplinary action by parent against offspring (mostly in mammals, usually associated with weaning).

6. Predatory aggressive behaviour. It includes the act of predation, possibly including cannibalism.

7. Antipredatory aggressive behaviour. It includes defensive attack by prey on predator, such as mobbing.

Most agonistic behaviour involves **competition** for some limited resource (food; water; access to a member of the opposite sex; or space for nesting, wintering, or safety from predators; see **Drickamer** *et al.*, 2002).

Causes of Aggression

The various causes of aggression have been categorized under the following two headings:

1. Endogenous factors of aggression

(*i*) Role of hormones. Aggressive behaviour in animals appears to be associated with the breeding periodicity. Males become comparatively more aggressive with the onset of the breeding activity. Androgen levels are responsible for aggressive behaviour in many species. The effects of androgen on aggressiveness are via its effect on lowering the threshold for aggression. Injections of **testosterone** restore aggression in castrated animals.

There are other hormones which influence aggressiveness in animals, because it has been found that male hormone does not completely regulate aggression. In fact, castrated individuals may often exhibit aggressiveness (**Davis,** 1957). In some species, **estrogens** too affect aggressiveness to some extent (**Guhl,** 1961). In fact, in some species *estrogens* enhance aggressiveness. LH (Luteinizing hormone) in starling and African weaver birds controls aggressiveness. Similarly **prolactin** and **progesterone** seem to manipulate aggressiveness. Levels of ACTH also influence aggression. High levels of ACTH (Adrenocorticotrophic hormone) reduce and low levels change aggressive behaviour.

The effect of hormone on aggression is mediated by the brain.

Invertebrate hormones are much different from those of vertebrates, and much less is known about how they work (**Breed** and **Bell,** 1983). Most of our information comes from crustaceans and insects. Males of many species of crustaceans have enlarged claws used for display and fighting. Development of these claws seem to be under the influence of a hormone produced by **androgenic glands** near the testes. Large glands produce more hormones, which leads to larger claws and greater fighting efficiency (**Nagamine** and **Knight,** 1980). In Juvenile lobsters (*Homaraus americanus*), levels of aggression are related to the moult cycle. When levels of the moulting hormone **ecdysterone** are high, aggression is low. This makes sense because after a moult the exoskeleton is soft and the animals are vulnerable. Once the shell hardens, and the animals compete for a shelter for the next moult, aggression is high (**Tamm** and **Cobb,** 1978).

Among insects, the paired **corpora allata,** neurosecretory glands in the head, produce substances called **juvenile hormones.** In grasshoppers, females shift from responding aggressively to males to become sexually receptive under the influence of juvenile hormones. If the *corpora allata* are removed, females continue to reject males. In male field crickets (*Gryllus* sp.), however, removal of the corpora allata had no effect on agonistic behaviour (**Adamo** *et al.,* 1994). **Roseler** *et al.,* (1986) implicated both juvenile hormone and ecdysteroids in the establishment of dominance hierarchies of female paper wasps (*Polister gallicus*), as they initiate nests in the spring. Subordinate females have lower levels of juvenile hormone and reduced gonad development.

***(ii)* Neural mechanisms.** The vertebrate brain structures most involved in aggression are part of the limbic system which includes structures such as the amygdala and the hypothalamus.

***(a)* Role of hypothalamus in aggression.** The hypothalamus is involved in defense and escape behaviour in a wide variety of vertebrates, ranging from fish to primates (**Huntingford** and **Turner,** 1987). Using lesions, electric stimulation, and single neurons recordings from specific areas of the brain, researchers have found that different brain sites are responsible for different types of aggression. For example, in cats, electric stimulation of the **ventromedial nucleus** of the hypothalamus produces growling, hissing, and attacking with claws (defensive attack) (**Flynn,** 1967). Stimulation of the **lateral hypothalmic area** produces a biting attack with no defensive elements.

***(b)* Role of amygdala and midbrain in aggression.** Areas of the brain that are involved in predatory aggression are the amygdala of the forebrain and the **central gray** of the midbrain. These areas are connected by nerve pathways, and they interact. For example, electric stimulation of certain areas in the thalamus causes cats to attack rats (**Bondler** and **Flynn,** 1974).

***(c)* Telestimulation.** The use of radio transmitters permitted electrical stimulation of specific brain areas in seminatural social groups (**Delgado,** 1967; **Herndon** *et al.,* 1979). Monkeys with electrodes implanted in certain parts of the thalamus, hypothalamus, or central gray, become aggressive when the electrodes are activated. In some cases, when stimulation is applied to the hypothalamus, lower-ranking monkeys become dominant as a result (**Robinson** *et al.,* 1969).

***(d)* Role of Fos protein in aggression.** Immunological techniques can be used to trace the paths of nerve axons to different brain areas. **Kollack-Walker** and **Newman** (1995) found in Syrian hamster brain that mating and agonistic behaviour are controlled by separate areas by the expression of **Fos protein.** Adult males were allowed to interact either with a sexually receptive female or an intruder male. They were then sacrificed and their brains sectioned and stained immunologically for Fos protien. In this way, active sites in the brain could be seen microscopically. The results show that some areas of the limbic system were activated by both mating and agonistic behaviour, others were selectively activated.

(*e*) Role of neurotransmitters in aggression. Chemical messengers, or neurotransmitters, are involved in the transmission of nerve impulses across synapses or across nerve-muscle junctions. These can produce *excitatory* or *inhibitory* effects depending on the nature of the post-synaptic receptors. Best known perhaps is **acetylcholine,** which is involved in the neuromuscular junction and in the autonomic nervous system. **Norephinephrine** (noradrenaline) is wide-spread in the pons and medulla, with fibers projecting anteriorly to the mid-and forebrain. **Dopamine** and **serotonin** are found in the midbrain, with fibers projecting anterirorly to the hypothalamus amygdala and straitum. These and other, such as **substance P (Shaikh and Siegel,** 1997), have been specifically implicated in the control of aggression.

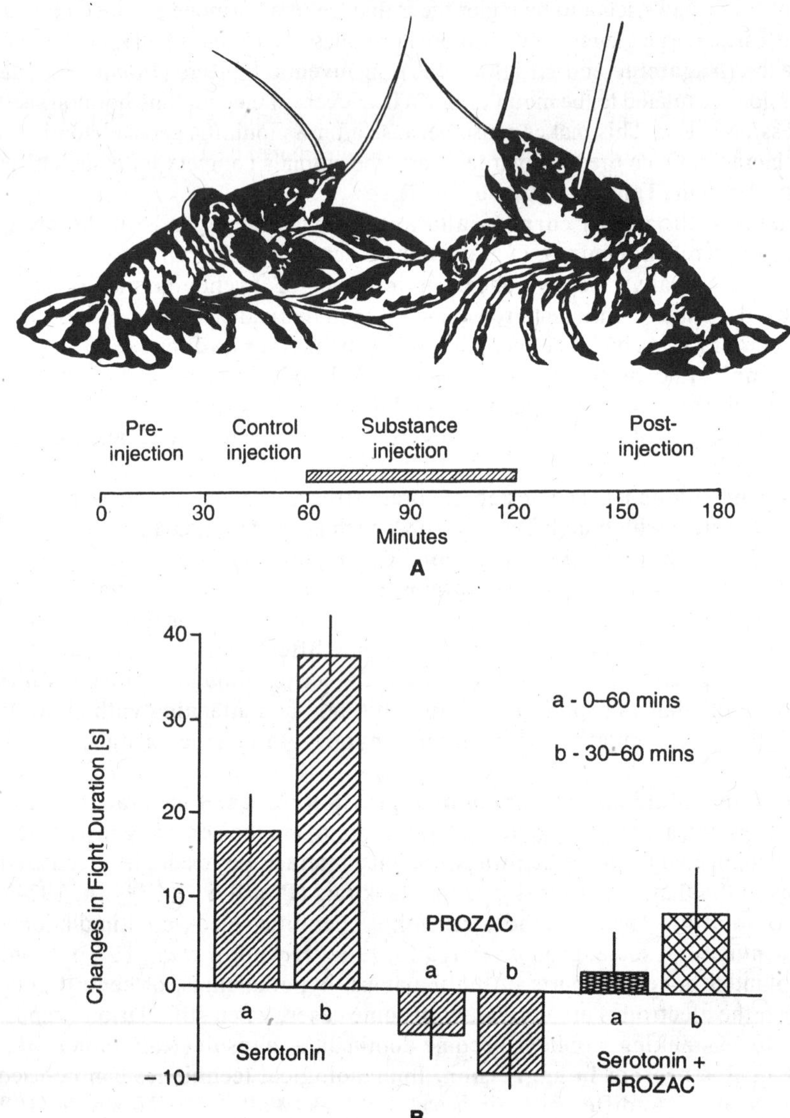

Fig. 8.4. A – Delago's experimental setup for delivering substances into freely moving crayfish. A cannula connected to a syringe pump is implanted in the pericardial sinus of the crayfish on the right. B – Changes in fight duration with serotonin and Prozac treatments. Each pair of bars represents the mean fight duration during the first and second 30 minutes of treatement with the agent (after Drickamer *et al.*, 2002).

Ample evidence links serotonin to aggression : a rise in brain serotonin level accompaines lowered aggression in a variety of vertebrates, from rainbow trout (*Oncorhynchus mykiss*) (**Winberg** and **Lepage,** 1998) to lab rats (**Olivier** *et al.,* 1995). Accordingly, administration of drugs that lower brain serotonin is accompained by increased aggression. Among invertebrates such as crustaceans, however, the effect is the opposite : serotinin typically increases aggression (**Huber** *et al.,* 1997). When subordinate crayfish (*Astacus astacus*) were given injections of serotonin into their haemolymph, they increased contests, and fights were more intense and lasted longer than prior to adminstration (**Huber** and **Delago,** 1998). When given **Prozac,** a selective serotonin reuptake inhibitor, along with the serotonin injections, there was no increase in fighting (Fig. 8.4).

***(iii)* Role of genes on aggression.** The synthesis of nerve structure, neurosecretions and other compounds is under genetic control, and agonistic behaviour is shaped by natural selection, as is any other behaviour. Agonistic behaviour has long been known to have a heritable basis (**Maxson,** 1981) : artificial selection can lead to significant changes in levels of aggression within just a few generations. For example, **Ebert** and **Hyde** (1976) tested wild female house mice for aggressiveness, and by selecting high-and low - scoring mice, they produced two lines: one with highly aggressive females and one with passive females. The unselected control lines were, as expected intermediate. Domestic laboratory strains of mice differ widely in aggressiveness (**Southwick** and **Clark,** 1968), as do dog breeds (**Scott** and **Fuller,** 1965). Siamese fighting fish (*Betta splendens*), fighting cocks, and even crickets have been artificially selected over the years for performance contests with large sums of money riding on the outcome.

Genes controlloing agonistic behaviour may be localized on certain chromosomes. Offensive behaviour in male mice, which involes bite-and-kick attacks on the flanks and rump of the opponent, is influenced by a region of Y chromosome. Genes in this area are hypothesized to affect synthesis of testosterone-dependent phenomena as well as the perception of olfactory stimuli triggering attaks (**Maxson,** 1996; **Monahan** and **Maxson,** 1998).

2. Exogenous Factors of Aggression

***(i)* Starvation** is one such factor. Hungry animals fight vigorously. Chaffinches and yellow buntings fight more when hungry; fighting is over once the food becomes available to these birds after a period of starvation.

***(ii)* Frustration** is another cause which elicits aggression. Frustration is defined as interference with the occurrence of an instigated goal response at its proper time in the behaviour sequence. Frustration is motivation for aggressive behaviour and leads to fighting in hungry animals.

***(iii)* Role of experience in aggression.** Aggression is influenced by experience. Experience shapes levels of aggressiveness. Animals reared in isolation are relatively more aggressive. Absence of litter males in the preweaning stage and of peers in the post-weaning stage stimulates aggression in mice.

***(iv)* Role of appetite.** Aggression is regarded as one of the motivated behavioural types. As such, there is a drive toward a goal. An animal which has been aroused demonstrated aggression rapidly. It appears that the animals become more aggressive if reared under some situations. For example, a winning animal fights readily than those with more succession of losses.

***(v)* Sex.** The relationship between sex hormones, particularly testosterone, and aggression is striking in seasonally breeding species. As the gonads increase in size in response to environmental changes in photoperiod, rainfall, vegetation, and so forth, fighting and wounding increase as well. Most of this increase is related to competition for breeding territories, social rank, or access of females. Key stimuli lying outside the body usually provoke aggression in animals. For example, a male stickleback fish defending its territory is aroused on seeing a rival male which it distinguished by a red spot on its belly. The aggression is so tightly associated with this red spot that a faithful model lacking the red spot on its belly becomes unable to elicit the aggression. Contrarily, crude model having a red spot provokes aggression. Similarly, aggression in robin is elicited by the red tuft of feathers in rival. A bird devoid of it will not provoke aggression.

In some monogamous species (*e.g.*, sea gulls) fighting occurs during the early phases or pair formation before a sexual bond develops. In others, courtship and copulation seem to have aggressive components, as evidenced by the appalling racket mating cats make, which makes the listener wonder whether it is a fight to the death or courtship. Other carnivores also have violent mating behaviour. When polecats mate, the male grabs the female by the scruff of the neck and drags her back to his nest. Some of this aggression at the time of mating probably stems from the fact the males tend to court females rather indiscrimately, often making advances when the females are not sexually receptive. Some mammals, mainly carnivores and lagomorphs, are **induced ovulators,** and vigorous courtship plus copulation may be needed to trigger the release of an egg into the oviduct as copulation occurs (see **Drickamer** *et al.,* 2002).

Adaptive Significance of Aggression

Agonistic behaviour has the adaptive significance of reducing **intraspecific conflict** and avoiding overt fighting which is not in the best interest of the species. Most species channel their 'aggression' into **ritual contests** of strength and threat postures which are universally recognised by the species. For example horned animals such as deer, moose, ibex and chamois may resort to butting contests for which 'ground rules' exists. Only the horns are allowed to clash and they are not used on the exposed and vulnerable flank.

Aggression in Fishes and Tetrapods

Fishes living in a dense yet buoyant medium, typically utilise in locomotion a complex powerful lateral musculature, controlled and toned by the action of paired and median fins. Spines and other ornaments can be used as defensive weapons in some groups, but most fish can attack only by **biting, butting** or **ramming** some part of body against the opponent. Conversely, most *tetrapods* have moved into a medium 800 times less dense than water and have been forced to develop a supportive appendicular skeleton, a largely nonmetameric parietal musculature in order to support the individual during locomotion or other activities. The addition of well-developed, operationally flexible appendages permit many tetrapods to increase the variety of movements that can be used in signaling, attacking and defending themselves.

Despite this acute distinction, it appears that remarkable similarities exist in the contexts associated with agonistic behaviours in the two vertebrate types (fishes, tetrapods) and that analogous behaviour often occur in analogous situations. This is apparent in the role of visual and acoustic stimuli in agonistic encounters. For example, a wide variety of visual fishes and tetrapods utilize autonomic responses as signals to convey rather precise information about motivation in fighting contexts (*i.e.,* colour changes in fish and reptiles; pilorection in birds and mammals). **Hostile encounters** in fish often contain various combination of autonomic and nonautonomic responses, such as fin-spreading, fin-fighting, opercle-spreading, tail-beating, mouth-gaping and pendulum movements that seem to have analogs in behaviours such as wing-raising, feather-flashing, gaping, pecking, neck-stretching and pendulum movements in birds and back-arching, rearing, tooth-baring, head-raising, tail-rattling and foot-thumping in mammals. These behaviours and the less ambiguous movements such as biting, butting, slashing, grasping, kicking, pecking, wing-beating, chasing, fleeing, freezing and others, all occur in various combinations in situations in which individual priorities or needs must be determined.

Many of the behaviours found in territorial fights, dominance encounters and other competitive situations seem to involve movements that serve either to *increase the apparent size* of the actor or to expose specially coloured structures or potentially dangerous weapons. Conversely, behaviours occurring in 'submissive' or 'subordinate' individuals often tend to decrease apparent size or obscure dangerous or strikingly coloured structures. For example, **Darwin** (1872) pointed to the extreme difference in *posture* between a hostile dog and the same animal "in a humble and affectionate frame of mind" (Fig. 8.5). A similar contrast is seen in gulls, where the beak is shown off in aggressive displays but is hidden, by turning the head away, in appeasing ones (**Tinbergen,** 1959).

Fig. 8.5. Aggression and submission in dog as explained by principle of antithesis of Darwin. The same dog is portrayed above approaching another dog with hostile intension and below in a humble and affectionate frame of mind.

Aggression and Animal Husbandry

The knowledge of aggression and its causes in animal is of a great value from the view point of the farm livestock. A knowledge of stimuli which provoke aggressive behaviours in different animals helps humans for domestication. A careful planning with a view of stimuli that elicits aggression will help keep animals such that they do not fight with one another. For example, the animals not able to live close may be separated by partitioning through opaque sides.

8.4. GAME THEORY MODEL

(Game theory and animal contests)

Maynard Smith and **Price** (1973) independently developed a model of how displays, rather than all out fighting, could be selected for at the level of the individual. During agonistic encounters, the decision about whether to attack, display, or flee depends on what the other participants does. Various alternative courses of action are called **strategies.** The participants choice of strategy is assumed to be heritable. The outcome of each game is determined by a **payoff matrix,** or a predetermined set of outcomes, the costs and benefits that result from different strategies. These costs and benefits are given in terms of expected Darwinian fitness measured by reproductive success (the number of offspring produced), which is the currency of the model. An **Evolutionarily Stable Strategy (ESS)** is a strategy, which, if adopted by most members of the population, cannot be successfully invaded by any rare alternative strategy (**Maynard Smith,** 1994).

One version is the **Hawk-Dove game (Krebs** and **Davies** 1987). There are just two strategies : **Hawk,** attack immediately, flee only if injured, and **Dove,** display to another Dove, but flee immediately without getting injured if attacked by a Hawk. If Hawk encounters Dove at a resource, Hawk always wins and Dove always loses. If the resource has value *V*, Hawk has no cost and obtains + *V*. Dove, which flees immediately does not gain or lose anything. If Hawk encounters Hawk at a resource, and ethologists assume that both Hawks are identical, each has a 50 percent chance of getting the resource and a 50 percent chance of being injured. So, each Hawk gets on average $(1/2\ V - 1/2\ I)$ or $(V - I)/2$. Finally, what happens it two Doves meet ? Assuming that they are identical and that both display, each has a 50 percent chance of getting the resource. Half the time, a Dove will win $V - D$, and half the time, it will lose and get $-D$. Thus, on average, each Dove will get $\frac{1}{2}V - D$. The payoff matrix for this game would be as follows :

	Against :	
Payoff to :	**Hawk**	**Dove**
Hawk	$(V - I)/2$	V
Dove	0	$V/2 - D$

When ethologists substitued some hypothetical numbers for these variables, they could compute the ESS (= Evolutionary Stable Strategy).

Let the resource value or $V = 50$;

Costs of injury , $I = 100$;

Costs of displaying, $D = 10$.

Plugging these values into the matrix above, they obtained :

	Against :	
Payoff to :	**Hawk**	**Dove**
Hawk	– 25	+ 50
Dove	0	+ 15

If all individuals in the population played Dove, the average payoff would be + 15. But what if a mutant that plays Hawk enters the population? It would do well against Dove, gaining + 50 fitness units. Hawk mutants would spread. Clearly, under these conditions pure Dove is not an ESS because it is invaded by another strategy. But as Hawk increases, the chances of Hawk meeting Hawk increase as does the cost of injury, and so the payoff approaches – 25, pure Hawk. At point Dove could invade, since Dove gets O against Hawk, not very high, but better than – 25. So pure Hawk is not an ESS either. The ESS in this case is some mixture of Hawks and Doves.

Ethologists calculate the stable mixture as follows: let p equal the proportion of Hawks, in the population, so $1 - p$ would be the proportion of Doves. The average payoff for each strategy is the payoff for interacting with each type of opponent times the probability of meeting that type. For Hawk, this is

$$p(V - I)2 + (1 - p)V$$

For Dove, this is

$$p(0) + (1 - p)(V/2 - D).$$

At equilibrium the average payoff for Hawk equals that for Dove. So, setting these two equations equal to each other and solving for $p = 35/60$, or 58 percent Hawk for the ESS, and 42 percent Dove.

In general, if $V > I$, pure Hawk is the ESS. If $V < I$, a mixed ESS results. Pure Dove is never an ESS. One conclusion ethologists could draw from this model is that, depending on the various costs and benefits, noncombative individuals that display and retreat may be following an adatptive strategy; they are not simply the losers or less fit victims of the aggressive dominants.

Development of Complex Hawk-Dove Models

More realistic or complex models have been developed that include situations where the contestant differ in some quality. Three kinds of asymmetry have been considered : (1) cases where one individual already owns the resource or occupies the area, when leads to home-field advantage; (2) cases where contestants differ in **resource holding power (RHP)**, as which one is larger or stronger than the other; (3) cases where the resource has different values for each of the contestants.

Maynar Smith (1976) added a conditional strategy to the Hawk–Dove model called **borugeois** (= Conservative) to deal with case (1). These individuals play a Hawk strategy if they are on home ground (or already own the resource in dispute) but play as a Dove if it is someone else's territory. Assuming that there are no other asymmetries, bourgeois turns out to be a pure ESS : Male speckled butterflies (*Pararge aegeria*) occupy mating territories on the forest floor; the resident male always wins when intruders enter, and the outcomes reverse if ownership changes (**Davies,** 1978). As predicted by the model, both contestants behave like hawks if they can be tricked into ownership of the same territory.

Asymmetries in RHP (= resource holding power) and resource value have been manipulated in laboratory studies of spiders. Working with female funnel–shaped spiders (*Agelenopsis aperta*), **Riechert** (1984) found that residency was usually unimportant in determining the outcome and that assessment of relative weights occurred early in the contest. Subsequent behaviour depended on the results of the assessment. If a spider has a large weight advantage, it immediately increased the fight, whereas smaller spiders tended to retreat. If they were the same size, then the resident won. **Austad** (1983) studied contests between male bowl and doily spiders (*Frontinella pyramitella* over access to a female. He tested his data against a model referred to as the **"war of attrition"** (**Maynard Smith,** 1974). In this game, two individuals display or fight continuously until one wins; costs are assumed to increase linearly with persistence time. **Austad** varied resource value by introducing an intruder male at various times while the resident male mated with the female. As more and more of her eggs were fertilized over time, her value to the resident male declined, and he was less inclined to fight. Males varied in RHP in terms of size : larger males usually won. As predicted by the model, fights were longest when males were of similar weight, and also when the female was of greatest value to the resident.

QUESTIONS

Long Answer Questions

1. Describe aggressive behaviour in animals.
2. What is conflict behaviour ? Describe different types of conflict behaviours.
3. What do you mean by displacement activities? Discuss various hypotheses regarding causation or displacement activities.
4. Give various conflicting situations that may arise during conflicting behaviour.
5. What is aggression? Mention different types of aggressive behaviours and discuss various causes of aggression.

6. Give an account of causes of aggression.
7. Give examples of behaviour displays of aggression in fishes and tetrapods.
8. Describe in detail the game theory model.

Short Answer Questions

Write short notes on the following :
(1) Hawk–Dove model
(2) Vacuum behaviour
(3) Displacement activities
(4) Prolonged conflict.

Multiple Choice Questions

Choose the correct answer from the four alternatives given

1. When an animal attacks another animal, this behaviour pattern is called
   (*a*) aggression (*b*) threat
   (*c*) appeasement (*d*) avoidance
2. The reaction which allows animals to escape from actual or potential danger in the environment is called
   (*a*) shelter (*b*) appeasement
   (*c*) avoidance (*d*) threat
3. The one of the advantage of living in groups or social living is
   (*a*) helps in locating food resources
   (*b*) increased competition within the group
   (*c*) increased risk of exploitation
   (*d*) fitness due to inbreeding reduced
4. Any behaviour affecting another animal, usually of the same species is called
   (*a*) motivated behaviour (*b*) ritualized behaviour
   (*c*) aggressive behaviour (*d*) social behaviour

Answers

1. (*a*); **2.** (*c*); **3.** (*d*); **4.** (*c*).

9

Communication in Animals

Social interactions in animals involve communication. Each social grouping must have a precise system of communication, for it is the only mode through which an animal alters its behaviour in response to the performance of others (Box 9.1).

Box 9.1.

Various important external stimuli impinging on an animal are those emanating from other animals. Predators, mates, flock companions, parents, offsprings and so on transmit signals to which animals respond in particular ways. In many cases these signals are the result of deliberate and complex actions which have evolved specifically to influence the behaviour of other animals. When we consider the relationship between a signal transmitted by one animal and the response it elicits in another, we consider an act of communication. **Communication** can be said to have occurred when an animal performs an act that alters the behaviour of another animal.

When animals communicate with each other, information of various different sorts passes between them. In the general sense of the word, a *signal* may incorporate details of species, age or sex of the signaller, the group to which it belongs, its family or exactly which individual it is. The signal may transmit information about the outside world, as in the alarm call of birds, vervet monkeys and in the bee dance which indicates the location of food, or about the signaller's state of motivation, as when a threat display shows that it is likely to attack (*e.g.*, male stickleback, Fig. 9.1). These and many other points of information may be gleaned from the signals an animal produces. In the strict scientific sense, as defined in **information theory,** many 'bits' of information pass from one animal to another as a result of signal. The more information received, the more uncertainty of the recipient as to what will happen next is reduced.

9.1. COMPONENTS OF COMMUNICATION

Communication among animals is complex and can be analysed into following components:

1. **Sender.** The member which gives off a signal.
2. **Receiver.** The individual who is affected by the signal.
3. **Channel.** It is the pathway of passing signal.
4. **Noise.** Background activity which is emitted, transmitted and received along with signal.
5. **Context.** It is the particular setting in which a signal is emitted, transmitted and received.
6. **Signal.** The information in question, *i.e.*, the stimulus to be transmitted.
7. **Code.** It includes complete set of languages of possible signals and contexts.

Evidently, it is the signal which travels, between the sender and the receiver.

Analysis of a Communicative Act

An act of communication can be viewed from two different vantage points: that of the sender and that of recipient. This point has been made most clearly by **Smith** (1977) who made the important distinction between *message* and *meaning.*

1. Message. Message is what the signal encodes about the sender and in some way describes the state of that individual. It may, for example, indicate that the individual is anxious, that it is an adult male in breeding condition or that it has just seen a predator.

2. Meaning. It is what the recipient makes out of the message. This can only be inferred from the response if any, that the recipient shows. It can differ very greatly from one recipient to another and is quite distinct from the message. The meaning can also vary according to *context* so that the same signal may elicit different responses in different circumstances. Because of this a comparatively small repertoire of signals may come to elicit many different responses, the one shown depending on the nature of the individual perceiving the signal and on its experience as well as on the immediate context and on variations in the signal itself.

Fig. 9.1. The head-down threat display shown by a male three-spined stickleback to other males that intrude in his territory.

The distinction between message and meaning can best be explained by the reference of following example. Consider a male bird singing at the top of tree (Fig. 9.2). In many birds species, *only males sing,* they only do so during the breeding season. They require **testosterone** (hormone) to be circulating in their blood and they sing much more when they are unmated. The message in a particular species song may therefore be "*I am an unmated male Dissimulatrix spuria in breeding condition*". The meaning of the song will vary according to the listener. To a predator it may mean "*Approach stealthily and attempt to catch*"; to a song bird of another species it may simply mean, "*Ignore*"; to a male of the same species it may mean "*Kept out. This territory is already occupied* "; to an unmated female it may mean "*Approach and attempt pair formation*". The identity of the listener and the *context* determine the response that will be shown and from this different meanings that the signal can have may be inferred. Some of these meanings may actually be disadvantageous to the sender, as when a predator is attracted. But for the signal to be produced, the net result must be advantageous (Fig. 9.2).

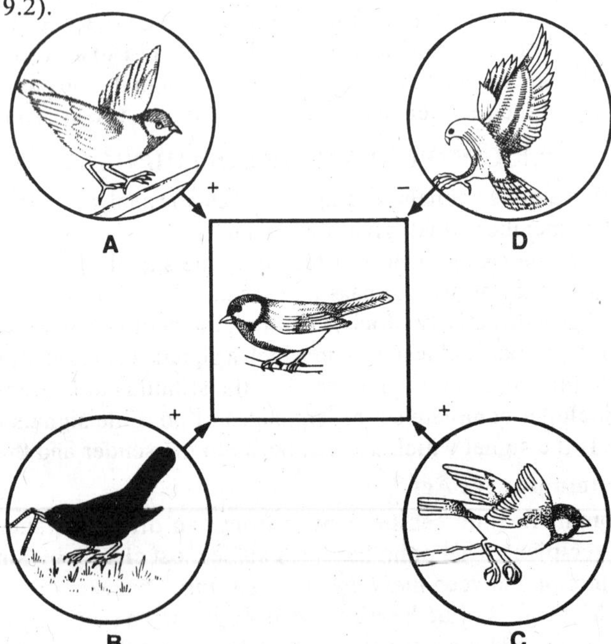

Fig. 9.2. The advantages and disadvantages of communicating. The male bird in the centre is singing on his territory. Advantages increase to him if his song attracts a female (A) or repels rival males (C). His song may be ignored by other species (B) but may also attract predators (D) which is a possible disadvantage of singing. The production of a signal will only be selected for if the net advantages outweigh the disadvantages.

When an animal communicates, the message is encoded in a signal. Signals are rather stereotyped, although they may form graded series so that slight nuances (hints) can be transmitted, whichever is the case, the message in some way defines the probability of what the signaller is doing or may do.

Types of messages. A set of 12 major classes of messages have been recognized by **John Smith** (1969). They include all types of message possibly involved in communication among animals. These are:

Class I. Identification messages. They identify the sender.

Class II. Probability messages. They indicate cause of action by the sender that may follow a signal.

Class III. General messages. They are diverse according to the need.

Class IV. Locomotion messages. They indicate movement.

Class V. Attack messages. They point to probable attack.

Class VI. Escape messages. They point to an escape attitude.

Class VII. Nonagonistic messages. They indicate improbability of agonistic acts.

Class VIII. Association messages. They indicate closeness among individuals.

Class IX. Bond limited messages. They indicate communication from bonds; *i.e.*, mates, parents and offsprings.

Class X. Play messages. They indicate playful actions.

Class XI. Copulation messages. They indicate copulatory behaviour.

Class XII. Frustration messages. They indicate conflict and frustration of individuals.

9.2. TYPES OF COMMUNICATION

Animals communicate some message, an encoded signal. The signals used for communication include vision, sound, olfaction, tactile, etc. Some species have unique signals, *i.e.*, electric discharges from the body of some fishes help them in locating food, water and

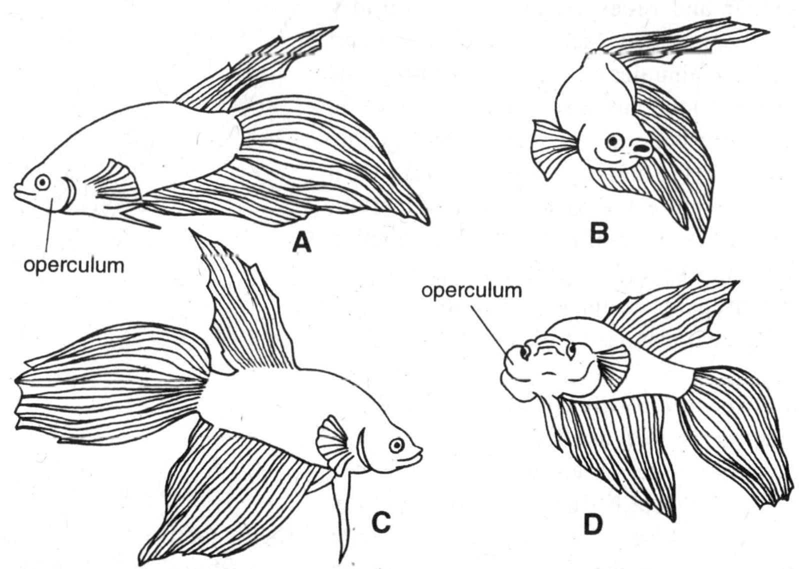

Fig. 9.3. Stages in threat display in Siamese fighting fish. A— and B—fish not showing threat displays. C— and D — Operculum and fins erected to increase their apparent size during threat displays.

resources. Which sensory modality is used for a particular signal depends to some extent on the information it conveys. In advertising for a mate, for example, larger animals tend to use sounds, as these travel rapidly, move round obstacles and can be detected at great distances. Smaller animals cannot achieve such high sound intensities and may rely instead on *pheromones,* smells which travel downwind and can, under favourable conditions, attract mates from remarkable distances.

According to the types of signals, the following types of communication have been recognised:

1. Visual communication. Changes in posture and colour are the main ways that animals communicate through the visual channel. Vision is the most important signal used in private and short-range communication. Such communications are meant for communication between two rivals in a flight for a territorial dispute. The communication between the male Siamese fighting fish (*Betta splendens*) at a territorial boundary has been shown in Fig. 9.3. The fish raises fins when it lies broadside to its rival (Fig. 9.3C) but closes its gill-covers when face to face. Thus, during visual communication both the sender of a signal and receivers display special facial expressions or postures to convey the message to conspecifics.

2. Olfactory communication. The olfaction (*i.e.,* smell, pheromone) as a signal of communication is generally used between sexual partners, between predator and prey, in territorial limitations, in locating food and water resource and also in the advertisement of the information among members of a society. Olfactory communication is not the best mode of communication, for smells diffuse only slowly through the environment, their speed and direction of travel being highly wind dependent and they can carry very little information (because after one smell is released time must elapse for it to disperse before another signal can be employed). However, there are situations where smells are ideal, *e.g.,* pheromones (*e.g.,* bombykol) produced by a female moth are detected by its male several hundred metres away. Territorial marking is a very good example of this type of communication.

3. Auditory communication. In auditory communication the signal used will not remain limited between sender and receiver, but it will rapidly traverse round corners and across the environment. This pattern of communication is thus advantageous in the sense that (*i*) the message can be sent; (*ii*) the frequency and speed of auditory signal is changeable as desired, and (*iii*) the message can be conveyed to a longer distance unlike the visual communication. Sound signals are very specific. A particular function is associated with a specific signal. For example, the mole cricket prepares a burrow in a horn-shaped design to enhance transmission of sound waves (Fig. 9.4).

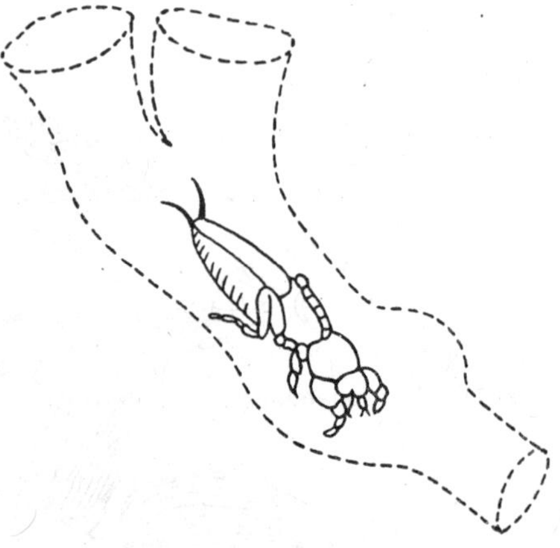

Fig. 9.4. Auditory or vocal communication in mole cricket.

4. Tactile communication. This mode of communication is not used commonly. Many invertebrates and few vertebrates have been found to transfer message through **physical contacts.** For example, blind workers of termites communicate through tactile communication in their subterranean tunnels. Elephants use tactile signals during courtship. The trunk plays an important role.

5. Animal communication as language. Most animal communication is very different from human language, as the message it transmits are not at all precise and word-like. They are much more akin to the signals we communicate to each other with our faces: the yawns, smiles, laughter and frowns which indicate other people what our feelings are (This is called **nonverbal communication**). **Verbal language** is different from this in many different ways. However the difference is not so great as might appear at first sight, because some animals are capable of transmitting detailed and accurate information just as we are. This is well exemplified by the dances of honey bees.

9.3. DANCES OF HONEY BEES

Honey bees have an elaborate communication system to signal to the other bees of the hive the location of the food. The Austrian ethologist **Karl von Frisch** (1886-1982) and his colleagues have deciphered the language of bees. They found that honeybees leave the hive and forage for food. When they return, other bees gather around and detect the odour (smell) of the nectar source the scout/forager has discovered. The scout performs a **dance** on the walls of the hive (comb) which indicates the distance and often the direction of food (**Von Frisch,** 1967; **Gould** 1976). This dance of bees is of following two types:

1. Round dance. If food is near the hive (less than 90 metres from hive), a scout bee of the species *Apis mellifera* performs the **round dance** (Fig. 9.5). In a round dance the forager or scout bee turn in circles first to the right and then to the left. Round dance intimates the recruited workers of hive that the source of food is less than 90 metres from hive but *gives no indication of direction.* Round dance facilitates fellow workers to search the correct food source by comparing the two smells (*i.e.,* sample of smell provided by the forager bees and the smell of target flower).

2. Waggle dance. If food is farther than 90 metres away, the scout/forager bee performs **waggle dance** (or tail waggling dance). Waggle dance indicates both *direction* and *distance* of the source discovered by the forager or scout bees and is regarded as **language of bee.** Waggle dance might be performed on the walls (**vertical surface**) of the hive or on **horizontal surface** at the entrance of the hive.

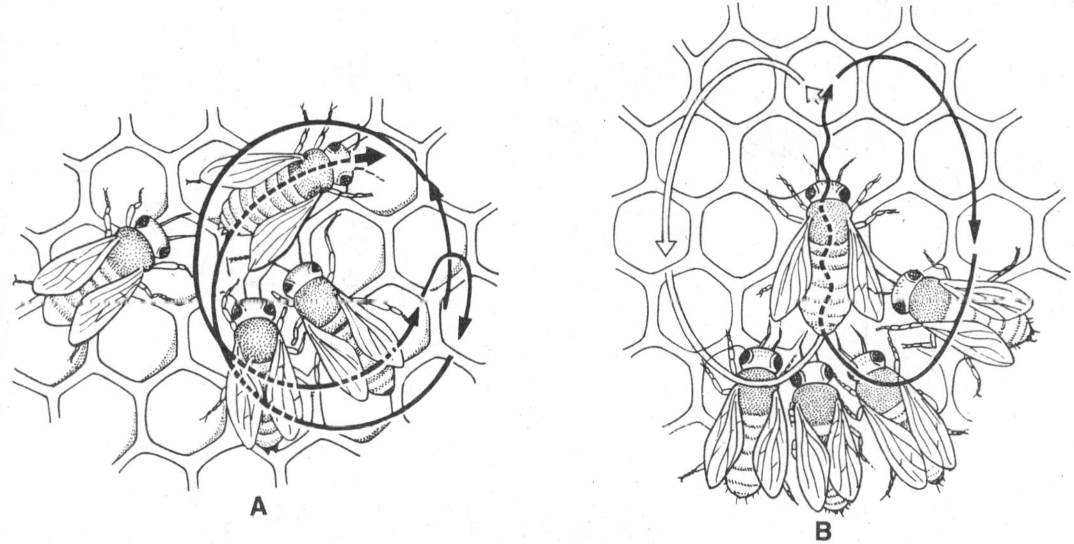

Fig. 9.5. Round dance (A) and waggle dance (B) of honey bee.

The waggle dance is shaped in a figure of eight (*i.e.,* 8) looping first to left and then to right with waggling of abdomen. Thus, a typical waggle dance consists of a **middle** or **straight run** *while the abdomen is waggled vigorously,* then the bee makes a *semicircular turn* and waggles in straight line again. This is followed by another semicircular turn in the opposite direction and another straight waggle run (Fig. 9.5). The dance is repeated many times. **Von Frisch** noted that *the direction of the straight run indicates the direction of food and tempo of the dance signals distance* (Box 9.2). The foraging bees use the sun as compass and when there are clouds they can tell the location of the sun by the pattern of light (*i.e.*, polarized light) in the sky. When the food is towards the sun, the dance will be in tune to keep the middle run of eight *vertically up* the hive. If the food is in the opposite direction of sun, the middle/straight run will go *straight down;* otherwise, the angle of waggle run to the vertical line equals to the angle formed between position of sun and food zone (Fig. 9.6).

Box 9.2.

The waggle dance involves the scout bee walking in a figure-of-eight and waggling of her abdomen during which according to **von Frisch** *the speed of the dance is inversely related to the distance of the food from the hive.* If food is 335 metres from the hive, the scout bee makes 30 runs per minute; if it is greater than 670 metres away, only 22 runs per minute are made. Several other elements of the dance vary with distance aside from the time needed to complete the entire circle: the duration of each waggle of the abdomen, the length of the run and the number of waggles per minute.

Bees may dance for long periods if the food source is rich. During this time, the sun changes position in the sky. Like the birds, bees have a built-in **biological clock mechanism** which compensates for such changes. As the sun's apparent position changes so does the direction of the waggle dance, so accurate flight directions are always given.

Since the waggle dances are generally performed within the darkness of the hive where vision is impossible, the worker bees must follow the dancing pattern of the scout by tactile sensation; they feel her position with their antennae.

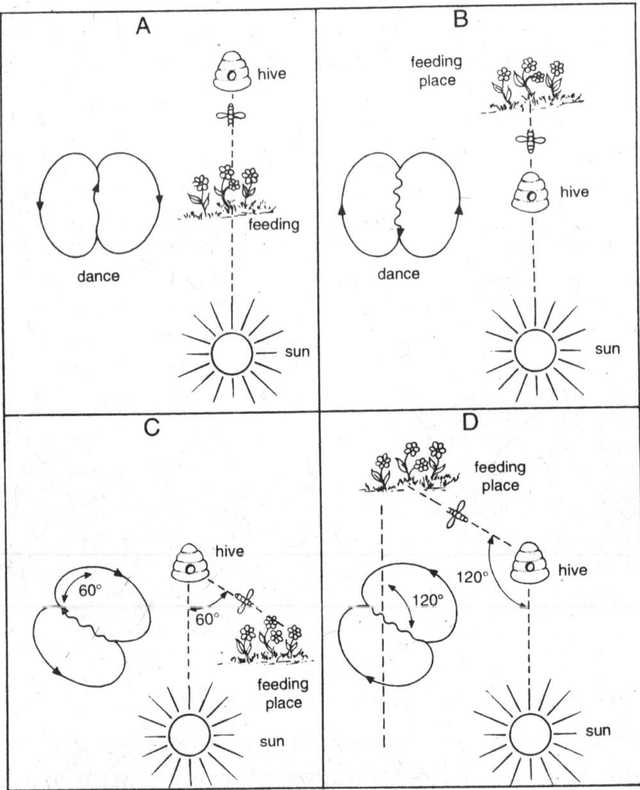

Fig. 9.6. Waggle dances of honeybees as they change in relation to the direction of the food sources from the hive relative to the sun.

The conclusion that the dance displays of bees encode fairly specific information about the distance and direction to good foraging sites was reached by **Karl von Frisch** after 20 years of experimental work. His basic research protocol involved training bees (which he daubed with dots of paints for identification) to visit particular feeding stations which von Frisch had stocked with rich sugar solutions or honey. By watching the dances of these trained bees in specially constructed **observations hives**, he saw that their behaviour changed in highly predictable ways depending on the distance and direction to a food source. More importantly still, his dancing bees were able to direct others to a food source they had found.

Some believe that recruited workers might rely exclusively on the flower odour present on the body of dancing bee as a guide for their search (**Olfaction hypothesis** of **Wenner** *et al.*). Some recent evidence too have suggested that bees may use high-frequency sound to communicate sources of food to other workers. However, the findings that a mechanical or **robot "bee"** can apparently induce recruits to search for food in a particular direction offers evidence in favour of the proposition that workers derive useful information from the dances (**Michelsen** *et al.,* 1989, 1992).

The bee dance has several notable attributes, two of which are features one might otherwise think of as unique to human languages. One is that it is **symbolic language**, since information about distance and direction are encoded in features of the dance in a stylised way. The other point is that the bees are **communicating,** as we often do, about events which are distant in time and space from where the communication is taking place. In other words, in the darkness of the hive they are telling each other where the food is even though this may be some kilometres away.

9.4. ALARM CALLS

In many passerine birds, individuals which detect a predator, such as a hawk, produce a distinctive **alarm call** (**Marler,** 1955). This call is very similar in several species of birds (Fig. 9.7) and consists of a **short-duration** (c.0.5 s), **high frequency** (7 kHz) note. A sound with these characteristics is very difficult to locate and this appears to be the selection pressure that has led to *convergent evolution* in several species. When they hear an alarm call, other birds fly into the nearest cover.

Some birds have more than one alarm calls, each used in response to a different kind of predator. For example, many birds use an alarm call in some contexts but others use an easily located **mobbing call** that attract other birds who together attack the predator.

The vervet monkey (*Cercopithecus aethiops*) has several distinct alarm calls, each signalling the presence of different predator (**Seyfarth** *et. al.,* 1980). Animals hearing these calls make appropriate responses, for example rushing up into trees in response to a **leopard call,** looking up in response to an **eagle call** and looking down in response to **snake call** (Fig. 5.15).

9.5. DIRECTIONAL INFORMATION OF SOUND AND HUNTING BY BARN OWL

Hearing as a substitute for vision occurs in a number of species, some of which have developed highly sophisticated and specialized extensions of normal hearing. All these specializations depend upon an ability to locate sound accurately. Comparison of the sounds reaching the two ears is the most important method of **determining the direction of sound** or **sound localisation** in vertebrates. A person with one ear can turn the head and scan for the direction of maximum sound intensity because the head blocks some of the sound and causes a **sound shadow.** With two ears, simultaneous comparison is possible, and this is much more rapid and accurate. If the ears are sufficiently far apart, there will be differences in arrival time and phase of the sounds coming from a particular directions. Thus, small animals rely on intensity comparison alone, whereas humans use both **monaural** and **biaural** (aural = pertaining to the

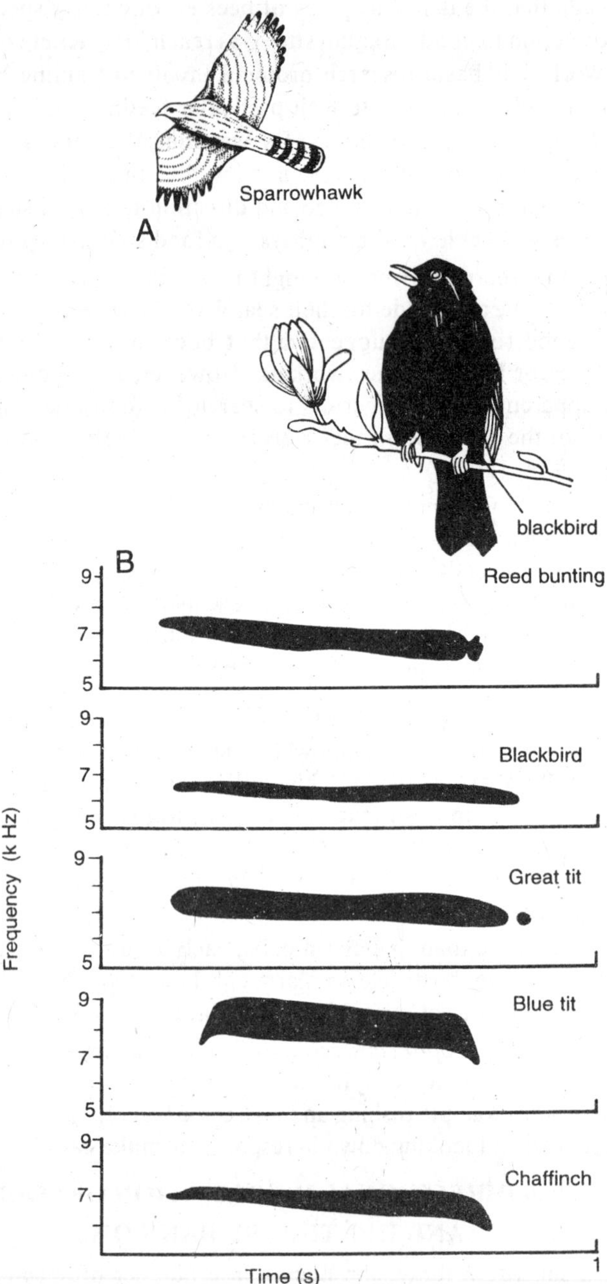

Fig. 9.7. Alarm calls in birds. A— A blackbird keeps very still and produces a "seet" call as a sparrowhawk passess over. B — Sonagrams of these calls as produced by several species of birds, showing how they are closely similar.

ear) methods. The pioneering work of **Roger Payne** (1962) has shown that the hearing abilities of the **barn owl** (*Tyto alba*) far exceed those of humans.

Barn owls are **nocturnal** hunters. They do not possess the mobile pinnae of mammals, yet they are fantastically good at locating a source of sound. Even barely audible sounds such as a leaf being pulled across a floor can be pin-pointed. Likewise, a barn owl can locate and capture freely moving mouse in total darkness. It can even determine the direction of the animal's movement and thereby align its claws with the long axis of the mouse's body. The barn owl is particularly sensitive to the differences in time of arrival of sounds at the two ears. This helps it to determine the azimuth (direction of the horizontal plane) of the sound. Sound intensity differences provide further hints to horizontal location. In this respect, the barn owl attains accuracy similar to humans, but it is nearly three times as accurate in determining elevation (direction in vertical plane) of the sound source. The barn owl appears to achieve this accuracy be means of the orientation of the auditory meatus (ear) and the arrangement of feathers around the face ("facial disc"). Together these design-features block all sounds except those arriving along the line of vision. All the barn owl has to do to locate a sound source is to move its head until the sound is loudest in both ears. It will then be facing the source. To aid this process, the barn owl's ears are placed asymmetrically on its head (Fig. 9.8). The right ear is directed slightly

Fig. 9.8. Facial structure of owl with some of the surface feathers removed to show the asymmetrical arrangement of the ears.

upward and the left ear slightly downward. The right ear is more sensitive to high frequency (3 to 9 kilohertz) sounds from above the midhorizontal plane of the owl's head and the left ear to high-frequency sounds from below. As the sound source moves downward, the high-frequency components of the sound become louder in the left ear and softer in the right. When the sound source moves upwards, the opposite occurs. This gives precise information about the elevation of the sound source because the owl uses low-frequency components of the sound to locate sounds in the horizontal plane and high-frequency components in the vertical plane. It does not confuse the two types of information, even though both depend upon comparison of the sounds reaching the two ears. *Barn owls locate and catch* its prey only when sounds of movements of prey can stimulate the two ears equally (Fig. 9.9). Thus, auditory stimulus filtering in the owl mainly involves the removal of directional redundancy.

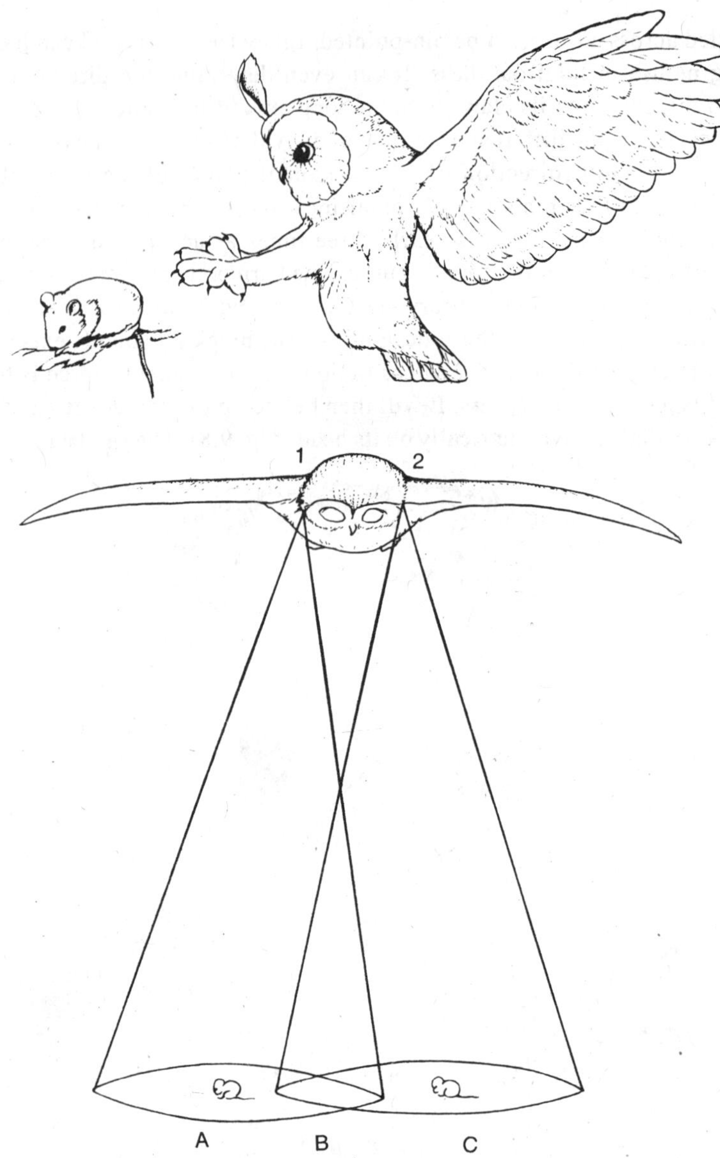

Fig. 9.9. Barn owls match the sound input to each ear to locate and catch prey (*e.g.*, mouse) in the dark. A mouse at the position A will stimulate ear 1 not ear 2 and vice versa at the position C. A mouse at position B will stimulate both ears.

9.6. ECO-LOCATION OR SONAR IN BATS

Owls use sounds coming from the prey to determine its location, but some vertebrates obtain information from the faint reflections or echoes of sound preys themselves produce. Such an **echo-location** or **animal sonar** is a good substitute for vision for those animals (*e.g.*, bats) which have to hunt in darkness. In eco-location, the animal emits high-frequency sounds and detects the presence of objects by the echoes produced. The principle is the same as that used in military sonar (radar). Simple forms of echo-location occur in shrews (*Blarina sorex*), South American oil birds (*Steatornis caripensis*) and Himalayan cave swiftlet (*Collocalia brevirostris*) that roost and nest in dark caves. More advanced forms of eco-location are found in whales, dolphins (*Torsiops*) and other marine mammals, but it reaches its peak in bats (Chiroptera).

The American zoologist, **Donald Griffin** (1958) first of all showed conclusively that the bats emit pulses of high-frequency sounds which strike objects and reflect back to the animals, which are able to detect and use the information in the echoes to locate, identify and capture prey while avoiding obstacles in their flight paths. The eco-locating bats emit bursts of ultrasonic sound pulses of *short duration* (5 to 15 milliseconds) and *high frequency* (12 to 120 kHz) which are beyond the auditory range of humans. The brief pulses permit accurate timing of the echoes so that the bat can determine the distance of the object producing the echo. Natural sound produced by wind and other animals are usually of low frequency, so by emitting high-frequency cries, the bats are unlikely to be subjected to interference. Another advantage of high frequencies is that they can be beamed precisely, thus allowing resolution of small objects. Bats produce their ultrasonic cries from a specialised larynx and emit them from the lips, as in naked-backed bat (*Pteronotus*), or from specially-shaped nostrils, as in the horseshoe bat (*Rhinolophus*) and leaf nosed bat.

Detection and analysis of sounds. The nervous system of bats has many specialised components for the detection and analysis of auditory information. Sound waves in the environment, including echoes from vocalisations produced by bat, pass through the ear openings and along the middle ear canal to the inner ear. There these waves strike a membrane covering the end of the ear channel, causing the membrane to vibrate. Ultimately these vibrations move specific portions of the basilar membrane of cochlea in tune with the frequency of sound waves entering the ear. The mechanical energy present in these movements deforms receptor cells or hair cells attached to basilar membrane; this energy is transformed into a receptor signal that is relayed to sensory neurons associated with each receptor when these neurons fire, their messages are carried away from the cochlea along the fibres that constitute the **auditory nerve** (VIII nerve). This nerve runs to specific regions of the bats brain, including the **lateral lemniscus** and **inferior colliculus.** The brain cells analyze input from the auditory sensory system and make decisions that control the movements of bat which determine its success in capturing of the prey.

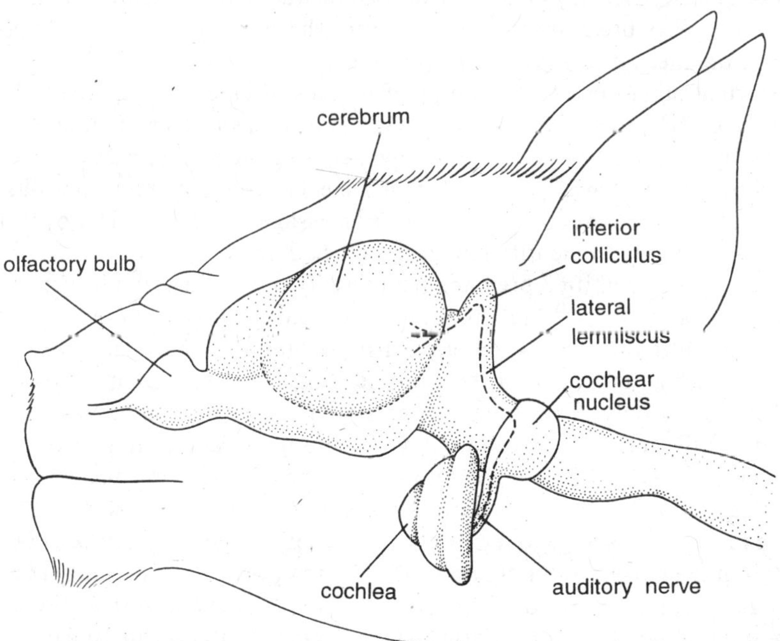

Fig. 9.10. Auditory system of bat, *Hipposiderus commersonii.*

Detection of echoes by bat. If a bat is to catch a flying insect, its first problem is to detect the echoes reflected from the potential prey. The trouble is that the echoes are very weak and return very soon after the bat produces a loud orientation vocalization. An auditory neuron that has just been exposed to a loud sound becomes temporarily insensitive to soft ones; the orientation cries are about 2000 times as intense as the echoes (**Griffin,** 1958). Bats preserve the sensitivity of auditory receptor cells in part because of a muscle in the middle ear. Just before the neurons in the bats brain order a vocalisation cry, they simultaneously send messages to the middle ear causing it to contract. The muscle diminishes the vibrations of the middle ear bones that transmit the energy in sound waves to the cochlea. The muscle relaxes 2 to 8 milliseconds after an orientation pulse, permitting a full response to the weak reflected sound waves coming from objects in bat's auditory field. Thus during the relatively quite interval between cries, the receptors are not only recovering from the recent vocalisation blast but are adjusted for and can detect sounds of very low intensity.

The protective mechanism of middle ear muscle decreases the loudness of sound passing down the ear canal to 1 per cent of their intensity. In addition to this mechanism, there are certain neurons in the lateral lemniscus of bat's brain that block transmission of auditory messages to higher regions of the brain during the period of a vocalization. The brain of bats also contains neurons, called **echo-detector units.** Within both the lateral lemniscus and inferior colliculus of some bats (*e.g.,* Mexican freetail bat), cells (neurons) have been discovered that respond more intensely to the *second* of two separate pulses of sound, one following the other in close succession, just as an echo would follow the vocalization pulse.

Identification and interception of prey by bats. To identify an object that is sending back echoes, the decoder neurons in a bat's brain must be able to discriminate between slight differences in the patterns of sound reflected from a leaf of a tree trunk or a moth. The little brown bat (*Myotis lucifugus*) can capture very small flying insects such as fruit flies and mosquitoes at a rate of two per second. Since airbone sounds do not readily penetrate water, fish-eating bulldog bats (*Noctilio*) do not locate fish that are entirely underwater, but capture those whose dorsal spines break the surface or those whose movements set up ripples. They scoop their prey from the water with their feet.

Bats use certain other cues to identify their victims. Flying insects, simply by moving, provide a distinctive cue for their bat predators. Moths offer a special signal; not only do they move, but as they fly their wings swing up and down, altering the surface area exposed to sound waves generated by hunting bats. This means that the echo from a moth changes in a regular, cyclical fashion. Bats capable of comparing sensory messages from an object over time could make use of this information to identify moths and other flying prey.

In order to catch an identified prey, the bat must (*i*) know where the insect is relative to itself (bat) and (*ii*) know where the insect is going so that bat can intersect the path of its victim. Bats have highly **directional ears,** which are sensitive only to sounds arriving from certain points in space. Most bats that catch insects on the wing have large external ears shaped to enhance directional sensitivity. A bat's external ears are flexible and can be moved about in an attempt to maximize the intensity of echo. [Bats waggle their ears during sound emission to scan the environment for echoes and to build up an acoustic picture in the manner of a line scan on a TV screen (**Land,** 1983)]. When the bat has succeeded in doing this for both ears, they will be pointed at the prey. The distance to the insect can be determined by brain cells that measure the interval between the pulse and the return echo from the pulse. For example, if the outgoing pulses are very short, as is vespertilionid bats (*Myotis lucifugus* and *Eptesicus fuscus*), there is no *overlap* between the end of outgoing pulse and the beginning of its echo. Since echoes arrive more quickly from nearby objects, the pulses are shortened progressively as an object is approached, so that zero overlap is preserved.

To capture a flying insect, the predator should not fly directly to the spot where the insect (*e.g.*, bug, moth) was first detected. Instead, the bat increases its rate of pulse production. In this way, it increases the rate at which it receives information about the prey's location and direction of movement. This permits bat to plot the flight path of the insect and to move to a place where the prey is likely to go. If the calculations are correct, the bat will arrive at a point in space at the same time as its victim.

The navigation of bats are also regulated by the following two effects:

1. Andrea Doria effect. A beeping bat leaving or entering its cave will often crash into a net that it is perfectly capable of detecting and avoiding. **Griffin** (1958) has entitled this the "**Andrea Doria effect**", in honour of the Italian cruise ship that crashed into the Swedish liner, Stockholm, off Nantucket some years ago despite having one of the most advanced radar system of its time. Evidently bats learn the route they normally follow when moving to and from a roosting site in their caves and are not rely listening to their sonar in this familiar area.

2. Doppler effect and its compensation. For a bat flying toward a stationary object, the perceived frequency of the echo is always higher than that of the outgoing sound. This is due to the **Doppler effect** that results from the relative motion of the bat and the object: the faster the bat flies, the higher the apparent frequency of the echoes. To compensate for this effect, the bats alter their outgoing frequency in order to maintain the perceived echo frequency as constant as possible. By this means the bat can obtain an estimate of its flying speed and the direction and relative speed of flying prey.

9.7. VOCALIZATION BEHAVIOUR : NEUROBIOLOGY AND DEVELOPMENT OF BIRD SONGS

Neurobiology of Bird Song

Birds produce song during exhalation, and they require muscular action to completely exhale : this gives them great muscular control over their vocalizations. The junction of the two bronchi from the lungs is modified into specialized **syrinx.** Muscles surrounding the syrinx control the tone and timing of the song.

Nottebohm and his colleagues have worked on various aspects of birdsong, especially in canaries (*Serinus canarius*), for nearly 30 years. They were first to discover that the avian brain has discrete **nuclei** responsible for the brain has discrete **nuclei** responsible for the production and processing of song (**Nottebohm,** 1975, 1980, 1981; **Nottebohm** and coworkers, 1976, 1981, 1984, 1986). The neural pathway for song production originates in the **HVC** or **higher vocal center.** There is a pathway from the HVC to the **RA (robustus archistriatum)**. The RA sends projections to the **DM (dorsomedial nucleus)** and to the hypoglossal nerve, which in turn innervates the syringeal muscles. Other centres control breathing and timing of signals. Lesioning any connection distorts or eliminates song output. The HVC is also involved in a second loop that deals with song perception and development.

Male canaries sing (Fig. 9.11); females generally do not sing. When a young canary matures at one year of age, it learns a song repertoire. Each succeeding year, the bird learns a new repertoire. Most songbirds show increased **testosterone** (hormone) around the time when they are singing. When birds are experimentally treated with testosterone, song nuclei show increased dendritic growth. In males, the HVC nucleus is 99 percent larger in the spring (when they sing) than in winter (when they do not sing), and the RA nucleus increases by 76 percent. Female birds treated with testosterone show anatomical changes similar to those observed in males and they start to sing. Experiments with adult canaries using radioactively labeled thymidine (a marker for DNA synthesis) indicate that new neurons are being formed in one of several brain nuclei involved in song control. Electrodes detected impulses resulting from auditory stimuli and confirmed that these new neurons are parts of functional circuits.

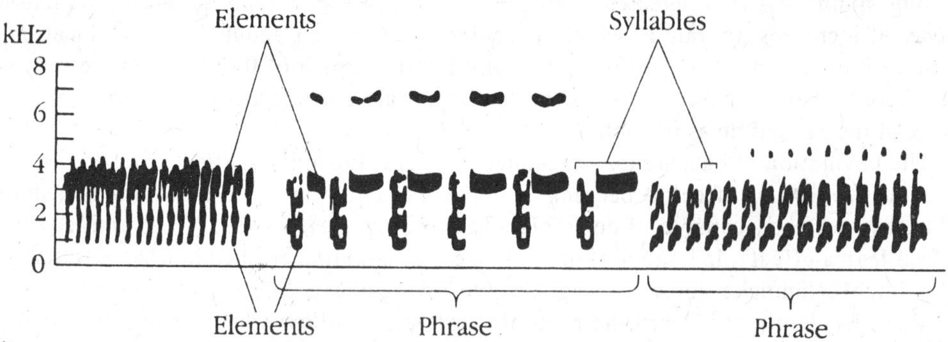

Fig. 9.11. Fragment of a song from an adult male canary. Canary song includes 20–30 syllable type, each repeated several times forming a phrase. This figure shows three phrases. Each syllable in the first and second phrase is composed of two elements, whereas syllables in the third phrase consists of a single element. The vertical axis represents sound frequency (in kHz) and the horizontal axis is in second (after Drickamer *et al.*, 2002).

Thus, neuroanatomical changes occur on a seasonal basis that corresponds to the annual behavioral cycle of the birds. As a new song repertoire is learned each spring, testosterone increases and new connections involving newly generated neurons are formed in brain areas associated with song control. These new neurons appear in response to a protein called **BDNF** (Brain-derived Neurotrophic Factor) : BDNF receptors are found in both sexes, but BDNF increases in response to testosterone, so it is normally found in males. When BDNF is infused into the HVC of adult females, the number of new neurons triples (**Rasika** *et al.,* 1999). At the end of the breeding season, there is a reduction in the size of the nuclei responsible for controlling the canary's song (**Nottebohm** *et al.,* 1986), and this corresponds to the bird's forgetting the songs of that year. Neuronal genesis in adults has now been found in a variety of taxa (**Tramontin** and **Brenowitz,** 2000).

Development of Bird Song

Birds learn to sing based on both a **genetic template** that varies with the species and **learning** from conspecifics, which may take the form of imitation and improvisation. There is a diversity of developmental strategies that led to the variety of songs and calls we are familiar with (**Kroodsma** 1978, 1981; **Nowicki** *et al.,* 1998). Comprehensive analyses of many bird species show that there are at least two major strategies for song development : ***(i)* imitation** of the songs of others, particularly of adult conspecifics; and ***(ii)* invention** or **improvisation.**

For example, the song learning of the wren (*Cistothorus palustris*) has been studied extensively. In nature, male marsh wrens sing more that 100 types of songs; neighbouring males generally sing identical song types; they often interact by counter-singing with one another using the same song type. To test their song-learning development, males were reared in special housing conditions where the songs they heard could be completely controlled. Males were played specially prepared tutor takes containing nine different songs each. Males were exposed to one tape for age 15 to 65 days, a second tape for age 65 to 115 days, and a third tape the following spring. By imitating them, the males learned the nine songs on the tape they were exposed to before 65 days of age, but they did not learn the songs on the subsequent two tapes. The males never song any invented songs. Further investigation revealed a more refined estimate of the peak sensitive period, which falls between about age 35 and 55 days (**Kroodsma,** 1978). In addition, males learned song tapes played for either 3 days or 9 days. Some males improvised some songs in these additional tests, presumably from elements of the songs on the tutor tapes.

The learning situation used for these marsh wrens involved only tutor tapes played over loudspeakers. In another test, young wrens were exposed to a number of song types on tutor tapes and then were given a period of social interaction with adult males with varied song repestroires. The **period of social interaction** occurred in the fall of the first year for some birds, and not until the following spring for other. Data of song repertoires for these birds indicated that song learning can occur early from the tapes, or at either of periods of exposure to adult males. Hence, birdsong learning is somewhat flexible, and social interaction may be an important feature of the process. These findings also make it clear that we should be careful when using only artificial stimuli such as loudspeakers. Social interaction has also been shown to be critical for language acquisition in human children (**Jerison,** 1973; **Freedle** and **Lewis,** 1977).

Longitudinal studies of song development indicate that some birds progress through a series of **stages** in song acquisition : **subsong, subplastic song, plastic song** and finally, **crystallized song.** Various environmental factors can influence the song development process, including social factors, such as the presence of conspecifics, and photoperiod, in relation to the time during spring or summer when the birds hatch.

9.8. LANGUAGE ACQUISITION BY HUMANS (AND APES)

Communication using **language** has tradionally provided a clear-cut separation of humans from other animals. By true language we mean both the use of symbols for abstract ideas and the understanding of syntax, so that symbols convey different message depending on their relative positions. **Syntax** means the arrangement of words and phrases to create sentences.

An example of complex visual communication is sign language in great apes (*e.g.,* Chimpanzees, *Pan troglodytes*). Although chimps lack the motor ability to produce the sounds of human language, they can acquire vocabularies of several hundred words by means of hand signs or substitute symbols. Apes can use nouns, pronouns and verbs to converse about their immediate needs; they can be taught to communicate with each other, but their ability to create sentences and use true symbolism still being debated.

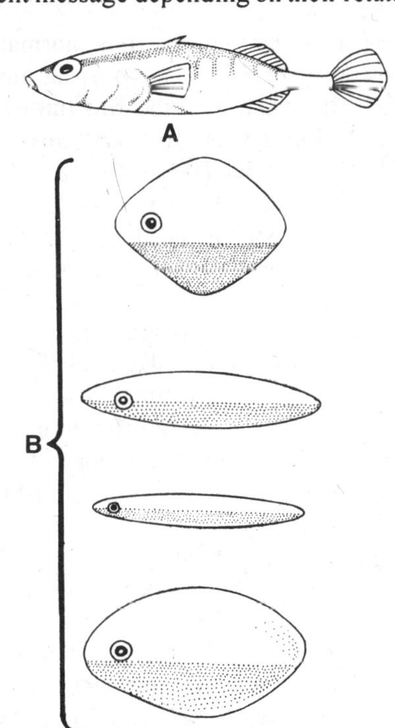

Fig. 9.12. A life-like model of a stickleback and four crude models. The crude models with red undersides elicited more aggressive responses than the life-like model lacking red.

9.9. RELEASERS (SIGN STIMULI)

Ethologists studying relatively stereotyped behaviour patterns (including instincts) have been interested in the nature of the stimulus that triggers the behaviour. Among the early researchers in this area were **Nikolaas Tinbergen** and **Konrad Lorenz.** By using dummies or physical models of animals, they tried to isolate the particular or characteristic of a species that triggers a response. For example, **Tinbergen** examined the aggressive and reproductive behaviour of the three-spined stickleback fish (*Gastrosteus aculeatus*). He found that dummy fish would stimulate male fish to fight as long as they had the red belly characteristic of male fish (Fig. 9.12). Thus, appearance of a characteristic red belly in male three-spined stickleback during its breeding condition serves as a **releaser** or **sign stimulus,** since it elicits aggression in other territorial males. The term releaser has been coined by **Tinbergen** for that specific stimulus which elicits a particular response pattern (*i.e.*, instinctive acts). Later on, the term **sign stimulus** has replaced the term releaser. Thus, a sign stimulus constitutes only a small fraction of the total environmental information available. It is part of a stimulus configuration and may be a relatively simple part. For example, as we have observed in the case of sticklebacks, the crude models suffice to elicit attack, provided they have a red underside. In contrast, a freshly killed male stickleback without a red belly is ineffective in provoking attack from other males. Thus, many of the details of the structure and texture of a male stickleback apparently are ignored by other males. In addition, **posture** was important in the sign stimulus (**Tinbergen,** 1951). This was shown by putting a male stickleback into a test tube, where it could not move. When the test tube was in horizontal position (normal swimming position), the fish did not provoke attack by other fish. But when the tube was turned so that the fish's head was directed downward, this stimulates attack (Fig. 9.13).

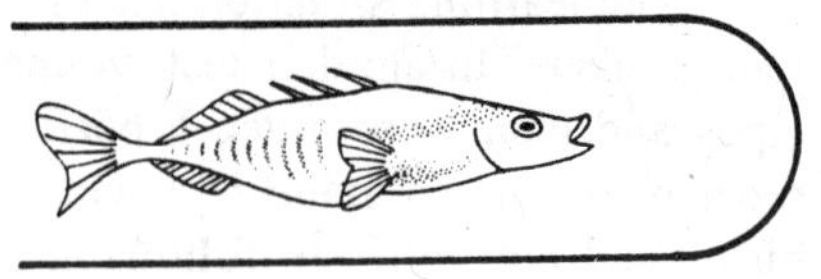

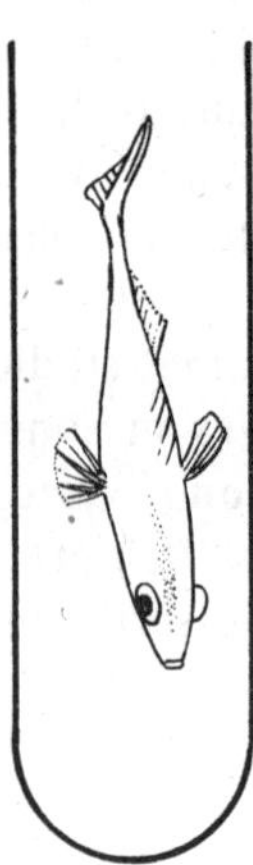

Fig. 9.13. Effect of posture on aggression in stickleback fish as experimentally shown by **Tinbergen** (1951). A fish inside a test tube stimulates much less fighting from the other fish if it is in the horizontal 'nonthreatening' position than the head-down 'threatening' position.

Similarly, **Tinbergen** found that the male was stimulated into courtship behaviour by two stimuli: the **swollen abdomen** of the ripe female and a particular posture when the female faced head-upward at a 45° angle. These stimuli encourage the male to go through a zigzag swimming pattern.

There are many more examples of sign stimuli among different animals:

1. The egg placed outside the nest of the greylag goose is the specific stimulus that releases egg-retrieval behaviour. However, the releasers do not trigger a fixed action pattern in all members of a species. The animal must be in the proper motivational state. Only a *female* greylag goose in *nest* responds to a wayward egg outside the nest. But once she is properly motivated and in the 'right' surroundings, the greylag goose is not seriously discriminating about what she retrieves. **Lorenz** and **Tinbergen** showed that almost any smooth and rounded object placed outside the nest would trigger the egg-retrieval behaviour of this species; even a small toy dog and a large yellow balloon were dutifully retrieved.

2. If a baby bird is banded with a shiny ring while it is in the nest, the parents will sometime treat the band as if it were a piece of faecal matter, a substance that is always removed from the nest and dropped some distance away. An adult bird may pull vigorously the ring despite the fact that its offspring's leg is attached to the band and despite the cries of distress from the nestling (**Welty,** 1975). The parent may actually succeed in throwing the baby out. It is as if the visual stimuli from the band triggered a certain behavioural response that is played out once it is "turned on".

3. A male English robin (Fig. 9.14A) will furiously and repeatedly attack a tuft of reddish feathers glued on to a stick and placed in his territory by an experimenter. The bird reacts to this stimulus as if it were another adult male with a reddish breast intruding on his holdings (**Tinbergen,** 1951).

Fig. 9.14. Various examples of sign stimuli in birds A — A territorial male European robin will attack a tuft of red feathers while ignoring a more complete model of a robin that lacks the red breast. B — Willow warblers attacking the stuffed head of a cuckoo. C — A male red-winged black bird copulating with a mount consisting of the tail of a female raised in the precopulatory position. D — A group of jackdews attacking the arm of Konard Lorenz as he carries a pair of black bathing trunks in his hand. However, these birds react indifferently to a human who is holding a nestling in his hand-provided the nestling has not yet acquired the first black feather.

4. European warblers will likewise peck and rip at the stuffed head of the parasitic cuckoo if it is placed in their nest (Fig. 9.14B).

5. Male red-winged blackbirds can be induced to copulate with a few feathers plus the tail of a female blackbird provided the tail is raised in the position females customarily adopt prior to mating (Fig. 9.14 C).

6. Lorenz's tame jackdews, which normally trusted him completely, attacked him violently when he happened to pull a pair of black bathing trunks from his pocket in their presence (Fig. 9.14D). The stimuli provided by the trunks triggered the same sort of protective attack that is normally generated when jackdews see a captured companion in a predator's clutches.

Law of heterogenous summation. It is important to understand that each instinctive act is not summoned by a single releaser. Instead, a particular response may be elicited by a number of releasers. For example, it has been found that fighting behaviour in the male cichlid fish is elicited by five stimuli: 1. silvery blueness, 2. dark margins, 3. highness and broadness, 4. parallel orientation to opponent, and 5. tail beating. It was found that any one of these stimuli would elicit hostile behaviour and that any two would elicit twice the reaction of one (although the relative strengths of reactions would be very hard to measure precisely). Therefore, responses to releasers appear to be additive, so that the whole is equal to the sum of its parts (not greater than the sum of its parts). The additive effect is called the **law of heterogenous summation.**

The problem of studying releaser, or their specific components, the sign stimuli, are also compounded by other factors. For example, in some cases, a *set* of releasers is necessary to elicit an entire adaptive response. Bees may initially be attracted to a coloured paper flower, but they will not land unless the flower scent is also there. So in this case the visual releaser alone will not initiate a behavioural pattern that has survival value.

Innate Releasing Mechanism (IRM)

The sign stimulus, whether a sound, colour, appropriate structure, odour or movement, releases a highly predictable response. The strict correlation between stimulus and appropriate response led ethologists to conclude that the nervous system is organised into discrete 'centres' capable of filtering the sensory input, correctly selecting the releaser stimuli and releasing an appropriate motor response. **Lorenz** (1950) and **Tinbergen** (1950) called this organisation the **innate releasing mechanisms (IRM)**. There are following three important aspects of this concept. *First,* the mechanism is envisaged as being innate, that is, both the recognition of sign stimulus and the resulting response to it are inborn and characteristic of the species. *Second,* the IRM has the role of releasing the response to the sign stimulus, implying that the IRM holds back the pent-up action-specific energy, or drive, until the appropriate sign stimulus is recognised, upon which the energy is released in the form of appropriate behaviour. *Third,* the response released by the IRM is stereotyped and part of the animal's innate repertoire of fixed-action patterns (FAP). **Lorenz** has used an analogy to illustrate how an IRM operates: the sign stimulus is a key that unlocks only one of many possible doors; once the door (the IRM) is opened, the specific response inside is released.

FAPs, as originally conceived by **Lorenz** (1932), were activities with a relatively fixed pattern of coordination, somewhat akin to reflexes. **Lorenz** drew several distinctions between fixed action patterns and reflexes. *First,* FAPs can be released by a variety of stimuli, whereas reflexes are elicited by specific stimuli. *Second,* whereas animals are motivated to perform fixed-action pattern, this is not true of reflexes. *Third,* FAP can appear in the absence of external stimulus and are then called **vacuum activities.** Some of these points would be disputed now. For example, the startle reflex occurs in response to a variety of stimuli.

Strictly speaking, a FAP is a motor response that can be performed in the absence of any guiding sensory feedback. Once released, the behaviour is produced in its entirety without alteration, as if there was a programmed motor message to be played back in response to specific sensory signals. Although both the IRM and FAP are hypothetical constructs and have been modified in response to vigorous criticism over the years, there is growing evidence that there are physiological mechanisms that resemble IRMs and that generates FAPs. For example, **Willows** (1969, 1971) has shown that special cells within the brain of the sea slug, *Tritonia* do have a programmed score of motor messages that they play back without receiving sensory feedback when other cells detect the presence of chemical substances from their arch enemies, the predatory starfish. **Willow's** model of nerve action has been modified slightly as a result of later discoveries (**Getting,** 1975), but it is clear that the slug's brain has special units that help it to perform a highly patterned set of thrashing swimming movements that make up a FAP. Likewise, a male cricket's nervous system can produce a complete calling song even if the sensory components of its neural machinery have been severed. The cricket does not have to hear itself in order to sing a perfect calling song.

The essential property of an innate behaviour or FAP is that the response is performed in a completely functional manner *the first time* an animal of a certain age and motivational state encounters the correct sign stimulus. The *context* in which the sign stimulus appears and the *motivational state* of the perceiver are both important in determining what the response of an animal will be. A very good example of this point is the reaction of a mature female herring gull to a herring gull egg (**Tibergen,** 1960). The stimuli associated with an egg will *release feeding behaviour* if the female is not breeding and not incubating eggs herself or if the egg is located well outside her nest. The same egg will *release incubation behaviour* if it is in her nest with at least one other egg, provided the female has not incubated the clutch for some time or has had to leave her nest because of a disturbance in the colony. The female 'knows' when and how to incubate her eggs properly the first time she settles down on her clutch. The same egg will release *egg retrieval behaviour* in which it is rolled back into her nest, if it lies just outside the female's own nest (having been pushed there during a hasty departure or having been placed there by a gull ethologist). Again the female uses a functional retrieval response the first time she encounters her egg just outside her nest.

Baerends (1959) has provided another typical example of IRM concept. The fact that every releasing mechanism has its own sign stimuli is well illustrated by the digger wasp *Ammophila adriaansei* that catches caterpillars and drags them to its nest as food for its larva. The wasp may respond to the perception of a caterpillar in different ways, all depending on which instinct is of it activated. When it is hunting, a caterpillar is caught and stung; when it is found near the nest opening, just after the wasp has opened the nest, it is drawn in; but when it lies close to the nest when the wasp is filling the nest entrance, it may be used as filling material. Finally, when we put it into the nest shaft when the wasp is digging out the nest, then it pushes the caterpillar away, exactly as she would deal with another obstacle for instance, a piece of plant root. It is, therefore, the same object which with different conditions of the animal, releases different responses. Still in every situation the caterpillar is always sending visual as well as chemical stimuli to the sense organs of the wasp where they will always be transformed into impulses. But then it depends on the instinct activated in the wasp that which of these impulses will be intercepted somewhere and which can pass along a still unknown way in the nervous system finally to stimulate the principal motor centre of the reaction. There are indications that in each case different stimuli are working. When hunting, *Ammophila* very likely becomes aware of the presence of a caterpillar by its odour, but when it loses the caterpillar during the transport to the nest optical stimuli are used to find it.

8.8. SUPERNORMAL RELEASERS

It has been found that, in fish the numerous sign stimuli composing a releaser could be segregated and analysed independently and that a deficiency of one part could be compensated by an increase in another part (**Seitz,** 1940s). This obviously means that the releasing value of any sign stimulus is not fixed. If blue is good, bluer is better. Since the releasing value of any sign stimulus is modifiable, then a question arises regarding the extent to which a releaser may increase its effect. Interestingly, sign stimuli or releasers can be artificially 'boosted' so that they are preferred over their normal condition by the animal. Incubating herring-gull, greylag goose and oyester-catcher, all three prefer eggs larger than normal and will try to sit on 'eggs' that are normally marked but too large to incubate (Fig. 9.15). Ringed plovers (*Charadrius hiaticula*) prefer a more strongly spotted egg over their own eggs. Male sticklebacks preferentially court females which are large and have distended abdomens indicating that they have a lot of eggs (**Rowland**, 1989).

Such exaggregated releasing stimuli that are 'preferred' over normal stimuli are called **supernormal releasers.**

Fig. 9.15. An oystercatcher (*Haemotopus ostralegus*) shows a preference for a giant egg rather than a normal egg (foreground) or a herring-gull's egg (left).

There are many other examples of supernormal releasers/stimuli. For example, herring-gull chicks will peck at models of the parent's bill from which they are normally fed with regurgitated fish. The adult herring gull has a red beak with a yellow spot on the lower mandible at which the chick pecks. The artificial bill (Fig. 9.16) is thinner than real one, coloured red with three white bars at the tip. This 'supernormal' bill attracts more pecks from naive chicks than does a realistic copy of the natural bill and head (**Tinbergen** and **Perdeck,** 1950).

Another supernormal stimulus was revealed by the work of **Magnus** (1958) with the silver-washed fritillary butterfly, *Argynnis paphia L.* Males are attracted by the flashing yellow-orange wing patterns of females as they fly by. Magnus could attract males to a revolving drum which bore yellow-orange and dark stripes and 'flashed' a wing pattern at

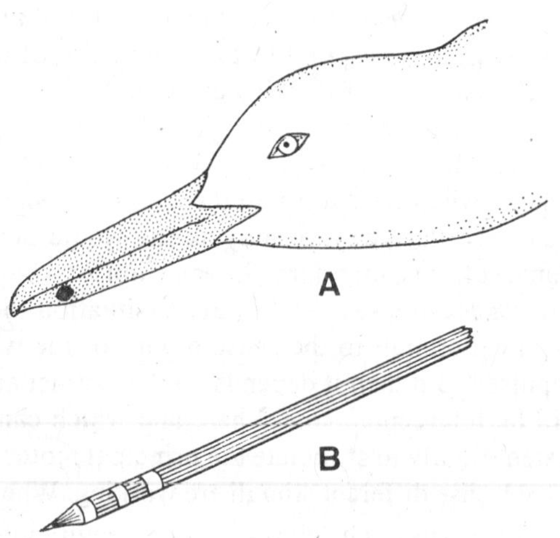

Fig. 9.16. A—Accurate three-dimensional model of a herring-gull's head; B—'Supernormal' bill. The later received 26 per cent more pecks from young chicks.

any required speed. The normal wing-beat frequency of a female fritillary is about 8 per second, but males show stronger responses the faster the wings of models are made to beat, upto as high as 75 per second.

Evolutionary constraints on supernormal releasers. In all these example there is no difficulty in understanding why the natural stimulus has not evolved further towards the supernormal condition. Female fritillary would attract more males if they beat their wings faster, but wings are not only for this purpose, they have to move insects through the air and there are severe mechanical limitations on their speed. Similarly a herring-gull's bill would probably be highly inefficient in all but attracting the pecks of its chicks, if it were as long and narrow as the supernormal model.

QUESTIONS

Long Answer Questions

1. Write an essay in communication in animals.
2. Explain the communication behaviours in honey bee. (*Bangalore 1994*)
3. Write a short note on waggling dance. (*Purvanchal 1997, 98, 99*)
4. Write a short note on the alarm calls.
5. Describe how does barn owl hunt in darkness.
6. Give an account of ecolocation in bats.

Short Answer Questions

1. Write a short note on sign stimuli. (*Allahabad 1994, 96*)
2. Write a short note on IRM.
3. Write a short note on supernormal releasers.

Multiple Choice Questions

Choose the correct answer from the four alternatives given.

1. The dancing language in honey bee is used between
   (*a*) queen bee and drone (*b*) drone and drone
   (*c*) drone and worker (*d*) worker and worker
2. The honeybees exhibit a type of dance to communicate the location of food. This is known as
   (*a*) round dance and waggle dance (*b*) break dance
   (*c*) waggle dance (*d*) tap dance
3. The biologist who discovered the meaning of the dances performed by returning honeybee foragers was
   (*a*) Karl von Frisch (*b*) Niko Tinbergen
   (*c*) Konrad Lorenz (*d*) Ivan Pavlov
4. If the honeybee is performing round dance it is conveying the information of food source from the hive at a distance of about
   (*a*) 50 metres (*b*) 100 metres
   (*c*) 1000 metres (*d*) 2000 metres
5. Which of the following is not true singing in male birds?
   (*a*) it is done to claim a territory (*b*) the typical song in characteristic of a species
   (*c*) all songs are learned from their parents (*d*) they generally sing at dawn or dusk
6. The faster a waggle dance is completed, the
   (*a*) farther away is the food (*b*) closer is the food
   (*c*) greater is the amount of food available (*d*) higher is the food off the ground

ANSWERS

1. (*d*) **2.** (*a*) **3.** (*a*) **4.** (*a*) **5.** (*c*) **6.** (*b*).

Pheromones
(Chemical Communication)

A **pheromone** (Gr., *pherein* = to carry; *harman* = to excite) is a chemical that is produced and released into the environment by one organism and affects the behaviour of another organism of the same species. Pheromones have a wide range of functions. For example, they are present on faeces or in urine and are used to mark territory in badgers or foxes; they provide information about sex and reproductive state of dogs and can delay and hasten sexual maturity in mice (**Indge,** 1997).

Pheromones are species-specific and produced by **exocrine glands** that are often hormone dependent. They are used as **chemosignals** in intraspecific communication, *i.e.,* in conveying information between individuals of a species. **Karlson** and **Luscher** (1959) coined the term 'pheromone' for substances that are secreted by an animal in the external environment and cause a specific reaction in a receiving individual of the same species, *i.e.,*an occurrence of certain behaviour or determination of physiological development.

Pheromones are also called **ectohormones** and released in minute quantity bringing about major effect by functioning as chemical messengers (**Novak,** 1975). They can be differentiated from hormones on the following bases:

1. Pheromones are released by the exocrine glands and transmitted externally through the environment.

2. Pheromones are species-specific and produce specific behavioural, reproductive and developmental responses in the bodies of other members of the same species. On the other hand hormones are released by endocrine glands and bring about changes in physiology of same organism.

Pheromones differ from some other types of chemicals used as signals among animals. **Allochemics** is the term given to those chemical signals that convey information between members of two different species (interspecific communication). Allochemics have been recognised into two groups: allomones and kairomones. While **allomones** adaptively favour emitter, **kairomones** favour receiver. Examples of allomones include repellents of certain insects, spray of skunk, etc. Example of kairomones include signals that enable the predator to locate its prey (**Dominic,** 1978).

10.1. DISCOVERY OF PHEROMONES

The discovery of pheromones took place by chance. A French biologist **Bonnet** in 18th century discovered the long trails of ants. He placed a colony of ants at one end of a table and at the other end he placed small pile of sugar. He observed the ants coming across the table to the heap picking up sugar and going back with it. The ants moved to and fro along a definite track, none of them diverting from it (Fig. 10.1A). Bonnet wondered why the ants followed such a definite path? To break the continuity of this invisible path he rubbed his finger across the path of ants running towards the sugar and saw that the ants stopped and searched about when they reached the rubbed place (Fig. 10.1B). They waved their antennae in the air and tapped

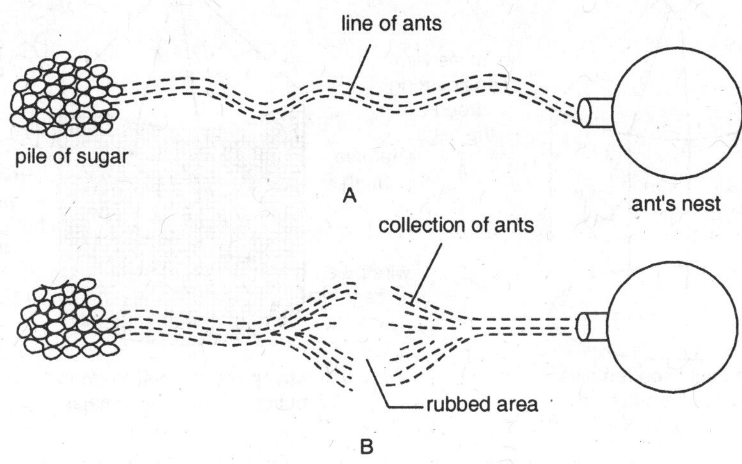

Fig. 10.1. Bonnet's experiment with ants to discover trail of chemicals.

them on the ground. There was a gradual crowding of ants on both sides of the line — some coming from the nest, some from the food, until a few adventurous ants run on the place where Bonnet had rubbed. Ants from both sides met, recognised each other and continued their journey once again. Bonnet thus assumed that the path of ants was actually a **trail of chemicals** which the ants could sense and follow.

Jean Henri Fabre (1823 1915) was a French entomologist who used direct observations of insects in their natural environments in his pioneering research into insect behaviour. **Fabre** one day placed a female moth in a cage near a window and to his surprise within 15 hours about 60 males of the same species came out of the woods and collected around it. Fabre began a series of experiments to try to discover what had attracted the males? When he placed the females in a tightly closed transparent container (Fig. 10.2A), where the female could be seen, no male appeared. On the other hand, a container made up of wire mesh (Fig. 10.2 B) occupied by the female moth which could not be seen but smelt. From this classical experiment **Fabre** concluded the presence of scent (or pheromones) in female moths.

10.2. TYPES OF PHEROMONES

Bruce (1970) had recognised following three types of main pheromones:

1. Releaser pheromones. These pheromones induce immediate and reversible bahavioural responses mediated directly by the central nervous system through first acting on neurohumoral pathways. These are used mainly in recognition of species members, sex, sexual status, aggression, ejection of milk, etc.

Bronson had substituted the term releaser pheromone by "**signalling pheromone**" with the reasoning that pheromones only transfer information and never dictate the type of response in the receiver.

Examples of releaser pheromones. Releaser pheromones are represented by **sex attractant, trail** and **alarm substances** of insects, fish, toad, mammals, etc.

1. In mammals, releaser, pheromone is present in the **urine** and **foot pads (Dominic** 1978). The pheromone present in footpads initiates the **aggressive behaviour** in a mouse on encountering an unfamiliar male. In mice, urine of male contains releaser pheromone for attracting females or organising aggressive activities and communicating fear between individuals.

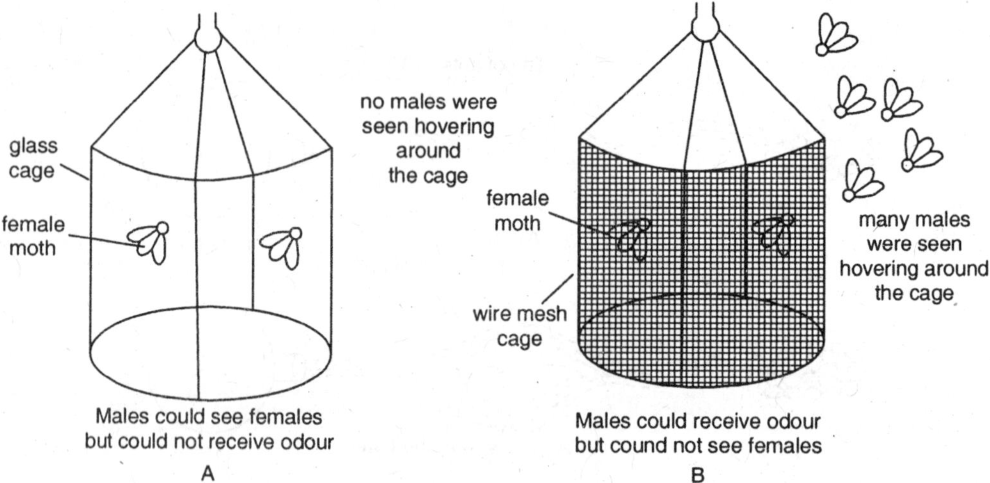

Fig. 10.2. Fabre's experiments. A — Males could see females but could not receive pheromones; hence no males were seen hovering around the case; B — Males could not see females but could receive pheromones, hence many male moths gather around the case.

In many other animals such as deer, dog, horse and sheep, etc., urine contains releaser pheromones which helps to elicit sexual attraction and arousal. **Vaginal secretions** of sheep, hamster and rhesus monkey act as releaser pheromone.

2. Whenever an *intruder* approaches a fish, **alarm pheromones** are produced in special flask-shaped cells of the skin. For example, the fish is frightened by the odour of the intruder — tadpole of toad. Such alarm pheromones are produced both by marine and freshwater fishes.

Alarm pheromones are also produced by certain skin cells of a toad tadpole. Whenever the tadpole is injured, all other in the medium move either to the bottom or run away from the injured tadpole.

Ants produce alarm pheromones in the form of **formic acid** from the abdomen to protect themselves from enemies. *Honeybees* and *wasps* also release alarm pheromones. In honeybee, while stinging she releases alarm pheromones along with poison and immediately after stinging she will move to other fellow members of the colony to show her sting and flatter her wing to aggravate the fellow members.

Alarm pheromones are also produced in many mammals such as pole cat, antelope, etc. Whenever an antelope is frightened, it will produce alarm pheromones from large glands concealed in the fur on the sacral region of the body by contracting sacral muscles.

2. The primer pheromones. These pheromones evoke a prolonged or long term endocrine or physiological responses in receivers mediated by the neuroendocrine pathways or through direct effects on the target organs.

Examples of primer pheromones. The common example of primer pheromones are those involved in suppression and induction of estrous cycle, termination of pregnancy and sexual maturation cycle.

1. If several female mice are caged together they affect each other's estrus cycles (sexual receptivity). The effect seems to result from an odour passed on from one female to the other. However, if a male mouse is introduced in the same cage it will shorten the estrus cycle. The pheromone responsible for the latter effect is present in the urine of male mouse and it is thought to be an androgen (testosterone) metabolite.

2. A pregnant female mouse will abort the litter she is carrying if she is exposed to the urine of strange male mouse. This is an example of the **"Bruce effect"**, named after **Hilda Bruce** who discovered that female mice tended to abort if a strange male mouse (*i.e.*, not the

father of her offspring) was around. Although it seems a rather maladaptive response on the part of female mouse to abort a whole litter, it appears that she is "*making the best of a bad job*". Male mice tend to kill newly born mice that they have not fathered themselves, so a female that resorb the embryo and starts a new litter with the new male is more likely to have surviving offspring than one that carries the litter to full term and then have it killed (**Manning** and **Dawkins,** 1998).

3. Honeybees, termites and ants have been most widely studied for observing primer effect. Among social insects, members of the same colony share a common pheromone which differs from that of the other colonies within the same species. The queen honeybee preserves her monarchy by exuding a pheromone called **queen substance** that keeps worker bees sterile and inhibits them from making new royal chambers or queen cells.

In termites, the caste system and size of the colony (*i.e.,* total number of individuals in a colony) are regulated by a pheromone, called **social pheromone** which is produced by a pair of reproductive termites (queen and king).

3. Imprinting pheromone. These pheromones act at critical period of developmental age and cause permanent change in the adult behaviour. Such pheromones have been found in some rodents, *viz.,* mice and rats.

Volatile and Nonvolatile Pheromones

Chemical signals are particularly well developed in insects and mammals. *Some chemical signals are designed to last only a short time:* they are volatile and tend to have low molecular weights. The chemicals used by some ants to signal *alarm* fall into this category of being short-lived, if they did not, then it would be impossible for the ants to signal the precise localization of a new source of danger. Ant alarm pheromones disperse well over a short range of 3-5 m, but they usually fade below detection levels within a minute or even less. It is even possible to make use of volatility of scent to achieve *patterning in time*. For example, some moths release their **sex attractants** (pheromones) in pulses at approximately 1 second intervals (**Dusenberry,** 1992).

On the other hand, other type of *chemical signals are designed to last for a long time,* allowing a message to persist in the absence of the signaller. *Territory markers,* for example, need to be persistent and therefore constituents must have a fairly high molecular weight (**Wilson,** 1975). The molecular weight cannot not be too high or the substance will be difficult to secrete and may not disperse well, *e.g.,* territory marker pheromone of spotted hyaenas which is secreted by **sub-caudal scent glands.**

10.3. ISOLATION OR EXTRACTION AND CHEMISTRY OF PHEROMONES

Isolation of pheromones has been a tedious task. **Robert T. Yamomoto** identified the sex attractants of cockroach *Periplaneta americana* by isolating it from a vapour emitted by female cockroach. He ran a stream of air through large cans containing thousands of female for a period of eight months trapping the vapour by freezing it with dry ice. From the condensate he could obtain only 12.2 mg of the pure sex attractant; this pheromone was a volatile yellow coloured substance having a strong floral odour. Male cockroaches respond the odour with intense excitement, raising their wings and making mating attempts. This pheromone was identified as **2, 2-dimethyl 3-isopropylidene cyclopropyl propionate.**

Another team of chemists at the Institute of Chemistry in Munich isolated the sex attractant of the female silk moth, *Bombyx mori.* They extracted the abdomens of five million females to obtain a tiny amount (12 mg) of the pure pheromone which they named as **bombykol.** The other attractants which have been isolated are the following: **gyplure** (gypsy moth), **civetone** (civet), **muskone** (musk deer), **honeybee queen substance** (bee), etc. Skunk, carnivore found from Canada to northern Mexico, produces a very foul smelling secretion from its anal gland. This secretion has three sulphur compounds: *methyl crotyl sulphide, crotyl and isopentyl mercaptan* and its main function is to repel predators and human beings. The musk deer and civet cat emit *muscopyridine.* The other acids which form active ingredients of pheromones are volatile *carboxylic acids* (red fox), *unsaturated lactone* (black tailed dear), *phenyl acetic acid* (mongolian gerbil), *isovaleric acid, isobutyric acid, n-butyric acid* and *isovaleric acid* (female rhesus; Table 10.1).

Table 10.1. Source and chemical composition of some pheromones.

Species	Source	Chemical composition
1. Silk moth	Lateral glands in female abdomen	$C_{16}H_{30}$ (Bombykol)
2. Gipsy moth	Lateral glands in female abdomen	$C_{18}H_{34}O_3$ (10-acetoxy-1-hydroxycis-7-hexadecane)
3. Tropical bug	Special duct in male	2-hexenol acetate
4. Ant	Mandibular gland	Formic acid
5. Honeybee	Head glands of queen	$C_{10}H_{16}O_3$ (9-keto-2-decenoic acid)
6. Male mouse	Urine	2-sec-butyl 4,5-dihydrothiazole and dehydroexobrevivomin
7. Primate species	Vaginal secretion	Acetic, propionic, isobutyric, n-butyric and isovaleric acids
8. Black tailed deer	Tarsal gland secretion	Cis-4-hydroxydo dec-6-enoic acid lactone
9. Pronghorn	Màle subauricular gland secretion	Isovaleric acid
10. Antelope	Interdigital gland secretion	Methyl ketones
11. Mongoose	Anal gland secretion	Lipids
12. Rabbit	Secretions of chin gland	Proteins, carbohydrates and lipids
	Inguinal gland	Phenyl acetic acid
13. Mongolian gerbil	Ventral abdominal gland	S_2-androst-16-ene-3-one and 3-hydroxy-S_2-androst-16-ene
14. Musk deer	Side glands	Lipids
15. Indian sheath tailed bat	Gular glands	Lipids and proteins

10.4. INSECT PHEROMONES

The insects have been most extensively studied for their pheromones and structures related to their production and perception.

1. Pheromones of Orthoptera (Locusts and Cockroaches)

Mature male *Schistocerca* which are bright yellow in colour accelerate the maturation of other less mature locusts of either sex. The pheromone responsible is secreted by the epidermal cells. Its effect is to stimulate the activity of corpora allata probably via nervous system. Cockroach is known to produce **sex attractants.**

2. Pheromones of Lepidopterans (Butterflies and Moths)

The males of Lepidoptera often produce scent from glands which are commonly associated with scales are known as **androconia** and are located on wings. Androconia have an elongated form and terminate in row of processes of fimbriae (Fig. 10.3 A). The pheromones produced by these glands evaporate through these fimbriae.

Pheromones are used to induce mating as in the case of the queen butterfly *Danaus gilippus.* The pheromone is released by the male and brushed on the female by a pair of brush-like structures, called **hair pencils** which are everted from the tip of the abdomen (**Green** *et al.,* 1990).

Males of *Amauris* have a small scent patch on each of the hind wings. These patches contain highly modified structure called **scent cups** (Fig. 10.3 B). Each scent cup has a median pore and below each cup is found a scent gland. Scent from these **scent patches** is dispersed by scent brushes associated with genitalia. In order to disperse the scent the insect lands with its wings spread and the bends towards one side and the scent organs are brushed.

In certain insect such as *Plaodia,* there are invaginated glands which open on either sides of the last abdominal segment (Fig. 10.3 C). The scent is dispersed by evagination of the gland.

Female silk moth (*Bombyx mori*) emits a sex pheromone, called **bombykol** to attract males from a pair of sacs (Fig. 10.3 F), called **sacculi lateralis** that are found on the last abdominal segment. The chemical formula of bombykol is as under:

$$CH_3 — (CH_2)_3 — CH = CH — CH = CH — (CH_2)_8 — CH_2OH$$

Most acute sense of chemicals is exhibited in nature is that of male Emperor moth *Endia pavonia* which can detect the sex attractant of female as far as 13 km.

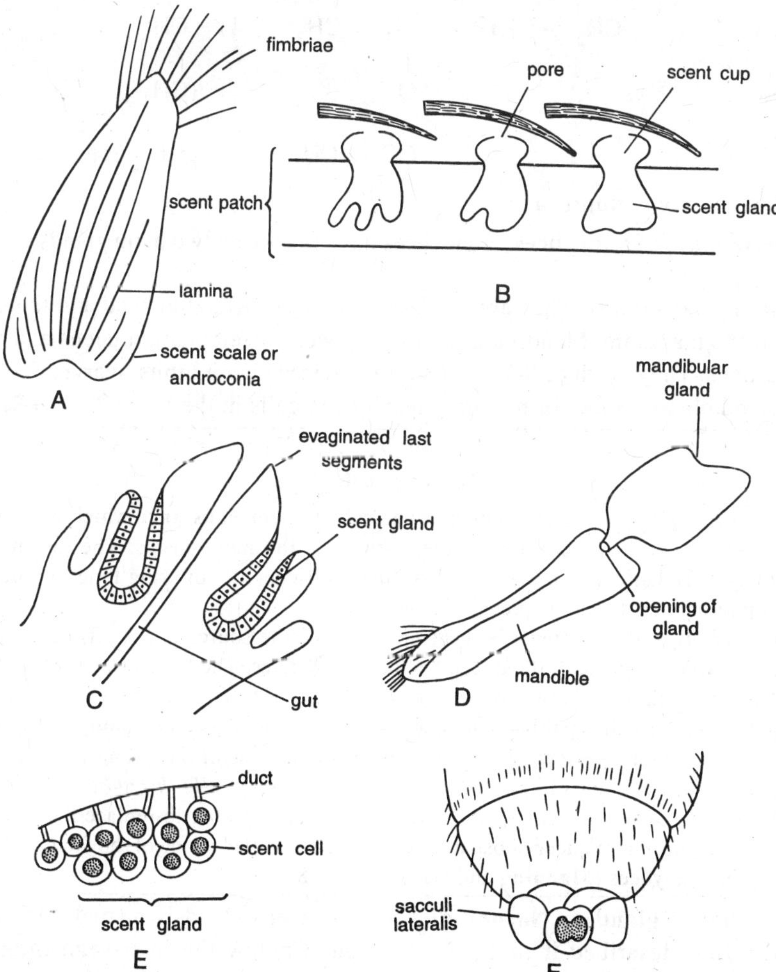

Fig. 10.3. Scent producing glands. A — Androconia; B — Scent cups; C — Invaginated scent glands; D — Mandibular gland; E — Nassanoff's gland; F — Sacculi lateralis.

Box 10.1.
Bombykol

We humans cannot smell bombykol, but it is potent in its effect on male silk moths. They can smell the bombykol released by a single female at a distance of over a mile. The male's olfactory organs are his antennae which are enlarged and have a fine mesh of side branches; they are so sensitive that according to experiments of **Dietrich Schneider,** a single molecule of bombykol elicits a response in the sensory neurone of the male. Only about 200 molecules need to hit the antennae of the male, for him to fly off in search of the female. The antennae of male moth also show a very specific response: they sense only bombykol and are not stimulated by even very similar molecules.

Once a male silkmoth has sensed bombykol, his only task is to fly in the correct direction; he must make the correct taxic response. He could achieve this by either of two methods. He could fly around, measuring the concentration of pheromone and then orient in the direction where the concentration increases; or he could simply turn upwind. Evidence indicates that moths follow the second rule: on sensing bombykol they fly zig-zag from side to side until they catch it again; and then they fly off upwind again. That set of response is enough to guide them to female.

The female gypsy moth *Porthetria dispar* attract males from over 4500 metres by releasing a sex attractant called **glypleure** whose chemical structure is.

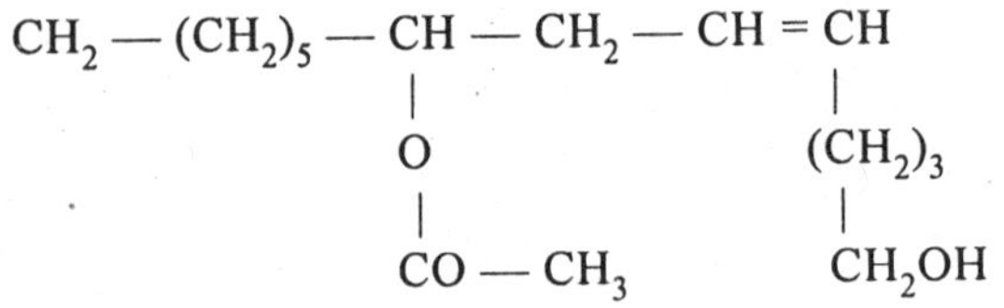

$$\begin{array}{ccccc} CH_2 - (CH_2)_5 - & CH & - CH_2 - & CH = CH \\ & | & & | \\ & O & & (CH_2)_3 \\ & | & & | \\ & CO - CH_3 & & CH_2OH \end{array}$$

3. Pheromones of Hymenoptera

***(i)* Pheromones of honeybees.** Honeybees have following two main glands which produce pheromones:

1. Mandibular glands. They are sac-like structures located in head and their ducts open at the base of the mandible. Mandibular glands are well developed in queens and workers but reduced in drones. They produce **9-keto-decanoic acid** which inhibits ovarian development of worker bees and obstructs the formation of new queen cells in the hive (Box 10.2).

Box 10.2.
Queen Substance

The queen honeybee secretes a pheromone called **queen substance** which both suppresses the ovaries of the workers and prevent them from rearing new queens. The queen is always surrounded by attendant workers who lick her body, subsequently offering food to other workers and with it pheromones from the queen.

The *level of pheromone* must be kept up and once the source is cut off, its concentration rapidly drops, which happens if the queen becomes ill or dies. The *effectiveness* of the incessant food sharing in circulating queen substance is shown by the fact that some workers in the brood area of the hive exhibit changed behaviour within an hour or two if the colony losing its queen. They begin to construct "emergency" queen cells in which some of the youngest larvae, destined in the normal course of events to become worker, *are fed royal jelly throughout their larval life and become queens.* These queens will eventually fight one another to replace their mothers. As colony grow, the dilution of queen substance below a critical level is one of the factors which leads to swarming in honeybees (**Manning** and **Dawkins,** 1998).

2. Nassanoff's gland or Nasnov gland. This scent gland is named after the Russian biologist who first described it in 1882. It is found below the inter-segmental membrane (arthrodial membrane) between 6th and 7th abdominal segments, which is exposed by bending the tip of abdomen of bee. This gland is made up of a number of large cells which open to exterior by small ducts (Fig. 10.3 E).

The queen bee attracts drones by a pheromone, the principal component of which is **9 oxodecenoic acid** produced in the mandibular gland.

(*ii*) Pheromones of ants. The social behaviour of ants is controlled mainly by pheromones: they also use visual and auditory signals, but most of their signalling is by means of smells (pheromones). Not only mating is controlled pheromonally in ants, but also finding and exploiting food, recruiting nest mates for battle, warning about enemies. The total number of scent glands employed by an ant such as *Iridomyrmex humilis* (Fig. 10.4) is enormous (more than one dozen) and all the ant's scent are meaningful. For example, if a lone ant finds a food source too large for it to bring back to the nest by itself, it runs back to the nest leaving a **pheromone trail** on the way. Different kinds of ants leave trails from different pheromone-releasing organs. *Solenopsis* releases its trail pheromone from its Pavan's gland; *Myrmica* from its poison gland; *Lasius* from its rectal gland. When the forager (food-finder) ant arrives at its nest it uses another pheromone to recruit other ants to come and collect the food; *Myrmica rubra*, for instance, having laid its trail from its poison gland, attracts its nest mates back to the food with a pheromone from its **Dufour's gland.**

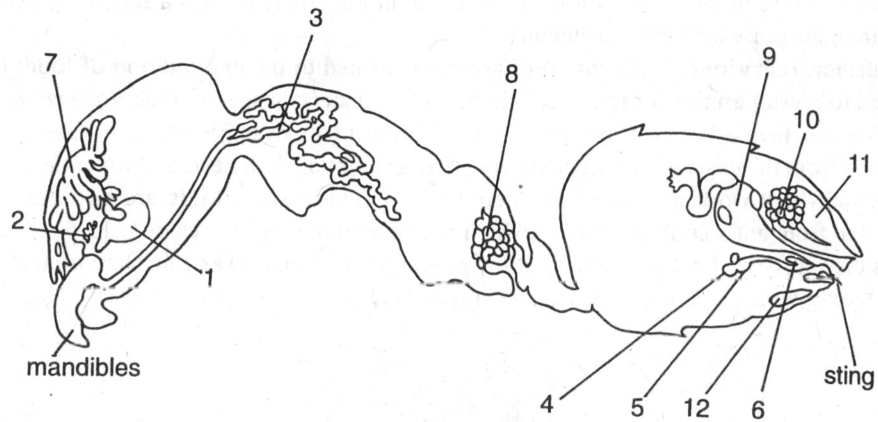

Fig. 10.4. Pheromones are manufactured and released from special glands. Here are the 12 main pheromone-releasing organs of the ant species *Iridomyrmex humilis*. 1. Mandibular glands; 2. Maxillary gland; 3. Thorax labial gland; 4. Poison gland; 5. Vesicle of poison gland; 6. Dufour's gland; 7. Post-pharyngeal gland; 8. Metapleural gland; 9. Hindgut; 10. Anal gland; 11. Reservoir of anal gland; 12. Pavan's gland.

Recruitment is not always affected by mass-acting pheromones. The food-finder might instead single out one another nest mate and then lead it alone back to the food source. *They match the recruiting party to the size of the food source.* **Bert Holldobler** has described how *Leptothorax* workers may recruit and lead a single ant when the food source is too big for one ant but only needs two to move it. When in *Solenopsis*, the food source is so large that many ants are needed, then **mass-acting pheromones** are released.

The scent trail of the ant is released as a volatile liquid (which evaporates) on the ground; but it is known that it is perceived by scent rather than taste. In an experiment, leaf-cutter ants of the species *Atta texana* are to follow a trail by walking along a plastic roadway placed just above the trail; they followed the trail as usual but must have done so by the airborne odour of evaporated scent. Like the silkmoth, ants sense pheromones through their **antennae**; but they make continual use of both antennae to keep them in right direction. Unlike the moths, which ignore relative pheromonal concentrations and simply fly upwind, ants steer by balancing the pheromonal concentration to the right and left. If the concentration increases to the right they turn in that direction until both antennae are sensing equal concentrations. This rule will guide the ants straight down the odour trail. In an experiment, **Hangartner** placed two trails in parallel; to begin with they were of equal concentration, but as they proceeded one of the trail (upper one in Fig. 10.5) became progressively

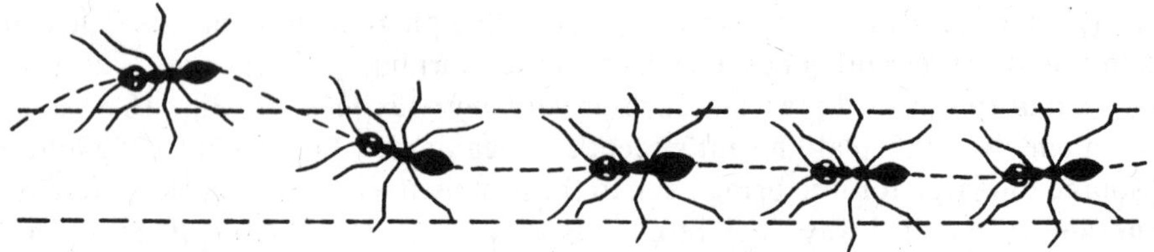

Fig. 10.5. Ants (*Lasius fuliginosus*) follow pheromone trails by balancing concentrations smelled through its two antennae.

weaker. The ant *Lasius fuliginosus* in the experiment started by walking between the two trails and then moved across to the stronger trail (bottom one in Fig. 10.5) in such a way as to balance the odour strengths sense by their two antennae.

Further, **recruitment pheromones** are not confined to the exploitation of food; they are also used to recruit armies for territorial disputes, or (in certain species) for *slave raids*. Not all ant species enslave other ants; but *Formica subintegra* (for example) does. It raids the pupae and larvae from the nests of other ant species, brings them back in nest, and when the pupae and larvae hatch they work as slaves for their captors. These slave-raids are coordinated by a pheromone from ant's Dufour's gland (which is enlarged in this species, Fig. 10.6 A). Dufour's gland is the source of the pheromones, called **propaganda substances** which are used to confuse the defending workers in the nest that the slave-makers raid (**Wilson,** 1971).

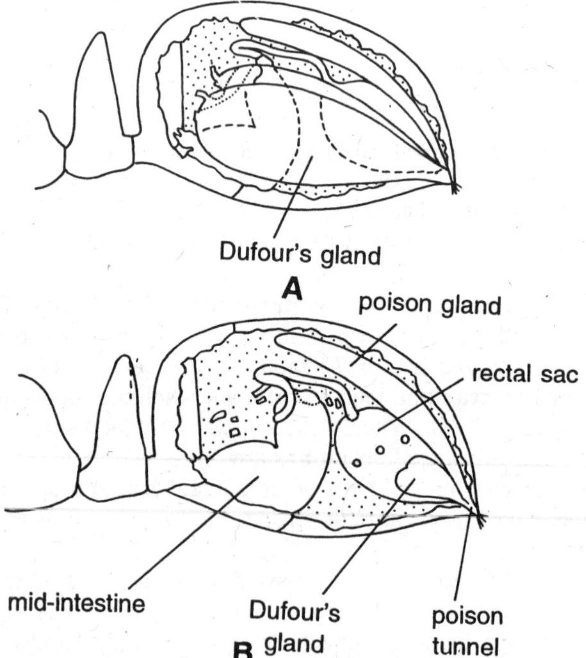

Fig. 10.6. The slave-making ant species *Formica subintegra* shown at the top has an enormously enlarged Dufour's gland (A) compared with a more typical ant species such as *Formica subserica* (B).

Moreover, with this kind of warfare going on between nests, the ants need to be able to distinguish their own nestmates from members of other nests. This distinction is made possible by other pheromones, called **colony odours.** Each nest of ants has its own distinctive smell.

Ants warn their nestmates about enemies with **alarm pheromones.** On sensing an alarm pheromone an ant may do any of a variety of things: it may run away from the source of the scent; it may freeze and play dead, or it may run towards the source of the scent and attack any nearby enemies. The different responses of ants are probably stimulated by different alarm pheromones. The *diversity of messages in an alarm pheromone* of an ant, the weaver ant *Oceophylla longinoda*, has been revealed by the chemical analysis of it. Its alarm pheromone contains over 30 different chemicals. The effects of only four of these were tested upto 1986. One makes the recipient ant generally 'alarmed', another makes it run towards the source and two others make it bite.

Lastly, in Africa, the larvae of **driver ants** secrete a **larval pheromone** which nurtures the larvae. According to **E. O. Wilson** in the ant nest, different pheromones are released to keep normal life of the ants in the nest. Such pheromones are called **organisation pheromones.**

Structures (Receptors) of Insects Perceiving Pheromones

Auguste Forel reported various structures on the ant's antennae and on other insects which seemed to be well adapted for receiving chemical substances. These structures were flat in some insects, cup shaped in others while in some they were in the form of short stubby hairs (Fig. 10.7).

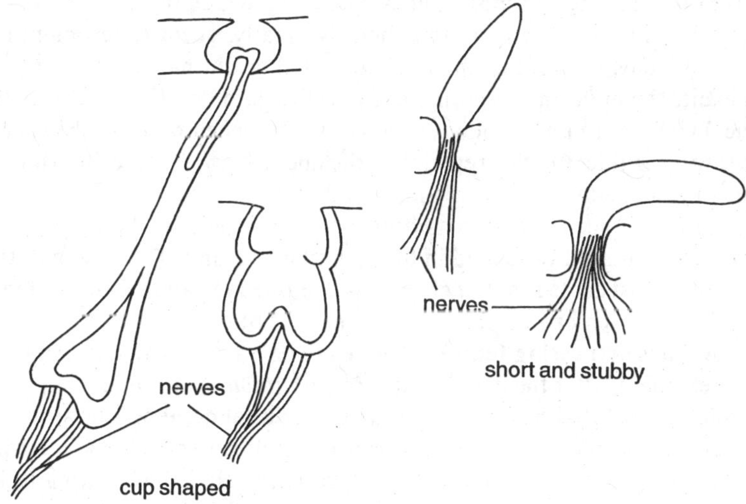

Fig. 10.7. Structures (receptors) perceiving pheromones in insects.

All these structures were supplied by nerves. The pheromones released by a conspecific are perceived by these specialized structures (receptors) and the messages are sent to control nervous system. Despite many variations, all chemically sensitive structures share a basic plan: the receptor cells and nerves. The receptor cells are positioned in such a way that stimulus-substance molecules from the surrounding medium can easily reach the special areas of plasma membrane. Certain animals have projecting antennae specially designed to receive pheromones.

German physiologist, **Dietrich Schneider** and his associates demonstrated that the male silk moth receives the stimulus of pheromones by specialized structure located on their antennae which act as **chemoreceptors.** They further discovered two specialized sensory structures on the surface of the antenna-**sensillae coeloconicae** which were pit-like and short rod-like **sensillae basiconicae** (Fig. 10.8).

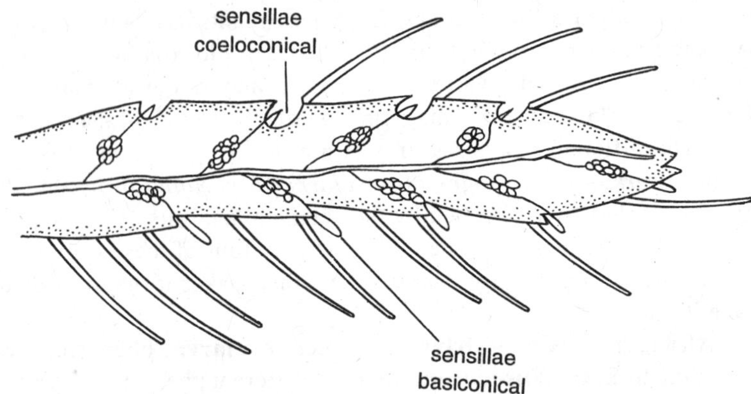

Fig. 10.8. Chemoreceptors on the antenna of male silkmoth.

Functions of Insect Pheromones

1. Insect pheromones as sex attractants. Pheromones are used by many insects to bring sexes together for mating. These pheromones are called **sex attractants.**

(i) Pheromones attracting males. Pheromones in female insects are usually released by exposing the glands by movement of abdomen. Normally, scent (pheromone) is released at particular times of a day, depending upon the nocturnal or diurnal nature of animal. The effect of scent is to excite the male and to promote take off toward her. Eventually as the male insect approaches the female, there is continuous increase in the concentration of the chemical attractant and this serves as a guide for the remaining distance. How far and in which direction the pheromone spreads, depends largely on the wind direction.

For example, the queen bee attracts drones by a pheromone, the principal component of which is **9 oxodecenoid acid** produced in the mandibular gland. The power of this pheromone is quite remarkable. Most acute sense of chemicals exhibited in nature is that of the male Emperor moth *Endia pavonia* which can detect the sex attractant of the female as far as 13 km.

(ii) Pheromones attracting females. There are only few examples of male producing sex attractants. For example, after the male beetle *Harpobittacus* has caught the prey and started to feed, two vesicles are everted from between the posterior abdominal segments. These vesicles are expanded and contracted and the scent is released which excites female for mating. On her arrival male present her the remains of prey and while female is busy eating it, he copulates.

2. Pheromones of social insects. The pheromones of social insects fall roughly into two categories according to their function:

(i) Those concerned with communication between conspecifics.

(ii) Those conserved with the maintenance of colony structure or caste system.

(i) Communication. (1) Many ant species lay scent trails by which they are able to find their way about. **E. O. Wilson** (1958) has reported that the trails of fire ant are formed by the pheromones of Dufour's gland. These trails can be of different types, one leading to food and other may just be a routine passage or it can be a trail of emergency exits. (2) If an ant is attacked in a colony it secretes **alarm pheromone** from mandibular or anal gland which attracts other members of the colony that quickly reach for the rescue of conspecifics. The alarm pheromones among social insects have three main functions: (i) to alert the colony, (ii) to release aggression and (iii) to mark the target to be attacked. (3) If you are stung by a bee and you reflexly kill that bee, the dead stinger might invite hundreds of bees; there are **alarm** and **distress pheromones** that are released from stinging or dead bees respectively which trigger a mass attack from the hive. Distress pheromones are very highly volatile to save the colony from a continuous disturbed state. (4) Worker bees are attracted to each other by a scent from Nasanoff's gland. The scent is released in various situations, *viz.*, at the time of feeding, during formation of new hive, to mark a source of water, to mark hive for recognition and for maintaining cohesion of swarm during flight.

(ii) Maintenance of structure of the colony. Queen bee produces pheromones from two glands *viz.*, mandibular and Nasnoff's, and she rules over a colony of 6000 to 8000 individuals for 5 to 7 years. To make those 8000 individuals work is controlled by producing varied concentrations of pheromones from just two glands. She decides the number in a caste, she prevents further construction of royal chamber and she passes orders to all the workers who execute work tirelessly. If a queen is removed from a colony her absence is perceived immediately by the workers who become restless, immediately they start building emergency **queen cells** or **royal chambers** by enlarging and reshaping the existing worker cells and the larvae within them are fed Royal jelly instead of Bee's bread. These larvae then grow into queens. The queen which comes out first of all locates sister queens with the help of pheromones and sting them to death before they even emerge.

10.5. PHEROMONES IN VERTEBRATES

A. Pheromones in Reptiles

Male garter snakes can follow **chemical trails** left by females and can even determine the direction in which the female was going. As the female moves along she pushes against certain objects, thus leaving more scent on the side in her direction of travel. The male tests each side of these objects with his forked tongue and recognizes which side the female has pushed against (**Ford** and **Low,** 1984).

B. Pheromones in Mammals

1. Scent marking by urine and faeces. Urine and/or faeces of many mammals have pheromones through which they can communicate with conspecifics. The most common use of urine and faeces is to scent mark a **core area** (an area which is used extensively and defended actively), **home range** (an area around the core area which is used actively but not defend actively) or **territory** (a bigger area around home range used but not defended). Some animals use urine and faeces for marking pathways, resting grounds, feeding grounds, and sleeping sites. Some animals urinate or defaecate on rivals, opponents, defeated conspecifics, etc.

(*i*) Male *Hippopotamus* marks its pathway between aquatic resting place and grassy feeding ground by the deposition of dung at places along the trail, usually close to some conspicuous objects. As hippopotamus defaecates while walking on the trail, it keeps moving its tail rapidly from side to side (Fig 10.9). In this way its faeces is splashed out and get deposited all over the vegetation above ground level. In this way it is placed at the nose height to make it more noticeable.

(*ii*) Similar to hippos, rhinos mark their pathway by defaecating and depositing piles of faeces at prominent landmarks so that it can be smelt and seen (Fig 10.10).

Fig. 10.9. Hippopotamus: its males use faeces for scent marking.

3. **Tigers** produce a pheromone called **tigeramine** which is a milky thick fluid with strong smell; it is passed out along with urine of tigers. Male tigers use this pheromone to mark their territories (**Brahmchay,** 1981). Tigeramine of different tigers contain **phenylamine** as its main and common constituent.

4. Main land **serow** (goat antelope *Capricornis*), found in northern India, has fixed pathways of movements. Initially these pathways are used visually and later the animal marks traditional spots by deposition its droppings there, so that it can just follow the scent and others are kept away.

5. **Polecat** and **civet** deposit urine and faeces at particular places along their trail.

Fig. 10.10. African rhinoceros: its male uses faeces for scent marking.

6. During their mating season, the adult males of **proghorn** (an antelope found in Canada, USA, Mexico) and **Sambhar** become territorial and mark their areas with their droppings and urine (Fig. 10.11).

Fig. 10.11. Sambhar uses urine and facial glands for scent marking.

Fig. 10.12. Skunk uses urine and faeces for scent marking at nose level.

7. The male **skunks** and **giant rats** adopt an unique style to mark the area at nose level for easy detection. They place their droppings above ground level by standing on fore limbs and hind limbs supported against some solid object as they defaecate and urinate (Fig. 10.12).

8. **Dogs, coyote** and **foxes** use just the urine to mark their territories (Fig. 10.13). The pheromone left with urine indicates minute details such as individual's identity, male or female, dominant or subdominant, adult or subadult and also how long ago that area was marked.

9. Like tigers, males of **lions, leopards** and all other big and small cats mark their home range with pheromones which is passed with urine. The cats select some suitable objects, back up to it, extend the back legs, raise the tail as the urine is passed, the tip of penis is curved backward so that the urine is given out as a fine jet between the hind legs.

Fig. 10.13. Fox uses its urine for scent marking. **Fig. 10.14.** Cheetah uses its urine for scent marking.

10. **Bears** use two different methods to mark their territory. They first mark a tree by scratching and chewing the bark, then they frequently urinate on such trees to consolidate the marking.

11. **Spotted hyaenas** (*Crocuta crocuta*) mark their clan territories both by smearing grass stems with paste from **sub-caudal scent glands** and by depositing faeces at latrines (*i.e.,* it carefully locates a suitable grass stem in the cleft of its extruded scent gland so that the paste-like exudate are placed conspicuously and at the right height above the ground). Where the territory is small, scent is placed strictly along the territorial boundary, but where small clan occupies a large territory and it is impossible to keep the boundary fully scented, both pheromones and faeces are deposited at strategic places within the territory (**Mills** and **Gorman,** 1987). Either way, other hyaenas are informed that a territory is occupied even when the owners are not around.

12. **Slow loris** occupies small territory which it marks with its urine. **Slender loris** sprinkles urine on tree trunks to mark its territory.

13. **Capuchin** (*e.g.,* white fronted capuchin) is found in southern central America and lives in large groups among the dense foliage of dense trees. Each group has its own home range in which the members move about. The adult members soak their feet and hands in their urine known as "**urine washing**" and smear it on the foliage. Urine washing is also seen in **slender loris.** Similarly, common **squirrel monkey** of northern South America, for demarcating its own territory 'leaves' scent by soaking its body specially tail in urine.

14: **Greater bushbaby** has a peculiar habit. While urinating one hand is half cupped in front of urine which is rubbed on tree branches. Here scent marking is done to identify the home or area where they spend maximum time. Some **lemurs** mark their territory by smearing their faeces on branches, quite like painting the house.

2. Scent marking by special glands. Many species of mammals have evolved special glands producing the scent. These have the advantage that their use can be restricted to necessity, *i.e.,* the animal could use pheromones as and when needed. This prevents their automatic and unwanted release with urine and faeces.

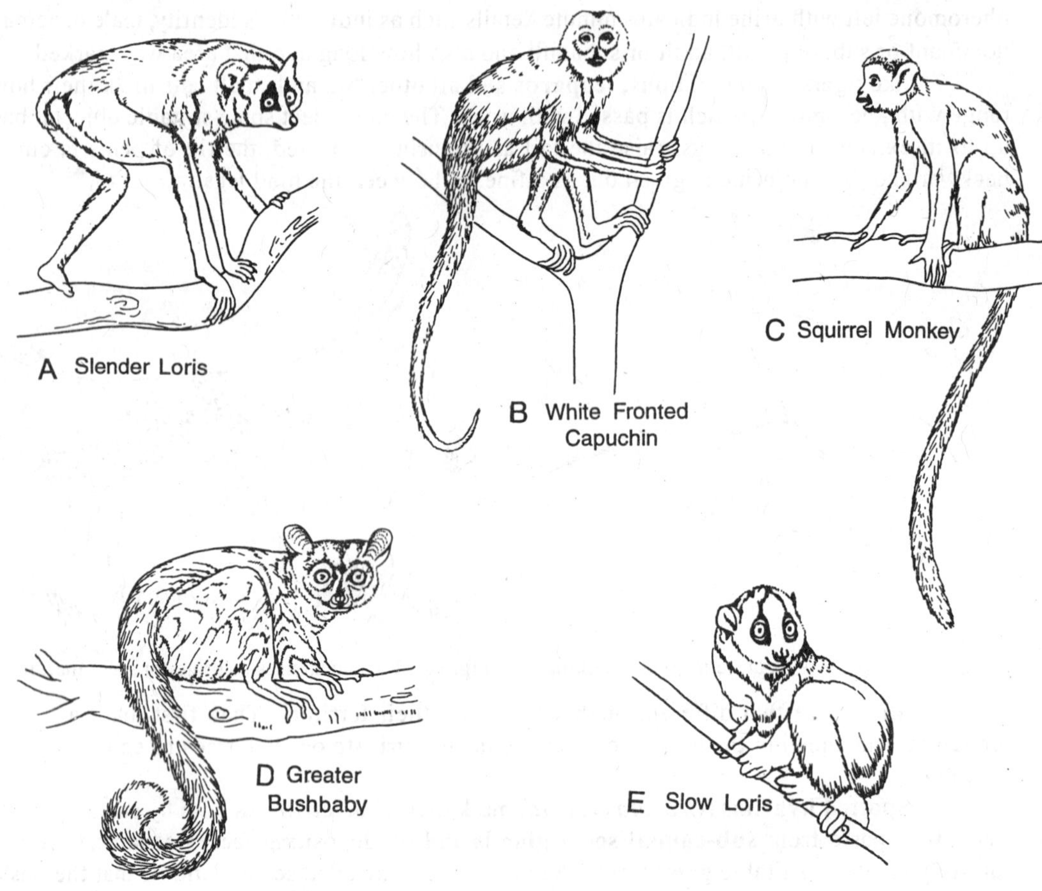

Fig. 10.15. Some mammals who use urine for scent marking.

(i) Anal glands. The commonest scent producing glands are the **anal glands.** In this case, to apply their secretions animals have to evert anal glands and rub them on the ground or on objects to be marked. Such method of marking occurs in monotremes, marsupials and in few placental mammals.

The **brown hyaenas** (*Hyaena brunnaea*) have two separate anal glands for different messages. They leave secretions as two separate pastes (one brown and other white) next to each other on stalks of grass. The male **beaver** marks dams, lodges and feeding trails by the pheromones produce by his anal glands and virtually keeps strange beavers out of his territory. He makes mounds by scooping up mud from his pond and on them he squats to deposit a smear from his anal scent glands. **Spiny ant eater** (Fig. 10.16) everts the cloaca and rubs it on the ground leaving a scent mark. The **brush-tailed opossum** and **sugar glider** have two types of **cloacal glands** with the help of which they mark their nests. In the **yellow footed marsupial mouse** cloacal secretions are rubbed on twigs and branches around their small territory. The anal glands of **black and red tamarin** (squirrel - like marmosets of Guiana and Amazon valley) produces a strong smelling scent which keeps others away. Male **tasmanian devil** has scent glands around its anus which it rubs on objects such as stones, branches, ground and grass to protect his female and young ones.

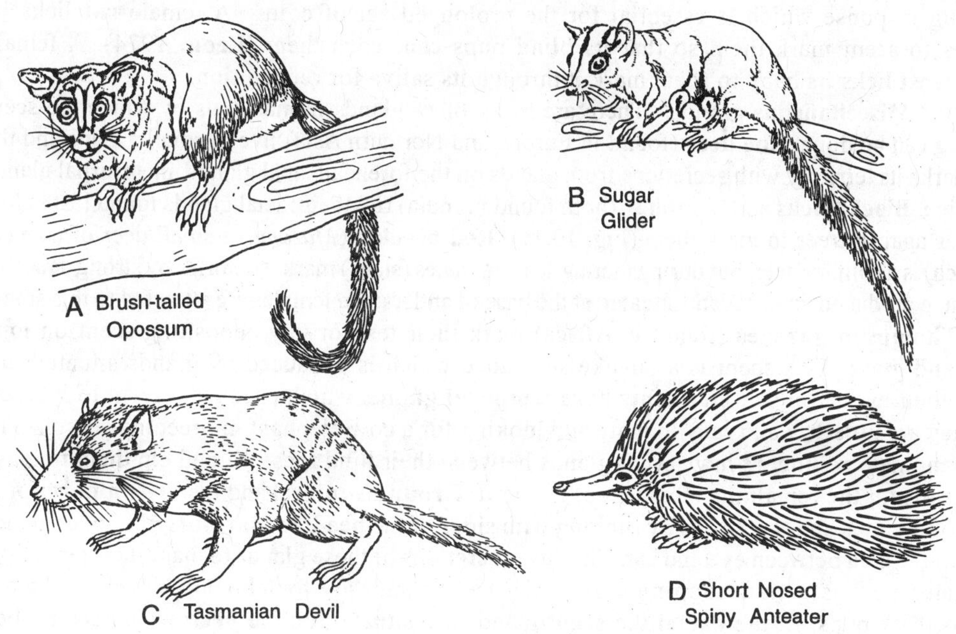

Fig. 10.16. Animals who use secretions of anal glands for scent marking.

In **ring tailed lemur** (Fig. 10.17) both male and female rub their glands to mark their territory. Male lemurs, however, contain additional scent glands under chin and on the penis which secrete pheromones to attract females. **Verreaux's sifakas** (Fig. 10.17) mark territories with urine and with secretions of male's throat glands. Brown hare of Europe marks its territory with strong smelling secretions produced from glands situated in its anal region, around its face and inside its cheeks. These secretions also help during its mating season to attract the mate.

Fig. 10.17. Animals who use secretions of glands found in forearm, throat and saliva.

(iii) Salivary glands. Saliva has been used as a marking agent by a variety of mammals such as bears, dogs, pigs, rats, etc. Most **marsupials** deposit saliva on twigs by chewing them and hence marking the foliage of their territory with saliva. **Hedgehog** first salivates on the object to be marked and then scratches it. The saliva of **boar** contains steroids that induce a standing response which is essential for the prolonged act of coitus. A female **rat** licks its nipples to scent mark them so that its blind pups can reach them (**Leon,** 1974). A female **wildebeest** licks its baby to scent mark it through its saliva for recognition.

(iv) Miscellaneous glands. There are many other glands in mammals by which the scent marking can be done. **Roe deer** (found in Europe and Northern Asia) live in small groups and the stag marks its territory with secretions from glands on the forehead, anal glands, metacarpal glands and urine. **Black bucks** and **Sambhar** (both found in India) rub their facial glands found just below the eyes against trees to mark them (Fig. 10.11). **Red brocket** (*Mazama,* a small deer of tropical America) is a solitary deer, but during mating season, males (stags) mark territories by strong smelling substance produced from a gland situated at the base of antlers; the females are attracted to this smell.

Thompson gazelles (found in Africa) mark their territory by depositing scent on long twigs and grass. This scent is a tar-like substance which is produced by glands situated just below their eyes. Bulls of **elephants** have **temporal glands** which produce a scent to indicate that they are **musth,** *i.e.,* ready to mate and looking for a cow elephant in breeding season. The dominant male **reindeers** have scent glands between their hind toes, which help them to leave scent trails for the remaining members of the herd. **Ground squirrel** and **American porcupine** mark their mates and territories by rubbing with sides of the head. In **marmots,** the scent glands occur in the area between eye and ear. They use secretions of these glands to mark their territories and mates. In **Australian hopping mouse,** the scent glands are found below the chin which it uses to mark pups. In the **camel** the scent glands are situated on the neck which are rubbed against the object to be marked and play an important role in mating. **Black tailed deer** has five different scent producing glands (Fig. 10.18) which are located at various places, producing different pheromones to carry out different functions. The male **musk deer** has a sac on the belly whose secretion attracts the female during the breeding season and it helps in marking scent at the territorial boundaries.

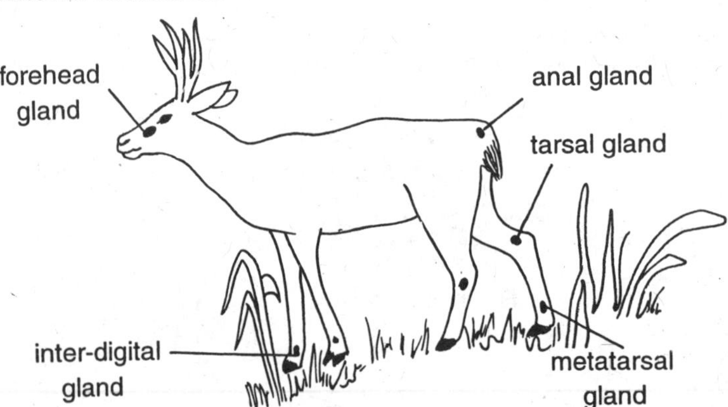

Fig. 10.18. Locations of different scent glands in the black tailed deer.

Functions of Pheromones in Vertebrates

In vertebrates pheromones have the following functions:

1. Scent marking provides sense of belonging. Koth Kolar observed that for animals, any new object was a source of unease until it had been safe by scent marking. The familiar smell gives a feeling of safety and sense of belonging. By receiving its own body scent from its dwellings, it gives them a feeling that "*this area belongs to me and I am safe here*".

2. Enhancement of self-confidence. Pheromones increases the self-confidence of animals. **Kuhme** (1963) recorded that the African elephant, when facing a rival, during the breeding season will sometime turn his trunk round and smell his own temporal gland. This appeared to increase elephant's confidence and make him more likely to attack.

3. Regulation of sexual behaviour. Pheromones have two types of roles in sexual behaviour: in marking the territories of male and in attracting the opposite sex for mating. The males mark their territories and ward off the rivals, *e.g.*, deer and antelopes. Or the rivals do not trespass marked territory, *e.g.*, tigers, other cats and dogs. Male dogs tends to urinate to mark all possible objects which fall in his territory — scooter, car, cycle, lamp-post, etc. Sex attractants brings two sexes together in most of the vertebrates.

4. Trail marking. Many animals mark their trail just to leave a record of their comings and going to be read by themselves and subsequent visitors in hippo and rhino. These sign posts are regularly visited and inspected by neighbours and thus each animal can keep a check on his fellows without actually seeing them. A new comer in the area will at once be detected and may be challenged. By marking the pathways animal does not lose path and *conserves energy* by not finding the way afresh every time it moves.

5. Territorial marking helps in avoiding recurring fights. Territory is first acquired by fighting, then the territory is loaded with animal's scent indicating that no more fights are required for this particular territory.

6. Miscellaneous functions. Sometimes males mark their mates with scent, *e.g.*, male rabbit marks his female with his chin glands secretions. Hamster mark a defeated rival and marked animal subsequently retreats. Mice and rats similarly mark a defeated opponent with urine. A wildebeest sniffs her newborn calf to form a permanent olfactory bond with her offspring. The calf also sniffs mother for her odour. Many mammals such as bighorn sheep and dogs often sniff and smell each other (Fig. 10.19) and when they do so, they can probably know the sex, the group it belongs to and its readyness for mating.

Fig. 10.19 Sniffing by bighorn male sheep of female's posterior.

Among many vertebrates the readiness of a female for mating is perceived by a male through pheromones produced from vagina. Therefore sniffing the genitalia is a common sight in ready-to-mate animals, *e.g.*, pheromone **copulin** is produced from the vagina of rhesus female. The male rhesus would inspect the hinder quarters of a female in heat before copulating.

7. Pheromone in avoiding inbreeding. Mice have been recently shown to use smell to distinguish mice that are genetically similar or genetically different from themselves in one particular genetic region. The **major histocompatibility complex (MHC)** is a group of genes that were originally thought to be entirely concerned with recognition of "foreign bodies" or tissues. It has now been established that *odour cues associated with differences in the MHC* *affect not just cellular recognition but behaviour as well.* Which companions a mouse nest with and which mate it chooses are both affected by its genetic similarity to those other mice at this particular region (**Lenington,** 1994). By smelling out mates that differ genetically from themselves, mice have a built-in way of avoiding interbreeding.

It has recently been discovered that similar effects may operate in humans, suggesting that our sense of smell may be more sensitive than we realise. Human females were asked to rate the smell of T-shirts that had been worn be six different men, they reported that the smells were pleasant from men most different from them in the MHC region and least pleasant from men who were similar (**Wadekind** *et al.,* 1995). What is more, the smell of the MHC-dissimilar men remained the woman's most of their own actual or former mates, implying **disassortive mate preference** in human (Box 10.3).

BOX 10.3.

The very word "perfume" has feminine over tones to many human male ears. Men can use "deodorant" and possibly "after shave", but the idea of all those dainty little bottles with their fussy parapheranilia is too much for the sensitive male ego. Yet no perfume industry can afford to neglect half its potential market, and perfume-markets are ever keen to crack the shall of male silence. **Craig Robert** of University of Liver Pool and his colleagues have been investigating this problem. They already know that appropriate scents can improve the mood of those who wear them. They have discovered that when a man changes his natural body odour it can alter his self-confidence to such an extent that it also changes how attractive women find them.

There are *three* broad theories of perfume use. One is that people employ it to mask body odours that they perceive as bad. The second is that some perfumes contain chemicals that mimic human pheromones – ambiguous, mysterious substances believed by some to play a role in mating. The third is that people use it to heighten or fortify natural scent, and thus advertise sexual attractiveness or availability.

All three theories could be true. In particular, the role of perfume as an olfactory disguise is obvious. Even here, however, there are some implied twists. Bad smells are not just a matter of poor hygiene. Illness and old age both brings characteristic odours of their own, and neither state makes people more attractive. Perfumes may bluff there messages. Hence the recent marketing a new scent called **Ageless Fantasy,** by Harvey Prince, which claims its product hide the "odour of ageing", suggested to be caused by the breakdown of a particular fatty acid in the skin.

As to pheromones, whether humans have these is questionable. A **pheromone** is a chemical that elicits a specific behavioural response at a distance. The most likely human candidate is a substance called **androstadienone.** This is a derivative of **testosterone** (hormone) that is found in men's sweat and is known, from brain-scanning studies, to promote activity in parts of women's brain. That this results in changes in behaviour has not, however, been clearly demonstrated.

The most interesting area, though, is the interaction between perfumes and natural scents that carry messages but do not have the specific properties of pheromones. Odours co-ordinate a wide range of human behaviour. Mothers can recognize their children by smell. Children can recognize each other. Relatives can be distinguished from non-relatives, even to the extent of understanding who is genetically different enough from the smeller to be a good choice of mate. The sexes themselves smell different, too. It is often claimed that women can learn information about a man's social status from his smell alone.

As long ago as the 1950s, a perfumer called **Paul Jellinek** noted that several ingredients of incense resembled scents of the human body. It was not until 2001, however, that **Manfred Milinski** and **Claus Wedekind** of the University of Bern wondered whether there was a correlation between the perfume a women preferred and her own natural scent. They found that there is.

The correlation is with the genes of what is known as the **major histocompatibility complex (MHC).** This region of the genome encodes part of immune system. It turns out that one of the most important aspects of mate choice in mammals, humans included, is to make sure that your mate's MHC is different from your own. Mixing up MHCs makes the immune system more effective. The MHC is also thought to act as a proxy for general outbreeding, with all hybrid then, vigour that can bring. Fortunately, then, evolution has equipped mammals with the ability to detect by small chemicals whose concentrations vary with differences in the MHC of the producer.

> This means that people are able to sniff out suitable MHC genomes in prospective partners. A woman, for instance, will prefer the smell of T-shirts that have been worn by man whose MHC genes are appropriately different from their own. Dr. Miliski and Dr. Wedekind also found an association between a woman's MHC genes and some of her preferences for perfume. Perception of musk, rose and cardamom is correlated with the MHC. Perception of castoreum (*i.e.*, odorous secretion of beaver) and cedar is not. (See, *The Scent of a Man*, The Indian Express, 24th Dec. 2008).

Pheromones in Humans

A case of pheromone base sex recognisation and sex attraction has been described in humans. Olfactory communication in humans is as important as in dogs (**Doty,** 1976), but its dominance is suppressed in adults due to psyco-sexual development. There exists a difference between the two sexes of humans in their ability to smell. Women are more sensitive. The smell-sensitivity also greatly varies during different phases of menstrual cycle and pregnancy in female, and following castration and male hormone supplementation in males (**Doty,** 1976). Women perceive musky odour of synthetic perfumes and sensitivity is highest during **ovulatory period**. Axillary sweating contains pheromonal substance in man. Similarly, vaginal secretion contains a pheromone called **copalin** containing volatile vaginal fattly acids with peaks at the time of ovulation.

Identification of human pheromone from male sweat has been carried out by **Dr. George Dodd** of Warwick University in U.K. It is identified to be **X-androstenol** (related to sex hormone) and its main function is to attract the females during sexual interaction.

10.6. MODE OF ACTION OF PHEROMONES

Insect pheromones exert their effect by **olfaction, physical contact** and **absorption (Berth,** 1970). Although, mammalian pheromones are olfaction sensitive, but physical contact is important for some primer pheromones in their influence on the ovarian cyclicity. Pheromones influence hormonal secretions which in turn directs behavioural change. While releaser pheromones directly affect neurohumoral activity of the central nervous system; primer pheromones influence neuroendocrine machinery through olfactory bulbo-nasal passage-vomeronasal organ-amygdala and stria-hypothalamus pathways. Hypothalamus triggers endocrine mechanisms and thus an effect is observed on the target tissue/organ.

10.7. ECONOMIC USES OF PHEROMONES

Pheromones such as **civetone** (from civet) and **muscone** (from musk deer) **are used** commercially in the preparation of perfume. They are also used increasingly as a **method of biological control** in insect pest species such as the gypsy moth. In these cases the artificial release of the pheromone **gyplure,** attracts males to the source of release where they can be captured and killed.

QUESTIONS

Long Answer Questions

1. What are pheromones ? Describe types and function of pheromones. (*Purvanchal 1993, 97*)
2. Write an essay on pheromones. (*Purvanchal 1995*)
3. Present an account of chemical communication in animals. (*Kochi 1999*)
4. Discuss pheromones and their action. (*Purvanchal 1994, 99*)
5. Write an essay on insect pheromones. (*Bharathiar (M) 1997*)
6. Describe pheromones of the vertebrates.

Short Answer Questions

1. Discuss pheromones in human beings. (*Purvanchal 1996*)
2. Differentiate between the following : Pheromones and hormones. (*Purvanchal 1996*)
3. Write short notes on the following : (*Purvanchal 1997*)
   (*i*) Dufour's gland
   (*ii*) Alarm pheromone.

Very Short Answer Questions

1. Which organ produces marker pheromones in ants. Discuss. (*Purvanchal 1999*)
2. Name of pheromones which affect behaviour directly through nerve supply.

Multiple Choice Questions

Choose the correct answer from the four alternatives given:

1. The volatile substance, secreted by organism into its environment which modify the behaviour of member of the same species is
   (*a*) pheromones (*b*) saliva
   (*c*) renin (*d*) noradrenalin
2. The first pheromone studied was
   (*a*) bombycol (*b*) muskone
   (*c*) geradiol (*d*) prostagandins
3. The term pheromones was given by
   (*a*) Starling (*b*) Bergstorm
   (*c*) De Vagnaud (*d*) Karlson
4. Pheromone is
   (*a*) a product of endocrine gland (*b*) used for animal communication
   (*c*) messenger RNA (*d*) always protein
5. Pheromones are secreted by
   (*a*) bulbouretheral glands (*b*) anal gland/perineal gland
   (*c*) prostate gland (*d*) cowper's gland

ANSWERS

1. (*a*) 2. (*a*) 3. (*d*) 4. (*b*) 5. (*a*)

Social Behaviour

(Also include Social dominance and Territoriality)

11.1. AGGREGATIONS

Few animals live solitary lives, most needing at least a mate from among their own species. Many animals live in groups which may be temporary or permanent. Frequently grouping together requires some behavioural adaptation which ensures the cohesion of the group. This may require a complex of special behaviour patterns such as those found in truly social animals, but the less permanent aggregations of animals involve simpler processes.

Here, a distinction can be made between an **aggregation** — a group of animals which came together for some external reasons and a **society** where the parents and offspring remained together for long periods kept in association by mutual behaviour. Thus, a *society* can be defined as a group of individuals belonging to the same species and organized in a cooperative manner. A pair of animals engaged in basic courtship and sexual activity is not included in this definition, nor are individuals who simply defend their territories from one another. Sexual behaviour and territoriality are important properties of societies and are correctly referred to as social behaviour, but having them is not sufficient to qualify a group as a society. Bird flocks, wolf packs, fish schools and locust swarms are good examples of true elementary societies. So are parents and offspring if they communicate reciprocally. Examples of animal societies based primarily on *structural specializations* are found in insects, while those based primarily on *functional specializations* occur in vertebrates.

11.2. GROUP SELECTION, KIN SELECTION, ALTRUISM, RECIPROCAL ALTRUISM, INCLUSIVE FITNESS

Social organization is most developed in insects. Insects are **eusocials** (*i.e.,* have true societies) and their eusociality is exhibited in terms of their following attributes:

(i) Cooperation. Members of same species show great cooperation.

(ii) Division of labour. There is distinct division of labour in some colonial insects. A group of individuals of the same colony become only workers and others only reproduce.

(iii) Overlapping of generations. At any stage, at least two generations overlap in a colony of insects.

Insect Society

A social insect has unique features. Although an insect society has thousands of members but they all constitute one family. Majority of the members are busy in activities that will end into the production of more individuals by the queen. These individuals are moulded into different casts by the "feeding action". Insect societies are termed **eusocial** for the following reasons:

1. They are essentially permanent.
2. They show mutual cooperation among inmates.
3. They contain offspring usually produced by a single female called as **queen.**
4. They exhibit the overlapping of generations — queen survives over her offspring generations, and
5. They include members which are so attached to one another that their existence depends upon the survival of entire colony.

Above described true societies are exemplified by only following two groups of the insects:

1. Order Isoptera, *e.g.*, Termites;

2. Order Hymenoptera, *e.g.*, ants, bees and wasps.

Origin and Evolution of Sociality in Insects

Eusocial behaviour seems to evolve through the following stages:

1. In the initial stage, called **quasi-social stage,** a communal nesting develops in which groups of female of same generation use common nest and cooperate among themselves in some important activities including care of the brood.
2. In the next stage, called **semi-social stage,** division of labour operates in reproduction. Only some individuals reproduce while majority of others assume the role of workers.
3. In last stage, called **eusocial stage,** the adult life of queen lengthens and thus overlaps with that of her offspring. This stage is related with development of **altruistic behaviour.**

Following two concepts have been forwarded to explain the way in which natural selection helps in the development of sociality:

1. Group selection concept. This concept is based on the view that the natural selection operates at group level. The whole colony or family group is selected for a social trait. However, since individuals form a group and so for a selection to become operative at group level it is obligatory that an allele must first become established by the selection at the level of individuals. This fact nullified the group selection concept and gave rise to a concept based on individual levels.

2. Kin-selection concept. An animal profitably lives in its environment, but its fitness is at not one but two component event. Eventually, individual's own altruistic sacrifice is counter balanced by the fitness of other in the population; the other may be close-relative. For instance, in hymenopteran insects (*e.g.*, wasp), sacrifice by a female in terms of reproductive success must enhance the reproductivity of sister or daughter.

Kin selection concept was proposed by **W. D. Hamilton** (1964). Its essence is that female wasps are really helping their reproductively competent sisters (*i.e.*, future queens) and only incidentally assisting their mother. That is, the *genetic goal* of their behaviour is to increase the chances of survival of their own genes by aiding very closely related siblings. The advantages of altruism are so great in this case that it can favour females that, so to speak, put all their eggs (*i.e.*, genes) in a sister's basket and forgo personal reproduction entirely.

Reciprocal Altruism

Helping behaviour existing between members of a social group is called **altruism. Reciprocal altruism** (a term coined by **Trivers,** 1971) means "*you scratch my back, I will scratch yours*". That is, one individual provides another with help, and the second subsequently reciprocates or pays back the first. In nature, reciprocal altruism is quite rare. However, it occurs in the vampire bats (*Desmodus rotundus*; **Wilkinson,** 1984).

Vampire bats are colonial mammals. During the day they roost together, often in hollow trees. At night they feed on the blood of domestic stock, such as cattle, horses, goats and pigs. Compared to other bats they have good eyesight and a keen sense of smell, which probably helps them to locate their hosts. Once a vampire bat has located a potential prey it often lands on the ground near the prey and then silently moves to the animal. The bat then painlessly inflicts a small wound with its needle-sharp incisors, secrete some anticoagulants in its salvia and laps up a meal of blood which may take up to 20 minutes to ingest.

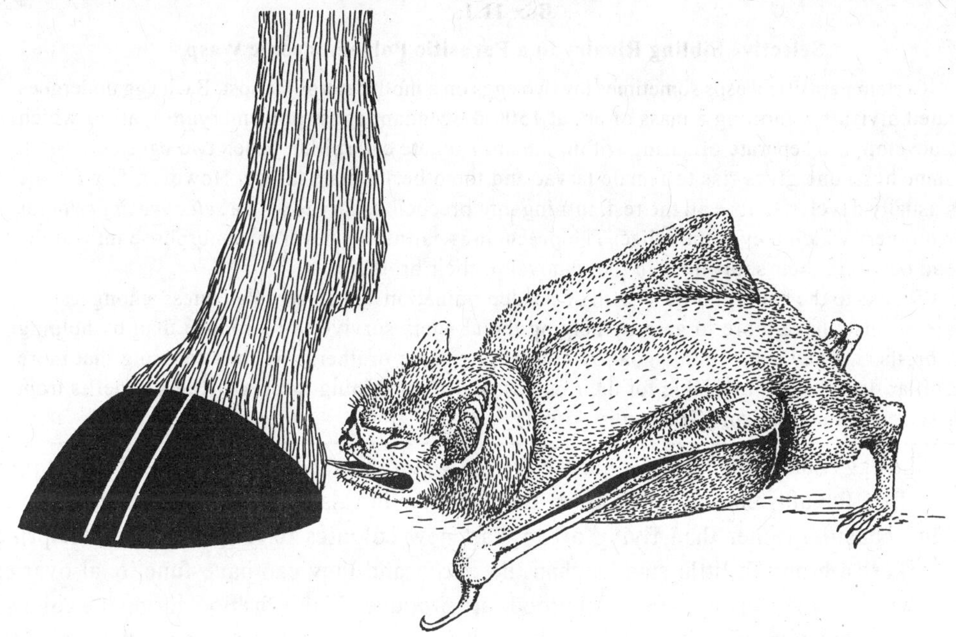

Fig. 11.1. A vampire bat laps at a tiny slit above the hoof of a horse.

If an individual fails to find a meal, it is in trouble. **Wilkinson** found that after about 50-60 hours without blood, a vampire bat starves to death. He found that a bat that had failed to obtain a meal was usually provided with regurgitated blood by a roost-mate which had successfully fed the previous night. Often such altruism was provided by a mother for her offspring. However, on a number of occasions, the bat receiving the regurgitated blood was either unrelated or related only distantly to the bat providing the food. Further, experiments on unrelated captive bats showed that there occurs reciprocation. It appears as though individuals remember from which individuals they have received blood and subsequently reciprocate if an opportunity presents itself.

11.3. SOCIAL ORGANISATION IN INSECTS

1. Social System in Paper Wasps (*Polistes*)

Paper wasps of the genus *Polistes* regularly built their nests in the shelter provided by an eave on a house. Reproductive females of temperate zone species emerge from cells in the nest late in the breeding season. They mate with males which are also being produced at this time and then spend the winter hibernating in a sheltered spot. In the spring, they rouse themselves and start a nest, which is constructed of chewed plant fibres. The nest contains a series of cells, each of which receives a single egg from the new queen. A foundress female may be joined by other overwintering females; this generates **dominance contests** and the formation of hierarchy, with the dominant female reproducing and the subordinates helping her to rear the larvae and protect the nest against predators and parasites.

The eggs that are laid early in the season are destined to become daughter wasps; the female controls the sex of her offspring by deciding whether to fertilize an egg with sperm she has stored from last fall's copulation or to lay an unfertilized egg. Young wasps reared from fertilized eggs become females; those that develop from unfertilized eggs become males (this is called **haplodiploid system of sex determination;** Box 11.1).

Box 11.1.

Selective Sibling Rivalry in a Parasitic Polyembryonic Wasp

Certain parasitic wasps sometimes lay two eggs on a moth caterpillar host. Each egg undergoes repeated divisions, forming a mass of about 1500 descendant eggs (polyembryony), all of which then develop into separate offspring within the unfortunate caterpillar. When two eggs coexist on the same host, one gives rise to female larvae and the other to male larvae. However, few female eggs usually develop before all the rest, turning into precocious larvae with a *selective appetite* for their brothers which they cannibalize. The **precocious cannibals** never metamorphose into adults instead devoting their short lives in the removal of their brothers.

Thanks to the haplodiploid system of sex determination and closer relatedness among sisters, female altruists have more to gain by helping their sisters survive to reproduce than by helping their brothers. They achieve this goal by eliminating some brothers and thus ensuring that more caterpillar flesh will be available for their maturing sisters, gaining **indirect fitness** benefits from their selective cannibalism (**Grbic** *et al.*, 1992).

As the eggs hatch, the newborn larvae are fed water, nectar and fragments of insect prey. When the first females emerge, they assist their mother in raising still more daughters (their sisters and cousins) rather than flying off to start new colonies to rear their own offspring. These workers are only a little smaller than the queen and they can have functional ovaries, although they rarely lay any eggs. Several broods are produced in this fashion during the colony's life, with more and more females (workers) joining the work force. They increase the size of the nest, add cells, feed the larvae, detect and drive off parasites and descend *en masse* (in group) on potential predators, stinging them to retreat. However, as the summer progresses in temperate regions of America, the queen produces increasing number of "lazy" females that do not join the workers but instead lounge about on the nest, stealing food from their working sisters. Later in the summer males emerge for the first time and these, too, do little to aid the welfare of the colony. Activity at the nest dramatically decreases at this time; the 'lazy females' and males fly off, *mating occurs with members of other colonies,* the males die and the mated females — future colony foundresses hibernate through the winter months to resume the cycle in the succeeding spring.

2. Social System in Honeybee

Honeybees live in colonies of hundred bees. Each colony has a **queen** which is somewhat larger than the worker bees. The queen lays the eggs which develop into new **workers, drones** (males) and **queens.** She also emits a complex series of chemical secretions, the **pheromones,** that regulate much of the behaviour of the workers (refer Chapter 10, Pheromones). Genetically both queens and workers are **diploid** and both are females. Queens, however, are fed a special rich larval food, a white, foamy, yoghurt-like **royal jelly** that is necessary for normal queen size and sexual development. In fact, workers can develop into sexually reproducing females but their reproductive organs are kept undeveloped through influence of the queen's pheromone. Drones are **haploid** genetically and are males that are produced by the laying of unfertilized eggs (*i.e.*, parthenogenetically). Drones are produced at the same time as new queens.

Box 11.2.

Unquestionally, the premier nonhuman architects are the social insects, whose beautifully designed homes provide protection from both predators and climatic dangers (Fig. 11.2). *Melipona interrupta* builds its nest in the hollow of a tree-branch.

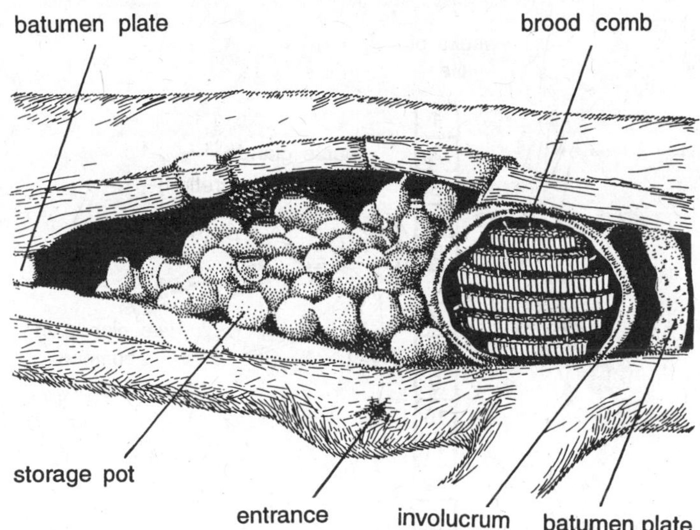

Fig. 11.2. Nest of a social bee, *Melipona interrupta,* in a hollow branch. As is true of most social species of bees, the architecture of the nest is elaborate.

When a hive of honeybees prepares to swarm or when an old queen becomes weak, the regulating pheromones of the queen become weak. This serves as signal for workers to begin raising new queen larvae. In the case of swarming, the old queen leaves with a group of workers and forms a new colony. In case of an aging queen, new queen displaces the old one. When a new queen hatches out and develops, she kills any other newly-hatched rivals and flies off to mate. By this time, drones have left the hive and have aggregated in traditional sites in large flying clouds. The drones from several neighbouring hives may all combine in such a cloud. When newly-hatched queens approach such a **drone cloud,** the drones rush at the queens avidly and several of them may mate with a queen in succession. The new queen then returns to her hive, lays her eggs and regulate the behaviour of workers (Box 11.3).

Box 11.3.

Except honeybees, in most social insects new colonies are founded by a single queen. She begins the construction of the nest and rears first batch of workers herself. These then take over the tasks of extending the nest and bringing food and the queen usually stays in the nest laying eggs from this point onward.

A honeybee worker lives for about 6 weeks as an adult and her activities are to some extent synchronized with her physiology (Fig. 11.3). Thus, she spends the first three days cleaning out cells and then begins feeding the older larvae a mixture of pollen and honey which she picks up from the storage cells in the hive. During this period the **pharyngeal** or **nurse glands** in her head have been developing. They secrete the so called **royal jelly** and from about the 6th to 14th day of her life the worker feeds this secretion to the younger larvae and any queen larvae in the hive. Royal jelly is fed to all larvae for a brief period early in their development, but those larvae intended to become queens develop in a larger cell and are fed royal jelly throughout. The worker's **wax-secreting glands** on the abdomen become active from the 10th day and at the same time the pharyngeal glands begin to regress. Worker gradually changes her behaviour from feeding larvae to cell construction. From about 18th day she may leave the hive occasionally for a few brief **orientation flights** (Box 11.4). At this stage she may be found guarding the hive entrance and inspecting incoming bees. From 21st day of age onwards the worker is primarily a

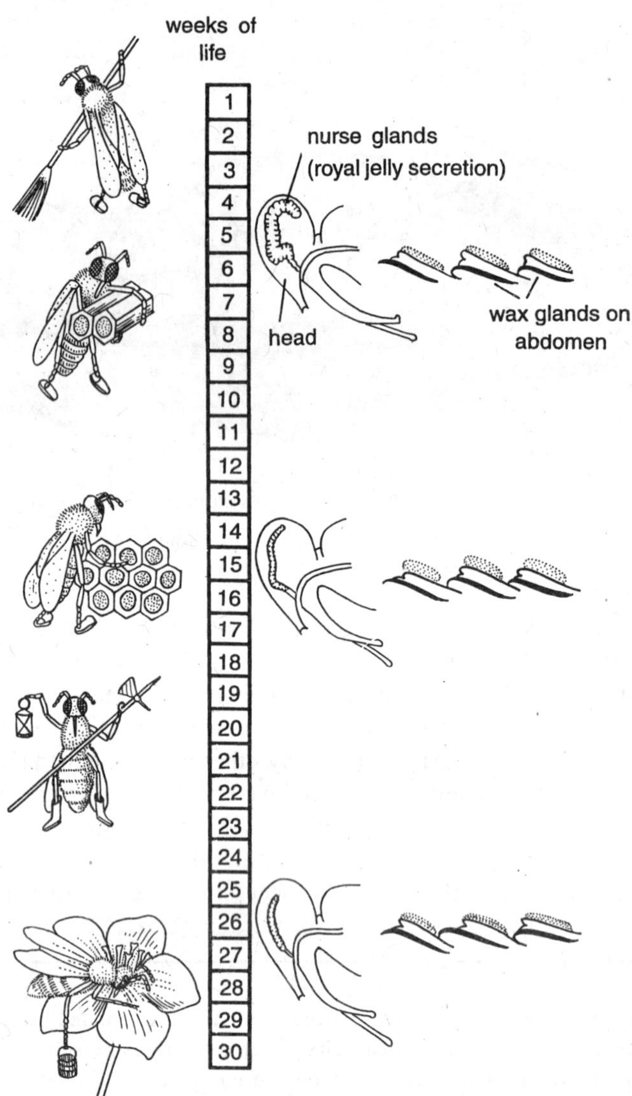

Fig. 11.3. Major activities of a worker bee as she matures. After emergence, an adult bee first cleans the hive, then tends the brood, builds more cells of comb, guards the nest entrance, and finally forages for pollen and nectar. This behavioural sequence of events is correlated with physiological changes in the nurse glands (which produce nutritive material for bee larvae) in the head and wax glands (which produce comb-building material) in insect's abdomen.

Box 11.4.
Orientation flights of Honeybee

Many insects make special 'orientation flights', during which they make a "fix" of the home area's position relative to sun and landmarks near by. If a honeybee colony is shut up in the hive and moved to a new site, a large proportion of the workers make orientation flights when they first leave the hive in its new position. They hover outside the entrance hole and then circle, gradually increasing their distance from it before flying off. During this orientation flight, lasting only 1 or 2 minutes, they learn the new position in sufficient detail to be able to return from long foraging flights.

forager bringing back nectar, pollen and water and usually remains so for the rest of the life (*i.e.*, 2 to 3 weeks). **Martin Lindauer** (1961, 1971) kept a record of the activities of one individual worker throughout her life. The sequence just described can be recognised, but the most conspicuous activities not previously mentioned are, *i.e.*, *resting* and *patrolling*.

A beehive is a marvelous society. Just as a mammalian body has elaborate homeostatic mechanisms that maintain a constant temperature, water balance and nutrient level, so a beehive has elaborate mechanisms that maintain hive homeostasis. When the hive is hot, worker bees fan air throughout it and cool it off. When the nurse workers attending the larvae are short of water, then turn to the nearest workers and signal a need for water. The latter give what water they have and then turn to their neighbours; thus, the shortage of water moves from bee to bee until it reaches a worker who leaves the hive and return with more water. When a predator or parasite enters the hive, workers rush forward and defend the colony, with the result they may die. It is almost as if the hive itself, and not the individual bees, were the organism.

3. Social Life in Termites (White ants)

Termites are social and polymorphic insects, living in large and well organized colonies. They differ from other social insects in having a larger number of castes, of which three are *reproductive castes* and two are *sterile castes.* The three reproductive or fertile castes of termites are as follows:

***(i)* Macropterous or winged forms.** These are normal winged (alate) males or females forming true **kings** and **queens**. They are characterized by a dark pigment; and possess well developed eyes, brain, sex organs and two pair of wings. The winged kings and queens leave the colony, mate, lose their wings and start a new colony.

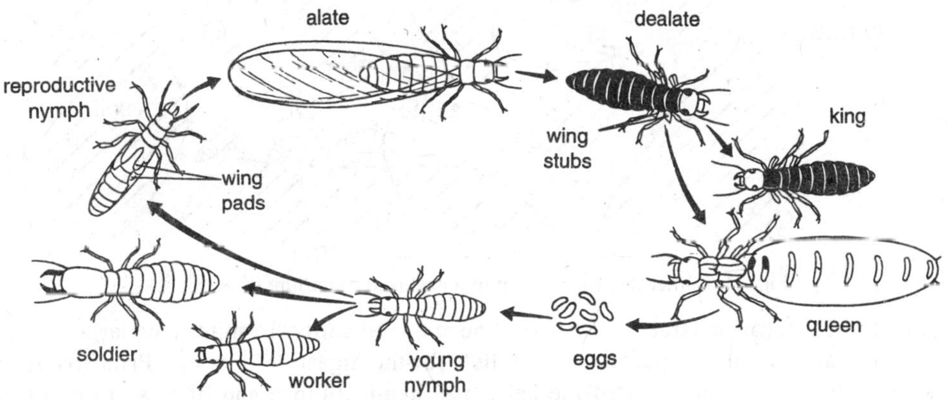

Fig. 11.4. Termites showing polymorphic individuals.

***(ii)* Brachypterous or short-winged forms.** These are short-winged males and females often called **supplementary** or **neotenic kings** and **queens**. If macropterous or primary king or queen dies, its place is taken by brachypterous individuals forming substitute king or queen.

***(iii)* Apterous or wingless forms.** These are worker-like individuals often called **ergatoid kings** and **queens.** They lack wings and possess less developed eyes and reproductive organs than in other reproductive castes. They are rare and found only in primitive termites, *e.g.*, *Leucotermes.*

The two sterile castes of termites are as follows:

1. Workers. These are small-sized wingless and include both male and female sterile individuals. The workers are most numerous than any other caste. In some termites (*e.g.*, *Odontotermes*), the workers are **di-** or **trimorphic.** They perform all the duties except reprodution and defence. They care for the eggs and young, search for and collect the food, feed and tend the queen, cultivate fungus in special chambers (*i.e.*, fungus gardens) in certain species, construct the nest or termitarium (*i.e.*, excavate tunnels and galleries and construct mounds) and perform other duties.

2. Soldiers. They are like the workers (*i.e.,* lack wings and reproductive organs) but are somewhat larger and possess stronger mouthparts. **Mandibulate soldiers** have larger body, strong head and huge mandibles for driving away intruders. **Nauste soldiers** are characterized by a median frontal rostrum on their heads and can eject a repellent fluid through their rostrum in warfare (**Singh,** 1997). Soldiers guard the colony (Fig. 11.5).

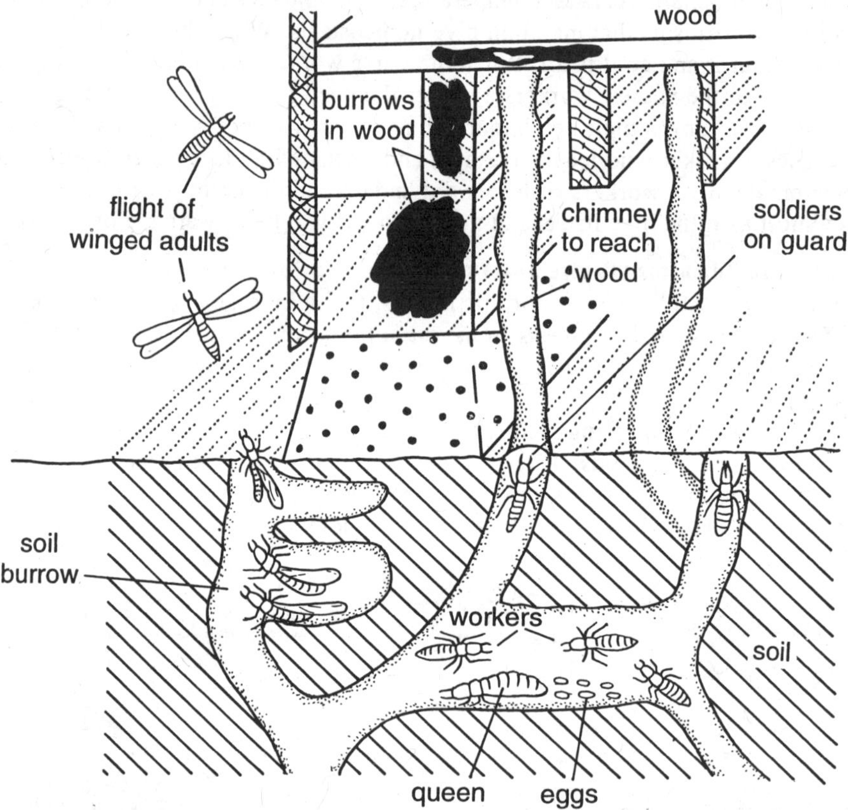

Fig. 11.5. Internal organization of a nest or termitarium of termites.

In termites, nutrition largely determines the major sociality. Termites feed largely on wood (cellulose). Many termites depend on **mutualists** for the digestion of wood. Primitive termites feed directly on wood, and most of the cellulose, hemicellulose and lignins are digested by mutualists in the gut where the paunch (a part of the segmented caecum) forms a **microbial fermentation chamber.** However, more advanced species of termites (75% of all the species) produce their own **cellulase** enzyme (**Hogan** *et al.,* 1988). A third group of termites (Macrotermitineae) cultivate fungi that digest wood and they then eat the fungus rather than the wood (see **Begon** *et al.,* 1996).

Termites refecate, *i.e.,* they eat their own faeces, so that food material passes at least twice through the gut, and microbes that are reproduced during the first passage may be digested in the second time round. The major group of microorganisms in the paunch of the primitive termites is of protozoans, consisting of anaerobic flagellates, such as *Trichomonas termopsidis* and representing unique genera that are found only in termites and in a closely related species of wood-eating cockroach (*Cryptocercus*). Bacteria are also present, but cannot digest wood. The protozoa engulf particles of wood and ferment the cellulose within their cells, releasing carbon dioxide and hydrogen. The principal products are volatile fatty acids (as in the rumen), but in termites it is primarily acetic acid and this is absorbed through the hind gut.

The bacterial population of the termite gut is less conspicuous than that of the rumen, but appears to play a part in two distinct mutualisms.

1. **Spirochaetes** are important members of the bacterial flora and they, together with rod-shaped bacteria, tend to be concentrated at the surface of the flagellates (protozoans). In the gut of one species of termite (*Mastoterma paradoxa*) the spirochaetes have been observed in synchronized movement actually propelling the flagellates. The association of spirochaetes and flagellate is mutualistic – the spirochaetes receiving nutrients from the protozoan and the protozoan gaining mobility from the spirochaete; so here we have a pair of mutualists living mutualistically within a third species.

2. Some bacteria in the termite gut are capable of fixing gaseous nitrogen (*e.g.*, *Citrobacter freundii* and *Enterobacter*). This appears to be the only clearly established example of nitrogen fixing symbiont in insects (**Douglas** 1992).

Wood in intestine of termite is also digested by *Trichonympha* (another flagellate) which are obligate anaerobes in the guts of insects, encysting prior to host moulting may allow the insects to maintain their protist symbionts (see **Brusca** and **Brusca** 2002). Protozoans (*Trichonympha*) enter young termites (nymphs) when they feed on fresh faeces of adult. This essentiality lays down the basis of evolution of social life in termites as it promised the overlap between two generations and thus, fulfilled one of the chief criteria of the sociality in insects.

Formation of new termite colonies. The new colonies are formed by normal macropterous individuals which come out of their old nest generally in rainy season in huge numbers on their colonizing flight, termed as **swarming.** After a brief flight these individuals alight on the ground and shed their wings. Each colony is formed by a royal pair; they mate and excavate a small burrow in the ground, called the **nuptial chamber.** The female (queen) lays about a million equipotential eggs per year. On hatching out from the eggs, nymphs may develop into one or more castes.

In the early stages of colony, nymphs develop into workers, soldiers and substitute reproductive castes which may become fertile and take the function of royal pair when the latter is destroyed. The production of different castes from similar diploid eggs has been explained by various theories based on a variable food supply (*e.g.*, king and queen have access of rich food around them); production of social hormone (pheromone) and group effect. For example, **pheromone** produced by queen and king at critical level suppress the ovarian development of newly reared workers. The egg-laying queen has a huge size possibly as a result of sluggish habits and rich food supply. She has tiny legs and a large egg-filled abdomen. When egg-laying capacity of the queen decreases her feeding is stopped; she dies of starvation and her fatty remains are devoured by other castes of colony.

The point is that the haplodiploid system of sex determination does not guarantee that worker's in eusocial hymenopteran societies will be very closely related nor do fairly low levels of r (*i.e.*, r = coefficient of relatedness) prevent eusociality in these insects. The termites, for example, are every bit as social as honeybees and paper wasps, despite the fact that both males and females are diploid. Termite colonies may have thousands of sterile workers that labour on behalf of huge, bloated queen in a nest chamber set in the centre of an immense nest mound riddled with tunnels.

Distinctness of Termite Colonies

Termites are diploid, so that males and females have the same number of chromosomes. Unlike workers in the Hymenoptera, termite workers include both males and females. Termites are also distinctive in that the primarily reproductive male (the king) stays with the queen after the nuptial flight, helps her construct the first nest and mates with her several times during her egg-laying period. Termite queen ends up as a huge and highly specialized egg-laying machine.

4. Social Life in Ants.

Ants are highly evolved social insects, showing polymorphism. A colony of ants consists of the following main castes or social types:

1. Queens or gynes. Unlike bees, a colony of ants contains several queens. They are fertile females having well developed sex organs. A queen ant is large in size than other members of the colony due to its large abdomen. The queens are also **dimorphic:** the larger forms are called **macrogynes** and smaller ones, **microgynes.**

Fig. 11.6. Food sharing between workers of the ant, *Formica fusca*. 'A' worker is receiving from 'B' worker which has regurgitated a drop of liquid from its crop and offers it between the outstretched mandibles.

Box 11.5.

Within ants there are **'army' ants** with huge colonies of upto 22 million individuals, which bestride the jungle floor eating everything edible in their path. There are "**fungus gardening**" species (Genus *Atta, Acromyrmex,* etc.) which grow fungus on specially prepared rotting leaves and live on the produce of the fungus (*e.g., Attamyces bromatificus*). Other ant species live by **milking honeydew** from herds of little insects called **aphids.** In yet others, workers form living "**honeypots**" or "**honey casks**" as they hang upside down from the roof of their nest, their abdomens hugely distended with honey (Fig. 11.7). Such workers remains motionless and are called **repletes.** Australian aborigine (natives) dig up the nests of these ants, take the head of repletes between their fingers and bite off their honeyed abdomens.

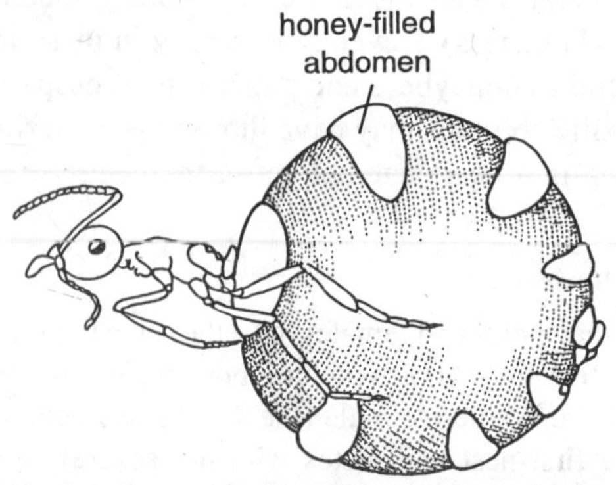

Fig. 11.7. Honeypots.

2. Males or aners. These are small, slender and fertile individuals. They have wings, smaller head, reduced mandibles, longer antennae and well developed sense organs and reproductive organs along with genitalia. Aners are also *dimorphic* having large sized **macraners** and small sized **micraners.**

3. Workers or ergates. Workers are normally sterile females. They are the smallest, wingless members of the colony having a reduced thorax, a small gaster (abdomen) and small eyes. They are also *dimorphic* including large sized workers called **macrergates** and small-sized workers, called **micrergates**. Worker ants collect food for the colony and they also care for the eggs and the young (Box 11.5). Food-sharing among workers (Fig. 11.6) is a method of communication in ants. It keeps each worker directly informed of the state of the food supplies within the colony.

4. Soldiers or dinergates. These are modified workers which lacks wings but have large heads and powerful mandibles. Their main social role is to protect the nest from enemies.

Formation of new ant colonies. The adult males and females swarm together in large numbers. Mating occurs in air during the **nuptial flight,** after which male ants usually die and females shed their wings. They may either return to their old nests increase their population or hunt a small burrow or excavate a small chamber in the ground to start a new colony. The first batch of eggs laid by the queen develop into wingless workers which then take over the charge of feeding the colony and tending the queen and the brood. The winged males and females are produced later (Box 11.6).

Sex Determination in Ants

Following two theories have been put forward to explain the determination of sex in ants:

1. Genetic theories. The queen lays two or more types of eggs. The wingless workers and soldiers are produced from unfertilized eggs (by parthenogenesis) while winged males and females from fertilized eggs.

2. Trophic theories. They refer to sex determination which depends on different kinds and amounts of food given to originally similar larvae or grubs.

Box 11.6.

Once the colony has reached a certain size the queen ant lays eggs which are reared as reproductives. The time when reproductives are first produced varies between species. In *Myrmica rubra*, a common garden ant in Europe, it is not until 9 years after founding, when the colony has grown to about 1000 workers (**Ridley,** 1986). In certain species queens have a life span of about 15 years.

Castes in Social Insects

All the social insects such as termites, honeybees, ants, wasps, etc., have **castes** (*i.e.,* they are **polymorphic**). A caste is a collection of individuals within the colony that are morphologically distinct from individuals in other castes and perform specific tasks. Castes allow colony members to specialise. The term *caste* is well suited to describe the division of labour within insect societies. It implies a rigid and limited role in society largely determined by one's upbringing (*e.g.,* diet). Thus, in a colony, there are always at least three *castes* — a **queen,** a **king** and lots of **workers.** Furthermore, in many species, the workers themselves may be subdivided into two or three castes (*i.e.,* they are either **dimorphic** or **trimorphic**). For example, in *Trinervitermes,* an advanced termite, the adult males may end up as **small soldiers** and **large soldiers** (Fig. 11.8). These have different functions and are physically distinct. As termites pass through their moults and gradually develop into adults, the juveniles may specialize and do different jobs from the adults. This is a bit like an assembly line in a factory where different people, or robots are specialized to perform different jobs with the greatest efficiency.

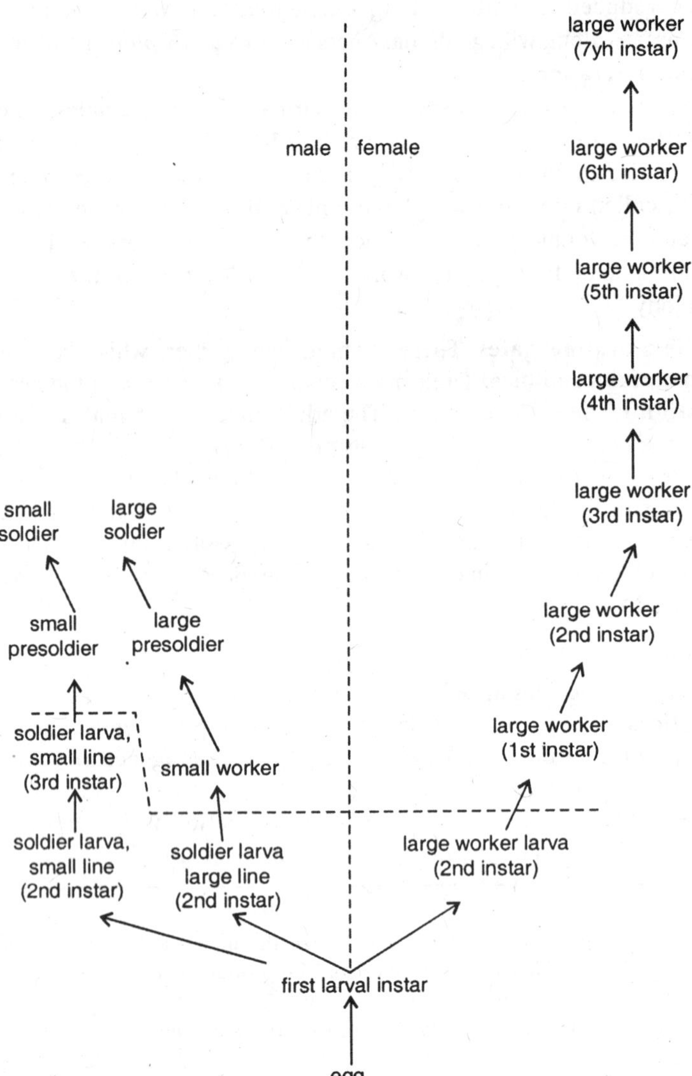

Fig. 11.8. Castes in the termite, *Trinervitermes.* Each time an individual moults it gets bigger. Individuals beneath the dashed line are helpless; those above it are capable of independent action and serve the colony.

Exception to Haplodiploidy Concept of Eusociality

Much evidence exists to suggest that very close relatedness is not essential for the evolution of eusociality. First the queen of many eusocial ants, bees and wasps mate more than once, a pattern that should tend to eliminate the close relatedness among the daughters of a queen. Queen honeybees, for example, often copulate with several dozen partners over a series of nuptial flights (**Koeniger,** 1986), a behaviour that could greatly reduce the coefficient of relatedness among the workers in a colony. On the other hand, female that mates with several males need not use the sperm from every male and evidence exists that some females of social hymenopterans are indeed highly selective (**Boomsma** and **Ratnieks,** 1996).

Further evidence that the workers in eusocial societies may not be especially closely related comes from the discovery that *many such colonies contain more than one queen ruling conjointly, each producing eggs cared for by the worker force at large.* In such a society, workers presumably help reproductive females that are not their sisters. Even in *Polistes* wasp colonies, which generally seem to be run by a single dominant female, **Joan Strassmann** and her colleagues (1989) have shown through genetic analysis that the actual average of *r* (coefficient of gene relatedness) of nest mates (females) almost never reaches the 0.75 maximum value, and often is less than 0.50. These results indicate that many paper wasp queens either mate more than once or share reproductive "duties" with other females in their nests.

11.4. SOCIAL ORGANISATION IN MAMMALS

Eusocial Behaviour in Naked Mole-rat

Eusocial behaviour lacks in vertebrates including birds and mammals except the rodent, the naked mole-rat (*Heterocephalus glaber*) of Kenya (**Jarvis,** 1981, 1985; **Gamlin,** 1987). The bizarre looking naked mole-rat is a little, hairless, sausage-shaped **diploid** mammal. It lives in colonies of 70 to 80 individuals (**Alcock,** 1998). Each colony occupies its own complex maze of underground tunnels, which may total 3 kilometers in length. The large size of their subterranean home stems from extraordinary **cooperation** among chain gangs of colony members, which work together to move tons of earth to the surface each year while burrowing in search of edible tubers (Fig. 11.9) Yet when it comes to reproducing, breeding is restricted to a single big "queen" and several "kings" that live in a centrally located nest chamber. Females other than the queen do not even ovulate, but serve as sterile helpers at the nest, consigned to specialized support roles for the queen and kings, as are most males in the colony (Fig. 11.9).

The breeding female (queen) may live for over 13 years and does little work other than suckle the young. She frequently leaves her nest chamber and patrols the colony. **Pheromones** (which are airborne chemicals that alter the behaviour of the other animals in the colony) are almost, certainly important. They are probably secreted by the breeding female in her urine which she leaves in special toilet areas. Experiments have shown that if these toilet areas are continuously cleaned be conveyor belts and automatic water flushing systems, several females comes into oestrus despite bullying by the queen. When this happens, fighting breaks out and females may kill each other. No colony ever seems to contain more than one breeding female.

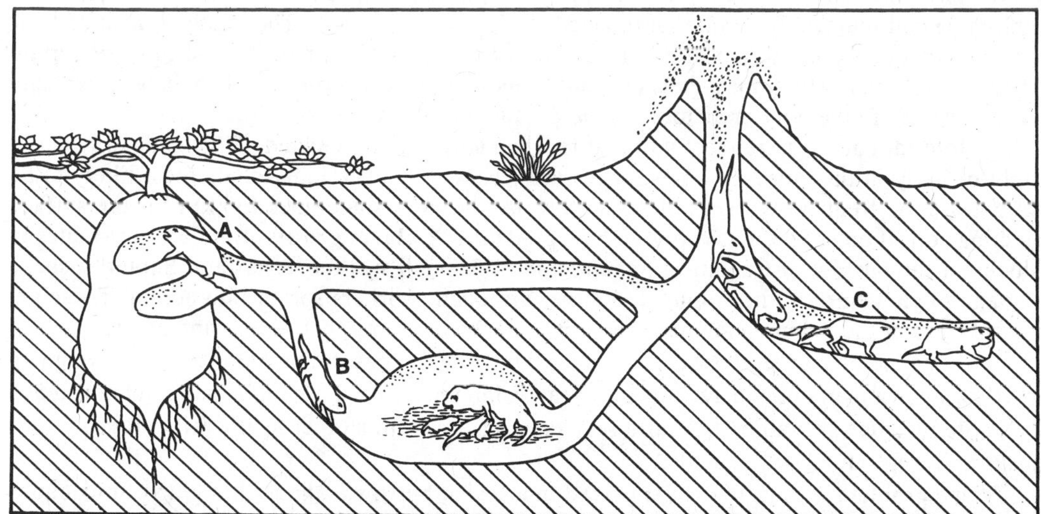

Fig. 11.9. Cross-section of part of the burrow system of a naked mole-rat colony. A — A frequent worker eating part of a growing tuber; B — The main chamber is occupied by the breeding female, subsidiary adults and young; C — A digging chain of frquent workers enlarging the burrow system.

In other species of mole-rats (family Bathyergidae) females usually have litters with between two and five young. In the naked mole-rat the average litter size is 12 and litters up to 27, the greatest number for any mammal, have been recorded. Naked mole-rats are also unusual in that their breeding is nonseasonal. The reproductive female of a naked mole-rats colony produces a litter as often as every 80 days.

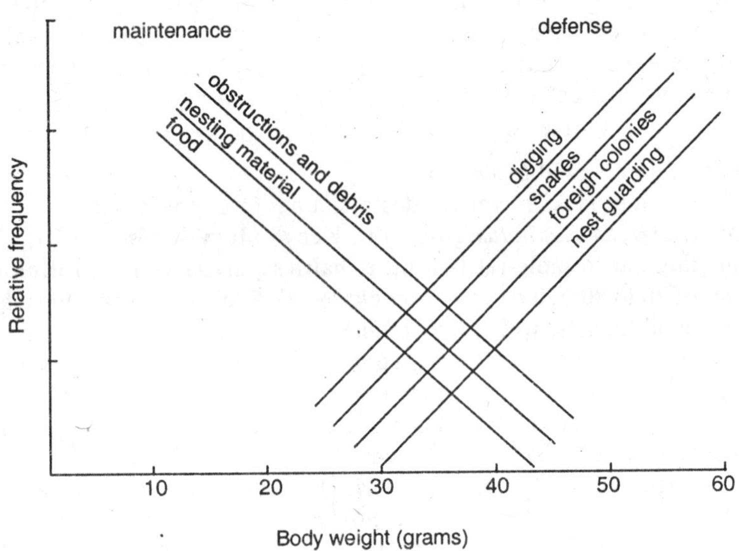

Fig. 11.10. Division of labour in naked mole-rat colonies. Small individuals engage primarily in maintenance activities; larger individuals undertake digging and defense duties. Only the very largest members of the colony breed.

Once a naked mole-rat is about three months old, it joins the "frequent worker" caste. This caste contains the smallest adults in colony. These do most of the day-to-day work such as **digging, foraging, transporting soil, keeping the tunnels clean** and **building nest for the breeding female.** Some individuals in this cast which include both males and females, grow very slowly and remain as "frequent workers" throughout their lives. Other individuals again both male and female, grow into large-sized "infrequent workers". These spend most of their time asleep but are important in defence. The main predators of naked mole-rats are certain carnivorous birds, which may grab the mole-rats as they kick soil out of their burrows, and snakes which are the only predators able to pursue mole-rats underground.

Mole-rat colonies are most vulnerable when soil is being ejected. This tends to occur at night or early in the morning when snakes are least active. The individuals most likely to be caught are the frequent workers because they are the ones sometimes found near the surface ejecting soil. If one of them is captured by a predator, its **alarm grunts** stimulate nest-mates to block off the burrow just behind it, abandoning the small worker to its fate. Should a snake manage to gain entry to the colony, the large infrequent workers spring into action. These are the **soldiers** of the colony and attack invading snakes by biting them again and again.

Naked mole-rats feed on the underground tubers of plants. Some of these tubers weigh up to 50 kg; by comparison, the average mole-rat weigh only 40 g. These tubers are not consumed by a colony in one go; rather, the mole-rats leave sufficient of a tuber to allow it to regrow, thus ensuring a continuous food supply.

One problem with living underground in a sealed burrow is the building up of poisonous carbon dioxide. *Naked mole-rats have the lowest metabolic rates of any mammals,* thus, minimising the amount of oxygen they use and carbon dioxide they produce. They are also the nearest thing yet discovered to a poikilothermic mammals. The temperature in a naked mole-rat colony remains a steady 29-30° C. At 15°C a naked mole-rat is so sluggish it can hardly move and its body fats begins to solidify. Their low metabolic rate may be the reason why they have lost most of their hair. Loss of hair may also reduce the risks of harmful ectoparasites which might be a particular danger to a highly social species living in a confined area.

Thus sterile workers of mole-rat usually do labour on behalf of relatives. In all of the highly social animals studied to date, colonies are composed at least in part of related individuals and workers may have the ability to partition their aid in accordance with the coefficient of relatedness. Indeed even diploid termite and naked mole-rat workers may be quite closely related to those they help *because of intense inbreeding involving repeated brother-sister or son-mother matings.* **DNA fingerprinting studies** of **Reeves** *et al.*, (1990) have shown that the members of any given naked mole-rat colony are extremely similar genetically while differing greatly from the members of other colonies, a condition that provides high potential indirect fitness benefits to helpers within a colony.

2. Social Behaviour in Lion (Pride formation, Cooperative hunting, Communal suckling and Mating Cooperation among Males)

Lions (*Panthera leo*) have been studied in the Serengeti and Ngorongoro Crater in East Africa continuously since 1966 and have been found to be most social of the cat family (**Schaller,** 1972; **Bertram,** 1978; **Packer** *et al.*, 1991).

Although the adult females, the lionesses, hunt in groups (called **cooperative hunting** or **social foraging; Alcock,** 1998), lions seem to get the most food *per individual* when they are solitary, or possibly when they hunt in pairs. However, there are other advantages of being social. For example, groups of lions can defend their kills against other animals (*e.g.*, a pack of spotted hyenas) and frequently take the carcasses killed by other carnivores, such as cheetahs.

The lionesses benefit in other ways from being in a group. Females remain in their natal groups, called **prides.** A pride typically consists of 4-12 adult females, 1-6 adult males and their offspring. The females in a pride are quite closely related to one another and lions are one of the few species to practice **communal suckling.** Cubs may suck from any adult female with milk. This means that, to some extent, females can take it in turns to go hunting and look after and feed the young.

Box 11.7.
Mating Cooperation among Males

Although parental care and altruism are restricted to females (*e.g.*, Belding's ground squirrels), male animals as different as dwarf mongooses, lions and bottle-nosed dolphins specialize in a different kind of cooperation, *the joint defence of groups of females.* In lions, for example, the males in a *coalition* living with a pride of females hurry to confront any opponents who dare to roar in their territory. When **Jon Grinnell** and his coworkers (1995) played taped roars in lion country, the resident males often approached the tape recorder and attacked a stuffed dummy lion placed nearby. The eviction of intruder males is essential for the maintenance of the resident's control of a pride and thus all members of a coalition benefits from a cooperative responses to potential threats to this control.

Craig Packer and his associates (1991) used DNA fingerprinting technique to determine the genetic relationships of pride members. As expected, they found that males in large coalitions possessed more similar DNA fingerprints on average than did males in smaller groups. Thus, duos and trios were often unrelated individuals that formed cooperative social teams to compete with other groups, whereas large groups were clusters of brothers, half-brothers and cousins that had stayed together after leaving their natal pride. In these large groups, a male that had little reproductive success was to use **Brian Bertram's** (1978) phrase "*reproducing by proxy*" through his companions, "because they all had a recent common ancestor and shared a relatively high proportion of the same alleles".

Males benefit from the presence of other adult males in the pride. This is because the adult males in a pride run the risk of being ousted by other males. As one might expect, the more adult males there are in a pride, the longer they tend to fight off other males. This means they can sire more offspring. It was used to be thought that the adult males in a pride were full or half brothers, and kin selection was invoked to explain the rather surprising observation that *there is no dominance hierarchy among the males within a pride, they all share fairly equally in the numerous acts of mating.* Now it is known that almost half of the breeding coalition of known origins contain no-relatives (Box 11.7). In fact fights do occur between the males in a pride and males can be badly hurt in these. Individual selection may be what is responsible for the relatively harmonious relationships between the males in a pride. After all if the length of time a male can expect to remain with a pride depends strongly and positively on the number of adult males, it may not be worthwhile for males to fight too much among themselves; cooperation may be the better strategy.

When a pride is taken by a new group of adult males, the new males seek out and kill as many as possible the young cubs (which are less than months old). About a quarter of all cubs die from such **infanticide**. The purpose of the males for this infanticide is that *it serves to bring the adult female back into oestrous* (**sexual selection hypothesis**). Thus, it is common with many mammals, including overselves that frequent lactation prevents ovulation (for 2 years in lioness). Infanticide appears to be a strategy used by males to ensure that they begin to reproduce as quickly as possible after taking over a pride (since males can expect to remain in a pride and have access to its females for just two years on an average; **Alcock,** 1998). Infanticide is now known to occur in several other species of mammals for similar reasons (*e.g.,* langurs).

One remarkable feature of lion reproductive physiology is that during oestrous, copulation occurs on average every 25 minutes. As oestrus lasts typically for days and female often take more than one oestrous to conceive, tremendous copulations are required to ensure one conception. To explain this peculiar behaviour the following explanations have been forwarded:

1. By requiring so many copulations for a single fertilization, female are reducing the value of each copulation with a male and thus, making it not worthwhile to a male in the pride to fight one another for access to oestrous females. According to this explanation, such fighting among males is to the disadvantages of females as should one or more of the resident adult males be injured, the pride is more likely to be taken over be new males which would, as we have described, kill many of the existing cubs.

2. Another, and almost exactly opposite argument is that this prolonged sexual activity by females, which is particularly apparent with that pride has been taken over, attracts other males to the pride. Such a coalition will be able to defend the pride for longest time against other males, which is to advantage of the females.

3. Social Behaviour in Primates and other Mammals

Sociality is very common among mammals. It predominantly occurs in deer, wolf, cattle, sheep, horse and primates. Of these, primates show an elaborate social life.

11.5. SOCIAL ORGANISATION IN PRIMATES

Primates show marked socialization. The sociality of primates has been enhanced by the following attributes: (1) enlargement of brain; (2) development of grasping hand; (3) great reliance on vision for exploration and communication and (4) diversity in their arboreal and terrestrial habitats.

Types of Social Behaviour in Primates

Southwick and **Siddique** (1974) have graded primate social behaviour into following *six* types:

1. Type I. Solitary. Examples. Orangutan, aye-aye, loris, etc. **Orangutans** (*Pongo pygmaeus*) lead an almost totally solitary life. They associate only for mating and an offspring is dependent on its mother. They are entirely arboreal and both sexes tend to stay in a home range over which they travel throughout the year following the fruiting of trees.

2. Type II. Monogamous. Examples. Gibbons, tree shrews, lemurs, marmosets, etc. Monogamy is rare. However it is perfectly exemplified by the **gibbons** (*Hylobates*) whose life-long pair bond between a male and female and their strict territoriality, maintained by elaborate 'singing' especially at dawn, show remarkable parallels to some birds. Gibbons are found in South America. Their groups include 4-8 individuals constituting an adult male and adult female and upto 4 young ones. There is not much difference in body size of male and female. They have equal dominance. Both of them (male and female) involve themselves in all activities with same intensity.

South American **marmosets** and **tamarins** (*e.g.*, silky tamarin *Leontocebus rosalia*) also have very small groups with often one, or at the most three adults of each sex in the group together with their young. Almost all other primates give birth to single offspring, but marmosets always have twins and males help with the care of infants, carrying them for much of the time.

Monogamous primates are usually smaller in size; they feed on high protein diet, *viz.*, insects, new leaves and ripe fruits. The parental investment is equal, the female feeds the young and male carries young on his back. Antipredatory device involves concealment.

3. Type III. Single male groups with bonded females and offsprings (Unimale bisexual groups). Examples. *Erythrocebus patas* (patas monkey), hanuman langurs, Fig. 11.11, red howler monkey, red tail monkey, blue monkey, etc. Type III primates typically live in unimale bisexual groups; their group may have 20 to 100 individuals, there will be just one adult; fully grown, big-sized, agile *dominant male* which is called **overlord** or **resident male.** Rest of the group is formed by adult females, subadult females, male and female juveniles, and infants. Adult male is the leader and coordinator of group activity. He initiates and determines the direction of group movement and activities such as where to go, when to feed, where to sleep, etc.

Fig. 11.11. Langur.

Thus, in male bisexual groups of **hanuman langur** (the entellus monkey; *Presbytis entellus*), it is usually the adult male who alone defends the territory. He herds females away from intruding males of all male group. Generally he alone indulges in fights. Very seldom females may also participate in the fights. The overlord (male) is much larger than females and is dominant over all the members of his group and there is no dominance hierarchy among females. Male parent investment is almost nil. The antipredatory strategy of langur involves climbing up the tree branches with all agility. Changing of overlord in a unimale bisexual group is of common occurrence. During interaction between adult male of unimale bisexual group and males of all male group, one of the adult males of all male group would chase the resident adult male and take possession of harem, this is known as **takeover.**

4. Type IV. Aggregate single male group with bonded females and offsprings. Examples. Baboons. The **Hamadryas baboons** (*Papio hamadyas*) are large sized primates found in Ethiopia and Somalia; the males have heavy mane around neck and have dog-like muzzle (Fig. 11.12). In their social organisation several females are more or less permanently bonded to a single male forming a so-called "**harem group**". A number of such groups band together, moving and foraging as a unit, perhaps 40 to 50 strong baboons. Thus, Hamadryas baboons form large troops on cliffs to sleep together. The troops separate into bands before travelling to foraging areas each morning; bands fragment into one male units while foraging and then reunite into bands to travel back to the cliff in evenings and nights to sleep together as a large multiple bisexual group (Fig. 11.12).

The other baboons, such as **olive baboon** (*Papio anubis*), yellow baboon (*Papio cynocephalus*), the chacma baboon (*Papio ursinus*) and common baboon (*Papio papio*), also have units of comparable size but here there are no persistent male/ female bonds. Adult males form temporary consortship with females as they come into oestrous, but otherwise move generally within the group.

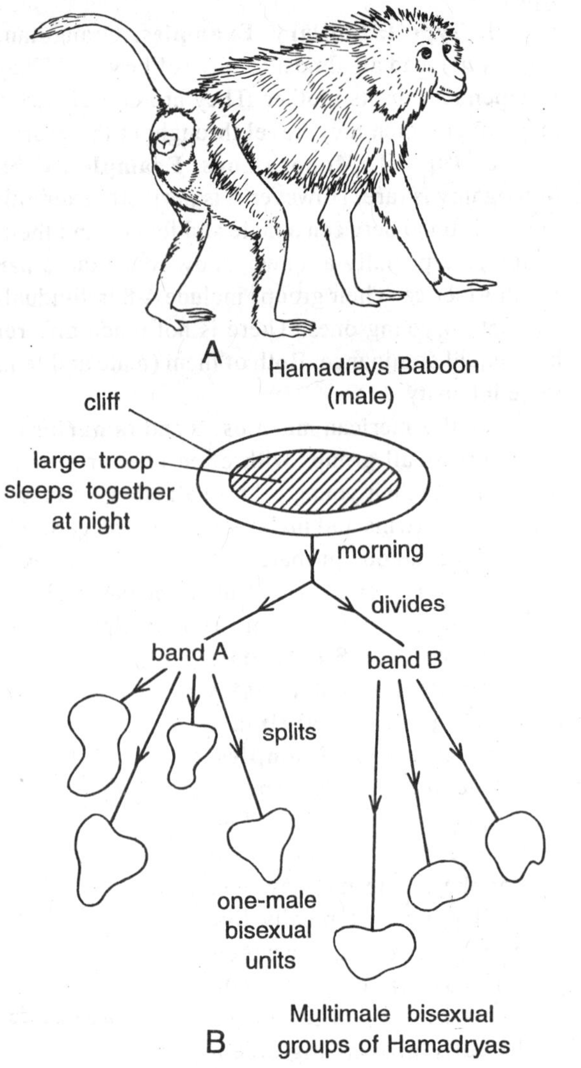

Fig. 11.12. Multimale bisexual group of Hamadryas baboon.

5. Type V. Multimale bisexual groups. Examples. Rhesus monkey, gorilla, spider monkey, squirrel monkey, etc. Typically there are 3-8 adult males in a group, each of which has 5-7 bonded females who remain with their infants. In a way, there are many small units living together thus forming a big group (sometime upto 180 or more individuals).

Sometimes there are following two major types of individuals within multimale bisexual grouping:

1. Those which do not divide in smaller feeding groups, *e.g.*, gorillas (Fig. 11.13).
2. Those which divide daily into smaller feeding groups, *e.g.*, rhesus monkeys.

In gorillas the group is typically a multimale bisexual type with several males and several females, all the members remain together. The males have dominance hierarchy. The most dominant is called **alpha,** then **beta, gamma** and so on. There is a simple linear dominance hierarchy among the adult males of a troop. But there is no clearcut dominance system among females. In a typical group of gorillas of 20 individuals, the oldest and largest male, develops gray hair on his back and is called **silver back.** He is also most dominant; rest of the males would be lower in status. Dominance consists of possession of right to the way on narrow path, or to a resting place or feeding site. Surprisingly, in contrast to all other primates, dominant males in gorilla is not very aggressive and all other males also have assess to receptive females. There is no conflict for the females.

Fig. 11.13. Gorilla.

Rhesus monkey (*Rhesus macques*) is widely distributed in India and lives in large multimale bisexual groups. The males have dominance hierarchy and in them the bonded females acquire dominance from the males they have been affiliated or bonded with. For example, if the alpha is most dominant male, his bonded females will enjoy high place in dominance hierarchy among females and rest of group members, even their infants acquire that dominance. The dominant males can be identified easily by their confident walk and by their long strides, they carry their tails up and a subordinate male walks carefully and tucks its tail between the hind limbs (Fig. 11.14). If the alpha male goes away from the group for a short while the beta male raises its tail and as soon as the alpha returns it again takes the tail down.

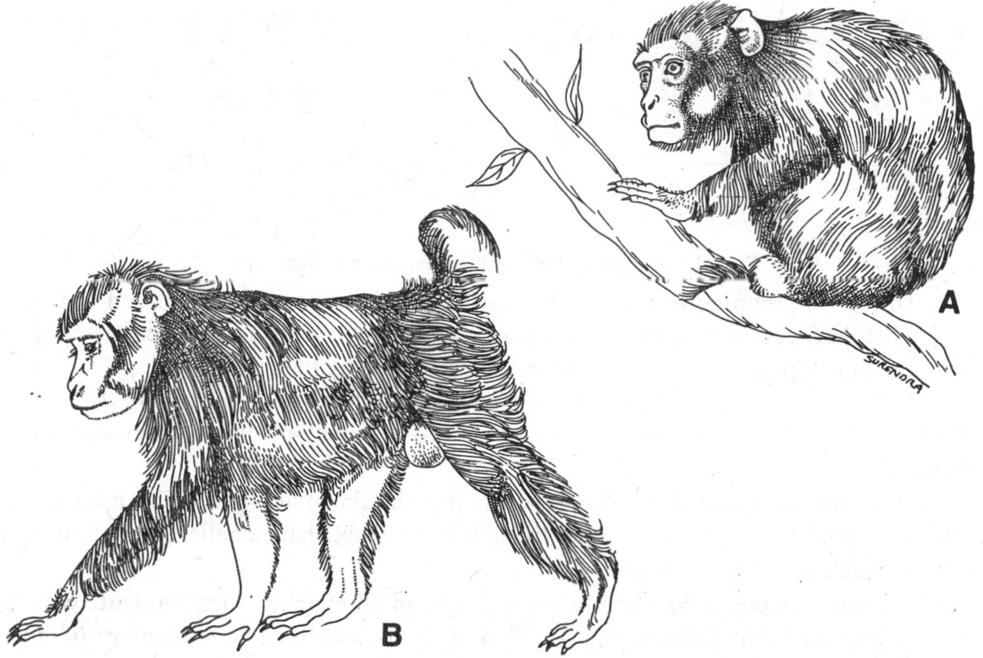

Fig. 11.14. Typical body postures assumed by a subordinate (A) and dominant rhesus monkeys (B).

The group may split temporarily into family units for foraging. All family units remain in near vicinity and can unite at the time of danger, for day resting and every evening for roosting.

6. Type VI. Diffuse social parties. Examples. Chimpanzee (*Pan troglodytes*) which are found in Guinea to Zaire and in Uganda and Tanzania and pigmy chimpanzee or bonobo (*Pan paniscus*) which is found in Zaire. The chimpanzees represent the living apes, the closest relatives of humans. Chimpanzees share about 99 per cent of their genetic material with humans. Chimpanzees are expert climbers, rest in sitting posture and walk on hind limbs but run on all four limbs (Fig. 11.15). Their social behaviour has been studied by **J. Van Lawick Goodall** (1968) and her group of University of California USA and **T. Nishida** of University of Tokyo, Japan. They found a remarkable and unparalled sociality among chimpanzees. Chimps usually live in groups. Males guard territory and restrict entry of males from other groups. The whole community searches food but breaks into smaller parties in case when food availability is less. Group size of chimpanzees is thus proportional to the food availability. Furthermore in small groups (n = 3 to 6), each member has responsibility of searching its own food.

Special Features of Primate Socialization

1. Primate groupings are close associations of conspecifics residing in a territorial limit. Each member has access to all kinds of information relating to food, water and danger, within the group and also to activities of neighbours.

2. Group size may vary from two to hundreds of members. But clustering is always avoided and optimum group size matches perfectly to the amount of food and also furnishes adequate sleeping site.

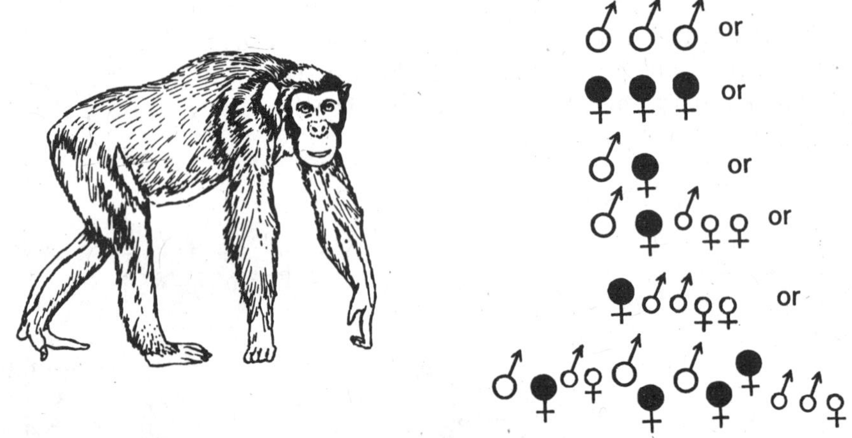

Fig. 11.15. Formation of diffused parties in chimpanzee.

3. The social grouping is based on the basis of rank relations, alliances and rivalries. A close tie exists between near-relatives and loose bonds are found between nucleus and peripheral individuals. This type of social organisation is exemplified by the monkeys.

4. Reproduction is all year phenomenon and young-ones have close relations with males and females of all ages, until they are mature themselves to play distinct ecological and social adult roles.

5. The group composition is adapted to life in a certain environment. This must suite the ecological demand. For instance, primates living in open tend to have smaller population density than those which are forest dweller.

6. Primate societies lacks blind following. Instead all members **communicate** with each other through sounds and gestures (Fig. 11.17) and also by watching the behaviours of other in the group. Each individual is constantly responsive to the posture, movements, gestures and

calls of other. This constant high level of **attention** to others is indeed dramatic.

7. In primates dominance and subordination are important features of their relationships with others (Fig. 11.15). Dominance always involves the threat of physical displacement or attack, even though it may be rarely necessary once rank is established.

8. Besides above negative interaction, positive interactions do occur in between individuals of a group of primates. There are many friendly contacts between animals as when they move and rest together, invite **grooming** or offer to groom another. Mutual grooming is very important as a appeasing gesture in primates. Often a dominant animal will "allow" itself to be groomed by a subordinate following a brief threat to which the subordinate has deferred. **Sexual presentation** as an appeasement gesture is very common in baboons and chimpanzees and is made by males or female towards a dominant animal who threatens or even if the subordinate wants to pass close to dominant (Fig. 11.16).

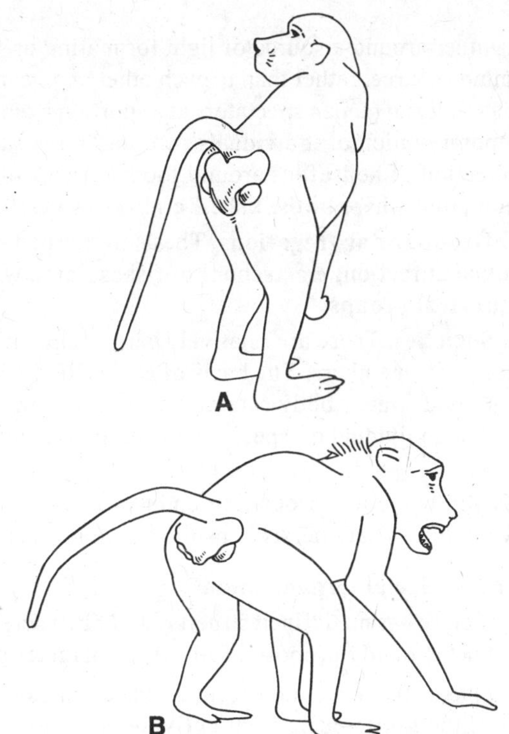

Fig. 11.16. Sexual presentation in the chacma baboon. A — 'Genuine' presentation by an oestrus female; she characteristically looks back over her shoulder toward the male. B — Appeasement presentation toward a threatening dominant animal; the general stance is the same but the facial expression is one of the fear.

9. Primates social organization is meant for constant exploration during daily walks on the basis of memory for the life purposes.

10. Primate societies have distinct division of labour and rank ordering. The task of decision making, child nursing, defense, foraging, exploration of new areas of food and sleep sites are performed by the animals according to status they enjoy in the group. Status, rank ordering or role distribution is based on age, sex and personality of the individual.

11.6. STRUCTURE AND ADVANTAGES OF SOCIAL ORGANIZATION

An animal **society** is a stable group of individuals intercommunicating with each other and catered by cohesion, cooperation and natural selection. **Eisenberg** (1965) has defined a social organization which contains (1) a permanent social structure; (2) a complex communication system; (3) specialization based on division of labour; (4) strict cohesion among the members, and (5) impermeability to conspecifics.

Social organization of a group depends on the total number of individuals, their age, sex ratio and number of adult males and these in turn depend on many important *factors* such as abundance, availability and dispersion of food, predation pressure, type of habitat and mating strategies. Not all groups of animals are truely social. There is clear distinction between mere **aggregations** or **associations** and true **societies (Alverdes,** 1927).

1. Associations. These are chance gatherings produced solely by external factors, *e.g.*, insects gather around a source of light for mating or groups formed by simultaneous attraction to a common source, rather than to each other, *e.g.*, earthworms collect under a rock for moisture and protection and human spectators at a sporting event or some cultural programme. Association is a temporary union of individuals reacting in the same way to similar environmental stimuli, *e.g.*, collection of houseflies around food source, collection of animals around a water hole, a migration point where birds, fishes, crabs or waterflies gather.

2. Groups or aggregations. These are formed largely by non-breeding individuals based on a mutual attraction, *e.g.*, school of fishes, large wildebeests and zebra herds. They are also called **survival groups.**

3. Societies. These are relatively permanent unions of individuals held together by mutual attraction of its members. The basis of social life is the interaction of individual members who exchange food, water, body care and sexual favours. According to **Tinbergen** any interaction between one individual of species with another member of the same species is known as **social behaviour.**

Lastly, while an association endures only while a specified set of external stimuli are effective, but a society persists inspite of great variation in external conditions.

Structure of Social Organization

A society has no defined structure but have certain properties. **Wilson** (1975) and **Brown** (1975) independently proposed following characteristics of social groups:

1. One of the most significant characteristics of social behaviour involves the number of animals of the same species that **actively come together** or **remain together** in a group. The minimum level of sociality (*i.e.*, the smallest social group) is found between a male and female who interact only during breeding season or between mother and infant.

2. Types and extent of connectedness between members within the group.

3. Reciprocal communication occurs for attracting and keeping the members of a group together. The communication may be visual, auditory, olfactory or tactile.

4. Much social behaviour is marked by **division of labour** and specialization of members according to their social roles, *e.g.*, queen, drone and workers in honeybee society.

5. A feature of social group in many species is an "**overlap of generations**", that is, families or part of families may stay together, *e.g.*, parents defending or protecting their young ones.

6. Eusocial animals have the highest level of social behaviour called **altruistic behaviour** or **aid-giving behaviour.** The most extreme forms of altruistic behaviour induces the sacrifice of one's life and reproduction for the sake of its colony. Origin and existence of sterile or non-reproductive castes among ants, bees, wasps and termites poses a difficult problem for natural selection theory as such individuals are not maximizing their own reproductive potential, instead they work for the benefit of the colony. They have probably evolved through **indirect** or **kin selection.** In a colony all the related animals live; the sterile castes by helping related animals ensure the perpetuation of some part of their own genes through their kins which is an extreme form of altruism.

Some other lesser significant attributes of social organization include identification of cohesiveness, degree of permeability to immigrants, extent of mutual cooperation, attention of the group and identification of distant and contact species.

Four Vital Qualities of Social Organization

Eisenberg (1965) has proposed the following four vital qualities of organized societies:

1. Communication. All organized societies of animals have some form of communication system. The members of a social group make gestures, postures, change colour, raise hair, they scent mark (by pheromones) and/or communicate through vocalization. They may convey messages by touching each other, or have some specialized forms such as ecolocation in bats, round and waggle dance in bees. Communication in higher animals is generally quite elaborate. For example, in chimpanzees, the facial expression alongwith vocalization communicate various forms of emotions and message (Fig. 11.17).

Fig. 11.17. Facial expression in chimpanzees.

2. Cohesion. The individuals constituting a society tend to remain in close proximity to one another. For example, all the bees of a group live in one hive, the individuals in a herd of deer, pride of lions and pecks of wolves remain in close vicinity in a given home range and the individuals in a troop of Hamadryas baboons while moving tend to remain in close proximity (Fig. 11.18). The adult dominant males (ADM) remain in the centre close to females, while adult subdominant males (ASM) take dangerous positions in front and rear.

Fig. 11.18. Cohesion between individuals of different age-sex classes in a Hamadryas baboon group. ADM = Adult dominant male; ASM = Adult subdominant male; AFI = Adult female with infant; I = Infant; J = Juvenile.

3. Division of labour. In organized societies, animals of different status, sexes or age groups have different functions in maintaining the society, *e.g.*, honeybees, wasps, termites, ants, baboons, macaque, etc.

4. Permanence and impermeability. The individuals making up a society tend to be same. There is little *migration* from the group. In most mammals the core of the group is formed by females who are related to each other, the males come and go, otherwise, the membership among females is permanent. Most organized societies resist **immigration** by outsiders.

Advantages of Social Behaviour

Sociality confers several advantages over an isolated mode of living. Some of the advantages of social behaviour are the following:

1. Antipredation

(i) Alarm calls. Sociality enhances defence against predators, for a number of individuals can be put on alert for detection of an enemy. Improved detection of predators — with more eyes and ears, there is an increased chance that one or more individuals will detect a predator before the others and be able to warn rest of the group. For example, in a group of deers or monkeys, one or two individuals will give **alarm calls** at the sight of approaching predator and the rest of the group members will take advantage of it. Likewise in a flock of birds alarm calls are given by few individuals and the rest get the advantage. Similarly, in prairie dogs which lives in large colonies, only one member would produce a loud alarm call when a coyote approaches, and rest of the praire dogs tend to dash into the burrow.

(ii) Guard behaviour. Animals also show **guard behaviour** in which one or a few individuals assume the role of watching for the entire group. They, thus, free others almost completely from the job of vigilance. For example, adult males of some of the monkey groups such as hanuman langur, occupy the highest canopy and remain vigilant for intruding members of all male groups and/or for predators.

(iii) Mutual vigilance. This behaviour has been observed among members of different species. Social animals living in the same habitat respond to the alarm calls given by the members of other species. For example, baboons, zebras, gazelles often forage together and each species respond to warning signals of each other.

(iv) Induction of startle response in predators. Erratic flight and explosive scattering form a vital antipredatory behaviour. Fleeing along an unpredictable path is a common defensive tactic in herding animals. Simultaneous eratic scattering of many individuals is likely to be more confusing to a predator than flight by a single individual. Most deer and antelopes use this tactic. Wildebeests form herds of 500 to 1000 individuals; they start running in all directions in a very confusing manner as the predator approaches.

(v) Geometrical effects. Social groupings provide a provision of formidable opposition against enemy. It is established that collection of weak forces prove to be winner against lonely mights. Predatory bird, hawk, prey upon wood pigeons. It was observed that attack success of hawk decreased as *the flock size of wood pigeons increased.* Further due to **geometrical effects** it is most likely that predators attack individuals on the periphery of a group. Prey animals in group should, therefore, gain protection by remaining in or moving towards the centre of a group. The animal is safer when it can keep another individual between itself and predator.

(vi) Mobbing. Running away or fleeing is the most commonly used antipredatory strategy among animals. However, there are examples when adult animals, generally males, **mob** a predator or form a defensive screen to scare it away. Soldiers or workers of social insects defend colonies by collectively attacking. **Musk ox,** a large Bovid found in North America, when attacked by wolves, adult female musk oxen form a defensive ring around their calves, while adult males form a ring around them and collectively chase wolves in attempt to drive them off. Elephants too form similar defensive formation. Groups of baboons have been observed chasing leopard, cheetah and lions (Fig. 11.19). Coatimunids, agoutis, California ground squirrels and various non-human primates have been seen mobing snakes. Large ungulates sometimes charge in a group to repulse lions, hyenas and wild dogs. Many species of colonial or flocking birds also mob predators.

Fig. 11.19. Defensive formation in baboon group.

2. Feeding Efficiency and Information Sharing

In groups, animals can pass on information to each other about the food or water source, preying and other skills.

(i) Cooperation. It is easier for a group of animals to catch a prey instead of catching it alone. Thus, **cooperative foraging** is beneficial, *e.g.*, for a group of monkeys it is easier to locate a bonanza of fruits. Early in the morning, members of a monkey group will spread and roam around in search of food and even if one individual locates the food it communicates vocally to other members of the groups which get the benefit of this information. When a baboon comes upon a water hole, it conveys this information to the others by its conspicuous drinking posture, with the hind part and tail sticking out.

In a similar way, by **cooperative hunting** even small carnivores such as wild dogs and wolves can hunt big animals such as elk, moose, mountain sheep and zebras. Lions, hyenas, killer whales and some dolphins also get benefit of cooperative hunting. Lionesses do most of the hunting together, often in groups in an organized way, some females will drive prey toward other lioness lying in wait.

Cooperation improves the ability of carnivores to protect carcasses from scavengers, a single lion usually cannot prevent a peck of hyenas or wild dogs from stealing a carcass, but two or more lions can.

(ii) Aunt behaviour. The animals of a group can leave their young ones with other group members and go foraging, *e.g.*, the pups of wild dogs, wolves and lions are often attended by few adults while other peck members can go out hunting. In a group of hanuman langurs, the related adult females temporarily take away the infant from mother to enable her to forage. This behaviour is called **aunt behaviour.** In elephants, for example, the related females, called **aunts** help in feeding the baby; related elephant help each other in acts such as guarding of the young while its mother drinks at a water hole (Fig. 11.20).

Fig. 11.20. Elephants aunt guarding.

(iii) Culture. A particular skill once discovered by an inmate suddenly or by trial and error persists throughout the group, if it is adaptive, despite that the discoverer has died.

Examples. 1. An experimental study involving chaffinches and house sparrows confirmed that flock members observe what others are eating and then begin eating those foods themselves.

2. In England just a few titmice birds started eating cream from milk bottles and gradually it became a major nuisance because hundreds of these birds learned by watching each other to open the milk bottles left on porches by milkman.

Fig. 11.21. Japanese monkey's food washing behaviour.

3. Japanese monkeys evolved a method that involved taking a pile of sand and grain down to the lake and drop it in resulting in the sinking of sand and floating of grain. This facilitates the scooping of grain from the water surface.

3. Facilitation of Reproduction

Group living improves **reproductive success.** For example, in solitary animals such as rhino and orangutan it is difficult for them to find a mate. They have to cover large areas in forest, spend time and energy to find a suitable mate but it may be easier to find each other in a larger social group. In a group, watching others courting and mating initiates behaviour in other members also.

11.7. SOCIAL DOMINANCE

Dominance is a common, but not universal kind of relationship between the members of a group in which some animals, the **dominant ones** have priority over other, the **subordinate ones.** The priority concerns access to such desirable resources as food, places to sit and mate. The Norwegian biologists **Thorleif Schjelderup-Ebbe** made the first important observations on dominance; he studied it in the common domestic hen (1935).

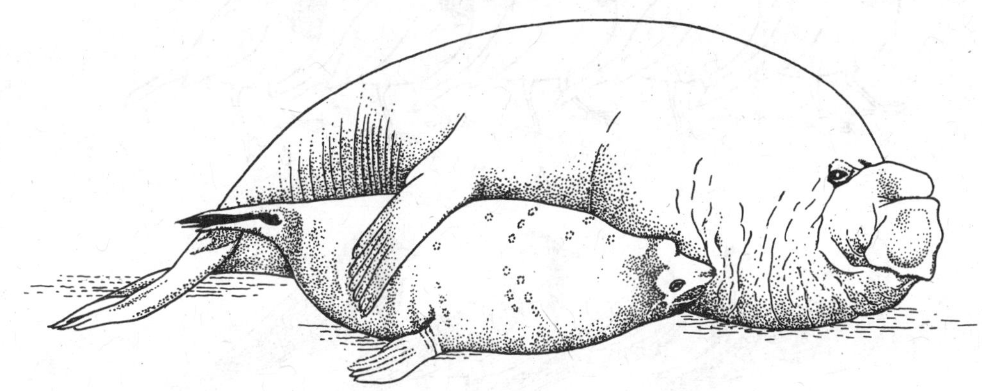

Fig. 11.22. Male and female elephant seals showing the great disparity of size which has evolved in these extreme polygynous animals. Mortality among males is commonplace and only a few achieve the size and strength to hold a harem of females on the breeding beaches.

Dominance involves a differential priority of access to resources which in some cases may include mating partners. If dominance confers a differential ability to inseminate females, it would be important that males fight to achieve dominance and attain the increase in fitness to be gained through siring (fathering) more offspring. For example, among elephant seals (Fig. 11.22), it has been estimated that less than one-third of the males in residence in a given season ever copulate (**Le Boeuf,** 1974). The higher the status of the male, as gained through male-male fighting, the more readily he approaches and copulate. In one study, 4 per cent of the males of elephant seals inseminated 85 per cent of the female (**Le Boeuf** and **Peterson,** 1969). Similar results have been obtained in a laboratory setting with house mice dominant or more aggressive males have been found to sire more offspring than subordinate or less aggressive males (**Horn,** 1994).

A weaker animal submits to the stronger animal by exhibiting a submissive pose. For example, carp (fish) closes its fins while submitting to **submissive behaviour.** A dog shows respect to its superior (*e.g.,* master) by licking the face of master because he thinks his master to be a dominant dog. Sometimes two animals fight in order to decide their ranking as to which one of them is superior. The fight goes on till one submits to the other and accepts its superiority.

1. Signals of Dominance

The males of Hamadryas baboons display their fangs for claiming superiority. Higher rank of the male is displayed by the bigger and larger fangs it possesses. In gorillas, the silver backed males remain at the top in ranking and females with young ones rank second. **Lebedev** and **Spanovskaya** deserved in the movement of the Malabar daanio (fish) that they show different angles in their hierarchy while swimming in water. The strongest fish swimes horizontally at about 2 degrees in relation to the water surface while the second in the hierarchy swims at 20 degrees and the third at 32 degrees and so on.

Badges of status. When animals encounter too many other animals to learn them all individually, they have a second method of avoiding unnecessary fights while keeping signal costs to a minimum. This is to have a dominance based not on individual recognition, but on

badges of status that can be instantly recognized even by unfamiliar individuals. Badges of status such as the *black bibs* of male house sparrows or the *black head feathers* of Harris' sparrows distinguishes dominant from subordinate animals (Fig. 11.23). What is striking about

Fig. 11.23. The most dominant male house sparrows (*Passer domesticus*) have the biggest black bibs (A) whilst in Harris's sparrows (*Zonotrichia querula*) the most dominant have the blacket heads (B).

them is that they are easy to be reproduced so can be used to cheat. However, it may be relatively easy for a weak animal to produce a status badge, but if it is constantly challenged by other dominant individuals, then it will pay a cost for its deception. The social consequence of cheating may be sufficient to keep cheats under control or at least at a low numbers.

In practice, status badges thus have a higher degree of reliability than they appear at first sight. They are found particularly in species such as sparrows and great tits where animals constantly encounter such large numbers that recognition on individual basis is difficult.

2. Dominance Hierarchies

Many social animals recognize pattern of authority; the two most basic being parent-child or provider-dependent and the relationship of stronger to the weaker or aggressive to submissive. The set of such relationships is often called **social** or **dominance hierarchy** (**Barnard** 1983).

The dominant hierarchy may be very simple involving a single individual ('despot') who is dominant to all others in the group but no rank relationship between subordinates (Box 11.8).

Box 11.8.

In naked mole-rat (*Heterocephalus glaber*), the colony contains only one breeding female, the "queen" who is dominant to all other females. In a captive group studied by **Susan Margulis** (1995), the queen was the only individual in the group to inspect the anogenital region of her female colony-mates and she was especially aggressive toward one female, the only one to have ovulated during her tenure. Upon the removal of queen, the victim of her aggression became one of the major contestants to become the sole breeding female in the group. The hormonal condition of the contestants changed markedly during this time, suggesting that the old queen could have learned about the hormonal state of reproductive capability of her fellow via anogenital inspection.

More often, there is a series of rank relationships such as that an "**alpha**" individual is dominant to all others in the group, a "**beta**" individual is dominant to all except the "alpha" and so on down to the lowest ranker who is subordinate to everybody. If it is possible to rank dominant males in a single sequence, the system is called **simple** and **linear dominance hierarchy.** The classical examples of linear dominance hierarchy are from so called "**peck orders**" of pigeons and chickens. The organization of an established flock of chickens is based on a set of dominance relationships. If food is given in a restricted location, it becomes apparent that one animal has priority of access to the food. It will exclude other chickens from the area of food, perhaps by delivering a peck at them. The other animals never return the peck. Such pecking is particularly prominent in newly established flocks; hence **Schjelderup-Ebbe** referred to dominance hierarchies as "**peck orders**". Typically in a group of dozen chickens one may recognize alpha, beta, gamma, etc., males in the linear dominance hierarchy.

Similarly **Bruins** and **Hediger** have reported *alpha, beta* and *gamma* ranks in the deer of the Basal Zoo where alpha dominated over others and when its antlers were cut off, it gives its privileges to the beta animals. Southern elephant seals and baboons provide other good examples of social dominance and dominance hierarchy (Box 11.9).

Box 11.9.

Male animals of all sorts and sizes, living and extinct, fight with bites, kicks, antler locks, jaw grapples, head butts, even neck slams in case of giraffes, in order to establish *dominance* over others. For example, when two male elephant seals throw themselves at each other on a beach, 4000 or so kilograms collide in a battle that will determine a winner, the one who stays, and a loser, the one who humps away across the sand. From his records of winners and losers in a population of southern elephant seals (*Mirounga leonina*) on South Georgia Island in the Atlantic ocean, **T. C. McCann** (1981) found that he could identify the number one male, an individual who had from 14 to 157 encounters with each of nine other males, and won them all. The number two male elicited submission from all but the top male and so on down the line. In other words, the elephant seals fought to establish dominance over others, forming a **linear dominance hierarchy,** with each male in his own place. In elephant seals, larger males generally have an advantage in their beach fights. The relationship between large body size and winning fights holds for a broad spectrum of animals (**Anderson,** 1994).

Baboons are another example of a species in which males are much larger than females, and here too adult males compete violently for *social dominance.* **Carlos Drews** (1996) found that males averaged one aggressive wound a bite—every 6 weeks, an injury rate about four times greater than that of females.

Sometimes as in rabbits, there are two linear hierarchies, one for males and another for females. In other cases, for example in certain monkeys, the dominance hierarchy for food may be different from that for an access to reproductive females. In certain animals, linear dominance hierarchies are complicated by triangular relationships such as that A beats B, B beats C, C beats A, but these are usually unstable.

Thus dominance hierarchy may be simple or intricate in animals. Intraspecific and interspecific hierarchies are found as seen in the mixed flocks of titmice. All great titmice rank superior to blue titmice, whereas blue titmice remains superior to the black-capped chickadees.

3. Types of Dominance Hierarchies

Barnard and **Burk** (1979) have distinguished the following *three* basic types of dominance hierarchies:

1. Statistical hierarchy. Individuals have no memory of past encounters and fight every time they meet. Because of variation in physical qualities some tend to win and others lose. Because of the costs of fighting and the consequent selection pressures for assessment, purely statistical hierarchies are likely to be rare in nature. However, there is evidence that they may exist at least in some lower invertebrates. For example, aggression between sea anemones (*Actinia* sp.) appears to conform to a statistical hierarchy where success depends on relative body size.

2. Confidence hierarchy. The act of winning a fight increases the vigour with which the victor enters its next encounter, so enhancing its chances of winning again and vice versa. Computer simulation of this 'confidence' effect ultimately produces a linear dominance hierarchy. The confidence dominance hierarchy has been recorded in crickets (Orthoptera: Gryllidae) and laboratory mice, where the outcome of fights can be predicted more from how individuals fared in their last few fights than from the characteristics of opponents.

3. Assessment hierarchy. Individuals use their memory of past encounters to decide which individuals they will fight and which they will avoid. Fights become calculated risks based on features of opponents (**assessment cues**) which correlate with having won or lost.

Assessment hierarchies are commonest type in nature. We would expect natural selection to favour individuals which are sensitive to any cue which predicts the outcomes of a fight because fights are likely to be expensive in time, energy and risk of injury. Indeed, the majority of recorded hierarchies in insects, amphibians, birds and mammals do appear to be of this type (**Barnard,** 1983).

4. Advantages of Dominance Hierarchy

The net effect of a hierarchy is that a small proportion of the individuals within a group gain priority access to limited resources such as food, water, roosting places and mates. Surviourship and reproductive potential may therefore be closely linked with an individual's dominance ranking.

For example, in a wolf pack during the consumption of the scare food, the dominant males eat first and other take what is left. Moreover, dominant males have sexual rights and priorities and they can interfere in the sexual behaviour of the weaker males. Such behaviour patterns are adaptive, ensuring that the strongest will be best fed and so remain strong and successfully reproduce while the weaker are less likely to survive. Thus, it is very desirable to a hen to be dominant. Dominant hens take the pick of the food and roost sites; dominant males also copulate more with dominant females than do subordinate males. Natural selection must therefore be favouring the dominant animals. *What kind of animals become dominant?* In hen dominant birds are usually larger in size. They also tend to have more of the male sex hormone, the **testosterone**, in their blood; a hen can even be made to ascend its dominance hierarchy by injecting some of this hormones in her blood. There is evidence, from other kinds of animals, of many other factors which affect dominance. *Parasites* , for instance, affect dominance, at least in mice. **W. J. Freeland** set up groups of three mice; the different mice having been injected with different quantities of parasites. The mice with the least parasites usually become dominant.

Box 11.10.

Spotted hyenas (*Crocuta crocuta*) live in large clans, hunt big game and compete fiercely with one another for meat from the animals they kill. **Dominant females** and their offspring gain more food than subordinate ones, and as a result have three times the reproductive success than their inferiors. The sons of alpha females "inherit" their mother's high dominant status, when fully mature, young male hyenas emigrate to another clan. If the son of a dominant female becomes the top male in his new clan, he will enjoy exceptional reproductive success, because only the dominant male mates with the many females in the clan.

Pseudopenile displays. The social system of spotted hyenas heavily rewards dominant females (**Frank,** 1986). In turn, the dominance interaction among females may be mediated by *pseudopenile displays* because the tendency of a female to present her erect pseudopenis to other in a greeting display is related to her social status. Subordinate females and youngesters are far more likely than dominant animals to initiate these displays. This greeting ceremony appears to promote cooperative behaviour among the participants. Cooperation between dominant and subordinates might be advanced if the dominant could check the physiological (*e.g.*, hormonal) state of a subordinate by inspecting its erect, blood-engorged penis or clitoris. Submissive subordinates may signal that they lack the hormones needed to initiate a serious challenge to the dominant. By this latter can afford to tolerate the *subordinates,* permitting them to enjoy the benefits of social life in the clan, since they do not pose an immediate social threat (see **Alcock,** 1993).

The dominance hierarchy of hens is also arranged by sex. Males are usually dominant to females. If there are many males and females in the group, the males form separate hierarchy about that of the females. In most species, which form dominance hierarchies, males are dominant to females, but this is not universal. There are some species such as hyenas (Box 11.10) and vervet monkeys, in which females are dominant to males. In many primates older animals are often dominant to juveniles.

5. Conclusion

The main reason why dominance reduce the amount of overt aggression within a social group is that weaker individuals come to learn that they are weak and therefore, avoid entering into fights that would be effortless to the stronger but dangerous to themselves. However additional mechanisms may be at work. In the pigtail monkey (*Macaca nemestrina*) the most dominant male reduces the fighting within his troop by policing any fights that breaks out. The establishment of a settled dominance hierarchy itself is the main limit on aggression.

11.8. TERRITORIALITY

Many species of animals establish "ownership" and defend specific areas of land called **territories.** This is called **territorial behaviour** or **territoriality. Kaufmann** (1971) has defined a territory as an area in which the resident animal enjoys priority of access to limited resources that it does not enjoy in other areas. Defence of a territory of some sort is in fact a common feature of the social organization of many vertebrates and invertebrates such as insects, fish, amphibians, reptiles, birds and mammals.

Box 11.11.

Terriotoriality is an important and widespread form of asymmetric intraspecific competition. It occurs when there is active interference between individuals such as a more or less exclusive area, the territory, is defended against intruders by a recognizable pattern of behaviour. Individuals of a territorial species that fail to obtain a territory often make no contribution whatsoever to future generations (**Begon** *et al.,* 1996).

Characteristics of Territoriality

Wilson (1975) has suggested that territories are more than simply defended areas. They are **dynamic** — often changing in size and shape with seasons, population density and animal's age. **Huxley** compared them with elastic discs having the resident animal at its centre. In conditions of high population density, the pressure along the disc's boundary is high and it contracts. However, there is limit beyond which further contraction is not tolerated by the resident. Either the resident aggressively resists further reduction in the size of its territory or the territorial system in the population breaks down. When population density and hence boundary pressure is low, territories expand. Once again, however, change does not go on indefinitely; very sparse populations are often characterized by territories whose boundaries are not cotiguous or touching (*i.e.,* they have a "*no man's land*" or "*buffer zone*" between them) or are so diffuse as to be indefinable. North American dunlins (*Calidris alpina*) provides a good example of elastic disc territories (Fig. 11.24). Sometimes the pattern of space use within the territory changes with its expansion and contraction. Tree sparrows (*Spizella arborea*), for instance, use the whole of their territory with equal intensity when it is fully contracted, but use only a central area intensively when it is expanded.

Territories with hexagonal boundaries. When several elastic discs are allowed to expand in close proximity to one another so that they meet, their originally circular boundaries distort to form hexagons. An important point is that perfectly-matching and contiguous hexagons leave no space between them. This feature would be of great advantage to territory owners in high density populations. By allowing their territory's boundaries to distort into mutually contacting hexagons, owners would maintain the maximum amount of space possible within their territories.

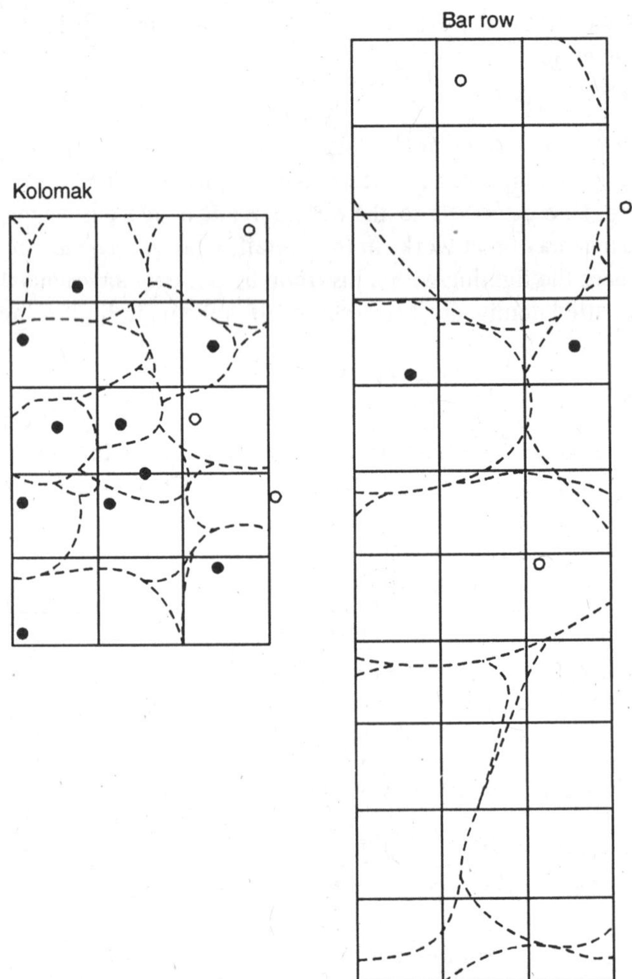

Fig. 11.24. Elastic disc territories in North American sandpipers. Territories are small in the high population density area at Kolomak, but large with "buffer zones" in the low density Barrow area.

Careful analysis of the dunlin territory boundaries (Fig. 11.24) revealed that at high population densities, these did indeed conform to hexagons. In the mouth brooder fish (*Tilapia mossambica*), the hexagonal territory boundaries of breeding males are sometime actually visible as sandy ridges surrounding the owners (Fig. 11.25).

Mudskipper is a small marine fish which has capacity to live out of water for periods when the tide is out. The males are highly territorial and build mud walls between the areas they occupy so that by the time the tide returns, the mud is a mosaic of areas, each fenced off like a suburban garden. When the tide is out, the walls preserve some water within the territory and this helps the animal to respire, but the main function of the territory seems to be to act as a **display area,** for the males leap in the air and this attracts females to mate with them **(Slater,** 1985).

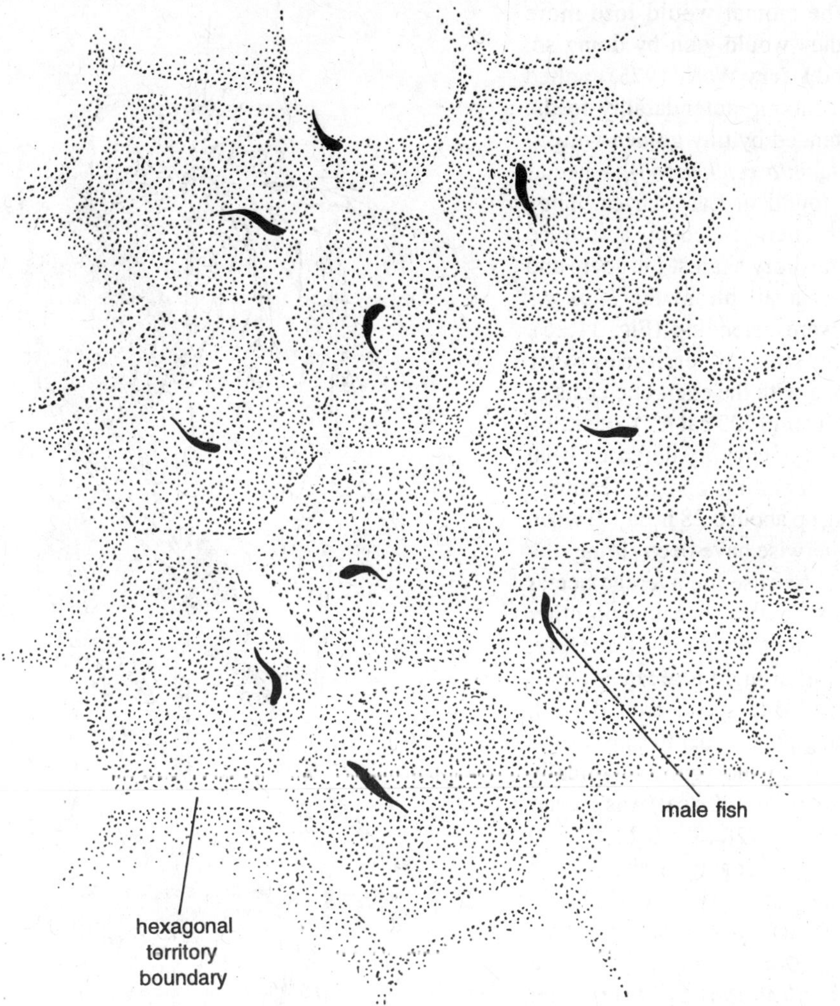

Fig. 11.25. Hexagonal territory boundaries in male mouth brooders, *Tilapia mossambica*.

Functions of Territoriality

Like other aspects of behaviour, the adaptive value of territoriality lies in the benefits it confers in increased survivorship and reproductive potential relative to the time, energy and risk costs involved. Thus any benefit that an individual does gain from territoriality must be set against the costs of defending the territory. In some animals this defence involves fierce combat between competitors, whilst in others there is a more subtle mutual recognition by competitors of one another's keep-out singals (*e.g.*, song, scent, etc). Yet even when the chances of physical injury are minimal, territorial animals typically expend energy in partolling and advertising their territories and these energetic costs must be exceeded by any benefits if territoriality is to be favoured by natural selection (**Davies** and **Houston,** 1984).

1. Territoriality and Calories

(i) Economic defendability. This idea was introduced by **Brown** (1969). It points out that animals should only go to the time and trouble of defending a territory if the resource they were defending (*e.g.*, food) was worth defending and could physically be defended. For instance, a very scattered food source might take so long to defend that the animal would lose more energy than they would gain by doing so. **Frank Gill** and **Larry Wolf** (1975) applied the idea of economic defendability to the territories defended by tiny golden-winged sunbird (*Nectarinia reichenowi*), a nectar feeding bird found in East Africa. They calculated the energy used by a sunbird defending its territory against intruders and also the energy available in the *Leonotis* flowers they were defending (Fig. 11.26). The nectar contents of territories were calculated along with the energetic costs of foraging for nectar (1000 cal h^{-1}), of sitting (400 cal h^{-1}) and of territory defence (3000 cal h^{-1}). By expending energy territory defence (taking up about 0.28 h day^{-1}) when they would otherwise have been sitting, the birds were able to raise the average nectar content of their flowers (because no other birds were foraging from them). Therefore, because foraging brought a greater rate of return, they needed to spend around 1.3 h less per day foraging, in order to satisfy their energy needs. What they gained typically (780 cal = 1.3 × (1000 – 400 cal) was greater than what they spent (728 cal = 0.28 × (3000 – 400 cal). Defending a territory was therefore energetically worthwhile.

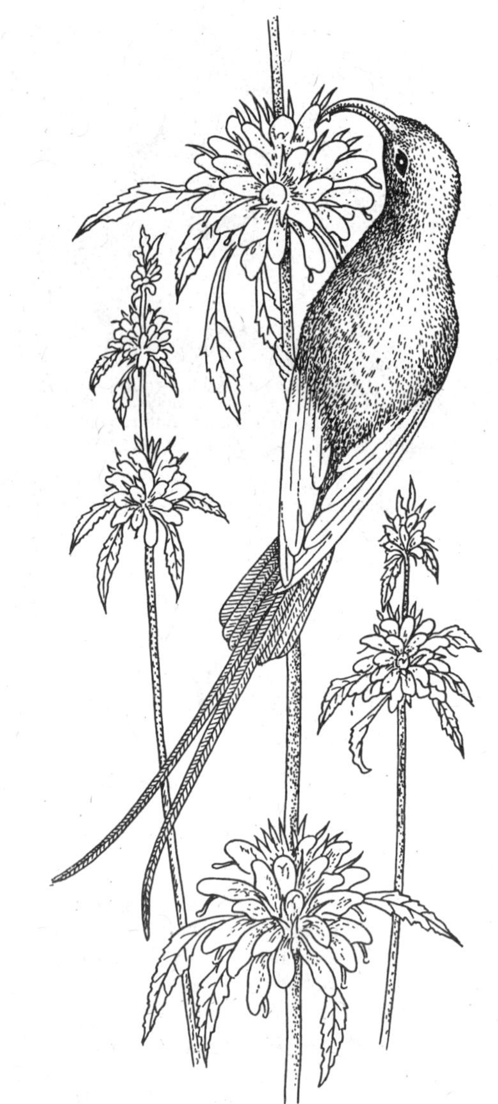

Fig. 11.26. The golden-winged sunbird defends nectar-filled flowers (of mint) against intruders if is energetically worthwhile.

(ii) Role of satellite males in economic defendibility. A study of pied wagtails (*Motacilla alba*) by **Davies** and **Houston** (1981) shows how finely tuned this energy balance can be. The wagtails were defending territories near a river bank (Fig. 11.27), where they fed on insects washed up on shoreline. They appeared to exploit this fluctuating food source extremely efficiently. Moving along the water's edge, a territory owner would patrol round the rest of its territory. Such a round trip takes approximately 40 minutes by which time more insects would have been washed ashore. Given that the advantage a wagtail gains from defending a territory is the exclusive access to any insects that are washed up, it is somewhat surprising to find that some territory owners tolerated another males, called **satellites,** in their territory for long periods of time. Such satellite males also walked around the territory following the same path as the owner but half a circuit behind so that each section of river bank was left for 20 minutes before being revisited by one of the birds. The owners appeared to be collecting only half the food they could collect if they had occupied territories on their own and had not allowed the satellites to feed. The solution seemed to be that the satellites 'earned their keep' by helping in defence of territory.

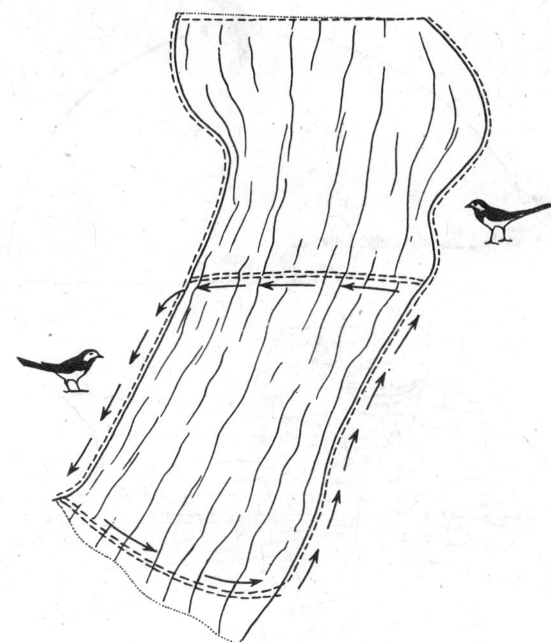

Fig. 11.27. Pied wagtails on riverside territories. The birds patrol around their territories collecting food washed up by the river.

Davies and **Houston** found that when owner was alone in the territory, 60 per cent intruders were spotted immediately. The remaining 40 per cent took some food from the territory before being chased. When a satellite was present, however, 85 per cent of intruders were spotted and chased immediately and only 15 per cent took any food. When food was sufficiently abundant, owners could increase their own feeding rate by upto 33 per cent by having a satellite to help them with defence, which was more than compensentated for the food eaten by the satellite. When food became scare, however, and there was not enough food for two birds, the owners became aggressive to the satellites and evicted them from their territories.

(iii) Variations in economic value of territories. In fact, describing territoriality only in terms of just 'winners' and 'losers' is an over-simplification. Generally there are first, second and a range of consolation prizes — not all territories are equally valuable. This has been demonstrated by a study of oyster-catchers (*Haematopus ostralegus*) on the coast of the Netherlands, where pairs of birds defend both *nesting territories* on the salt-marsh, and *feeding territories* on the mudflats (Fig. 11.28; **Ens** *et al.*, 1972). For some birds, called **residents,** the feeding territory is simply an extension of the nesting territory: they form one spatial unit. For other pairs, called **leapfrogs,** however the nesting territory is further inland and hence separated spatially from the feeding territory. Residents fledge many more offspring than do leapfrogs, because they deliver fair more food to them. From an early age, resident chicks follow their parents to the mudflats, taking each prey item as soon as it is captured. Leapfrog chicks, however, are imprisoned on their nesting territory prior to fledging, all their food has to be flown in. It is far better to have a resident than a leapfrog territory.

2. Territoriality and Breeding

Territorial behaviour plays a crucial role in the breeding biology of many species. The parcelling up of vital environmental resources (such as food or nesting sites) by members of one sex can secure their access to members of the other sex. Usually it is males who defend limited resources. Females who require the resources are then constrained to venture into male territories where they may be

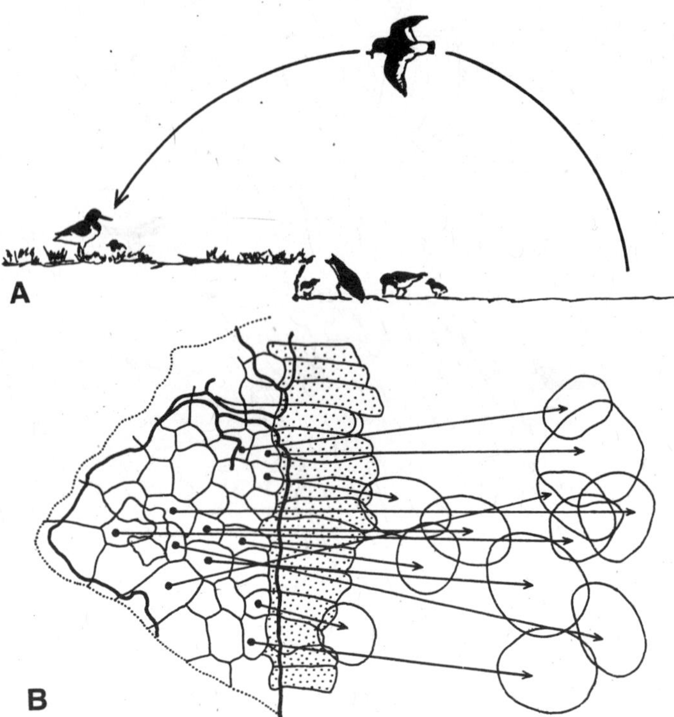

Fig. 11.28. A — A coastal area in the Netherlands providing both nesting and feeding territories for oyster - catchers. In 'resident' territories (dark shading), nesting and feeding areas are adjacent and chicks can be taken from one to the other at an early age. B — 'Leapfrogs' have separate nesting and feeding territories (light shading) and food has to be flown in until the chicks fledge.

mated. Depending on the distribution of resources in the territory and ability of males to defend them, females may be able to use some feature of the territory in choosing a mate.

One feature female might use is **territory size.** In long-billed marsh wrens (*Telmatodytes palustris*), female's choice appears to be related to the size and the quality of the food supply within a male's territory. Males which are **polygynous** have the largest territories, **monogamous** males have slightly smaller territories and unmated '**bachelor**' males have the smallest territories. Quite interestingly, males with large territories may accumulate a number of females while nearby owners of small territories remain unmated. Territory size also appears to be an important determinant of breeding success in male three-spined sticklebacks. In this case, large size of territory is negatively correlated with egg predation by intruder males.

In other cases, **territory qualities** other than size seem to be important. In red-winged blackbirds, the type of vegetation on male's territory influences the risk of predation of offspring. Males whose territories contain most effective type of vegetation attract the most females. In pronghorn (*Antilocapra americana*) it is the males whose territories provide the best foraging material which attract most females.

Female American bullfrogs (*Rana catesbeiana*) choose territories which are defended by **older** and **larger males.** Mortality among developing embryos is lower in these territories because they are subject to less extreme temperature variation and are infested by fewer predatory leeches.

Further, beeswax is an essential part of the diet of bird, orange-rumped honey guide (*Indicator xanthonotus*) and males maintain year-round territories at the location of bee's nests (Box 11.12). Since these nests are sparsely distributed on exposed cliffs a small number of males can monopolise all the food supplies. Courtship centres on the bee's nests and mating success is high for territory owners. One male, for instance, was known to copulate 46 times with at least 18 different females. Likewise, during period of female sexual receptivity male impala (*Aepyceros melampus*) and water buck (*Kobus defassa*) divide the habitat into defended territories. In this way a few males end up with all the matings when female herds move in.

Box 11.12.
Mutualism between Honey guide and Honey badger

An African bird, the honey guide (*Indicator indicator*) has formed a remarkable relationship with the ratel or honey badger (*Melliovora capensis*). A honey guide that has located a bee's nest leads the honey badger to it. The mammal tears open the nest and feeds on honey and bee larvae, and later the honey guide gets a meal of beeswax and larvae. The honey guide can locate a bee's nest but not break it open, while honey badger is in just opposite position. The reciprocal link in their behaviour brings mutual benefit.

Lek system. It is a type of male's territory which is observed in the "lek system" of birds such as white-bearded mankin (Fig. 11.29), sage grouse, ruff (*Philomachus pugnase,* (Fig. 11.30)) and prairie chicken and in mammals such as Uganda Kob (a large antelope *Kobus kob*) and the hammerhead bat. Here the males gather in a tighter group of **lek** and within the group, each

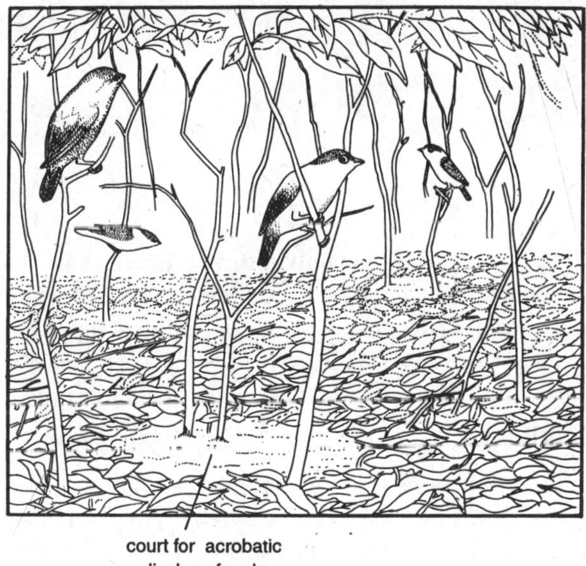

Fig. 11.29. Lek system of white-bearded manakin (A South American bird). Male defends tiny display sites; females (shown here on the far left) visit the lek to select a mate from among the several males displaying there.

male defends a small territory. There is no food in any of the territories and the function of these leks is not fully understood (**Balmford,** 1991). One possibility is that leks occur when females are very spread out, perhaps because their food is widely dispersed and the males are unable to defend a large enough feeding territory to attract them or keep them in one place (**Emlen** and **Oring,** 1977). The lek therefore becomes a sort of '**marriage market**' where the normally widely dispersed females come to choose a mate, perhaps because the light and sound of many males all displaying together attract females from a large area. **Rayan** *et al.* (1981) studied the Tungara frog (*Physalaemus pustulosus*) where the males aggregate into leks or **choruses** and showed that with bigger choruses, more females approach. Fortunately for the males, they are also safer from predation by frog eating bats in bigger groups.

Another possible advantages of leks is that females can make a direct comparison between different males on the basis of their displays or the sounds that they make. **Gibson** and **Bradbury** (1985) showed that it is the males of the sage grouse that display for the longest and in most vigorous way that are chosen by the females. In black grouse (*Lyrurus tetrix*), another lekking species, the female choose the males that have the highest chance of still being alive in six months time, suggesting that the females are using the lek to assess the health and viability of the displaying males (**Alatalo** *et al.,* 1991).

Fig. 11.30. Lekking in the ruff.

Superterritories. Quite contrastly, in the **polyandrous** American jacana (*Jacana spinosa*) females divide up suitable ponds and lagoons into "**superterritories**" within which a number of males set up their small breeding territories. Females holding superterritories often have several males incubating clutches simultaneously and readily replace clutches which are predated. Such polyandrous female birds are 50 to 70 per cent bigger than males.

3. Other Functions of Territories

The territories have other functions too, for example in ground-nesting sea birds such as gulls, terns and gannets (Box 11.13) they avoid **cannibalism.** These territories may be as little as 1 m across and contain no food at all. They consist simply of a small defended area around the nest which is always occupied by one member of the mated pair and are important because of the cannibalistic nature of other gulls in the colony, some of which are specialized in eating eggs and chickens.

Box 11.13.

The even distribution of the nesting territories of gannets (*Sula bassana*) is remarkable. The gannets obviously gain protection by nesting together on islands and from mutual defence against aerial predation, but each pair still requires a space, albeit little more than a square meter, which is defended against hostile neighbours. The result is an even spread birds across all the nesting areas.

Territoriality is a 'contest'. There are **winners** (those that come to hold a territory) and **losers** (that do not) and at any one time there can be only a limited number of winners. The contest nature of territoriality ensures a comparative constancy in number of surviving, reproducing individuals. One important results of territoriality is **population regulation** (*i.e.*, the regulation of number of territory holders). Thus when territory owner dies or are experimentally removed, their places are often rapidly taken by new comers. For instance, in great tit populations, vacated woodland territories are reoccupied by birds coming from hedge rows where reproductive success is noticeably lower (**Krebs,** 1971).

Territorial behaviour may enhance an animal's reproductive potential in other ways. Many of these involve reduction in the risk of predation. The defence of roosting positions in many species of bats, starlings, sparrows, and pigeons may result in successful defenders having priority access to the safer sites or those nearest to rich food supplies. In great tits, increased spacing between nest sites reduces the risk of egg and fledgling predation by rats and weasels. In ectothermic animals, such as lizards, aggressive spacing may decide access to the best sunny areas. Forest-dwelling anoles (*Anolis* sp.) fight for access to particular sunny perches in trees.

Secondary Functions of Territorial Behaviour

So-called secondary functions are also ascribed to territorial behaviour. For example, animals may become familiar with parts of their home range and thus be able to escape predators, find food and establish safe resting areas more effectively. Another suggested "secondary function" of territoriality is a reduction in the spread of parasites and disease between individuals. Certainly spacing reduces the amount of physical contact between animals, especially in colonially nesting birds such as gulls. By defending territories, males may also be in a better position to maintain established pair bonds. If the territory is large enough to encompass the female's normal range of activity, preventing access by other males is easier.

QUESTIONS

Long Answer Questions

1. What is meant by the term social organization ? Give an account of social organization in an animal which belongs to an insectan order Isoptera. (*Punjab 1999*)
2. Give an account of social organization in termites. (*Punjab 1997, Rohilkhand 99*)
3. Describe eusocial behaviour in animals. Discuss Hamilton's kin selection concept.
4. Write an essay on dominance and dominance hierarchy.
5. Differentiate between association, aggregation and society. Describe characteristics and advantages of social organisation of animals.

Short Answer Questions

1. What is meant by the term social organization ? Give an account of social organization in honeybee ? (*Magadh (H) 1992, 95, 97; Punjab 1998,99*)
2. Explain the kinds of social organization in primates. (*Bangalore 1995*)
3. Write a short note on "Leadership behaviours in primates". (*Bangalore 1994*)
4. Write a short note on dominance. (*Gorakhpur 1990, 95*)
5. Write a short note on dominance hierarchy. (*Bangalore 1995*)
6. What is territorial behaviour ? (Give a short account)
7. What is 'nuptial flight' ? Why does it occur in bee ? (*Calcutta 1991*)

Very Short Answer Questions

1. **Answer these questions in only 20 words.**
   (*a*) The different castes of honeybee exhibit division of labour. Justify. (*Kurukshetra 1996*)
   (*b*) What is Royal jelly ? (*Kurukshetra 1998*)
2. Name the haploid castes of honeybee. (*Purvanchal 1999*)

Multiple Choice Questions

Choose the correct answers from the four alternatives given:

1. Social behaviour in animals are of (*Bharathiar 1997*)
   (*a*) territorial rights (*b*) dominance and subordinance relationship
   (*c*) leadership (*d*) all the above
2. At levels of individuals, which of these is an example of "division of labour" ?
   (*a*) different parts of our brain
   (*b*) different kinds of receptors
   (*c*) polymorphic forms of a termite colony
   (*d*) variously adapted legs of honeybee
3. Which of these are colonial in habit ?
   (*a*) locusts (*b*) white ants
   (*c*) bed bugs (*d*) mosquitoes
4. Queen honeybee is adapted for
   (*a*) controlling other bees
   (*b*) laying eggs
   (*c*) laying eggs and rearing the youngs
   (*d*) preparing honey
5. Honeybee stores honey in
   (*a*) stomach (*b*) salivary glands
   (*c*) cells of comb (*d*) crop
6. If a worker bee of same species is transferred from one to another hive
   (*a*) it will be welcomed
   (*b*) it will be killed and eaten up
   (*c*) it will be shut in a special cell for becoming familiar with the smell of this hive
   (*d*) it will be killed and kicked away
7. In 1975, a Harvard University biologist published the book *Sociobiology,* in which the author attempted to show that characteristics of social behaviour evolved in the same way that anatomic and physiological traits evolved
   (*a*) Stephen Jay Gould (*b*) Ireanaus Eibel-Eibesfeldt
   (*c*) Edward O. Wilson (*d*) Eric R. Pianka
8. The hypothesis that altruistic behaviour is selected for to improve the fitness of an entire group of related or unreleated individuals is called
   (*a*) group selection (*b*) kin selection
   (*c*) inclusive fitness (*d*) reciprocal altruism

9. Altruistic behaviours between closely related animals are selected for because they
(*a*) reduce fighting between species
(*b*) increase the frequency of the altruistic individual's genes in the next generation
(*c*) ensure survival of altruistic individual
(*d*) none of the above

10. Which of the following sequences is most common in territorial behaviour?
(*a*) threat, combat, advertisement
(*b*) advertisement, threat, combat
(*c*) combat, threat, advertisement
(*d*) threat, advertisement, combat

11. Dominance hierarchies are most common when
(*a*) resources are concentrated in one part of environment
(*b*) males are larger than females
(*c*) there is no parental care
(*d*) resources are uniformly distributed throughout the environment

12. When a wolf exposes its throat ("shows the jugular vein") to another wolf, this means
(*a*) it wants to mate with other wolf
(*b*) it wants to be groomed by the other wolf
(*c*) it wants to fight with the other wolf
(*d*) it is subordinate to the other wolf

13. Drones are (*Magadh (H) 1997*)
(*a*) haploid fertile males (*b*) deploid fertile males
(*c*) diploid fertile female

14. Workers in beehive are (*Magadh (H) 1997*)
(*a*) diploid sterile females (*b*) diploid sterile males
(*c*) haploid sterile males

ANSWERS

1. (*d*)	**2.** (*c*)	**3.** (*b*)	**4.** (*b*)	**5.** (*c*)	**6.** (*d*)	**7.** (*c*)
8. (*b*)	**9.** (*b*)	**10.** (*b*)	**11.** (*a*)	**12.** (*d*)	**13.** (*a*)	**14.** (*a*)

12

Reproductive Behaviour Patterns
(Courtship, Mating Systems and Parental Care)

12.1. MATING SYSTEMS

In many land animals, fertilization involves mating or copulation. During mating, the females are in a defenseless position. In such animals, the mating involves the suppression of the escape behaviour. In many species the female takes the larger share than the male in feeding and protecting the young and is more important than the male. Also, one male can fertilize more than one female, another reason why the individual males are biologically less valuable than females. It is the reason why the female needs pursuation more than the male and courtship is so often the concern of the male.

There must be close coordination between the males and females, *i.e.,* they must find each other, during actual copulation they must bring their genital organs in contact with each other, and the sperm must find the egg cell. This orientation is one of the vital task of mating. Thus, reproductive behaviour of an animal involves the following components :

1. Choosing the Member of the Right Species

Since the genes are different in each species, mating between animals of different species bring widely different genes together. This easily disturb the delicately balanced growth patterns. Mating between different species, therefore, often results in fertilized eggs which are unable to live or die at the beginning of their growth. In certain cases, **hybrids** may live but are less vital or infertile. This has led to development of differences between the mating patterns of different species so that each individual can easily recognise its own species.

2. Sexual Selection

Within a breeding animal population, there is often intense competition between members of one sex for opportunities to mate. Typically, it is males that compete with each other for access to females because, whereas females need mate only once, males can enhance their reproductive success by mating with many females. Sexual selection is the evolutionary force that arises when some males are more successful than others in obtaining mating. Male competition may take one of the two forms.

(*a*) Males may fight over each other over females. In such cases **sexual selection** favours those that are largest, strongest or which have the most effective weapons (*e.g.,* horns). This form of sexual selection is responsible for the antlers of deer stags and for the larger size of males than the females.

(*b*) In another system, males may compete in a less direct way. In such cases, mating success depends on how effective a male is in attracting and stimulating females. Thus, variation in male mating, success depends on preferences among females for cetain males rather than other. This form of sexual selection favours elaborate male displays and the physical characteristics that are associated with them, such as bright colours and elaborate plumage. The courtship display and big tail of the peacock are thought to have evolved through sexual selection based on female preference.

The theory of sexual selection has been controversial since it was first formulated by **Charles Darwin.** While there has been no disagreement over the idea that male strength and weapons have evolved because of the advantages they confer in direct competition for females. A recent study has provided the evidence that female choice has been responsible for elaborate male displays and plumage. Further, the females do prefer those males that have the most exaggregated courtship characters.

3. Seasonality

Many species, particularly those living in temperate zones, breed during a part of the year only. The birds have a spring peak in their reproductive activity. In other species, the breeding behaviour takes place at other times of the year. Many species of deer and elk breed in the fall; wolves and coyotes breed in midwinter and many species of seals and sea-lion breeds in the late spring and early summer. Although different species have different gestation periods, most give birth in late spring and early summer. The major advantages to seasonality lies in timings of the birth of offsprings so that they appear when there is good weather and plenty of food available. Mating season and gestation period appear to interact in such a way that young take birth in late spring or early summer.

4. Mating Systems

The following mating systems have been observed in various species (Box 12.1) :

1. Monogamy. In **monogamous** mating systems, each breeding adult mates with only one member of the opposite sex. The 90% of the birds are monogamous. Many birds establish a territory during the breeding season. Both male and female ensure the food supply for the young ones. The formation of long-term pair bonds is advantageous because less time is spent in finding a mate during each reproductive cycle. In long-lived birds, such as sea gulls, those breeding with former mates have higher reproductive success because of less aggression between mates and greater synchronization of sexual behaviour. In birds one or both the parents are responsible for the parental care. In synchronization two animals come together and are physiologically ready to reproduce.

BOX. 12.1.

ECOLOGICAL CLASSIFICATION OF MATING SYSTEMS

Emlen and **Oring** (1977) developed following ecological classification of mating systems :

1. Monogamy. Neither sex is able to monopolize more than one member of opposite sex.

2. Polygyny. Male controls access to more than one female. It is of following four types :

***(i)* Resource-defense polygyny.** Male controls access to females indirectly by monopolizing critical resources, *e.g.*, male walnut flies (**Papaj,** 1994).

***(ii)* Female-defense polygyny.** Males control access to females directly, usually females are grouped for other reasons, *e.g.*, seals.

***(iii)* Male-dominance polygyny.** Mates or resources are not monopolizable, female select mates from aggregations of males, as in leks, based on the quality of the male's display or his territory, *e.g.*, sage grouse (*Centrocerus urophasianus;* **Wiley,** 1973; **Gibson** and **Bradbury,** 1985).

***(iv)* Scramble polgyny.** Males actively search for mates without overt competitions, *e.g.*, female woodfrogs (*Rana sylvatica;* **Berven,** 1981), horseshoe crabs (*Limulus polyphenus;* **Brockman,** 1990).

3. Polyandry. Females control access to more than one male. It is of following two types :

***(i)* Resource–defense polyandry.** Females control access to males indirectly by monopolizing critical resources, *e.g.*, American jacana (*Jacan spinosa;* **Jenni,** 1994).

***(ii)* Female-access polyandry.** Females do not defend resources essential to males, but they interact among themselves to limit access to males, *e.g.*, African wild dog (*Lycaon pictus;* **Moehlman,** 1986).

About 4% of the mammalian species show monogamy, *e.g.*, white-banded gibbon. In this case the males defend the territory and youngs remain with the parent until they became sexually mature.

2. Promiscuity. It refers to the absence of any prolonged association and to multiple mating by at least one sex, *e.g.*, deermice (**Birdshell** and **Nash,** 1973).

3. Polygyny. It is most common type of mating system. In this system the male has access to more than one female. In polygyny, males defend areas containing the feeding or nesting sites critical for reprodution. The female's choice of a mate is influenced by the quality of the male and his territory.

Territories that vary sufficiently in quality may attain the '**polygyny threshold**', the point at which a female may join an already mated male possessing a good territory than an unmated male with a poor territory. Thus some males may get two or moe mates while others get none. Some males monopolise females and exclude other males from their harems.

4. Polyandry. In polyandrous system females control access to more than one male. In most cases, females provide parental care while males seek new mates. If breeding success is low due to high rate of predation of the youngs or eggs, females produce many offspring.

For example, breeding sites of the American jacana, found in central and south America, are limited and are divided into small territories by males. Female jacanas control **super territories** that may encompass the nesting areas of several males. Frequently, several males incubate the clutches of one female. Usually, breeding females dominate the males and proved little parental care. In this case, the females are specialised only in egg production.

5. Courtship. Courtship behaviour functions to bring together two animals of different sexes of the same species for mating. The first problem is that of simply locating a potential mate. Obviously, the conspecifics must be of the opposite sex if reproduction is to be successful. Courtship entails a complex sequence of interacting signals to ensure that an animal mates with an appropriate partner. Timing is an important part for successful reproduction and both male and female must be in appropriate physiological condition for mating.

12.2. COURTSHIP BEHAVIOUR

The term **courtship** refers to the behavioural interaction that occurs between males and females before, during and just after the act of mating. **De Morris** has defined 'courtship' as the heterosexual communication system leading to consummatory sexual act. In some animals, courtship is brief and superficial but in other, it lasts for a long time and involves vigorous and elaborate displays. The male often plays an active role during courtship. It is also usually male that initiates sexual interaction and displays to the female whose behaviour is receptive or non-receptive.

Box 12.2.

Courtship behaviours are fixed action patterns that precede copulation. In courtship the males and females exhibit interacting sequences of stereotyped behaviours, in which a behaviour by male elicits a behaviour of the female, which in turn elicits another behaviour of the male and so on. Each component in the sequence is not unique to courtship (*i.e.*, it may be normal element of feeding, aggression and parental care) but the particular combination of behaviour is unique. The main function of courtship behaviour is to ensure that the two individuals are of the same species. Courtship behaviour informs a potential mate that the intention is breeding and not aggression.

A nervous organisation must develop within the animal prior to its sexual maturity. This determines both the form of courtship movements and their link with significant stimuli. Courtship is not induced with mechanical consistency. A male encountering a female may on one occasion court her intensively and on other occasion ignore her completely. Variations in responsiveness are often called **changes in motivation.** Many factors affect sexual responsiveness of the mates. It may decline temporarily just after coitus or after prolonged unsuccessful courtship. It may drop sharply if an enemy appears. It may also fall gradually if the animal becomes ill. Sexual responsiveness may disappear completely during the winter months.

Box 12.3.
Sexual selection

However skillful an animal is at repelling predators, foraging efficiently, or defending territories, these abilities have evolutionary consequences only if the individual succeeds in passing on more of its genes than other genetically different individuals in the same species. **Reproductive behaviour** is therefore the central focus of natural selection. In this regard one has to seek answer of the following questions:

1. Why the reproductive behaviour of males and females often differ dramatically?
2. Why do males so often take the initiative in courtship mates?
3. Why do females so often reject their suitors?
4. Why do males fight with one another over females?
5. Why do females prefer males that sport bizarre ornaments and perform strange behaviour displays?

Answering these intriguing questions require us to understand about **sexual selection** which is a category of Darwinian natural selection. Sexual selection occurs when one sex prefers the gametes received from certain members of the opposite sex.

Why Do Animals Court?

Sexual reproduction creates a social environment of conflict and competition among individuals as each strives to maximize its genetic contribution to subsequent generations. Males usually make many small gametes and try to fertilize as many eggs as possible, while providing little or no parental care. Because females make fewer, larger gametes and often provide additional parental investment in their offspring, they usually have a lower potential rate of reproduction than males. As a result, receptive females are scarce, and males typically compete for access to them, while females can choose among many potential partners. In a few species, however, these sex roles are reversed, providing an opportunity to test the theory that the differences between the sexes stem from differences in their relative parental investments (or in their potential reproductive rates).

In evolutionary terms, courtship provides chances for better survival. But it is generally believed that courtship is concerned with reproductive fitness rather than with survival. Even then courtship is quite important, *since a sterile animal, however long it lives, cannot influence subsequent generations.* Reproductive fitness too depends upon many factors; among them obtaining a mate, producing and rearing healthy offspring are significant.

Charles Darwin pointed out that the winning of a mate is an exceptionally complicated process in which *display gives one mate advantage over the other.* According to him courtship sometimes helps male and female to find one another, to indicate suitable breeding sites to one another and to synchronize physiological processes for fertilization. It may help to break down barriers which prevent the mates from coming together.

Causes of Diversity of Mating Systems

Mating in most vertebrates is basically an affair between two individuals. In fish such as herring, where several males may discharge gametes simultaneously, this is not so, but as a general rule sexual behaviour involves affairs between a single male and female. In spite of this basic similarity, there is enormous **diversity** in the way mating occurs. There are several reasons for this diversity. First of all, mating is rarely random in vertebrates. In most cases, their is a *competition,* either between females for nest sites and food, or between males for opportunities to fertilize the greatest number of females. This competition has led to courtship displays and sequences with elaborate eye-catching, stimulating or intimidating signals.

Secondly, to prevent mismatching with members of other species, *courtship behaviour must be* species-specific. This has also led to considerable diversity in displays, especially when related species share the same habitat. In birds and frogs, we find that **allopatric species** (species that normally do not overlap in their distributions) often have similar vocalization and that **sympatric species** (species with overlapping distribution) invariably have very different vocalizations. This indicates the importance of species distinctiveness in signaling.

A third cause of diversity in courtship behaviours is due to variations in the types of tendencies involved in a mating. For example, a species in which males are highly territorial is more likely to exhibit conflict behaviours stemming from both aggressive and sexual tendencies than in mobile herd or school where territories are not established. Since many displays have evolved from and reflect these motivational conflicts, we expect such differences to be reflected in courtship behaviour.

Finally, the need for temporal (momentary) synchronization in mating varies widely in different animals. In some fish, mutual simultaneous discharge of gametes is required to fertilize the eggs. This requires a careful preparation to discharge in which several display sequences by both male and female are performed to bring both into the same motivational state at the same time. Similarly, male and female doves exchange a complex sequence of signals that coordinate the equally complex sequence of mating behaviour necessary to raise their young. Repeated signals by one mate eventually trigger special behaviours in the other, which in turn evoke the next sequence in the first mate. By this means, doves go through a courting, copulation, nest-building, egg-laying and young-feeding cycle with each mate participating at the right time. Other birds do not require this synchronization and their courtship behaviour differs accordingly. If we take all these different factors into consideration, we can categorize most mating systems into one of the following categories: monogamy, polygamy (polygyny and polyandry including lekking).

Mechanism of Courtship

Courtship fulfils four major functions: 1. Mate finding; 2. Persuation; 3. Synchronization; and 4. Reproductive isolation.

1. Mate finding. For unisexual animals, the locating of a suitable mate is necessary for their survival. Recognition of a receptive partner is the first link in the chain of events leading to fertilization. In higher animals, mate finding is a highly organized process which involves one or more of the senses such as sense of sight, smell, sound, touch and even taste.

For various diurnal and some nocturnal animals *vision* is the primary factor which plays an important role in recognizing a mate. Owls and certain insects have special lenses which enable them to find a mate by straight site. Fire-flies and many of the inhabitants of the deep ocean have light producing organs (bioluminescence) which help in mate finding. Sea cows and whales rely on chemical trails for mate finding and on land scents (pheromones) are very important for bringing the sexes together. In many animals (*e.g.*, frog) sound produced by the males works as auditory cue by which females locate the males.

2. Persuation. In some animals, meeting of male and female leads almost immediately to mating. In many species, normally the male is more ready to mate than others. After recognizing a potential mate, the next barrier for the male is to bring the female into close proximity. Male performs certain behaviour patterns which stimulate the female until she becomes sexually receptive. The female is generally more valuable than the male because in most cases, she carries the eggs after mating and also have a greater role in protecting the young one (parental care). In addition, the male is usually capable of fertilizing more than one female. Therefore, the female requires some persuation and male plays more active role in courtship. Often such displays involve a variety of activities that stimulate the female in different ways.

In some cases, the female attacks the male and eats him (*e.g.*, spiders). In these circumstances, male courtship behaviour may serve not simply to stimulate the female sexually but also to suppress her non-sexual behaviour. In this way, male establishes his identity as a mate and makes necessary arrangements to remove temporarily the carnivorous instinct of female before he approaches her.

3. Synchronization. The occurrence of the same behaviour in different individuals (*i.e.*, males and females of a species) at the same time is called **synchronization.** Precise synchronization of courtship activities of male and female is especially important in species in which there is **external fertilization.** For successful synchronization both sexes respond to external clues such as day length, lunar cycle, the presence of predator, etc. In marine forms, synchronization is related with the tides and phases of moon.

4. Reproductive isolation. The role of courtship ensures that animals mate only with a member of their own species. It is typically achieved because courtship displays are highly species-specific. In this way, the signals used for attraction, persuation, appeasement and synchronization vary in different species. In addition, the partner that receives such signals is usually responding only to the displays of its own species. For example, in most frogs, males produce calls which are species-specific in terms of their pitch and timing. Female frogs approaches only to the calls of their own species. The antennae of many male moths selectively respond to the odours emitted by females of their own species. As a result, hybridization between two species is extremely rare.

Examples of Courtship and Mating in Animals

1. Mating in Arachnids

(*i*) Mating in spiders. Spiders are predatory animals. They are ready to kill and eat the animals of small size that come within their range. Female spiders are cannibalistic too. So matings in spiders is difficult and quite dangerous for the male who is smaller and weaker. Under these circumstances, male courtship behaviour has dual functions: it stimulates female sexually on one hand and suppresses female's non-sexual behaviour on the other hand.

Some species of spiders employ a mechanical mean of attracting the opposite sex. Male spiders approaches the web of a female sitting at the centre of the web and pluck a thread of the web at the species-specific frequency. The plucking '**serenades**' the female and reduces her natural aggressive manner so enabling the male to approach and mate her. Unfortunately if the male "woos" a female of the wrong species or 'plays the wrong tune' he is attacked and killed.

In some species, male spiders present the female with a **nuptial gift** in the form of an insect wrapped in silk (Fig. 3.17). While the female unwraps the gift and eat the insect, the male is able to mate with her, without being attacked. In spiders, the mating procedure is also quite unique. Male weaves a small pad of silk on which a drop of sperm is deposited and this is sucked up by specially modified pedipalps. In due course of time, this is inserted in the vagina of the female.

(*ii*) Courtship and mating in scorpions. In scorpions also, the male not only stimulates the female sexually but he has to suppress the non-sexual behaviour of the female so that she may not eat him. In ethological terminology, the function of courtship is to provide releaser stimuli which block hunger drive.

In scorpions, courtship takes the form of a dance called '*promenade a deus*' by **Fabre.** On finding a suitable mate, the two scorpions stand face to face with their tails upraised and move about in circles (Fig. 12.1); in some other species, the male rocks. The male then seizes the female with his pedipalps, and together they walk backward and forward (**promenade**). The behaviour may last from 10 min to hours, depending on how long it takes to locate a suitable site for spermatophore deposition (*i.e.*, firm surface). Eventually, the male deposits a spermatophore which he attaches to the ground. A wing-like lever extends from the spermatophore. The male then manoeuvers the female so that her genital area is over the spermatophore. Pressure on the lever of spermatophore releases the sperm mass, which is taken up into the female orifice. Thus, in scorpions, coitus does not occur.

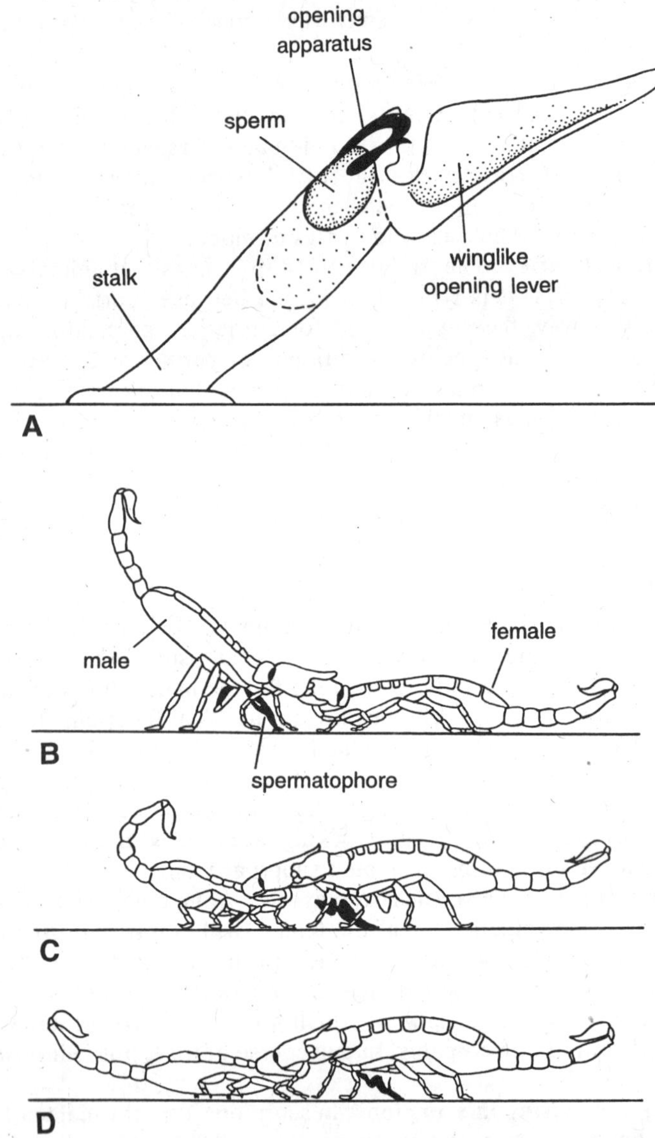

Fig. 12.1. Courtship and mating in scorpion. A — A spermatophore; B–D— Sperm transfer in scorpions. B—While holding the female's pedipalps in his own, the male (on the left) deposits the spermatophore on the ground. C — The female is pulled over the spermatophore. D — The spermatophore is taken up into the female's gonopore.

2. Mating in insects (*Drosophila*). In 1915, **A. H. Sturtevant** described and analysed the courtship of fruit fly, *Drosophila melanogaster.* Mating behaviour of *Drosophila* consists of species-specific fixed action patterns which are accompanied by orientation movements. Such patterns are known as **courtship displays** and involve a number of elements or signals which are performed sequentially. After initiating the courtship displays, the male terminates its actions at any point in the sequential performance of the signals or he may repeat the full pattern numerous times. If the potential mate is non-specific female, a conspecific male or a previously inseminated *conspecific* female then he usually terminates the coital action quickly. If the individual approached, is a conspecific virgin female, he usually persists until either copulation occurs or one or both flies terminate the encounter. Male actions are of two types: those involved in male-to-male encounter and those of male-to-female encounter. In most of the cases the forelegs, wings and mouth parts of the male serve as signalling structures. The female signals are more limited as compared to those of the male and are produced by the wings, legs, genitalia and movements of abdomen. Females perform these signals in response to courting behaviours of male fruit flies and are divided into two types: rejection response and acceptance response. During a orientation movement a male visually lives upon a potential mate. He slightly elevates his body, turns to face the potential mate and approaches it (Box 12.4).

Males make physical contact with female by **tapping** and quickly moves to her rear (abdomen). Copulation occurs only if the female responds by performing acceptance signals. The male mounts and copulates by curling tip of his abdomen under and forward, simultaneously lunging upwards and forward thrusting his head under or between her wings. If she has spread wings, male grasps her body with his fore legs and middle legs and then achieve intromission. The duration of copulation is species-specific. The shortest time recorded is 5 seconds in *Drosophila enigma* and longest in that of *Drosophila ancanthoptera* with 62 minutes.

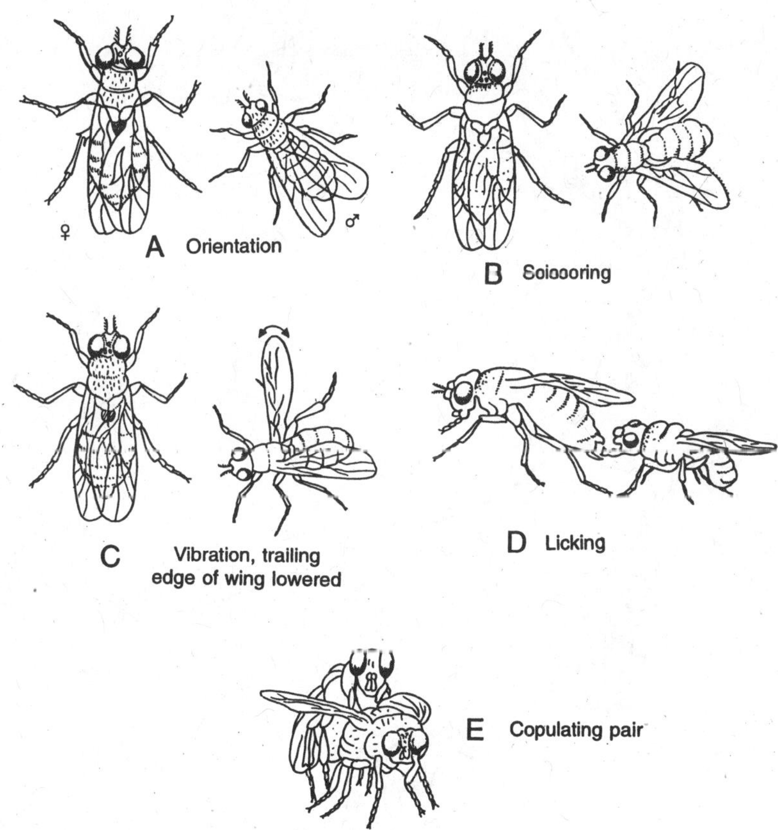

Fig. 12.2. Courtship and mating behaviour of *Drosophila.*

3. Courtship in fishes (three-spined stickleback). Tinbergen described the pattern of breeding behaviour in the male three-spined stickleback (*Gasterosteus aculeatus*) which is a common but fascinating inhabitant of European ditches. The reproductive behaviour is activated by the lengthening days of spring and that time the visual stimulus of a suitable territory may elicit either aggression or nest building. Nest building will start unless another male appears.

The male is stimulated into courtship behaviour by two stimuli: the swollen abdomen of the ripe female and a particular posture when the female faced head-upward at a 45° angle. These stimuli encourage the male to go through a **zig zag swimming pattern** (*i.e.,* he is torn between chasing her and leading her to nest he has built). This, in turn, stimulates the female to swim toward the male. The male then swim toward the nest and female follows. The male shows the nest to the female and she enters. The male trembles (*i.e.,* he pokes the tail of female with his snout and this induces her to lay eggs) and the female lays her eggs. He then promptly chases her off, return to nest to ejaculate his sperms over the eggs, producing fertilization. Then male spends days fanning oxygen-laden water over them until they hatch (Fig. 12.3).

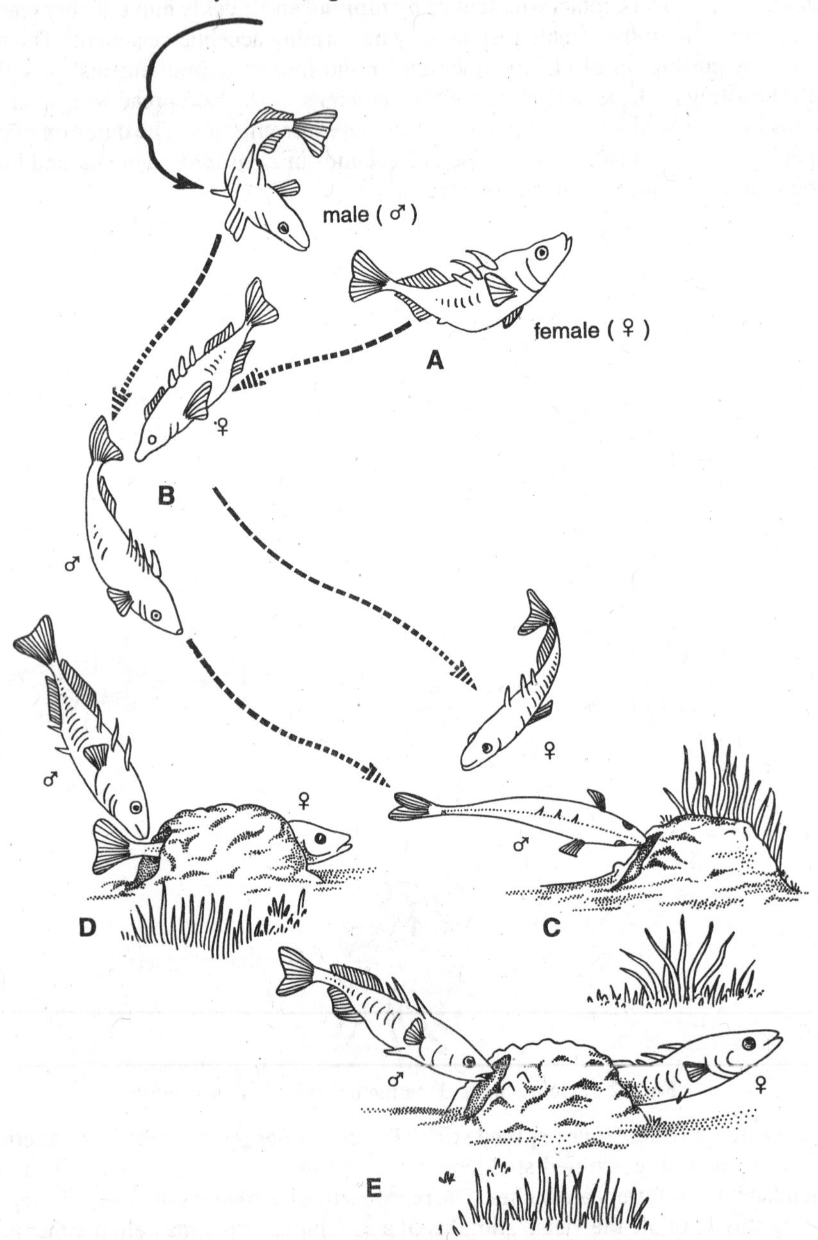

Fig. 12.3. Courtship and mating in three-spined stickleback.

Box 12.4.
Signals/Stimuli in Courtship of *Drosophila*

1. Male courtship elements. Tapping, wing flicking, wing fluttering, wing semaphoring, wing scissoring, wing vibration, leg vibrations and licking.

2. Female rejection signals. Abdomen elevation, abdomen depression, decamping flickering, fluttering, kicking, extrusion.

3. Female acceptance signals. Genitalic spreading, wing spreading and ovipositor extension.

4. During courtship, *Drosophila* males produce wing vibrations which result in the production of **courtship songs** which have been recorded and analysed. These songs include two elements: the **sine song** and the **pulse song.** The sine song of *Drosophila melanogaster* consists of a humming sound that is reminiscent of flight sounds. The pulse song is comprised of repeated sound pulses separated by time or interpulse intervals. Both males and females are capable of receiving auditory courtship signal via their antennae whose **aristae** serves as velocity sensitive receptors.

5. The role of olfactory stimuli in the form of pheromones in mating behaviour of *Drosophila* has also been suggested.

6. All species of *Drosophila* court during the period of daylight and some also court in darkness.

7. Chatterji and **Singh** (1988) have reported that in some species of *Drosophila*, males possess dark black patch on their wings which serve as visual stimulus to the females during courtship.

4. Courtship and mating in amphibians. In salamanders, scent is the main factor in recognition of sex and species. Males and females are of different colours. In the male, the Hedonic glands are located mainly at the base of the tail and on the underside of the head. Skin secretion of females has the odour that enables the male to identify them as to which species and sex they belong. After finding a mate, the male rubs its chin against her head, so that, she may smell his scent. Male captures female with his tail and invites her to dance. The dance may consist of the male transporting her on his back or joining together making a figure of eight around each other. In certain species, the female straddles placing her head on the base of the male's tail and then the two waddle (to walk with short steps, swaying from side to side) together in this position. The male of two-lines salamander (*Eurycea bislineata*) applies a secretion from glands on his head on to the female's skin. He then lacerates her skin with two protruding teeth so that the secretion enters her blood stream (Fig. 12.4).

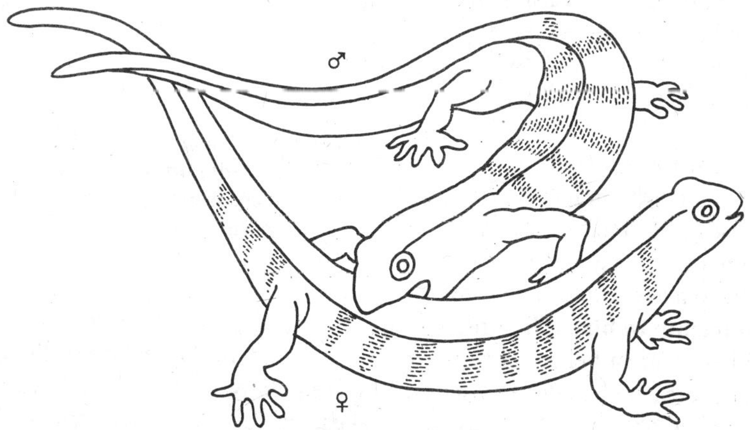

Fig. 12.4. Diagram showing courtship of two-lined salamanders (*Eurycea bislineata*) with the male lacerating the female's back.

In most salamanders, the fertilization is internal. Male salamander deposits a small jelly covered package of spermatozoa, the spermatophores, either on land or in water according to the habitat of species. Then female comes and picks up one of the spermatophores with the lips of her cloaca. It is placed inside her body in a special receptacle known as **spermatheca** where the sperms remain to fertilize the eggs.

The male **newt,** *Triturus,* develops a large crest having black and red spots and an orange belly in the breeding season. He performs a dance to attract the female. If she follows and touches his tail with her snout, he deposits spermatophores which are picked up by her cloaca. Just after, before his arrival to the surface of water for breathing, male again renews his dance and deposits two or three more spermatophores. She picks up these spermatophores once again. The female thus tests the vigour of the male. According to **Halliday,** *the chance of fertilizing the female by the male depends on how long he can stay submerged and the number of spermatophores he can produce.* Sperms may be stored for many months in a spermatheca. Fertilization takes place in the oviduct and then eggs are laid separately wrapped in a leaf.

Male **bullfrogs** produce advertisement calls by their vocal cords and amplify them by inflatable sacs present in their mouth because, probably louder and longer calls are more attractive to females and more inhibitory to other males. Females are attracted by such calls and clasped by males in a state termed **amplexus.** Ova and sperms are released simultaneously and external fertilization takes place in water.

4. Courtship in reptiles. Many reptiles exhibit well marked courtship behaviour during the breeding season. The males of certain **lizards** fight and display ritually either to terrify each other (Fig. 12.5)or to evoke a suitable response from the female. The male American chameleon, *Anolis carolinesis* displays a rhythmical up and down bobbing of his body to show his **dewlap,** which is a bright-red spot of skin beneath the skin. As the female approaches the male, the latter twists his tail around the female and inserts his hemipenis.

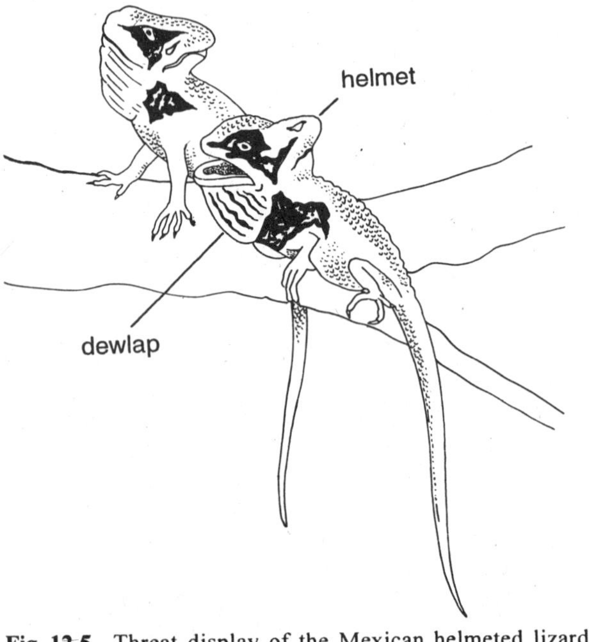

Fig. 12.5. Threat display of the Mexican helmeted lizard (*Corythophanes hernandezii*). A threatner's apparent size is increased by extending the dewlap and opening the mouth.

In case of the **spotted turtle,** when a male approaches a female, the usual response of the latter is to move away but often she looks coyly over her shoulders to make sure that he is following her.

5. Courtship in birds. In birds, phenomena of display and courtship attain a complexity incompatible to human beings. A bird's song is often an initial feature of the ritual which first attracts and excites the female. The male bird is frequently aggressive in defense of his territory. Consequently, courtship in females often takes the form of reactions which tend to avoid provoking of the attack. For example, female kestrel (a European falcon, *Falco tinnuculus*) show infantile behaviour reminding the appeasement employed by the young. Courtship ceremony begins by offering food to each other which is a behaviour pattern carried over from infancy.

In **avocetes** (Long-legged shore birds of Genus *Recurvirostra*) having webbed feet and slender up-curved bill, male and female both preen their feathers in a hasty fashion during courtship (Fig. 12.6). After preening when the female adopts a characteristic flattened posture indicating her readyness for the mating only, then the male mounts and copulates.

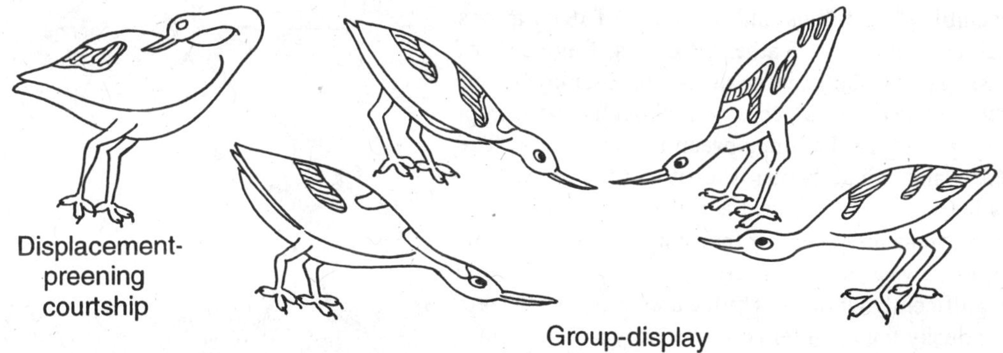

Fig. 12.6. Displacement-preening courtship and group display in avocete (shore birds).

The precoital displays of **herring gull,** *Harus argentatus* is also quite interesting. Both male and female bob their heads upward uttering a soft melodius call with each bob (Fig. 12.7). After a series of such mutual head tossing the male takes the initiative in copulation and suddenly mounts and mates.

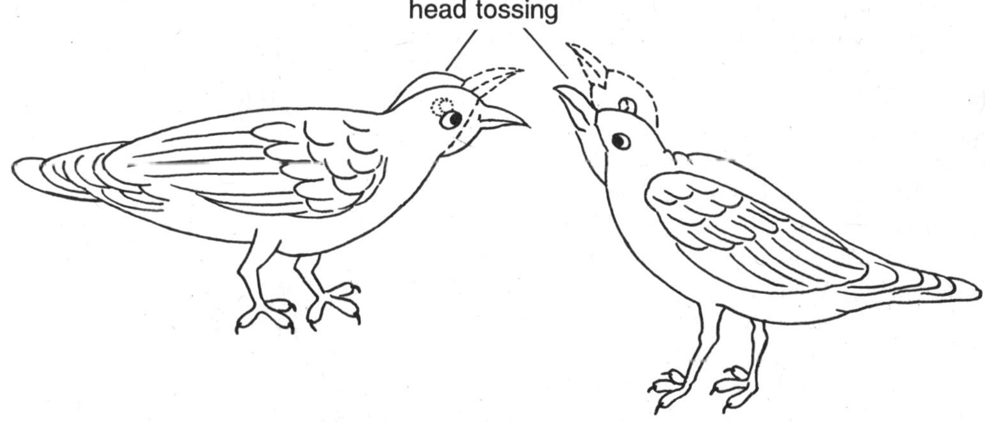

Fig. 12.7. Head-bobbing precoital display of herring gulls.

Pair-bonding and display of ritualised-courtship dances in the Great-crested grebe (*Podiceps cristatus*) are unique in animal kingdom. It was well studied by **Julian Huxley** (1914). The courtship ceremony includes a series of behaviours such as head shaking ceremony; dive and eat display, mutual greeting and eat displays; and penguin dance. In penguin-dance both birds dive and reappear with bunches of weeds (*i.e.,* nest material) in their bills. They swim towards one another, and then spring upright and move together shaking their heads from side to side, with crest and neck ruff raised. The nest material is held firmly in the bill. **Huxley** has coined the term **ritualization** for precoital displays of great-crested grebe.

The brilliantly coloured **bower birds** of Australia, New Guinea and neighbouring islands build display grounds and decorate them with various objects such as flowers, stones, shells, etc. The satin bower bird, *Ptylonoihychus violaceus* of East Australia constructs a **bower** (*i.e.,* rustic cottage) with two parallel walls of arched twigs. On one side of bower is a **display ground** in which is found an assemblage of things and varieties of decorations such as blue parrot feathers, flowers, fragments of glass, papers, shells, wasp nests, etc. Display begins with sunrise and may occur throughout the day except during feeding, bathing, preening or calling. The male stands squarely on his territory (bower) making a whirring noise, arches his tail in fan-like manner and stiffens his wings, at the same time keeping his neck low and erect. His plumage glistens magnificently while eyes bulge and become rose red. The dusky female utters convulsively a few guttural sounds. Occasionally, she arranges the disordered twigs of the bower. In the mean time the male bounds wildy about, producing an extraordinary medley of mechanical sounds, but most males are rejected by visiting females. After a female bower bird is copulated, she leaves to lay eggs and incubate them until they hatch, feed the hatchlings and shepherd the fledglings around the neighborhood. Her partner stays at the bower, working slavishly on its improvement and decoration. A few males with especially well decorated bower will copulate with more than one female (the record is 33).

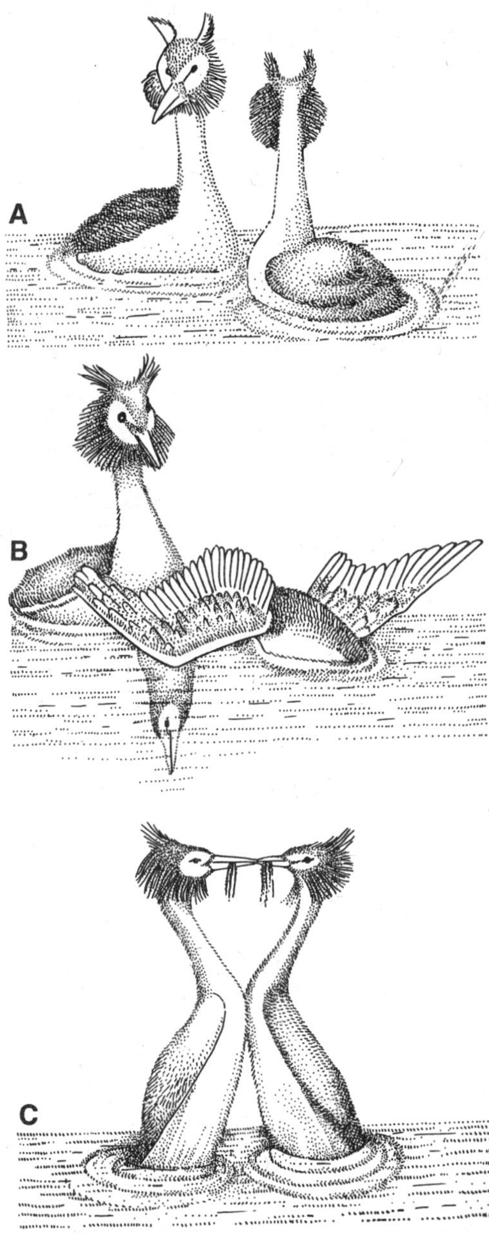

Fig. 12.8. Penguin dance of great crested grebe.

Indian peacock is a polygynous bird. During breeding seasons, cock forms a drove or harem of 4 to 5 peahens. The courtship behaviour of peacock is very fascinating. While courting, the peacock spread its beautiful tail whenever a peahen approaches but as she comes near him he takes an about turn showing her his rear portion. If peahen is ready for mating, she would run swiftly around the tail to be able to see him from front again. The peacock responds by rustling his tail feather. Then he will turn around again and this courtship game will be repeated several times. At last thc peahen will lie down infront of him giving signals for mating.

The male **Baya weaver bird** (*Ploceus philippinus*) builds several nests one after another, each of which is a swinging retort-shaped structure with a long vertical entrance tube, compactly woven out of strips of paddy leaf and rough-edged grasses, suspended in clusters from twigs of a babul or a palm tree usually over a stream or tank. Blobs of mud, collected when wet, are stuck inside the dome of nest near the egg-chambeı. In a breeding season, to attract females for mating purpose, male Baya birds (up to 10 to 50) make whistle-like mating calls in chorous, accompained by flapping of wings in unision while weaving their nests in a colony. In this case, incubation and all feeding of the chicks are done by the female alone.

Fig. 12.9. Various courtship displays of the green heron.

The **green heron** has an elaborate courtship behaviour that combines songs with stereotyped movements. This North American bird nests in tall trees along streams or marshes. Courtship begins when the male advertises himself by posturing on a treetop with up-pointed bill and giving a call that sounds like *skow* (Fig. 12.9). Females are attracted to the male's call and one female soon perches nearby and answers with a slightly different call, *skeow.* The duet continues for some time, until the male allows the female to approach. Both birds then perform an elaborate aerial dance above the nest, displaying the unique colours of the plumage of their species to one another. The pair then returns to the treetop, where the male enters the nest, displays his brightly coloured feathers, and sways from side to side, pointing his bill up and away from the female and uttering a soft *aaroo-aaroo.* The male then points his bill down, snaps it sharply, and bobs and bows. The female enters the nest, and the birds stroke their bills together and preen one another's neck feathers. These behaviours signal a readiness to accept physical contact and copulation takes place.

Mating in mammals. In mammals, olfaction plays a major role in the regulation of courtship behaviour. After smelling a female, the males of many species display a response in which neck is extended and the upper lip is curled (*e.g.*, males of cow, buffalo, ass, etc.). Females often ask for mountings. Sometime by approaching a male she muzzles or licks him. Often she runs away from the male when approached but appears again a solicitation behaviour starts and then she escapes.

In case of **red deer,** the female deer (hind) frequently runs away when approached by the male deer (stag) but she soon stops and wait for him. She licks him and then runs again only to wait once more for his approach. In **bottle-nosed dolphins,** vocalization, nuzzling of partner's genitalia, rubbing of bodies, stroking with flippers, displaying of the underside, chasing, and head butting occur, during courtship.

Courting in Elephants

The elephant's life span is comparable to our own and the key element of elephant society is the **matriarchal group** led by a mature female with her daughters and their offspring of all ages. A baby is born into a group where every individual knows all the others from long experience in their company. It is nourished and closely protected for several years by its mother and other relatives. Slowly it acquires the adult repertoire of feeding behaviour, learning how to select food and how the group migrate around its home range to match the seasonal changes of vegetation and water supply. Females do become sexually mature until about 20 and pubety is even later for males.

Males leave the family as they become sexually mature around 10–15 years of age, but they are unlikely to be able to mate with females until much later. They move in small bachelor groups, continuing to grow, and are unlikely to breed until they are 40 to 50 years old (see **Manning** and **Dawkings,** 1998).

The elephants use tactile signals during courtship. The trunk plays an important role in courtship behaviour. After chasing the female the male caresses her head, trunk and tusks, then strokes her back gently until she indicates her willingness to mate (Fig. 12.10).

12.3. PARENTAL CARE

Looking after the eggs or the young until they are independent to defend themselves from predators is known as **parental care.** In other words, parental care is any action or behaviour performed by the parents towards its offsprings that increases its chances of survival (**Trivers,** 1972). Parental care behaviour is any behaviour performed after breeding, by one or both parents, that contributes to the survival of their offspring (**Keenleyside** and **Reese,** 1978). Animals exhibit a great diversity in care of their eggs and youngs during development.

Fig. 12.10. Courtship in African elephant. A–The male chases the female in heat. He raises his trunk to receive her pheromone. B–Male catches female's trunk by his own. C–Lastly he comes back and gently strokes her back if she allows mounting.

1. Parental Care in Insects

Exclusive male parental care is very rare among animals but common in fishes and certain insects such as water bugs in which occur **back brooding.** In *Belostoma flumineum*, males take care of eggs that their mates (females) lay on their backs. Such males spend much of their time keeping the eggs near the water surface, pumping their bodies to keep relatively aerated water moving over the eggs, which they also stroke at intervals with their hind legs. Eggs that are separated from their brooding attendant fail to develop and die (**Smith,** 1997).

Box 12.5.
Parental care in Invertebrates

Except for the social insects, parental care is not well developed among invertebrates. Some retain eggs within the body until they hatch; others such as crayfish, carry eggs externally. Invertebrate parental care is most highly developed in social ants, bees and wasps. Social insects provide all **five** functions of parental care— **food, defense, heat, sanitation** and **guidance.**

2. Parental Care in Fishes

Parental care behaviour is universal among fishes. Of some 250 families described in **Breder** and **Rosen's** (1966) encyclopaedic treatise on fish reproduction, about 77 per cent fish show no parental care, another 17 per cent include fish species that care for the egg only and less than 6 per cent contain species that are known to care for eggs and newly hatched young.

Fish show all grades of parental care behaviour from random spawning and from deposition of large number of uncared eggs to the protection of young. The lack of parental care behaviour is correlated with the production of great number of eggs and sperms.

Two general types of variation in parental care behaviour exist among fishes. *First* either both parents, or one alone cares for the offspring. Thus, there are **paternal, maternal** and **biparental** species. *Second,* the eggs and newly hatched young are either maintained on the substrate — that is, on plants, under stones, in excavated pits and so on (these are called **substrate-brooders** or **guarders**) or carried about in the parent's mouth (these are called **mouth-brooders** or **incubators**). Fishes have evolved many means of affording care to fertilized eggs and young ones by one or both sexes.

1. Scattering eggs over aquatic plants. In some fishes such as pikes, *Esox lucius*; carps, *Cyprinus carpio*, *Carrassius auratus,* etc., eggs are scattered usually over aquatic plants to which they are attached.

2. Depositing eggs in sticky covering. In many carps, eggs are usually laid with some special sticky covering by means of which they are attached to each other and to the stones, weeds, etc. In yellow perch, *Perca flavescens* eggs are deposited in a rope of single mass.

3. Laying of eggs at suitable places. Suitable spawning grounds are selected by anadromous fishes such as *Salmo solar, Acipenser, Oncorhyncus,* etc. They dig excavation in gravel substrate, lay their eggs in the pits, cover them with gravel and desert them. The sand gobi *Pomatoschistos minutus* lays its eggs in some protected spot where they are guarded by the male who aerates them by his movements.

Fig. 12.11. Shallow basin-like nest of sunfish.

4. Nest building. Only a few fishes construct nest which may be complex like that of birds or simple. Simple nests of fishes are merely hollowed out depressions in the bottom as in the lung fishes. Males of many species such as darters (*Etheostoma*), sunfishes (Fig. 12.11) and cichlids prepare a shallow basin-like nest. All stones and rock crystals are carefully removed from the bottom. The eggs are laid in the nest and the male remains on guard till the young ones are hatched.

The male African lung-fish, *Protopterus,* prepare a simple nest in the form of deep hole in swampy places along the river banks. After spawing he guards the nest. The South African lung fish *Lepidosiren* also prepares a nest in the form of a burrow and the male developes highly vascularized filaments on its pelvic fins for aeration. The male bowfin, *Amia cal*va, of the great lakes of North America builds a crude circular nest among aquatic vegetation (Fig. 12.12). The male stands on guard till the young ones are hatched. The young ones leave the nest only under the protection of the father.

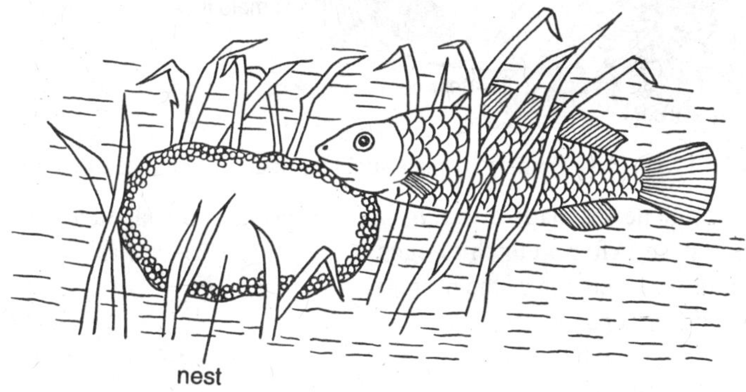

Fig. 12.12. Male *Amia calva* guarding its circular nest.

Before the onset of its courtship, the male stickleback (*e.g.,* Three-spined stickleback, *Gasterosteus aculeatus* and ten-spined stickleback, *Pygosteus pungitius*) builds a quite elaborate spherical or elongate nest. The nest is built by collecting plant fragments, rootlets and the like and then binding them together with adhesive kidney secretions. The various activities of male such as probing, boring, sucking and glueing, result in the formation of a compact nest with an internal chamber (tunnel) to receive the eggs. Male drives and induces the female into the nest for laying eggs, then chases her away, enters the nest, fertilizes the eggs and guard them from intruders (Fig. 12.13).

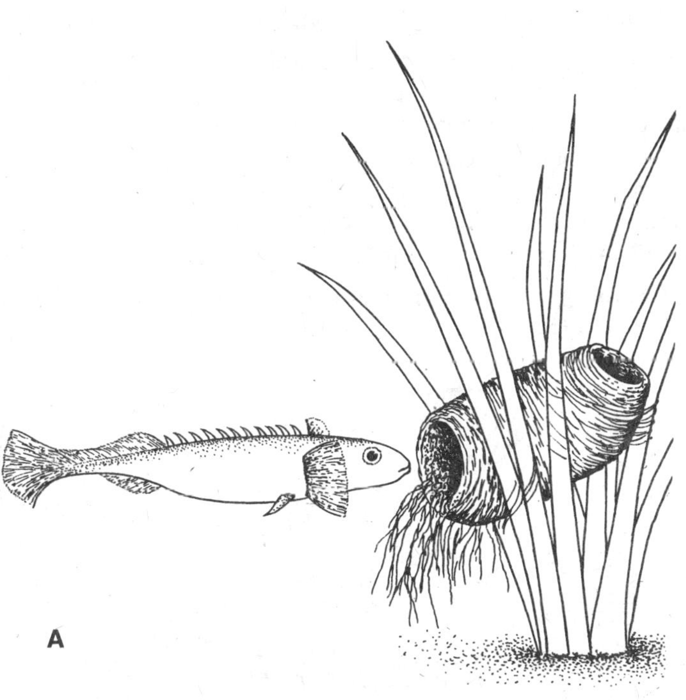

Fig. 12.13. A— Nest of ten spined stickleback;

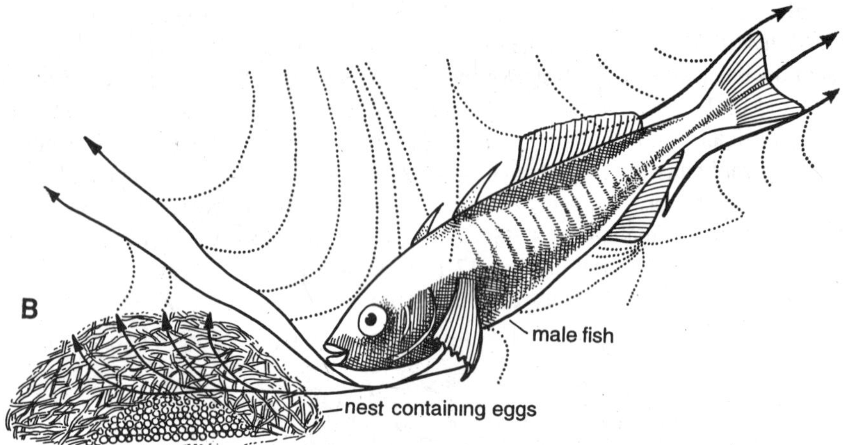

Fig. 12.13. B — Fanning by male stickleback at the entrance of its nest.

The most elaborate nest is made by *Apelts quadracus*; its cup-shaped nest is attached to rooted plants close to the bottom. After a clutch of eggs is laid, the male builds an extension of the nest up and over the eggs, with a concave upper surface to the extension. A second clutch of eggs is laid on the new nest floor and this procedure may be repeated several times, until the male has several clutches of eggs stacked vertically within a single multitiered nest (**Rowland,** 1974).

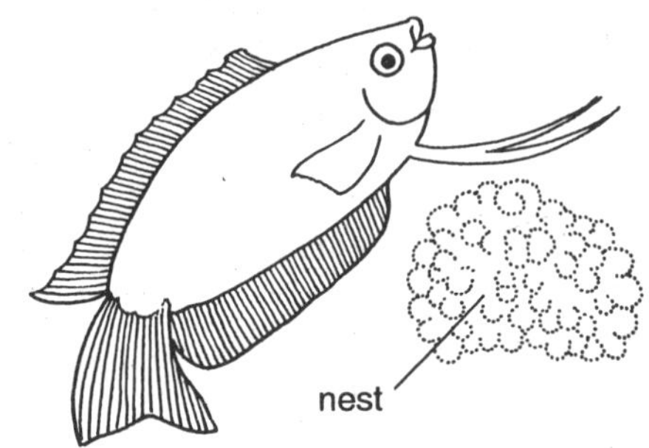

Fig. 12.14. Male Siamese fighting fish defending his floating nest.

Floating nests are made by American catfishes, in which the eggs are suspended in a mass of bubbles and mucus produced by the fish. The male Siamese fighting fish (*Betta splendens*) too builds a floating nest and sticks the fertilized eggs to the lower surface of foamy nest. He stays on guard on this nest and fights till death to defend it (Fig. 12.14). The male paradise fish, *Macropodus* also prepares a similar foamy nest.

5. Coiling round the eggs. The butter fish (*Pholis gunnellus*) rolls all eggs into a ball and curls around it (Fig. 12.15).

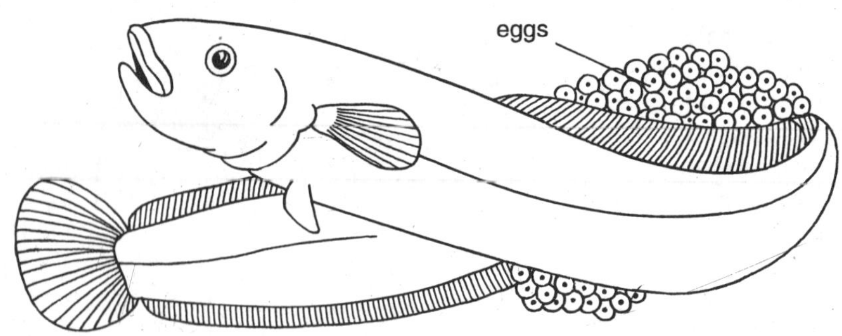

Fig. 12.15. A butter fish coiling around the egg.

6. Deposition of eggs by ovipositor in mussels. The female *Rhodeus amarus* (European bitterling) deposits eggs in the siphon of a fresh-water mussel (Swan mussel; Fig. 12.16) by means of very long urogenital papilla (ovipositor). Male immediately sheds the sperms on the opening (of mussel) over the eggs.

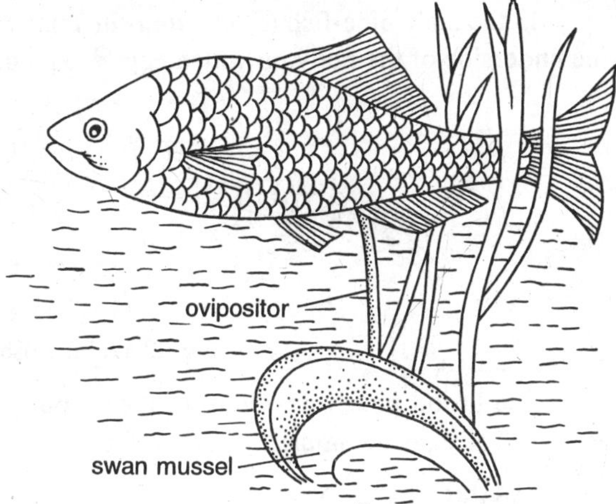

Fig. 12.16. Oviposition by European bitterling in swan mussel.

7. Egg brooding in mouth and intestine. The female *Tilapia mossambica* broods the fertilized eggs in her mouth (Fig. 12.17). She allows the young to take refuge in her buccal cavity in times of danger for some days after hatching. In the North American sea catfish (*Galeichthys felis*) the male carries eggs in the mouth for a period of nearly six weeks. The eggs of this oral incubating fish are large and relatively few in number. During this period the brooder fish do not take any food, thus, exhibiting great degree of self-sacrifice. *Tachysurus* keeps the fertilized eggs in its intestine till hatching occurs.

Fig. 12.17. Mouth brooding (oral incubation) in *Tilapia.*

8. Brood pouches. The male sea horse and pipe fish carry eggs in a brood pouch on the abdomen. In sea horse (*Hippocampus*) fertilized eggs are transferred by the female into the brood pouch on the belly of the male. These eggs are carried by males until their hatching. Eggs become embedded in the folds of the brood pouch and for the exchange of respiratory gases a sort of placenta is formed.

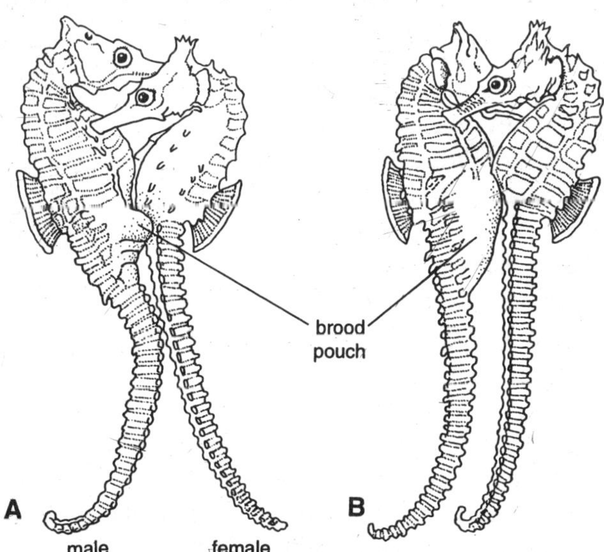

Fig. 12.18. A male sea horse (left) receives eggs from his mate, which fills his brood pouch and swell his abdomen substantially.

In the male pipe-fish (*Syngnathus acus*), a brood pouch is formed by two flaps of skin on the underside of the body on which eggs are placed by the female (Fig. 12.19).

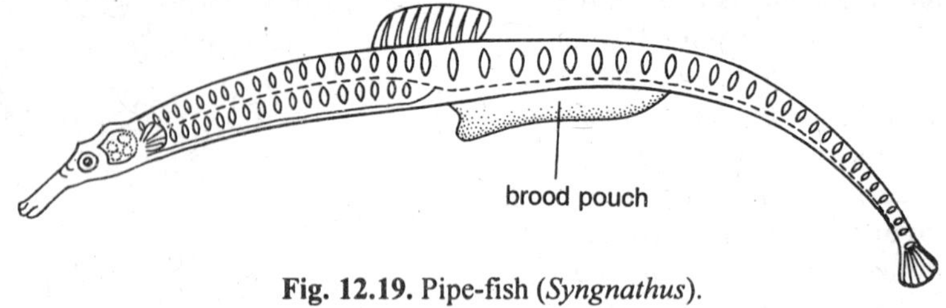

Fig. 12.19. Pipe-fish (*Syngnathus*).

The brood pouch develops an inner spongy lining which is ricnly supplied with blood vessels. The eggs get nourishment until hatching. Fry may return to the pouch when in danger.

9. Attachment of egg to body. The male nursery fish (*Kurtus*) of New Guinea, carries eggs held in a cephalic hook (Fig. 12.20).

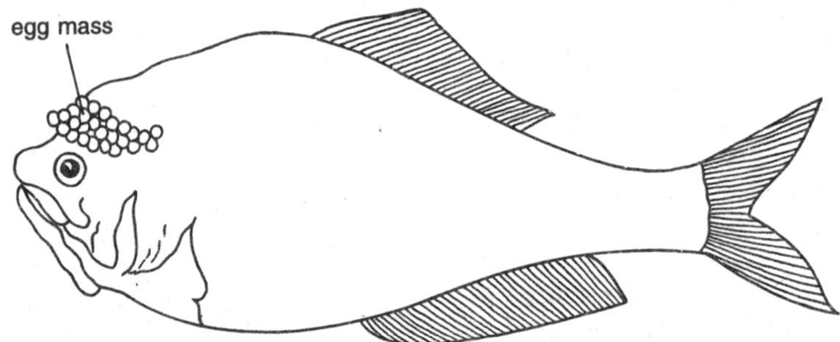

Fig. 12.20. Nursery fish (*Kurtus*) carrying egg mass entangled on a hook-like process.

10. By putting eggs in integumentary cups. In siluroids (*Aspredo* and *Platystacus*) the eggs are pressed into soft and spongy skin of female and carried about.

11. Egg capsules. In oviparous Elasmobranches such as rays and cat sharks (*Scyllium* and *Raja*) fertilized eggs are laid inside protective horny egg capsules, called **Mermaids purse.** This capsule remains attached to the aquatic weeds by their tendrils. The development proceeds inside the capsule until the yolk has been used up. The youngs hatch out after rupturing off egg case.

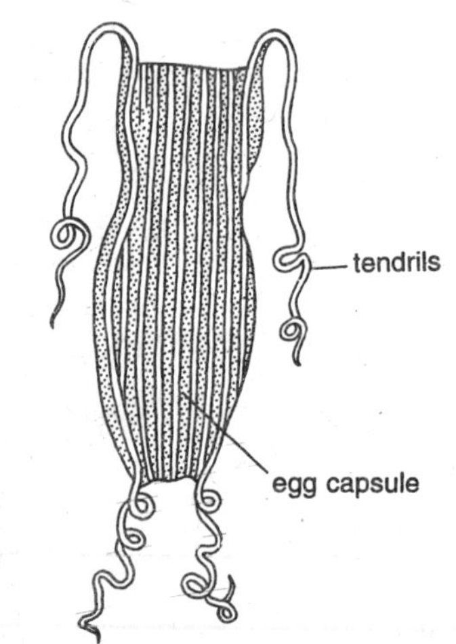

Fig. 12.21. A horny egg capsule of cat shark.

Feeding the young. Like birds, some fishes help young in feeding. During its **plunge-feeding,** *Geophagus* spp., thrusts its snout vigorously into the substrate, withdraws it and expels the mouth contents. Loose particles may then be ingested. If an adult *Geophagus* is foraging in this manner with a school of its progeny, the young fish also gather and feed on the expelled material (**Loiselle** 1978).

Another form of "feeding the young" also occurs among cichlids (*Symphysodon discus, Etroplus maculatus, Chichlasoma citrinellum*). As the youngs reach free-swimming fry stage of development, they begin to graze on the mucus on their parents' sides. Mucus-producing cells in the adult's epidermis are most numerous at this stage of the reproductive cycle.

Viviparity (True internal incubation). Viviparity describes the highest degree of parental care and provides maximum protection to the young ones. In viviparous elasmobranches (*e.g., Scoliodon, Mustelus*) eggs develop in the uterus. The mucous lining of the uterus forms fluid-filled protective compartments, one for each embryo. Each embryo receives nourishment from the uterine tissues through the yolk-sac placenta (Fig.12.22).

In two orders of Teleosts, Cyprionodonts and Perciformes some species (*e.g., Zoarces, Gambusia* and *Poicilia*) show **internal fertilization.** In them youngs develop within ovary but they are not attached to the wall of ovary. These embryos develop freely in the sac feeding upon an "embryotrophic" material, apparently produced by the discharged ovarian follicles, which become highly vascular and remain as such throughout the several months of pregnancy.

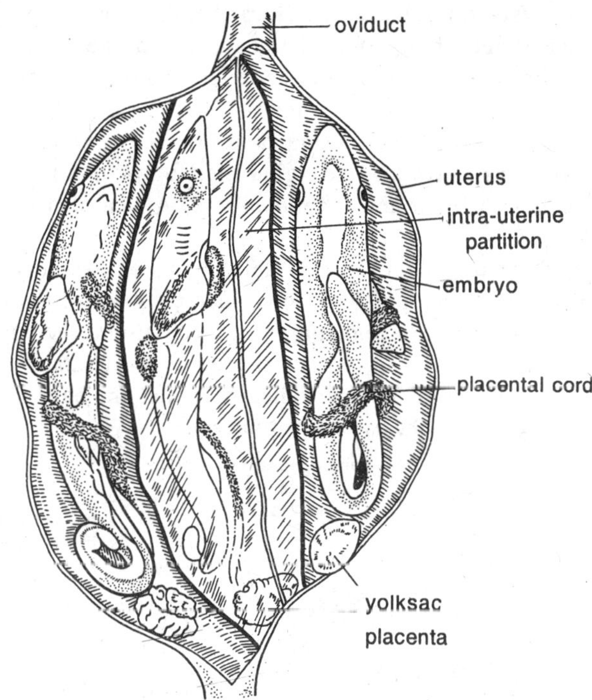

Fig. 12.22. Uterus of *Scoliodon* cut open to show the embryo with yolk-sac placentae.

In shiner surf-perch (*Cymatogaster aggregata*) also the eggs are fertilized in ovarian follicles but are soon released into the cavity and are nourished by a secretion from the ovary (Fig. 12.23). The males are retained in the ovary until sexually mature.

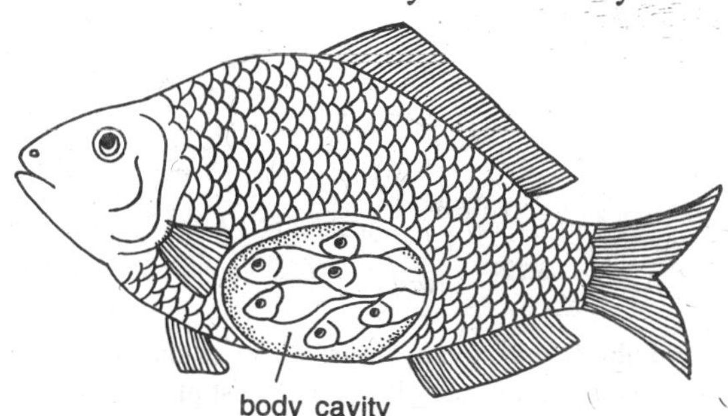

Fig. 12.23. Body cavity of *Cymatogaster aggregatus* cut open to show fully formed youngs.

In the nurse shark, *Gingly mostoma,* there occurs an intermediate condition between the oviparous and viviparous. It is **ovo-viviparous.** Its eggs are covered by a horny case and the development of the embryo occurs in the uterus. Fully developed youngs are hatched out by breaking the shell inside the uterus.

3. Parental Care in Amphibians

Perhaps no group shows more diversity in parental care than do the amphibians. Anurans show much greater diversity than urodeles and apodans. Parental care is associated only with those species that place their eggs in single clusters. The methods of parental care by amphibians generally fall under two broad categories: 1. protection by nests, nurseries or shelters; and 2. direct caring by parents. In apodans and salamanders, parental care consists only of attendance of the eggs, mostly by the female. In anurans the eggs are usually guarded by the male and the parental care involves transportation of eggs and larvae also.

A. Protection by nests, nurseries or shelters. Amphibians have evolved various curious methods of giving protection to their defenceless eggs and larvae from the predators:

1. Selection of site. Many amphibians lay eggs in protected sites most of which are (microhabitats) on land. Many tropical frogs and toads lay eggs on land near water. Many tree frogs lay their eggs not on land, but on leaves and branches overhanging water. Species of *Phyllomedusa, Rhacophorus, Hylodes,* etc., glue their eggs to foliage hanging over water. *Rhacophorus malabaricus* in India and *Chiromantis* of Africa also deposits their spawn on trees.

2. Defending eggs or territories. In Urodela, males of *Cryptobranchus* and *Hynobius* drive away all enemies except ripe females of own species. *Andrias japonicus* shakes the eggs for proper aeration. Either sex of *Proteus anguineus* may attend the eggs. It waves its tail to direct water currents over the eggs. In terrestrial plethodontid salamanders egg clutches are usually attended by females. The female periodically rotates the eggs in the clutch in order to increase aeration and to prevent adhesive malformation.

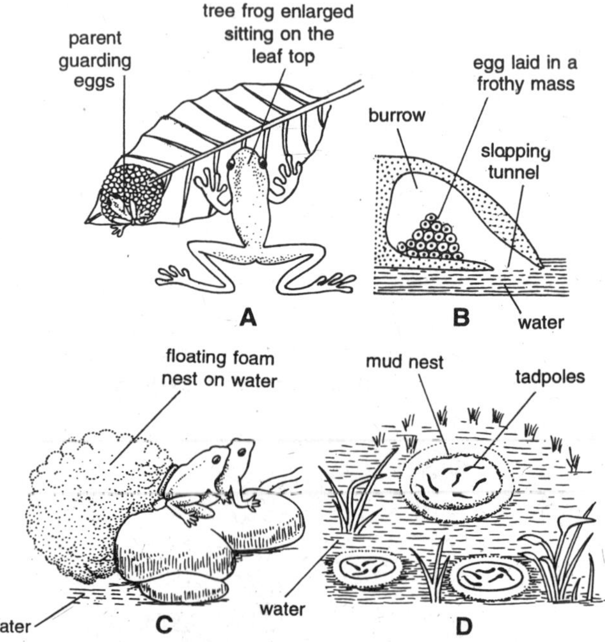

Fig. 12.24. Parental care in Amphibia. Protection by nests, nurseries and shelters. A — A tree frog guarding eggs glued to a leaf overhanging water. B —Foam nest of *Rhacophorus schlegeli* in a slopping burrow near water. C—Foam nest floating on water. D—Mud nest of *Hyla faber.*

In Anura, males of green frog *Rana clamitans* and other species maintain territories and attack small intruders to defend eggs. In *Manlophryne robusta,* the male actually sits over and holds with hands the elastic gelatinous envelope containing eggs numbering 17. Some tree frogs laying eggs above water may sit besides the eggs or rest on top of them. Males of *nest building* gladiator frog, *Hyla rosenbergi* are highly **territorial** and *aggressive*. They dig a shallow pit (a nest) near the water. Eggs are laid as a surface film on water in the nest. If the surface tension is broken, the eggs sink to the bottom of the nest and die. Male moves around the nest area and attack intruders whose entry into the nest may disrupt the surface tension. The terrestrial eggs of *Dendrobates auratus* are periodically moistened by males by periodically urinating on the egg clutch. Female of *Hemisus marmoratum* provides moisture to developing eggs. She sits on the clutch in an underground chamber and digs a tunnel leading to the water. The tadpoles escape to water from the nest chamber.

3. Direct development. In some terrestrial or tree frogs, such as *Eleutherodactylus, Arthroleptis, Hylodes* and *Hyla nebulosa*, the eggs hatch directly into little frogs thus avoiding larval mortality. In red backed salamander *Plethodon cinereus*, the hatchlings are miniatures of the adults.

4. Defending the larvae. Female *Leptodactylus ocellatus* provides protection to the tadpoles. The tadpoles move around in the dense masses. The female remains with the tadpoles and attacks birds attempting to feed on tadpoles. The tadpoles of *Dendrobates* develop in water-filled leaf axils of various kinds of plants or tree holes, etc. Females of *D.pumilio, D.histrionicus* and *D.lehmanni* carry individual tadpoles to separate water-filled leaf axils and protect them. Females of *Hyperolius obstetricans* remain with their eggs on leaves overhanging the streams. Upon hatching the female kicks tadpoles out of leaves into the water.

5. Foam nests. Many amphibians convert copious mucous secretions into nests for their young. In Japanese tree frog *Rhacophorus schlegeli,* the mating couples dig a hole or tunnel into which eggs are left in a frothy mass to avoid desiccation. During rains, hatching tadpoles are washed down the slopping tunnel into pond or river water for further development.

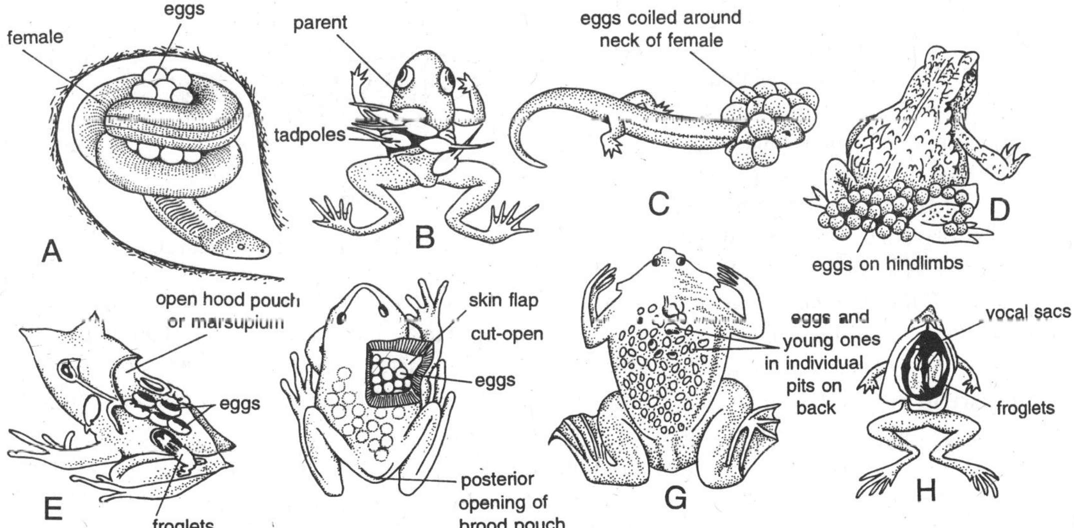

Fig. 12.25. Direct parental care in Amphibia. A— Female *Ichthyophis* coiling around eggs. B—Transportation of tadpoles attached to back of a parent. C — *Desmognathus fuscus* with eggs. D—Male *Alytes obstetricans* carrying eggs around his thighs. E—A marsupial frog with eggs exposed in open brood pouch on back. F—*Nototrema* (*Gastrotheca*) with flap of dorsal brood sac cut open to show eggs. G—In *Pipa,* eggs develop completely into individual capsule on back of female; H—Froglets inside vocal sacs cut open of *Rhinoderma darwinii.*

The female of South American tree frog, *Leptodactylus mystacinus*, stirs up a frothy mass of mucus, fills it in holes near water and lays eggs in them. The tadpoles developing in these foam nests can readily enter water. Some anurans lay eggs in nests of foam floatlng on water. Male frog *Adelotus brevis* actively defends the eggs laid in **foam nest.** The males contain long tusks on the lower jaw which are used in fighting.

6. Mud nests (Nurseries). In the Brazilian tree frog, *Hyla fabre,* the male digs a little crater-like hole or **nursery** in mud in shallow water, in which the female lays her eggs. Tadpoles hatch within these mud nests and develop upto a stage when they are large enough to fend themselves.

7. Tree nests. The South American tree frog, *Phyllomedusa hypochondrales,* lays eggs in a folded leaf nest with margins glued together by a cloacal secretion. The tadpoles when formed fall straight into water below. Another tree frog, *Hyla resinfictrix,* lines a shallow tree cavity with bees wax obtained from the hives of certain stingless bees. Females lay eggs when this cavity is filled with rain water. Here, the young develop relatively free from predators.

8. Gelatinous bags. In *Phrynixalus biroi* large eggs are enclosed in a sausage-shaped transparent gelatinous membranous bag, secreted by female and left in mountain stream. *Salamandrella keyserlingi* also deposits 50 to 60 small eggs in a gelatinous bag which is fastened to aquatic plants.

9. Communal nests. The toad *Nectophyrynoides malcolmi* prepare **communal nests** in which eggs are deposited by several females. This nest is guarded by a single male.

B. Direct Caring by Parents

1. Coiling around eggs. The apodans such as *Ichthyophis glutinosa* lays eggs in a shallow hole near the water and coils herself around the gelatinous egg mass (Fig. 12.25 A). The attendant female eats the infected eggs and thus saves other eggs from infection. Another apodan, *Idiocranium russeli* coils around the eggs in a dense mat of grasses.

In the salamander congo eel *Amphiuma,* the female lays large eggs in burrows in damp soil and carefully guard them by coiling her body round them until they hatch (Fig. 12.26). The female salamander *Plethodon* also coils round the eggs which are laid in small packages in the hollow of a rotten log or beneath a rock. In *Megalobatrachus maximus* (Japanese giant salamander), it is the male who coils round the eggs.

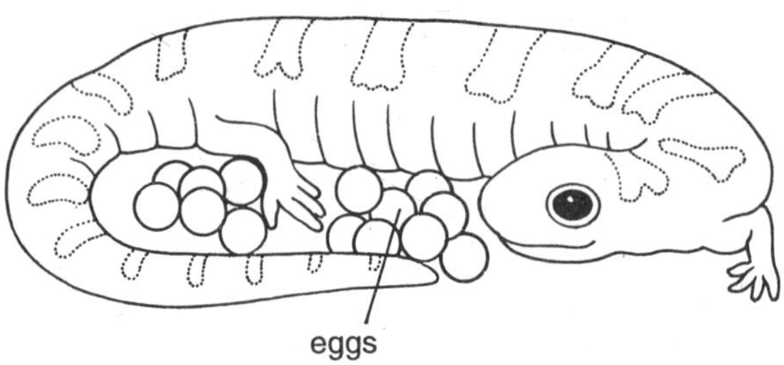

Fig. 12.26. Parental care in salamander.

2. Transferring tadpoles to water. Some species of small frogs such as *Phyllobates, Arthroleptis, Pelobates, Dendrobates, Sooglossus,* etc.) in both tropical Africa and South America deposit their eggs on ground. The tadpole hatching out fasten themselves to the back of one of the parents (*e.g.,* female in *Sooglossus seychellensis*) with their sucker-like mouth on their sticky ventral side and are transported to water.

3. Eggs glued to body. Many amphibians carry the eggs glued to their body. In the dusky salamander, *Desmognathus fuscus,* female carries the string of eggs coiled around her neck, until they are hatched. The female places her chin on the egg clutch thereby subjecting the egg to slight vibrations caused by pulsations of the throat. *Desmognathus* aggressively defends her egg. In Sri Lanka tree frog, *Rhacophorus reticulatus,* the eggs are glued to the belly of female. When female of European midwife toad, *Alytes obstetricans,* lays eggs, the male entangles them around his hindlegs. He carries them with him until they are ready to hatch. At that time he releases the tadpoles into nearest water.

4. Eggs in back pouches. In one group of tree frogs, called **marsupial frogs** or **toads,** the female carries the eggs on her back, either in an open oval depression, a closed pouch or in individual pockets. The eggs develop into miniature frogs before they leave their mother's back. In the Brazilian tree frog *Hyla goeldii* (=*Cryptobatrachus evansi*), the posterior part of the back of female forms a sort incipient brood pouch in which eggs remain exposed. In *Nototrema* (= *Gastrotheca*) *marsupium* the eggs are covered by skin forming a single large brood pouch on the back which opens posteriorly in front of cloacal aperture. The Surinam toad, *Pipa pipa* (which is completely aquatic) during breeding season, skin of female's back becomes thick, vascular, soft and gelatinous. The male presses fertilized eggs against female's back, where they sink into individual pits. A hinged cover forms over each egg enclosing it in a small capsule. Complete metamorphosis occurs within capsule and tiny tailless toads leave mother but do not enter water.

5. Organs of brooding pouches. Males of terrestrial South American Darwin's frog, *Rhinoderma darwinii,* pushes at least two fertilized eggs into his relatively large **vocal sacs.** Tadpoles complete their development and are metamorphosed in the vocal sacs and come out as mini frogs. In the West African tree frog, *Hylambates breviceps,* the female carries eggs in her buccal cavity. In *Arthroleptis,* it is male who keeps the larvae in his mouth.

The only known case of **gastric incubation/brooding** in vertebrates is found in the Australian frog *Rheobatrachus silus*. The female keeps the eggs in her stomach. The tadpoles are expelled through mouth after metamorphosis.

Feeding of larvae. *Dendrobates* is the only amphibian known to feed their youngs.

Viviparity. Some anurans are **ovoviviparous.** They retain eggs in the oviducts and the females give birth to living young. African toads *Nectophrynoides* and *Pseudophyryne* give birth to little toads. The European salamander *Salamandra salamandra* produces 20 or more small young while alpine salamander *S. atra* gives birth to one or two fully developed young.

Viviparity is wide spread in order **Gymnophiona** and three out of four families have this mode of reproduction, common examples include *Typhlonectes, Geotrypetes, Dermophis,* etc.

Role of Parents in Parental Care

1. The most common form of parental care is the attendance of the egg clutch by one parent.
2. Physical protection of eggs from predators either by driving them away or by eating them.
3. The attendant parent may also eat dead and fungus infested eggs to reduce spreading of fungus to other eggs of clutch.
4. Protection against drying of terrestrial eggs is provided by actively moistening them.

4. Parental Care in Reptiles

Internal fertilization and terrestrial reproduction are fully developed among reptiles. Few eggs well supplied with yolk may be carried inside the mother's body until they hatch or they may be placed in nests buried in the ground and given little subsequent care. Crocodiles are the exception in reptiles. They actively defend the nest and young for a considerable period of time.

5. Parental Care in Birds

Parental care behaviour of birds includes the following activities.

1. Nest construction. Most bird species build **nests** (Fig. 12.27) that is, *structure in which the eggs are laid and then incubated.* Some lay their eggs directly on sections of narrow rocky ledges or on patches of bare ground that have been scraped clean of vegetation and debris. Others make excavations in the ground or in trees, posts and the like, while still others use existing excavations. In either case, the birds may or may not build nests inside such holes. A majority of species of birds build **true nests** which range in complexity from simple collection of sticks to those requiring varying degrees of care and precision in the formation of completed nest cups.

Each species of birds builds its nests in its own way, in the same general type of location and of the same type of materials. The materials used in nests include grass, twigs, hair, feathers, strings, moss and clay. A nest may be a mere pit in sand (ostrich), a loose framework of twigs (dove), a cup woven of plant materials (many song birds), a hanging retort (Retort is a vessel with a bent tube and used, for the heating of substances) completely woven of strips of coarse saw-edge grass or paddy leaf (weaver bird) or a chimney made of mud (cliff swallow). Most birds conceal their nests from enemies. Woodpeckers place their nests in tree hollows, kingfishers excavate tunnels in stream banks for their nests and birds of prey build their nests high in lofty trees.

An extreme form is the large **colonial nest** of the social weaver (*Philetairus socius*), which builds first a roof of coarse straws in a large tree and under this accumulates a large mass of woven plant materials with many individual nest chamber in it.

BOX. 12.6.

SOCIAL PARASITISM

Brood parasitism or **social parasitism** (see **Begon, Harper** and **Townend,** 1996) is well developed in social insects. Here, the social insects use workers of another, usually very closely related species (**Choudhary** *et al.*, 1994) to rear their progeny.

Eggs of catfish *Synodontis multipunctatus* are incubated in the mouths of cichlid fishes together with cichlid's own eggs. As they develop they feed upon the fry of the host whilst still in its mouth and so almost entirely destroy the host's reproductive investment.

Bird brood parasites lay their eggs in the nests of other birds, which then incubate and rear them. They usually depress the nesting success of the host. Amongst ducks **intraspecific brood parasitism** appears to be most common. Females may lay a few eggs in the nests of neighbours of their own species, although more usually they rear their own young. Young ducklings are largely self-supporting soon after they hatch and there is therefore probably little cost to the host bird.

Most brood parasitism is **interspecific.** About 1% of all bird species are brood parasites — about 50% of the species of cuckoos, two genera of finches, five cowbirds and a duck (**Payne,** 1977).

2. Incubation of eggs. Incubation is the process by which the heat necessary for embryonic development is applied to an egg after it has been laid. This is done by sitting on the eggs. The incubating or brooding of eggs is almost always necessary in birds. Because birds are warm-blooded or homeothermic, the embryo will die if the egg is allowed to cool for long. In many cases, only the female incubates; in others, the male and female work in shifts; and in few cases, *e.g.*, ostrich, the male performs this duty. The incubation temperature of most birds is slightly above 100°F. The *duration of incubation* is in general related to the size of the bird. It varies from 12 to 14 days for the smaller birds and from 40 to 50 days for the large birds. Eggs of the royal albatros, have the longest incubation period, about 80 days. The eggs of hen hatch in 21 days. In many birds such as sparrow, hen, duck, quail, etc., incubation starts when the last egg is laid. Thus, their baby birds hatch at about the same time. Some birds such as hawk, owl, etc., incubates eggs as soon as each is laid. This produce 'stair step' family.

When the male and female share in incubating, the eggs may only rarely be left uncovered. When the female does all or most of the incubating, her mate may bring her food. Short period of cooling do not harm the eggs except in very cold weather, and they are often left uncovered while the female goes to feed. In some groups, such as some ducks, the female is deserted by her mate early in the incubation period, and she often carries on alone. In most grouse and pheasants, the female is deserted by the cock even earlier, immediately after the eggs are laid.

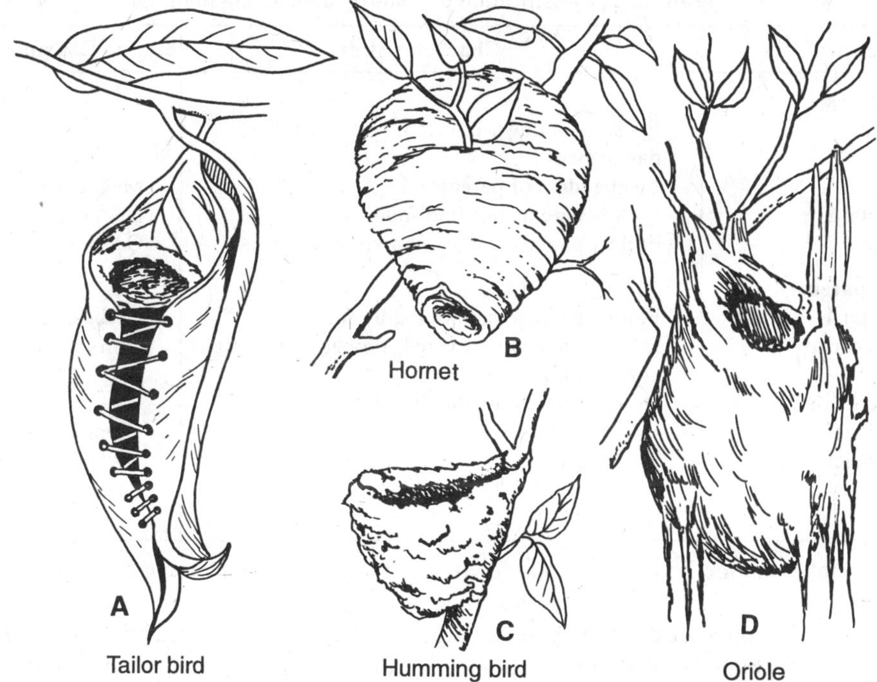

Fig. 12.27. Various types of nests in birds.

In some species when the eggs are periodically left unattended by the bird they are camouflaged in various ways. The female duck pulls the edges of the downy lining of the nest over the eggs when she leaves, and the grebe covers its eggs with material from the side of the nest. This covering is apparently for the concealment of the eggs. Egyptian plover nests on the sand and buries its eggs in the sand. The bird has been seen to sit on the sand above the buried eggs and to regurgitate water on the nest.

The time that individual birds of different species spend continuously on the nest incubating eggs may vary greatly. In many small, active song birds the incubating birds may leave the nest twice an hour or even often; in many birds (*e.g.,* hawks), periods of continuous incubation may last for hours. In some gallinaceous birds (turkeys, pheasants and quails) the bird may leave the nest only once a day. In pigeons, the two parents may incubate alternately; the male incubating during the mid-day hours, the female the rest of the time. In albatrosses, the attentive periods of each parent may last several days, during which time the sitting bird fasts completely. In some parrots, the female incubates almost continuously, being fed by the male. In some hornbills, the female remains in the nest and is fed throughout incubation by the male. In the emperor penguin, the male alone incubates continuously for the whole 60 day period and eats nothing.

During brooding, the parent occasionally turns the eggs with its beak. This not only warms them evenly but prevents adhesion of embryonic structures of the inside of the egg shell.

3. Care of the young. The young birds at the time of hatching are of two types: **precocial** and **altricial** (Table 12.1). The *precocial young birds* such as quail, fowl and duck are covered with down (feathers) at hatching and can run or swim as soon as their plumage dries up. The *altricial young birds* such as swift are naked, blind and helpless at hatching and remain in the nest for a week or more. They grow rapidly and mature early. Both types of young birds need care by the parents for some time. Their care includes *feeding, guarding* and *protecting* them against the sun and rain. Either one or both the parents may perform these duties. The food of young birds, depending upon the species, comprises worms, insects, fish, seeds and fruits.

Table 12.1. Classification of maturity at hatching in birds.

Type	Characteristics	Example
A. Feed selves		
1. Precocials	Eyes open, down-covered, leave nest first day or two	
Precocial 1	Independent of parents	Megapods
Precocial 2	Follow parents but find food	Ducks, shore birds
Precocial 3	Follow parents and are shown food	Quail, chickens
B. Fed by parents		
Precocial 4	Follow parents and are fed by them	Grebes, rails
2. Semiprecocials	Eyes open, down-covered, stay at nest though able to walk	Gulls, terns
3. Semialtricials	Down-covered, unable to leave nest	
Semialtricials 1	Eyes open	Herons, hawks
Semialtricials 2	Eyes closed	Owls
4. Altricials	Eyes closed, little or no down, unable to leave nest	Passerines

The chief avian methods of parental feeding of the young are carrying food to the nest in the bill and placing it in the open mouth of nestling; *regurgitating food*, which the juvenile picks up from the ground or from the parents' mouth and throat; and uncovering food items on the ground (by pecking or scratching), after which the juveniles pick it for themselves. The pigeon is unique in feeding its young with the "**pigeon's milk**", which is the sloughed-off epithelial lining of the crop.

While guarding the young, birds show overt aggression towards potential predators, and in the use of signals to stimulate the young to take evasive action. The specialized form of parental protective behaviour known as "injury feigning" is common among shore birds and water fowl, where its chief functions is probably to distract predators from eggs or juvenile birds.

Lastly there are several more specialized forms of avian parent-young interaction. These include long term social bonding based on imprinting of the young on their parents, feeding of the female on the nest by the male, and communal nesting and rearing of the young.

6. Parental Care in Mammals

Most mammals are altricial or semiprecocial. Among the most precocial mammals are the seals.

1. Parental care in Prototheria. Female duck-bill or platypus (*Ornithorhynchus*) is oviparous. Eggs are large with much yolk and plastic shells. There occurs no uterine gestation. Eggs are incubated by mother and newly hatched youngs are very immature. They are fed on milk in abdominal pouch till fully developed (Fig. 12.28).

2. Parental care in Metatheria. Female metatherians are viviparous. Gestation period for uterine development is small, two weeks in opossum and five weeks in kangaroo. The youngs are born in a highly immature stage (*i.e.*, tiny, naked and blind) but contain clawed fore limbs. The baby clutches the fur of mother through the claws and reaches the pouch of the mother and take a nipple of the mammary gland in the mouth. The niple swells within the mouth of the baby and the baby cannot leave it now. The baby completes its development lying attached to nipple. The sides of nipple extend into his pharynx. Milk is pressed out by a special muscle attached to the epipubic bone.

The mother remains unaware of the birth of a baby and does not recognize young one. After birth the baby enjoys long protection in the pouch but the mother does not look after it. The fully developed baby leaves the pouch but may return into it for safety.

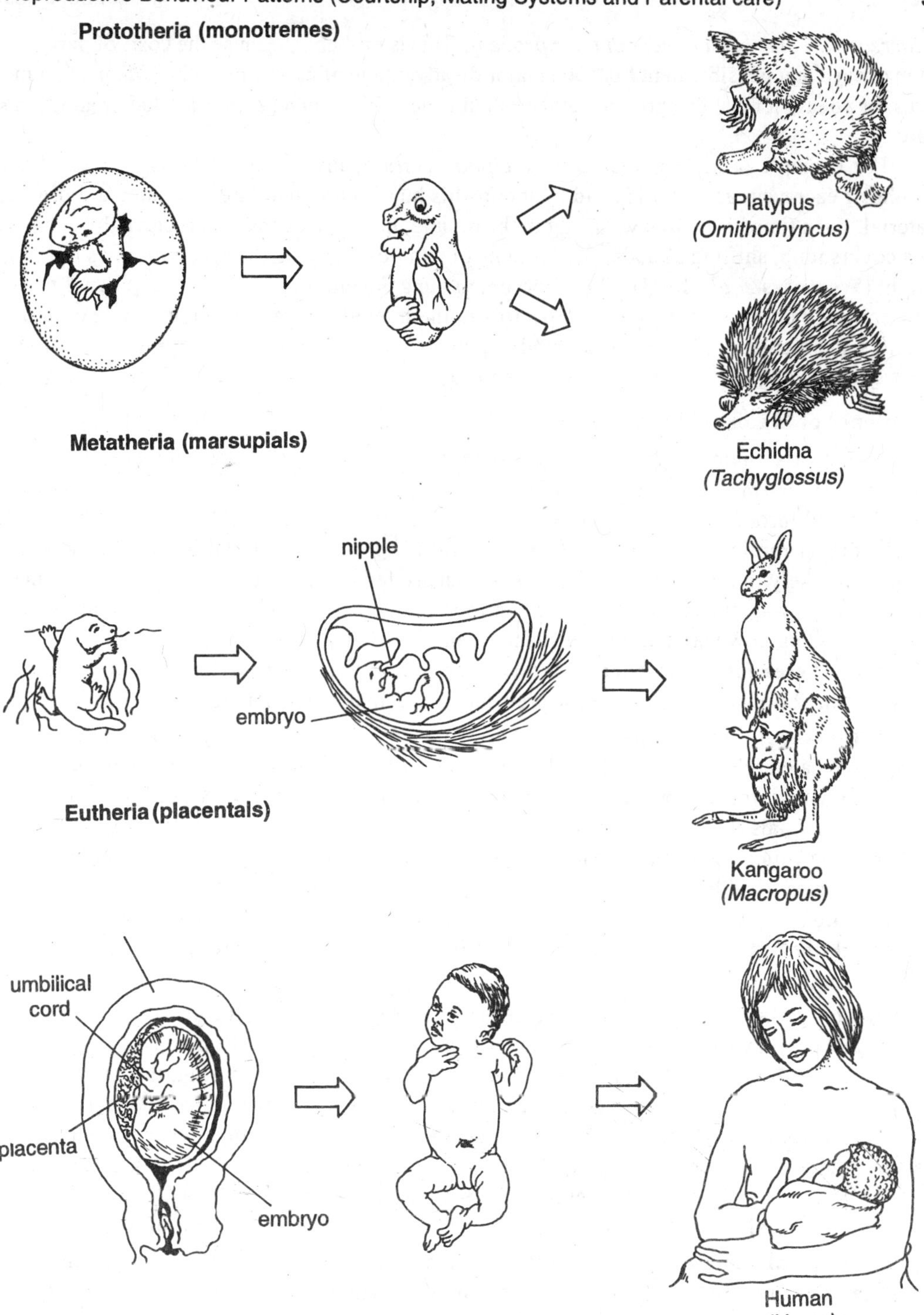

Fig. 12.28. Parental care in three major groups or existing mammals.

12.4. COST OF PARENTING

Once animal has mated, it faces yet another decision, namely, whether to provide parental care for any offspring that result. *Most animals do not invest time and energy in efforts designed to increase the survival chances of their progeny.* This is probably because the costs of parenting often exceed any possible benefits. One major disadvantage of parenting is that when an animal cares for young, it must forego some other valuable activities, such as finding food or additional mates.

For example, a male mallee fowl (*Leipoa ocellata*) invests about 5 hours a day for 6 months of each year as he builds and maintains his huge "**compost** " **nest** of sand and organic material. Each time his mate lays an egg in the mound, he digs a deep pit to receive the egg, and then covers it up, shifting about 850 kilograms of compost and sand, nearly 500 times his body weight (**Weathers** *et al.,* 1993). The embryonic young developing inside the eggs benefit from their concealment from predators and heat from the mound's decay, which the male carefully monitors and regulates. This costly parental investment presumably help the eggs he has fertilized survive to hatch. However, when the chicks appear, they are left completely to their own devices.

Prevalence of Maternal Parental Care

In almost all animals, males have the potential to help rear the kids. However, *extensive paternal care is rare and exclusive male care of the young is almost nonexistent.* In most mammals and throughout the rest of animal kingdom females do much more than their "fair share" of the parenting. For example, female California ground squirrels with young nearby are much more likely to kick sand at rattlesnakes than are fathers of those youngsters. While male squirrels drift away, the mother squirrels continue to interact with the enemy, moderating their risk be assessing how warm and large snake is, in part by listening to its rattle.

Hypotheses for prevalence of female versus male parents. Gross and **Shine** (1981) proposed the following hypotheses for prevalence of female versus male parental care.

1. Investment hypothesis. In just described example of California ground squirrel, most probably female ground squirrels have a greater interest in the welfare of their young since they have already invested so much energy in making eggs and nourishing embryos. But females of many other animals abruptly terminate their parental investment after laying their eggs or giving birth to their young (This shows that past investment in an offspring does not automatically prompt females to additional parental care). In fact, given the trade-offs involved in parental care, selection ought to favour added investment in an offspring *only* when its improved chances of survival (the benefit) compensate the parent for its loss of future reproductive opportunities (cost).

2. Low reliability paternity hypothesis. This hypothesis links the prevalence of female versus male parental care to the fact that females are more likely to be the parent of an offspring than are males. If a female lays a fertilized egg or gives birth to an offspring, this progeny will definitely have 50 per cent of her genes. A male has no such assurance, especially if his partner practices internal fertilization, in which case she may have mated with another male and used his sperm to fertilize her eggs. Therefore, to the extent that a male runs the risk of caring for progeny other than his own, his potential gain from parental care falls, thus affecting the cost-benefit ratio of male parenting and making its evolution less likely than that of maternal care.

3. Order of gamete release hypothesis. This hypothesis for female bias in parental care is that females are more likely to be left holding the fertilized eggs than their male partners, enabling males not willing to invest in parental care to "desert". By this argument, **internal fertilization** should be linked with exclusive maternal care, because after a male inseminates a female, he can depart at once, leaving his mate to make the decision whether to provide parental care or not. In cases in which such care was advantageous the female would be "forced" to provide it, given the absence of her mate. In contrast, when **fertilization is external,** females often deposit their eggs before males shed their sperm, giving the females a brief moment to flee, leaving the males to confront the parental dilemma.

Data on relationship between mode of fertilization and parental care in fishes and amphibians (Table 12.2) are consistent with the order of gamete release hypothesis.

Table 12.2. Relationship between the mode of fertilization and the sex providing parental care among families of fishes and amphibians (**Gross** and **Shine,** 1981).

Sex provides parental care	Families with internal fertilization	Families with external fertilization
Male	4 (14%)	75 (70%)
Female	25 (86%)	32 (30%)
Total	29	107

This hypothesis was discarded by the following cases: paternal care is common in frogs, in which males release their sperm into a nest before the female deposits her eggs there. Among some shorebirds, such as spotted sandpiper, paternal care is also more common than maternal care, even though males donate sperm to fertilize eggs long before egg laying, giving them ample time to desert (**Szekely** and **Reynolds,** 1995).

4. Association with young hypothesis. This hypothesis holds the following argument: because females carry the eggs, mothers are more often in a position to do something helpful for their offspring when eggs are laid or infants born. In contrast, fathers usually have no direct control over the birth of their offspring and may not even be around when their progeny appear.

Male fish (*e.g.*, 'fanning' by male three spined sticklebacks, Fig. 12.13) are very unusual in often providing **uniparental care.** In many species (mouth brooding cichlids) however, male fish may pay only a small penalty in lost reproductive opportunities for caring for eggs laid in their territories, whereas the cost of parental care to females may be much greater in immediate losses of fecundity and reduction in growth rate, which carry additional penalty in terms of long-term reduction in fecundity.

QUESTIONS

Long Answer Questions

1. Write an account of reproductive behaviour in animals.
2. Describe different ways of courtship behaviour met within animals. *(Purvanchal 1993)*
3. Write an essay on courtship behaviour. *(Purvanchal 1995)*
4. Define courtship and mating behaviour, and give an account on the role of stimuli in mating behaviour. *(Purvanchal 1999)*
5. Give an account of parental care available in various groups of animals. *(Punjab 1998, 99)*
6. Write an essay on parental care in animals. *(Punjab 1995)*
7. Write an essay on "Parental care in fishes". *(Purvanchal 2002, 03)*
8. Give an account of different modes of parental care observed in fishes. *(Purvanchal 1997)*

Short Answer Questions

1. Write a brief account of mating behaviour in scorpions and spiders. *(B.H.U. 2001; Kanpur 2000)*
2. What is courtship behaviour? *(Kerala 1997)*

3. Discuss the following :
   (*a*) Courtship display among birds *(Purvanchal 1995, 98)*
   (*b*) Mating behaviour *(Purvanchal 1995)*
   (*c*) Nest building habits in fishes. *(Purvanchal 1999)*
4. Write a short note on parental care of eggs in vertebrates. *(Punjab 1996)*
5. Discuss the following
   (*a*) Parental care in amphibians *(C.C.S. 2001)*
   (*b*) Parental care in fishes *(Puranchal 1995, 98)*
6. Give a brief account of parental care in birds. *(Punjab 1995)*
7. Give a brief account of cost of parenting.
8. Why parental care is more often provide by females?

Very Short Answer Questions

1. Who makes the "Mermaid's purse" and why? *(Purvanchal 1999)*
2. Write short notes on the following :
   (*a*) Nesting and brooding in birds.
   (*b*) Courtship behaviour in spiders *(Purvanchal 2003)*
   (*c*) Brood pouches in fishes. *(Purvanchal 2000, 02)*

Multiple Choice Questions

Choose the correct answer from the four alternatives given.

1. Courtship behaviour is a form of
   (*a*) taxis (*b*) kinesis
   (*c*) fixed action pattern (*d*) imprinting
2. A pair bond is most common among animals in which
   (*a*) there is elaborate courtship
   (*b*) there are fewer males than females
   (*c*) the female cares for the young by herself
   (*d*) both the male and the female care for the young
3. Parental care behaviours are selected for because they
   (*a*) reduce fighting among the young
   (*b*) increase the frequency of the parent's genes in the next generation
   (*c*) ensure the survival of the parents
   (*d*) force the parents to cooperate with one another and so decrease competition
4. In a particular behaviour pattern, the female prefer those males that have the most exaggragated courtship characters. This is known as
   (*a*) choosing the member of the right species (*b*) monogamy
   (*c*) courtship (*d*) sexual selection
5. The mating system where the male has access to more than one female is known as
   (*a*) monogamy (*b*) polygyny
   (*c*) polyandry (*d*) none
6. The process where by two animals come together and are physilogically ready to reproduce is known as
   (*a*) courtship (*b*) synchronization
   (*c*) copulation (*d*) reproduction

ANSWERS

1. (*c*) **2.** (*d*) **3.** (*b*) **4.** (*d*) **5.** (*b*); **6.** (*b*).

13

Biological Rhythms

(Circadian, Circalunar and Circannual Rhythms and Biological Clocks)

Most living things are exposed to the rhythms of the solar system. Because the earth revolves around the sun with regularity, rotates at a constant rate on its tilted axis and has a predictable and recurring position in relation to moon, life on earth is subjected to cyclic fluctuation in its environment. Thus, almost all animals and plants have to live in an environment whose properties are not constant over time, but instead show considerable periodic fluctuation. In this way, *most organisms are exposed to drastic changes resulting from the rhythms of night and day or summer and winter.* In addition, there are *tidal* and *lunar fluctuations* that exert greatest influence on organisms inhabiting the tidal zones. In the tune of environmental rhythms, animal behaviour too exhibits a marked periodicity in its pattern of performance. This is due to the fact that *the animals, during the course of their evolution have acquired a variety of endogenous rhythms or biological clocks whose periods are matched with those of rhythmic events in the environment.* Consequently, some rhythms of organisms are matched to the daily or 24-hour cycle (**circadian rhythm**), others to the 12·4 or 24·8 hour lunar-day or tidal cycle (**circatidal rhythm**), the 29-day lunar cycle (**circalunar rhythm**), the yearly seasons (**circa-annual rhythm**) or the time (14·7 days) between successive spring low waters (**semilunar** or **circasyzygic rhythm**). The remarkable thing about these rhythms is that they persist even when the environment cycles (**zeitgebers**) to which they are **entrained** are artificially excluded. They therefore seem to form a time base for behaviour or "**biological clock**" within the animal. Such a biological clock is assumed as analogous to human-made (physical clock) while appears to be integrated functional network of some structure located in 'higher' centres of animals.

Box 13.1.

Biological clocks imposes a kind of organizational structure on behaviour. It helps animals to change their behavioural priorities in relation to the time of day, month or year.

In most instances, behavioural rhythmicity can be explained in adaptive terms, *i.e.*, animals typically perform an activity at those times when it is best to do so (**Daan**, 1981). For example, most terrestrial reptiles inhabiting the temperate zones are active only in the summer months, and even then around midday. This is because these animals are poikilothermal (cold-blooded) and need high external temperatures for optimum activity levels. Many bird species reproduce mostly in the spring and summer because at any other time of the year there would not be enough food to raise the young. Many invertebrate animals living in the tidal zones will confine their predatory activities either to high tide—depending on whether they naturally hunt on land or in the water.

13.1. TYPES OF BIORHYTHMS

1. Circadian Rhythms or Biological Clocks

Regularly occurring daily cycle of light (day) and darkness (night) have been known to exert a far-reaching influence on the behaviour and metabolism of many organisms. The response of different organisms to environmental rhythms of light and darkness is termed **photoperiodism.** Each daily cycle includes a period of illumination followed by a period of darkness and is called the **photoperiod.** The term **photophase** and **scatophase** are sometimes used to denote the period of light and the period of darkness respectively.

Animals are not active continuously throughout each 24-hour period. Some such as birds and butterflies are most active during the daylight hours. These are called **diurnal animals.** Other animals such as crickets, cockroaches and many insects, owls and bats including many other mammals are active at night; they are called **noctural animals.** There are still certain species such as rabbits which reserve their active times for the twilight hours of the day, in early morning (dawn) or early evening (dusk). These animals are called **crepuscular animals.** Thus, the 24-hour cycle is marked by alternating periods of waking and feeding during daylight and sleeping or resting at night in diurnal animals. This alternation is reversed in nocturnal animals. Seasonal changes in the number of hours of daylight during the daily cycle regulate various activities of animals such as reproduction, hibernation and migratory movements.

Such regular day-night approximately 24-hour cycles, daily periodicities or rhythms have been discovered for almost all animals belonging to various major taxonomic groups. Such regular repeatable rhythms correlated with the earth's 24-hour rotation are called **circadian** (L., *circa,* about + *di(em),* day) **rhythms** or **circadian clocks.** This term was first coined by the American ethologist, **Franz Halberg** in 1959, for repeating pattern of activity that runs about 24 hours in length. Circadian rhythms are popularly called as **biological clocks** since their property is understood analogously to the physical oscillators of 24 hours.

Historical

Study of biological clocks (*i.e.,* biorhythms) is called **chronobiology** and it is of recent interest. However, the knowledge of biological rhythms/periodicity is ancient. As early as 4th century BC, the opening and the closing of leaves of *Tamarindus indicus* (the terminal tree) according to day/night sequence was noted in the writings of **Androsthenes.** Periodic sleep movements of *Mimosa* rhythms were investigated in 1829 and 1832, respectively by **de Mairan** and **de Candolle. Charles Darwin** (1880) and **Wilhelm Pfeffer** (1845-1920) emphasized about innateness of biological rhythms in plants.

The problem of rhythmicity of different physiological processes in different organisms was started to be investigated in detail nearly 75 years ago. The first report of biological clocks in USA came from **John Welsh** at Harvard (working with crustaceans) and **Orlando Park** at Northwestern (working with insects) in 1934. In Germany **Erwin Buenning** demonstrated in 1935 in a plant free-running rhythm. The scientific community greeted their reports with distinct coolness and ignored them. Then, in **1948, Dr. Frank A. Brown** had courage to announce—"*Organisms time themselves by environmental cues*". These cues, he theorised, are not the distinct such as light and temperature, but indirect environmental influences.

Aschoff in 1954 introduced the **zeitgeber** concept, which underlies that **the environmental factors drive endogenous biological rhythms.** His school then developed and acquired the knowledge of the functional properties of the biological clocks in humans and other animals under normal and experimental conditions. Mammalian biological clocks were, however, first investigated by **Curt Richter** (1922) of United States. The school of **Aschoff** (Germany) and **Halberg** (USA) gave a significant boost to the chronobiological study.

Examples of Circadian Rhythms

1. De Coursey (1960) provided an experimental proof of existence of a circadian rhythm in flying squirrel (*Glaucomys volans*) in the form of its wheel-running activity (Fig. 13.1). Flying squirrel had been entrained to a 24-hour light/dark cycle and was then kept in continuous darkness. The Figure 13.1 shows the remarkable precision in the onset and cessation of wheel-running in the complete absence of the environmental cues to which it is related. It also shows that like the most circadian clocks, the squirrel's clock is not calibrated to exactly 24-hours. Wheel-running started slightly earlier each day.

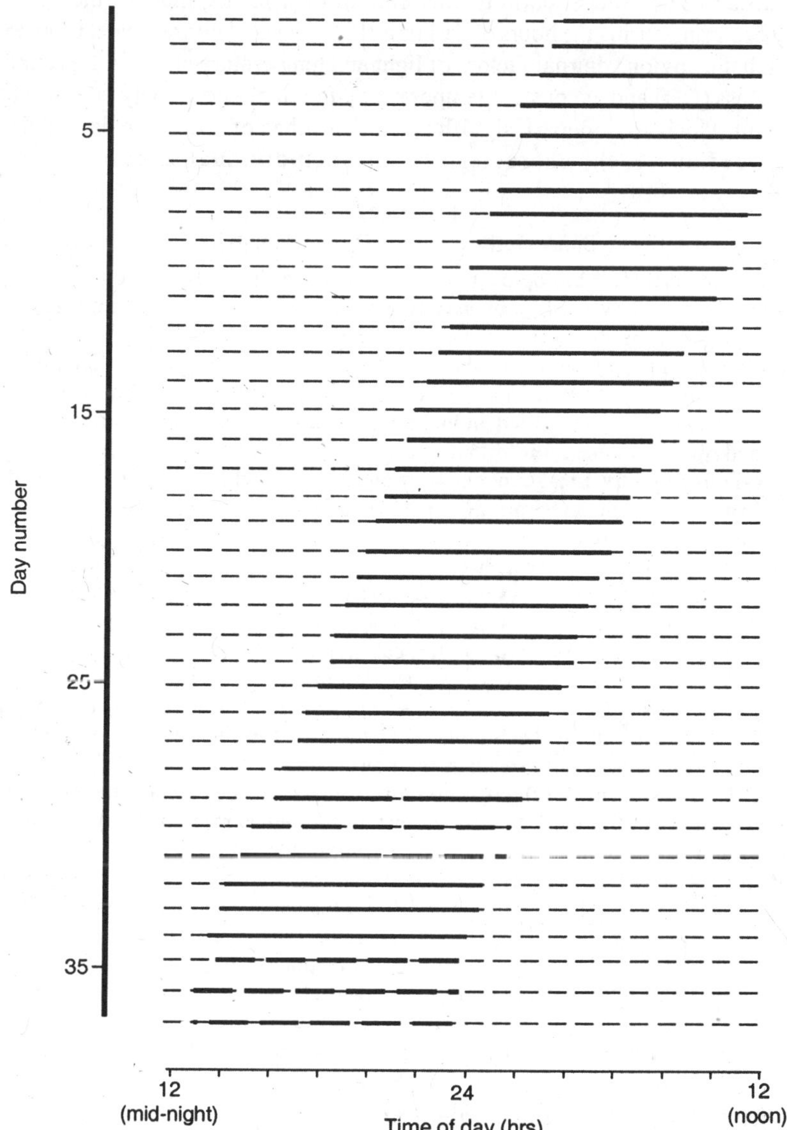

Fig. 13.1. Circadian rhythm of wheel-running activity in a flying squirrel under conditions of constant darkness.

2. Among invertebrates, circadian rhythms occur in behaviours ranging from vertical migrations by marine plankton and periodicity of pupal eclosion in dipterans to the sophisticated dance 'language' of foraging bees.

(*i*) Algae and phytoplankton photosynthesize during the daylight hours in the upper regions of a lake or ocean. At midday, when the sunlight is most intense, many mobile **zooplankton** are well below the surface. As darkness approaches, these creatures swim upward to feed upon the phytoplankton and each other. They sink back into deep waters after sunrise. The distance that a species migrates is variable; protozoans move only centimeters whereas larger species travel many metres.

(*ii*) Female crickets usually hide in burrow or under litter during the day and move about only after dusk, when it is relatively safe to search for mates. In response, male *Teleogryllus* crickets start calling to attract mates in the evening of each day (**Loher** and **Orsak,** 1985).

(*iii*) Cockroaches are almost entirely nocturnal in their habits, most of their locomoter and feeding activity occurring during the hours of darkness. If, however, they are placed in an experimental situation in which the obvious diurnal factors of light and temperature changes are removed [*i.e.,* in continuous darkness (DD) and at constant temperature], the rhythmic activity is observed to persist but with a period which may depart significantly from that of the solar day. This natural or endogenous circadian clock, characteristic of an individual cockroach in specified conditions, may show surprising accuracy in its onsets and in different individuals may vary between 23 and 25 hours.

Box 13.2.
Some Definitions of Chronobiology

1. Biological rhythm. A biological rhythm in organisms is the expression of their recurring life processes. A rhythm is wave-like or **oscillatory** in character, *i.e.*, maximum and minimum states of an event appear at identical (regular) intervals of time.

2. Exogenous and endogenous rhythms. A rhythm which is dependent on environmental stimuli is called **exogenous rhythm.** In contrast, a rhythm that does not require environmental input to maintain its periodicity is called **endogenous rhythm.** It is self-sustaining *oscillator* but it may be entrained by environmental information.

3. Free-running rhythms. An endogenous rhythm which is not entrained by any environmental information demonstrating its period spontaneously under constant conditions is known as **free running rhythm.**

4. Period. The distance between two identical positions of a self-sustaining rhythm (*i.e.*, two maxima and two minima) is the **cycle** or **period** of the rhythm. A period thus includes two phases: a **rise** (*i.e.*, strengthening of a process) and a **fall** (*i.e.*, weakening of a process). It is designated τ (Greek alphabet *tau*) for the biological rhythm and T for **zeitgeber.**

5. Phase. It is the on the spot state of an oscillation, *i.e.*, any point within the cycle.

6. Zeitgeber. It is a German term, *zeit* = time; *geber* = giver. Zeitgeber is a periodic environmental factor that entrain the biological rhythm. For example, the light and dark (LD) cycle is the permanent *zeitgeber* of circadian rhythms.

7. Amplitude. It is a range of fluctuation indicating the degree to which the process deviates from some mean value. Amplitude explains the level of biological rhythm in action. Any effect on biological rhythm is realised on its period, phase or amplitude.

8. Entrainment. It is the coupling of an endogenous rhythm to a *zeitgeber* in a way that both of them express the same period (Synchronization). For instance, if a circadian rhythm is entrained by a 24 hour cycle, it becomes a daily rhythm of 24 hour.

Presence (P) and absence (A) of mother is taken by pups as night and day, respectively. P:A cycle entrains circadian rhythms of mouse pups (**Viswanathan** and **Chandrashekran,** 1985).

(*iv*) The fruitfly *Drosophila pseudoobscura* shows a circadian rhythm of **pupal eclosion.** Unlike the locomotor activity of a cockroach which recurs nightly in the same individual, emergence of a fruitfly from its puparium occurs but once in its lifetime. It can be a periodic event, therefore, in a mixed-age population of insects. Such rhythms of pupal eclosion (or egg hatching, moulting from one instar to another, pupation, mating and oviposition) are widespread in insects. In *D. pseudoobscura* the adults emerge in well-defined pulses every 24-hour either just before "dawn" if the light component or *photoperiod* is less than about 7 hour, or after 'dawn' in a longer photoperiods (**Pittendrigh,** 1965).

3. Human body too goes through regular biological cycles every day. Many of these circadian rhythms reach a peak at some regular point during the day or night. **G.G. Luce** (1973) has suggested that our blood-sugar level, pulse-rate and blood-pressure also reach peaks around

5 to 6 p.m. According to him, our sensory acuity (*i.e.*, keenness of sense perception) such as hearing, vision, taste and smell, also are at their best in the late afternoon. Such a sensory acuity reaches a second peak around 3 a.m.

Diagnostic Features of Biological Clocks/Circadian Rhythms

A circadian rhythm has the following properties:

1. It must persist in laboratory under conditions of constant light (LL) or dark (DD) and constant temperature.
2. It must express itself in LL or DD as free running rhythm, showing period close, but not equal, to 24 hours.
3. Its free running period must be compensated for changes in the surrounding temperature. That is **temperature quotient** (Q_{10}) for circadian rhythms should be close to 1. (*Temperature quotient* is defined as the ratio between the period of circadian rhythms before and after a 10°C increase (Q_{10}) in the temperature).
4. A circadian rhythm is known to be *entrained* by the light-dark (LD) and temperature cycles.
5. A circadian rhythm must be able *to shift and resets its phase* in response to a change in light and dark conditions, and temperature or chemical disturbance.

Thus most significant features of circadian rhythm or biological clock are their persistance within free running period of approximately 24 hour and the demonstration of entrainment phase shifting and resetting properties of them in LD cycles.

Entrainment and Related Phenomena

Entrainment is a process by which an endogenous rhythm (biological clock) is linked with an external geophysical rhythm. A rhythm or clock in LL or DD and in constant temperature free runs with a period approximating 24 hours. A clock when presented with 24 hour LD *zeitgeber,* it sets and expresses a 24 hour rhythm. This is entrainment.

Entrainment is an important property of the biological clock, as it accounts for diurnal or nocturnal habit of the animals while diurnal species rise with the sunrise, nocturnal species start their activity with sunset. Despite the coupling of endogenous clock to an exogenous agency (*zeitgeber*), there still remains a difference between the periods of onset of physical event and actual biological process. This difference that occurs in the beginning between the entrainer and the biological response is called as **phase-angle difference** and it is denoted by the symbol (psi ψ). This property is called **phase-angle property** of the biological rhythm.

The phase-angle difference is **positive** when animals start activity before the onset of the *zeitgeber;* it is negative when onset of the *zeitgeber* precede the onset of the activity, shifting from one LD cycle to another shift of the phase of the circadian rhythm and the clock is re-entrained to new *zeitgeber.* It is termed as **phase-shifting property** of the biological clock. The amount of phase-shift is described as positive or negative according to the direction in which it has taken place.

Structure of Biological Clock

Chronobiological investigations aim on asking the following sort of questions: *where does the circadian clock exist?* Or, *What is structural complexity of the biological clock?*

1. Structure of biological clock in insects. In crickets (*Teleogryllus*) each day's bout of calling could get underway because the crickets possesses an internal timer or clock that measures how long it has been since the last bout began; they could use this environment-independent system to activate the onset of a new round of chirping. Alternatively, the insect's neural mechanisms might be designed to detect declining light intensity, or some environment cue, and to activate signaling only when the critical cue appears. Evidently, complete control system for cricket calling has both environment-independent and environment-dependent components: an environment-independent timer or clock, set on a cycle that is not exactly 24 hour long, and an environment-activated entrainment device for synchronising the clock with local conditions.

Experimental proofs for location of clock in insect's brain. If one cuts the nerve carrying sensory information from the eyes of a male cricket (*Teleogryllus*) to the optic lobes of his brain, it deprives him of his vision and he enters a **free-running cycle.** Visual signals are evidently needed to *entrain* the daily rhythm, but a rhythm persists in the absence of this information. If, however, one separates both optic lobes from the rest of the brain, the calling cycle breaks down completely; the cricket will now call with equal probability at any time of the day. From these results it was concluded that in cricket **master clock mechanism** (Fig. 13.2) resides within the optic lobes, sending messages to other regions of the nervous system (**Page,** 1985; **Johnson** and **Hasting,** 1986).

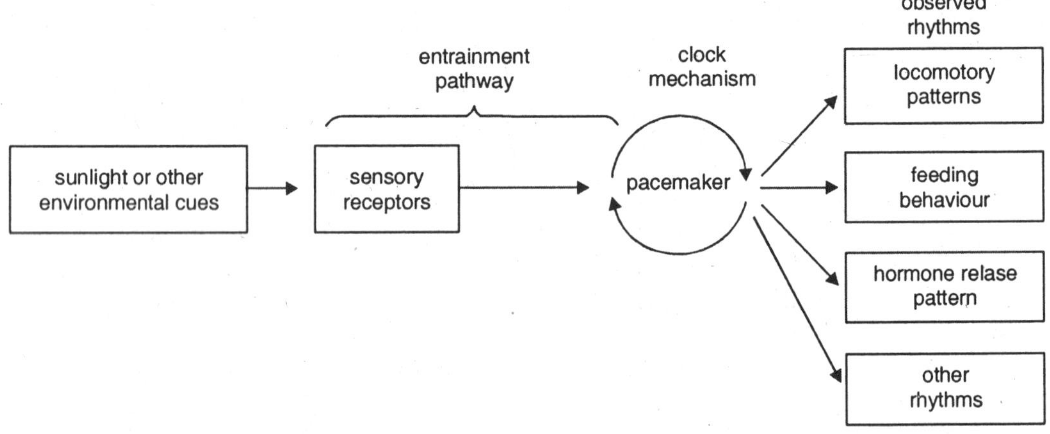

Fig. 13.2. A master clock may act as a pacemaker to regulate the many other mechanisms that control the circadian rhythms of an individual.

To investigate the nature of biological clocks in two species of silk moths, **Trumen** and **Riddiford** (1970) used surgical techniques. Adults of one of these silk-moth species usually emerge from their pupal cocoons at dawn; the other species enters the world as adults in the middle of the night. The removal of the brain from a silkmoth pupa of either species does not kill the creature or prevent metamorphosis, but it does destroy the emergence pattern. If the brain is transplanted from the head of the pupa to its abdomen, however, the animal is likely to emerge at the customary time for its species. This suggests that *the clock mechanism is contained within the silkmoth's brain.*

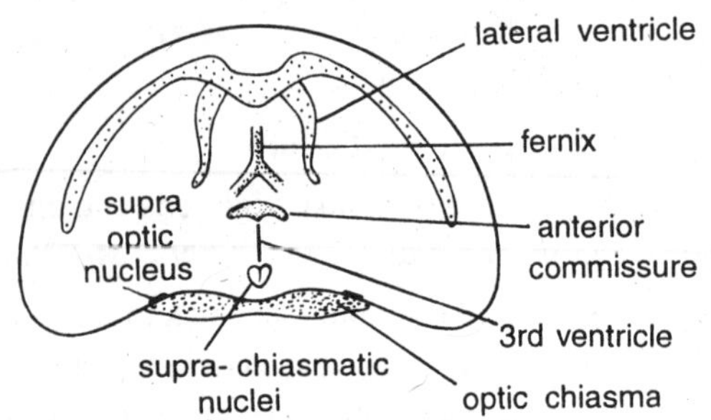

Fig. 13.3. A diagrammatic location of the SCN or suprachiasmatic nucleus in a coronal section of golden hamster brain.

2. Structure of biological clock in vertebrates. Stephen and **Zucker** (1972), **Moore** (1974) and **Rusak and Zucker** (1975) have demonstrated that the **biological clock** *is neural* and that in mammals the probable site of it is the **suprachiasmatic nucleus (SCN)** of the hypothalamus (Fig. 13.3). SCN is a pair of cell clusters in the hypothalamus that receive inputs from nerve fibres originating in the retina. In mammals, SCN lesions lead to complete destruction of circadian rhythm. Experiments with SCN of hamsters and Norway rats have suggested that SCN contains a **master clock** mechanism in these animals (**Decoursey** and **Buggy,** 1989).

Box 13.3.
Pineal Gland and Biological Clock

Chemical signals from SCN neurons activate a specific gene (CREM) in pineal cells, a gene whose protein product (ICER) is presumably involved in some way in melatonin manufacture. As the ICER protein builds up in pineal cells, however, it begins to inhibit activity in the very gene needed to make more ICER (*i.e.,* it indicates the presence of self-regulating negative feedback of genes in SCN cells).

This means that the pineal gland has a way to alter its biochemical output in relation to the recent history of photoperiods that the animal has experienced. This ability enables the pineal to adapt to seasonal changes in day length, thereby helping the animal adjust its daily schedule to these changes.

Thus, in some mammals, the SCN of hypothalamus contains the central pacemaker that sends signals to the pineal gland. The pineal gland cyclically changes its production of **melatonin** with adjustments for shifts in photoperiod length (Box 13.3) integrating the environment-independent and environment-dependent elements that regulate daily changes in behaviour. The adaptive value of the environment-independent components of such mixed systems may be that they enable individuals to adjust the timing of their behavioural and physiological cycles without having to constantly check the environment. At the same time, the presence of an environment-dependent element permits individuals to fine-tune their cycles in keeping with the subtle (delicate) variations in their particular environment. As a result, a rat will become active at about the right time each night, but will accommodate to change in day length as spring becomes summer, or summer becomes fall.

Light as Synchronizer of Biological Clock and Aschoff Rule

Light affects the biological clock through the photoreceptors/transducers/photopigments, etc., in different animals. Light is perceived through extra-ocular (extraretinal) and ocular (retinal) photoreceptors. In mammals, it is mainly ocular, but in birds it is extraocular and ocular.

Box 13.4.
Naked mole-rats lack biological clocks

In contrast to rats and mice, naked mole-rats live in colonies that occupy elaborate underground tunnels. They almost never come to the surface. When do they open a burrow to the outside, it is usually just to throw out dirt from fresh tunnel excavations. They have no special dependence on what is going on above them during the day or the night. And as one might predict, *naked mole-rats exhibit no circadian rhythm of any sort.* Instead individuals **scatter generally** brief episodes of wakefulness among longer periods of sleep, with the pattern changing irregularly from day to day (**Davis-Walton** and **Sherman,** 1994).

Jurgen Aschoff (1960) formulated a rule for the effects of light on the biological clock. As it states, *the period length of night-active animals increases with increasing intensity of light in LL conditions, whereas that of the day-active animal decreases with increase in light intensity in same LL condition.* Thus, on increasing the intensity of light in constant illumination condition (LL), the frequency of the free-running rhythm increases in diurnal animals and decreases in nocturnals. This rule is also called **circadian rule** and still holds true for majority of the animals in which biological clocks has been studied.

Functional importance of biological clocks. Biological clocks seem to occur universally in temperate and tropical organisms including fungi, algae, plants, insects and all vertebrates. These may have evolved in animals to time their behaviour in relation to the external environment. They provide the following adaptive advantages to the animals:

1. Bed bugs are nocturnal predators. Their activity rhythm coincides with the diurnal habit of the humans; they are active from sunset to sunrise, a period that is rest period of humans.
2. Filarial worms (*Wuchereria bancrofti*) infest the human lymphatic system; their larvae parasitize the blood. The worms increase in number during night and vanish from blood during day (by 9.00 a.m.). If the infested person changes its habit, from diurnal to nocturnal, the worm will change its periodicity accordingly.
3. Eclosion activity in *Drosophila* takes place before dawn so as to coincide the emergence of the adult at the optimum humidity level in environment and at low temperature. This is vital because when these otherwise fully developed individuals emerges, their skin is not yet completely hardened (*i.e.,* developed into a protective armour) which means that the insect would dry up very quickly.
4. Development of the timing mechanism sets a predator to hunt when the prey is most active.
5. Related species living on same diet feed at different hours of day to avoid competition.
6. Camouflage in ostriches is a function of the daily (circadian) clock. While black plumed male guards the nest in night, females with pale-brown garb sits around the nest in day.
7. Daily clocks measure the length of the day and thus determine the seasonal phenomena in animals which use day-length as environmental information.
8. The synchronization of an enzymatic activity and the synergism of the hormonal effects on the body's environment is possible only by the principle of biological clock.

2. Circatidal Clocks (Tidal Rhythms)

Lunar day is greater than solar day. It is of 24·8 hours and is due to the revolution of the earth on its own axis, but relative to the moon. The lunar day becomes 50 minutes longer, as moon rises 50 minutes late every day.

Tides are sea water movements which are caused by some astronomical factor, *i.e.,* due to gravitational pull of moon and sun. They represent a rhythmic rise and fall of water level and often waves of long wavelengths characterize the process. In a lunar day two tides occur at about 12·4 h difference. These two tidal floods may or may not be of the same height depending on the position of moon in relation to earth. If the moon is at equator, the tides will be at the same height but if it is north or south of the equator the height of the two tides will be unequal.

Types of tides. Tides are of following *three* types:

1. Diurnal tides. A diurnal tide is characterized by one high and one low tide per lunar day. Such tides occur in the Gulf of Mexico. These tides result when moon revolves the earth over the equator.

2. Semidiurnal tides. In this type of tides, one tidal height is greater than that of the other tide. Such tides occur in Atlantic Ocean, North Sea and Indian Ocean. These tides are caused when moon revolves the earth the north or south of the equator.

3. Mixed tides. This type of tides is characterized by the occurrence of tides of unequal strength in indefinite spatial and temporal pattern. Mixed tides occur in Pacific Ocean, Australian Coasts, Carribbean and Arabian Seas.

Spring and neap tides. A lunar month is of 29·5 days. As a result of alignment of earth, moon and sun, two types of tides occur at two times in a lunar month. **Spring tides** are extreme tides and occur when the earth, moon and sun come in a line (*i.e.,* the gravitational pull of sun and moon work together). Thus, spring tides occur two times in a lunar month—one at the time of new moon and another at the full moon time. Likewise, weak **neap tides** occur at first and quarters of the month when positions of earth, moon and sun is set at right angles. In this case due to right angle arrangement the gravitational pulls of the sun and moon work against each other.

Tidal rhythms. These are common in marine animals and have periods extending between two tides (12·4 h apart) and occur in animals exposed to tidal floods. Any organism living in the intertidal zone of the sea shore are alternately submerged by water and exposed to air. This results in change of several types of environmental factors such as **pressure, salinity, food supply, temperature, predation risks,** etc.

A French scientist, **Charles Bohn** (1903, 1904) first described tidal periodicity in *Convoluta rescoffensis,* an intertidal flat worm. Further studies on tidal rhythms were performed by **M.K. Chandrashekaran, F.H. Bornwell** and **H.W. Honegger.**

Examples. Clear-cut examples of tide-entrained cycles which persist under laboratory conditions are the vertical migration cycles in sand-dwelling platyhelminthes, polychaetes and diatoms, expansion and contraction rhythms in sea anemones, filteration rates in bivalve molluscs and swimming activity in fish. In an experiment of **Palmer** (1973), shore crabs (*Carcinus maenas*) were raised in the laboratory from eggs to adulthood under a day-night regime without tidal influences. These crabs showed only a circadian rhythm of activity. However, a tidal rhythm appeared after the crabs were given a cold-shock treatment. It seemed that the endogenous tidal clock had been dormant until started by the cold shock.

Similar phenomena occur in the tidal rhythms of other crabs. Fiddler crab (*Uca*) show tidal rhythm of activity that may persist for up to five weeks under constant laboratory conditions. The crabs emerge from their burrows at low tide and actively forage, court, etc. As the tide floods, they retreat back into their burrows.

3. Lunar Rhythms (Circalunar Clocks)

The lunar cycle of 29·5 days is known to influence a variety of aspects of animal behaviour. In these cases behaviour is said to show **circalunar** or **circasynodic periodicity.** Such circalunar cycles are common among marine invertebrates and insects.

1. In the annelid palolo worm (*Eunice virids*) of the Pacific Ocean (*i.e.,* Polynesian Islands), reproductive activity occurs only during neap tides of the last quarter moon in October and November. The palolo worms live in tunnels in the coral reef and their posterior parts, containing the genital organs, become detached from the anterior parts, swim to surface of the water, where they swarm and release eggs and sperms. This swarming phenomenon occurs seven to nine days after full moon. Fertilization occurs in the midst of this swarm of gametes and new individuals formed return to the coral reef below after a few days of development. The palolo worms provide abundant food for sharks and other fish, but by synchronizing their reproductive activity the worms are able to overwhelm these predators and a proportion of the gametes are always able to survive.

2. Likewise, the Mediterranean polychaete *Platynereis dumerlii,* an inhabitant of sea bottom, transforms into its sexual phase (heteronereis) and swarms at the sea surface in synchrony with the full moon. The people on the Fiji Islands know when these annelid worms swarm; they consider them delicacies and they catch them with nets to eat at feasts. *Platynereis* perform their swarming activity during the days around the new moon and during swarming worms release their gametes and then die. Such a lunar periodicity of these worms can be maintained under constant laboratory conditions although it appears to be labile to several zeitgebers such as light.

3. The California grunion (*Leuresthes tenuis*) is a small fish that seasonally spawns three or four days after a new or full moon, a time corresponding to the highest tides of the month. One to three hours after the tides begin to recede, the females, hotly pursued by the males, ride to the beach, bury their tails in the sand and deposit their eggs a few centimetres below the surface. While a female is depositing her eggs, one or more males encircle her and release sperm, which flow down the wet side of the female and fertilize the eggs. All fish return to the ocean on the next sweep of a high wave. The buried eggs are thus protected from the dangers of the sea until the next spring tide, a month later, brings high seas to carry the young out to sea.

4. Lunar rhythms are also known in terrestrial animals. For example, pit-building activity in the insect-ant lion (*Myrmelon obscurus*) whose larvae are called **doodlebugs** reaches a peak at around full moon. Lunar rhythms are characteristics of many insect species from Placoptera, Ephenaeroptera, Lepidoptera, Diptera and Trichoptera.

5. Jamaican fruit bats have a pattern of feeding that involves leaving the day roost in the evening and feeding throughout the dark nights during the period of the new moon. During the full moon, however, they depart from their day roost at the normal time in the evening, but return there when the moon is high, even when obscured by clouds, suggesting that they make use of an endogenous lunar clock to time their foraging and feeding behaviour.

4. Semilunar Rhythms (Circasemilunar or Circasyzygic Clocks)

Some seashore animals show semilunar or circasyzygic rhythms which are synchronised with fortnightly cycle of spring and neap tides. The periwinkle (*Littornia rudis*), for example, shows a marked 15-day periodicity in its locomotory activity. The species lives high up on the shore and is only covered by the high water of spring tides.

5. Circannual Rhythms

The term circannual rhythm is coined for rhythms that persist and recur annually with the free-running period of approximately 365 days (one year). Annual periodicity in reproductive cycles have been observed in many insects, fishes, birds and mammals. For example, in its natural habitat the 'carpet' beetle *Anthrenus verbasci* feeds on the organic matter in old birds nests and takes one, two or even three year to complete its development. **Blake** (1959) showed that the development of this insect comprised successive periods of **diapause** and **growth,** one diapause for each winter passed. The natural period of this circannual rhythm (41-46 weeks) could be entrained to an exact annual cycle by the naturally occurring changes in the day length.

Among fishes, brook trout spawn in the fall; bass and blue gills in late spring and summer. Among birds, some of the best examples of circannual clocks are shown by warblers and white-crowned sparrows. Among mammals, circannual rhythms control activity and hibernation cycles, *e.g.*, hedgehog (*Erinaceus europaeus*), rodents, ground squirrels (*Spermophilus* spp.) and chipmunks (*Eutamias*).

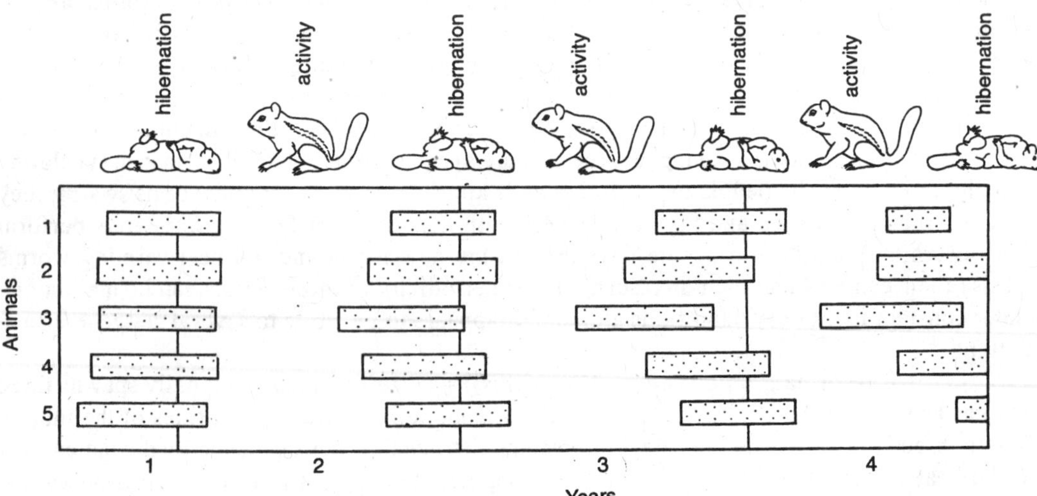

Fig. 13.4. Circannual cycle of the golden-mantled ground squirrel. Animals held in constant darkness and at a constant temperature nevertheless entered hibernation at certain times year after year.

Testing the hypothesis that an animal has a circannual rhythm is technically difficult because individuals must be maintained under constant conditions for at least two years after their removal from the natural environment. One successful study of this sort involved the **golden-mantled ground squirrel (Pengelley** and **Asmundson,** 1974) of north-temperate North America. This squirrel in nature spends the late fall and winter hibernating in an underground chamber. Five members of this species were born in captivity, then blinded and held thereafter in constant darkness and at a constant temperature while supplied with an abundance of food. Year after year, these ground squirrels entered hibernation at about the same time as their fellows living in the wild (Fig. 13.4).

Several behaviours of animals, thus, have been cyclic or rhythimic, *e.g.*, 1. eating and drinking; 2. sleep and activity; 3. learning and memory.

QUESTIONS

Long Answer Questions

1. What is biological clock ? Explain its various types giving examples. (*Magadh* (*H*) *1990, 94, 98*)
2. Discuss the phenomenon of biological clocks. (*Calcutta* (*H*) *1991, 96*)
3. What is meant by the term 'biological clock'? Elaborate your concept on 'circadian clock'. (*Calcutta* (*H*) *1992*)
4. What is circadian rhythm ? How circadian rhythm influences the endogenous and exogenous system of animal behaviour ? Add a note on its significance. (*Calcutta* (*H*) *1994*)

Short Answer Questions

1. Explain circadian rhythms. (*Calcutta* (*H*) *1991, 93*)
2. Describe briefly the following :
   (*i*) Biological clock (*Magadh* (*H*) *1992, 96* ; *Kanpur 1995, 97*)
   (*ii*) Biorhythms (*Lucknow 1995*)
   (*iii*) Circadian rhythm. (*Calcutta* (*H*) *1999*)

Very Short Answer Questions

1. What is lunar periodicity ? (*Kerala 1999*)

Multiple Choice Questions

Choose the correct answer from the four alternatives given

1. Which of the following is not true of biological clocks?
   (*a*) they are exogenous components of rhythmic behaviour
   (*b*) they are endogenous component of rhythmic behaviour
   (*c*) they are entrained by zeitgebers
   (*d*) they are entrained by light cycles
2. Endogenous rhythm of 24 hours is called (*Magadh (H) 1998*)
   (*a*) circalunar (*b*) circadian
   (*c*) circannual (*d*) circasynodic

ANSWERS

1. (*a*); **2.** (*b*).

Migratory Behaviour

(Also Includes Orientation and Navigation)

14.1 MIGRATION IN ANIMALS

The migration is movement of a population of organisms from one environment to another. The word "**migration**" has been derived from Latin *migrare* meaning to go from one place to another. It is a sort of periodic and directed travel due to changes in environmental conditions. Thus, we can define migration as mass directional movement of large numbers of a species from one habitat/location to another (Box 14.1). The term therefore applies to classic migrations (the movements of locust swarms, the inter continental journeys of birds, the transoceanic movement of eel) but also of less obvious examples such as to-and-fro movements of shore animals following tidal cycle. The study of periodic phenomena (such as migration) in animals in relation to changes, climatic and other ecological factors is called **phenology.**

Box 14.1.

The process of migration in some animals was initiated as means of survival but later on by virtue of natural selection it became genetically fixed pattern of behaviour and differs from species to species. This behaviour was selected since it provided better survival, easy availability of food, better habitat, etc.

Evolutionary model of migration. According to evolutionary model developed by **Baker** (1978), animals tend to assess the utility of their present habitat (h_1) relative to that of another potential habitat (h_2). They migrate only when the utility of h_1 drops below that of h_2 multiplied by a **migration factor** (m) (*i.e.*, when $h_1 < h_2m$). The factor m is an expression of the cost/benefit ratio of the migratory act itself. **Baker** (1978) proposed the following "**hierarchy of migration**".

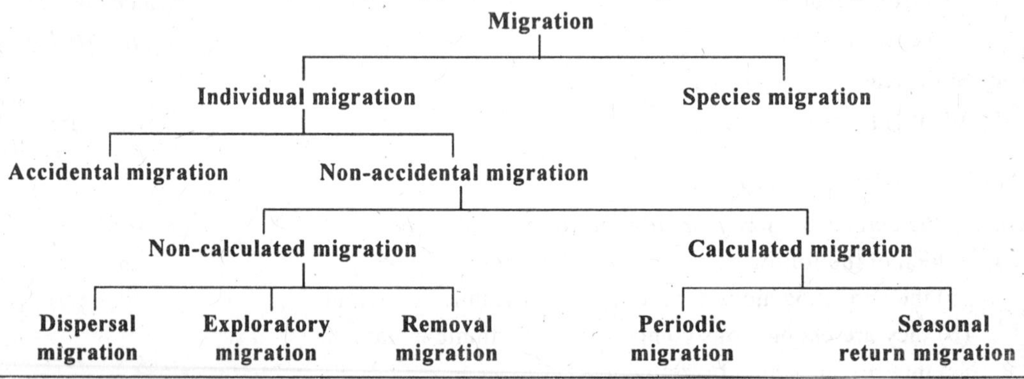

Patterns of Migrations

1. Diurnal and tidal movements. Individuals of many species move from one habitat to another and back again repeatedly during their life. The time scale involved may be hours, days, months or years. In some cases, these movements have the effect of maintaining the organism in the same type of environment. This is the case of movement (called **tidal migration**) of crabs on a shoreline. Crabs move with the advance and retreat of the tide. In other cases,

diurnal migration may involve moving between two habitats (or environments), each of which can supply a *limited resource.* For example planktonic algae both in sea and fresh water lakes descend to the depths at night but move to the surface during the day. It appears they accumulate phosphorus and perhaps other nutrients in the deeper water (the *hvpolimnion*) at night before returning to photosynthesize near the surface (the *epilimnion*) during day light hours.

2. Seasonal movement between habitats. Many motile organisms (animals) make seasonal moves between habitats. The paths of the environment in which resources are available change with seasons and populations move from one type of patch to another. The **altitudinal migration** of grazing animals in mountaineous regions is one example, where, for example, the **American elk** and **mule deer** move up into high mountain areas in the summer and down to the valleys in the winter. Thus, by migrating seasonally the animals escape the major changes in food supply and climate that they would meet if they stayed in the same place.

In amphibians such as frogs, toads and newts, a sort of migration occurs between an aquatic breeding habitat in spring and a terrestrial environment for the remainder of the year. The youngs develop (as tadpoles) in water with a different food resource from that which they will later eat on land. They will return to then same aquatic habitat for mating, aggregate into dense populations for a time and they separate to lead more isolated lives on land. In this case an individual may make several return journeys in its lifetime.

3. Long-distance migration. The most notable habitat shifts are those that involve travelling long distances. Most terrestrial birds in the Northern hemisphere move north in the spring when food supplies will become abundant during the warm summer period, and move to south to savannahs in the autumn when food becomes abundant only after the rainy season. In this case, migration seem to involve transit between areas that supply abundant food, but only for a limited period. *They are areas in which seasons of comparative glut/excess and famine/ scarcity alternate and which cannot support large resident populations all the year round.* For example, swallows (*Hirundo rustica*) which migrate seasonally in South Africa, vastly out number a related dense species. The all-year round food supply can support only a small population of resident, but a seasonal excess exceeds what the residents can consume. Long distance migrations may be *one-return journey migrations* or *one-way only migrations* as shown by the following examples:

Examples

1. Migrations in birds. Of 1200 Indian birds, about 900 are residents and remaining 300 are migrants. Migratory birds come across the Himalayas from northern part of Asia and Europe to escape from the severe winters prevailing there and enjoy the much warmer and longer days in central and southern India. Ducks, geese, wagtails, flycatchers and ruffs are important examples of winter migrants of India. They arrive in India in September or October and return in March or April.

Tiny ruby-throated humming birds fly nonstop across 800 kilometres of the Gulf of Mexico twice a year. Migrating hoope (*Upupa epops*) is found to hop up a Himalayan pass at 20,000 feet. Studies have been made of songbird species within the genera *Phylloscopus* and *Sylvia* that exhibit a range of migratory habits. It appears that in the typical long distance migrants such as garden warbler (*Sylvia borin*), the subalpine warbler (*Sylvia cantillans*) and willow warbler (*Phylloscopus trochilus*), there are marked seasonal changes in body weight, moult, testes size, nocturnal restlessness and food preferences. The European population of these species winter in Africa and migrate across the Sahara Desert.

Middle-distance migrants such as black cap (*Sylvia atricapilla*) and the chiffchaff (*Phylloscopus collybita*) winter in Europe and Africa and show moderate seasonal changes in body weight and other migratory indices. The Sardinian warbler (*Sylvia melanocephala*) and the Dartford warbler (*Sylvia undata*) are partial migrants that winter within the Mediterranean breeding areas, while Marmosa's warbler (*Sylvia sarda balearica*) is resident the year round and endemic on the Balearic Isles and the Pithyuses in the Mediterranean sea. These species have a fairly constant body weight throughout the year, and they have a moult of body feathers of long duration which alternates with periods of nocturnal restlessness.

Some birds migrate many hundreds of kilometers over the ocean, where there are few landmarks. For example, the arctic terns (*Sterna paradisaea*) breed in the northern arctic latitudes, for example, in northern Canada, flying east across the Atlantic in the fall (autumn) changing course to the south off the coast of England, and finally spending the winter far in the south reaching Antarctica. In the spring they reverse the migration, though probably by a different route. Thus Arctic tern travels a 35200 km round trip course each year. Similarly, the slender-billed shearwater (*Puffinus tenuirostris*) breeds in southeast Australia and Tasmania, and migrates to Alaska via Japan. Its return journey takes it down the western coast of North America. Thus, during its annual flight in a large clockwise loop around the pacific, it crosses the equator on two occasions, and in this way it is exposed to a variety of changes in length of day, apparent path of the sun, climatic conditions and so on.

2. Migration in mammals. Among the mammals, wildebeests, caribou, zebra, gazelle, bison, seals and whales perform arduous journeys each year. For example, in the Serengeti Plain of Tanzania, East Africa, long **seasonal migrations** are performed by zebra (*Equus* spp.), wildebeest (*Connochaetes taurinus*) and Thomson's gazelle (*Gazella thomsoni*). These species spend the dry season (July to December) feeding in open thorn woodlands where, owing to the low rainfall, they are limited in their movements to the vicinity of water holes. In December, when the first rain come, the herds move into the central plains of the Serengeti and then migrate in a predominantly anticlockwise direction following the rains and new vegetation. Most apparently wildebeest can sense where rain is falling by using their eyes and ears (but not by their noses) and then migrate in that direction.

Although the direction of migratory movement is purely opportunistic (animals simply move towards areas where rain can be seen or heard falling), the sequence in which the different species move round is determined by specific food requirements. In the Serengeti, the sequence is generally zebra first, wildebeests second and gazelle third. Stomach analyses have shown that zebra take a high proportion of grass stem material. By trampling the grass and removing the tough stems, zebra open up the herb layer so that wildebeest, which feed mainly on grass leaf, can then move in. The joint action of the zebra and wildebeest exposes the broad-leaved plant layer preferred by the gazelles. By following the rains, the zebra continually move from a depleted habitat to one with fresh, abundant vegetation. In their turn, the wildebeest and gazelle capitalize on the selective feeding of the previous species and also move from poor to good habitats. The survival value of seasonal return migration is illustrated by the mortality figures for migrant and non-migrant species. The most important mortality factor in the migrant zebra, wildebeest and gazelle populations is predation. Few migrant animals die of starvation. In non-migrant species such as impala (*Aepyceros melapus*) and wart hog (*Phacochoerus oethiopicus*) the opposite is true. In them starvation often outweighs predation as a cause of death.

3. Migration in reptiles. Migratory reptiles include the leather-back turtles, which move vast distances (upto 3000 miles of oceanic travel) between a breeding area in French Guieana and points near North America and Africa (**Pritchard,** 1976). One population of the **green sea turtle** — *Chelonia* nest on Ascension Island, a tiny speck of land in the centre of the Atlantic ocean between Africa and Brazil. The adult female turtles visit the island only to deposit their eggs in beach sands. They then swim 1000 miles or so to warm, shallow water off Brazil, where they feed on marine vegetation for several years before returning, usually to the same beach, to lay another clutch of eggs (**Carr,** 1967).

4. Migration in amphibians. Amphibians too show long distance migrations. A salamander, **the California newt,** breeds in pools in mountain streams in the spring and then spends the summer months underground. In the rainy fall and winter, the newts wander widely over the forest floor in search of insect food. In the spring, the adults migrate back to the stream, often to the pool where they bred the previous year. Newts experimentally displaced hundreds of metres upstream or downstream had no difficulty relocating their home pools.

5. Migration in fishes. Some fishes show mainly **one-return journey migrations,** *i.e.*, they make only one return journey during their life time. They are born in one habitat, make their major growth in another habitat, but then return to breed and die in the home of their infancy. For example European eel (*Anguilla anguilla*) travels from European rivers and ponds across the Atlantic to the Sargasso sea, where it is thought to reproduce and die (although spawning adults and eggs had never actually been caught there). The American eel (*Anguilla rostrata*) makes comparable journeys from areas ranging between Guianas in the south, to south-west Greenland in the north.

The **salmon** makes a comparable transition but from a freshwater egg and juvenile phase to mature as a marine adult. From marine habitat, it then returns to freshwater sites to lay eggs. After spawning, all Pacific salmon (*Oncorhynchus nerka*) die without ever returning to sea. Many Atlantic salmon (*Salmo salar*) also die after spawning, but some survive to return to the sea and then migrate back to the home of their infancy.

Migratory fish return, often with very high precision, to the home of their infancy and some are known to use information from the **sun, polarised light** and **geomagnetic fields** as generalized cues that increase the "homeward-bound" direction of migration. But the more sensitive control of the direction of long-distance migrations appears to be determined by **olfactory cues** and when closer to home **memory of local topography.** There is strong evidence that each river and tributary has a characteristic odour and that young *fish imprint to this odour* before they migrate to sea. On the homeward journey they track the changing concentrations of the chemicals responsible for the odour. The eel can apparently detect concentrations of **β-phenyl-ethyl alcohol** (that represent only two or three molecules on the olfactory epithelium). There is some evidence that salmon remember separately, and in order, each odour along the seaward migration and *react to each in reverse order when they return home* (**Leggett,** 1977). An important effect of "homing" in migrating animals is that mating will be largely restricted to geographically localized populations within the species.

6. Migration in invertebrates. There are many examples of migratory invertebrates. For example, **Atlantic lobsters** living off the continental shelf of the United States apparently move back and forth from deep water to coastal areas, a one-way trip frequently exceeding 80 km (**Cooper** and **Uzmann,** 1971). Caribbean spiny lobsters have been observed marching along in a **conga line** of dozens of individuals with each crustacean holding on to the one in front of it. This formation results in drag reduction and conserves the energy of migrating individuals.

Among the migratory insects, the monarch butterfly (*Danasus plexippus*) has a 'return journey' pattern of migration. Its populations travel north in the USA and Canada to breed in summer, and south to Florida and California to overwinter, but not to breed. The same individuals may return to the north in the spring, but it appears that each individual makes no more than one return journey. On a much smaller scale, **one return ticket migrations** can also be seen in the life cycles of many insects, for example, butterflies, moths, caddish-flies and stoneflies. The individual spends the early part of its life in one habitat (the larval food plant for the butterfly, the aquatic stage for the caddish fly) migrates to another habitat as a sexually mature, reproductive adult and return to the juvenile habitat only to lay eggs.

In some migratory species, the journey for an individual is on a strictly one-way ticket. In Europe, butterflies such as the clouded yellow (*Colias croseus*), red admiral (*Vanessa atalanta*) and painted lady (*Vanessa cardui*) butterflies breed at both ends of their migrations. The individuals that reach Great Britain in the summer breed there and their offspring fly south in autumn and breed in the Mediterranean region—the offspring of these in turn come north in the following summer.

Ecological factors regulating migration. Most migrations occur seasonally in the life of individuals or populations. They are usually seen to be triggered by some external seasonal phenomenon (*e.g.*, changing day length) and sometimes also by an internal physiological clock. They are often preceded by quite profound physiological changes such as **accumulation of body fat.** They represent strategies evolved in environments where seasonal events such as **rainfall** and **temperature cycles** are reliably repeated from year to year. There is, however, a type of migration that is tactical, forced by events such as **overcrowding** and appears to have no cycle or regularity and these are most common in environments where rainfall is not seasonally reliable. The economically disastrous migration-plagues of locusts in arid and semi-arid regions are most striking examples.

Locusts (*Schistocerca gregaria*) change their behaviour when they become crowded, during periods of peak population growth. There is a 'phase' change that occurs over three generations, from a **solitaria phase** (when they behave as individuals), through a generation in a **transience phase** to a third generation in a **gregaria phase** when they start to band together and behave as members of a herd and they then fly as powerful flying machines, forming immense swarms in the daytime. Dense swarms of locusts migrate downwind into areas of low barometric pressure, where they are most likely to encounter rain and abundance of food plants. They fly by day and stop migrating when they encounter wet conditions. Sexual maturation, copulation and deposition of eggs then occur. Locust-like behaviour is also shown by other swarming pests such as army-worm (*Spodoptera exempta*), senegalese grasshopper (*Oedaleus senegalensis*), spruce budworm (*Choristoneura fumiferana*).

14.2. FISH MIGRATION

Generally fishes live in a constant habitat and restrict their movement within particular territorial limits, but there are many others which migrate from one type of habitat (*e.g.*, fresh water) to another (*e.g.*, sea water or vice versa) and travel long distances. In most cases this movement of a large number of fishes may occur for the purpose of feeding, spawning or shelter and is called **migration.**

I. Types of Migratory Fishes

1. Diadromous fishes. These fishes migrate between fresh water and sea water. Diadromous fishes are subdivided into following three types:

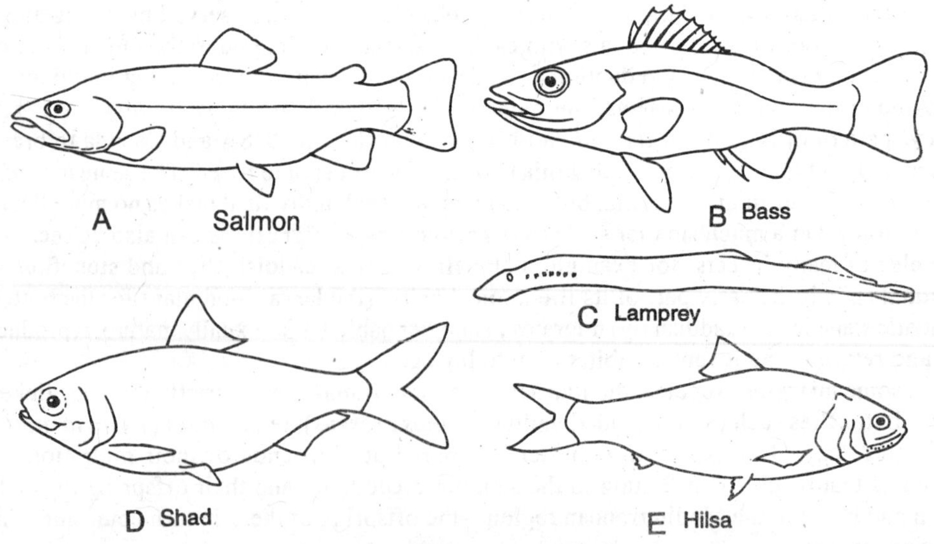

Fig. 14.1. Anadromous fishes.

(*i*) Anadromous fishes. (Greek *Anadramein* = to run upward). These fishes migrate from sea to fresh water for spawning (breeding purpose), *e.g.*, sturgeon (*Acipencer*), Atlantic salmon (*Salmo salar*), pacific salmon (*Oncorhyncus nerka*) and *Hilsa*.

(*ii*) Catadromous fishes. These fishes migrate from freshwater to the sea for spawning, *e.g.*, the common or freshwater eels—*Anguilla anguilla* and *Anguilla vulgaris*, and the American eel—*Anguilla rostrata*.

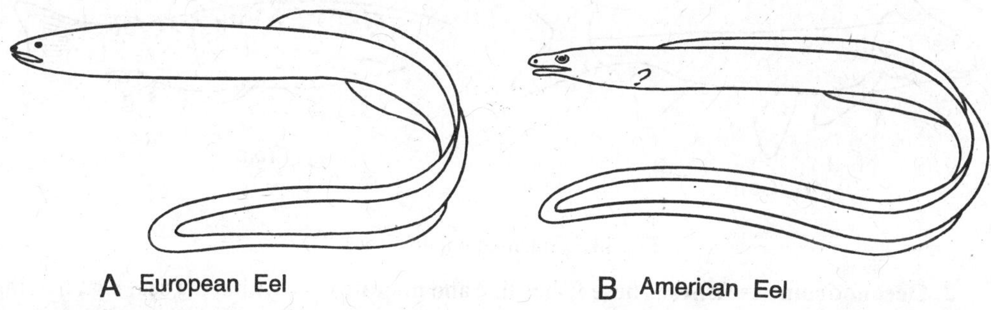

Fig. 14.2. Catadromous fishes.

(*iii*) Amphidromous fishes. These fishes migrate from sea water to fresh water and *vice-versa*. This type of migration is not for the purpose of breeding but for other purposes (*e.g.*, food). This movement may occur regularly at some definite stage of life cycle, *e.g.*, gobis undertake migration for food.

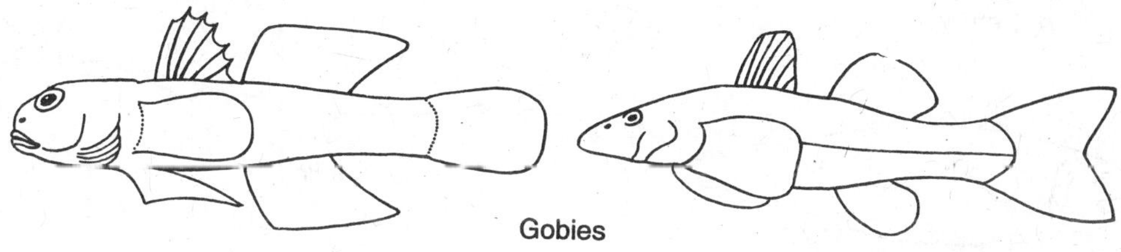

Fig. 14.3. Amphidromous fishes.

(*iv*) Semi-migratory fishes. These fishes migrate from sea water to estuaries and are not marked by migratory changes, *e.g.*, roaches, white fish and asiatic milk fish (*Chanos*).

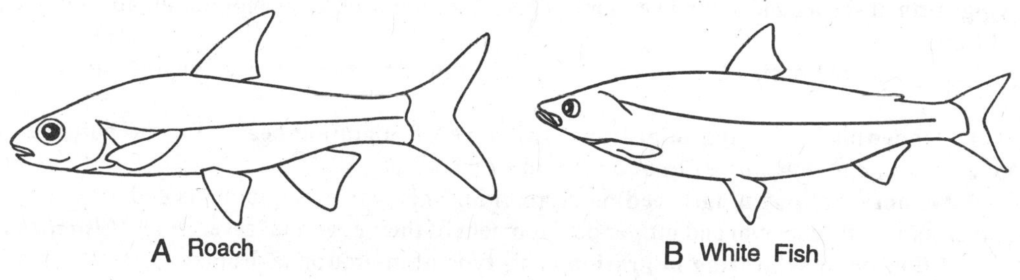

Fig. 14.4. Semi-migratory fishes.

2. Potamodromous fishes. These fishes live and migrate only within fresh waters, *e.g.*, carps, trouts. Carps and trouts travel long distances in large rivers in search of suitable spawning grounds. These fishes return to feeding areas after spawning.

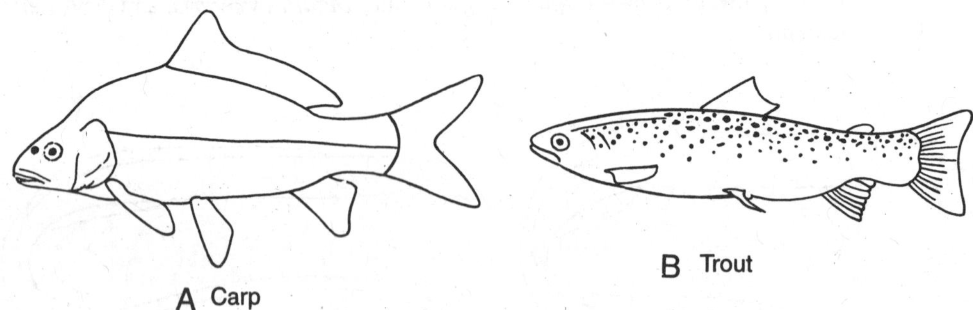

Fig. 14.5. Potamodromous fishes.

3. Oceanodromous fishes. These fishes live and migrate only within the sea, *e.g.*, herrings (*Clupea*), mackerel (*Scomber*), sardine (*Sardinella*) and tuna (*Thunnus*). These fishes travel long distances in the sea from deeper ocean waters to the shallow in shore spawning areas at the time of breeding. After spawning they return to their feeding grounds.

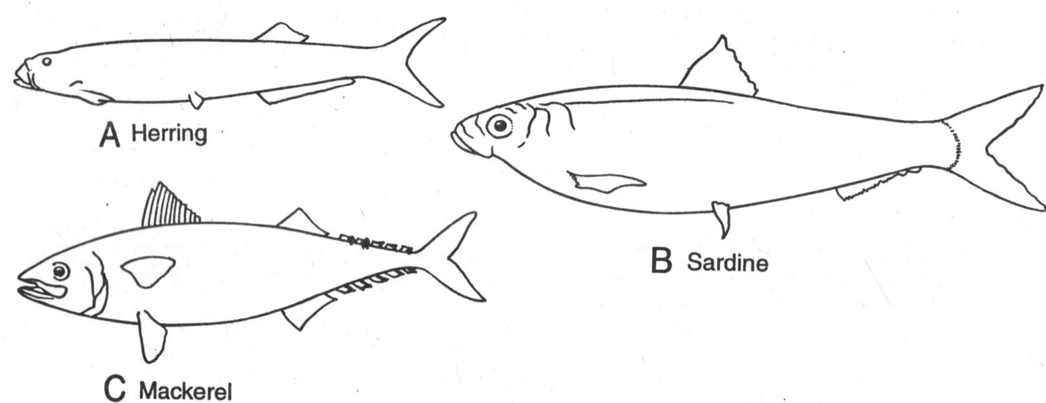

Fig. 14.6. Oceanodromous fishes.

Many marine fishes also perform seasonal migration. For example, swordfish (*Xiphias gladus*) and barracudas (*Sphyraena*), living in warm tropical seas perform latitudinal migration, moving north in spring and south in autumn. This migration occurs only for the suitable climatic condition.

Further, on the basis of the purpose, migration in fish has been classified into following types:

1. Alimental or feeding migration. This type of migration occurs for the purpose of search of food, *e.g.*, *Chanos, Harpoaon* (Bombay duck), etc.

2. Gametic or spawning/breeding migration. This type of migration is undertaken by a fish to ensure better survival and proper development of their eggs and larvae, *e.g.*, *Hilsa ilisha.*

3. Climatic or wintering migration. This type of migration takes place to secure more suitable climatic condition, *e.g.*, sturgeons, Atlantic salmon, etc.

4. Osmoregulatory or protective migration. This type of migration takes place for osmoregulation (*e.g.*, water and minerals).

II. Methods of Migratory Movements

1. Denatant and cotranatant. Migratory movements are of two types: (*i*) along with the water currents, called **denatant** (e.g., movement of pelagic eggs and young salmons) or against the water current, called **contranatant** (*e.g.*, migration of adult salmons towards spawning ground).

2. Drifting. By this method fishes are carried passively along with the water current. The young fishes migrate to nursery grounds by drifting.

3. Swimming. This is of following two types:

(*i*) Random swimming. It leads to uniform distribution or to an aggregation.

(*ii*) Oriented swimming. The oriented swimming takes place either towards or away from the source of stimulation or at some angle to an imaginary line running between them and the source of stimulation.

III. Speed of Fish During Migration

Speed is an important factor in migrating fish. Speed at which different species of fish move during migration is influenced by a number of physiological and environmental factors:

1. Maximum speed. The maximum speed at which a fish can move is ten times a body length per second (body length × 10/sec). This speed can not be maintained by the fish for more than one minute; they have to slow down to regain stamina and may again attain the maximum speed.

2. Maximum sustainable speed. The speed can be sustained by the fish over long periods of time. It is three times a body length per second (*i.e.*, body length × 3/sec.), *e.g.*, 1. In herring of 25 cm size, the maximum sustainable speed will be 25 × 3 = 75 cm/sec.; 2. In a cod of 80 cm the maximum sustainable speed will be 80 × 3 = 240 cm/sec.

Spectacular migrants such as salmon, cod and eel breed in one area but grows up and feed in another area. Interestingly, the distance between the feeding and spawning grounds may be over 700 miles.

IV. Examples of Fish Migration

1. Migration of salmon. In fishes, migration of salmon is a well-known and classical example. Salmon migrate in schools and are **anadromous**, *e.g.*, they spend their adult lives at sea but return to fresh water to spawn (Box 14.2). There are nine species of salmon belonging to two genera, Atlantic salmon, *Salmo* (*S. salar, S. gairdneiri* and *S. trutta*) and Pacific salmon, *Oncorhyncus: O. tschawytscha* (Chinook or king salmon), *O. nerka* (Sockeye salmon), *O. kisutch* (Coho salmon), *O. keta* (Chum salmon), *O. gorbuscha* (Pink salmon), *O. masu* (Japanese masu salmon). All these nine species of salmons perform migration but there are important differences among these species. The Atlantic salmon make upstream spawning runs (breeding migrations) year after year. The Pacific salmon make a single spawning run, after which they die.

Box 14.2.

A stream in which a salmon was born has proved itself suitable for reproduction. A mature salmon will therefore endure a long and difficult journey from oceanic feeding grounds to freshwater spawning grounds, risking death from exhaustation or predation, in order to have offspring in or near its birth place (**Alcock, 1998**).

The **Atlantic salmon** spawn mainly during November and December. They ascend the rivers as the breeding period approaches, travelling several thousand kilometers in sea and then inland. On entering freshwater this *contranatant* migrant stops feeding and loses weight. As a result, the energy requirement for up river movement is provided by accumulated nutrients chiefly in the form of fat deposits in body. Their bright silvery colour changes to dull reddish-brown shade. The body of the male salmon becomes spotted with red, orange and large black

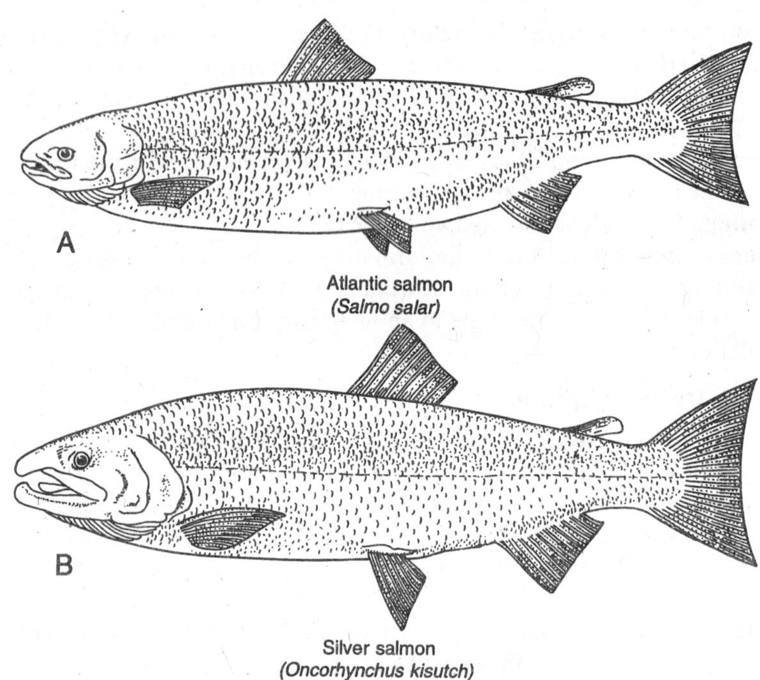

Fig. 14.7. Two common species of salmon. A—Atlantic salmon (*Salmo salar*); B—Pacific silver salmon (*Onycorhynchus kisutch*).

spots. Such male breeding salmons are known as **"red fish"** and the darker ripe females as **"black fish"**. After selecting suitable spawning grounds, the fish segregate into pairs. Shallow saucer-like depressions are prepared in the river bed by the females where spawning takes place (Box 14.3).

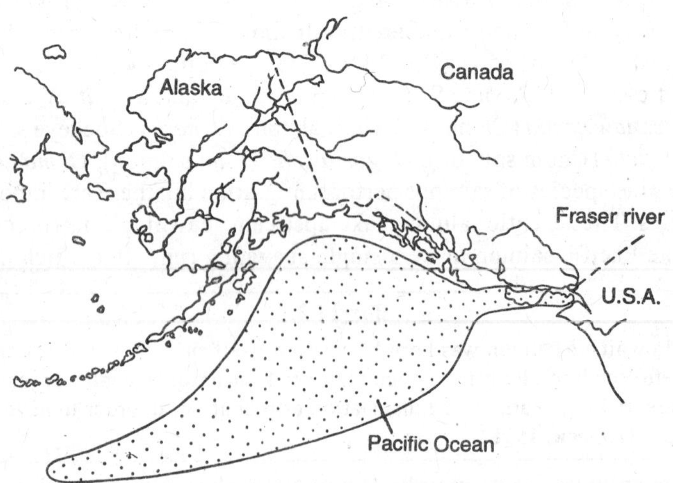

Fig. 14.8. Salmon that are born in British Columbia's Fraser river move far out in the northeastern Pacific ocean (shaded area) before returning to breed in the river where they began life.

Box 14.3.

The fertilized eggs of salmon hatch into small salmons towards spring. A large amount of yolk is retained by the young salmon on emerging from the eggs; this stage is called the **'alevir stage'**. After further growth, the yolk is reabsorbed and a size of 3-5 cm is attained. This larva is now called a **"fry"** which grows further to form a **"fingerling"** and remains in the river for an year or more. During this period the young fish continues to feed and grow and transforms into the **"smolt"** (stage after attaining a certain amount of maturity). Smolts stop feeding and then starts their journey towards the sea. Some species may migrate after hatching (*e.g.*, pink salmon *O. gorbuscha*) while others take as long as 7 yrs (*Salmo salar*) before starting migration.

After spawning, the return journey to the sea begins, but few males survive to breed a second time. However, many females are able to reach the sea and start feeding. They soon recover their normal condition and silvery colour. The salmon does not usually spawn more than three times in its life span of eight or nine years.

The **king salmon** of the Pacific coast of North America travels about 3600 kilometers from the sea in order to reach the rivers to spawn. However, all of them die after spawning. No male or female succeeds in returning to the sea, as they are already too much exhausted to undertake a return journey.

Arthur Hasler (1960) has studied the migration of **sockey salmon** (*O. nerka*). This salmon hatch in streams of the United States and Canada. After the smolt stage of development, they swim down-stream to the Pacific ocean. After two or three years in sea, they become sexually mature and migrate back to the exact stream where they were spawned. The return journey to coast is probably accomplished by means of celestial cues (*i.e.*, stars or azimuth position of sun) and internal biological clocks. Once at the coast, however, they have to select the correct river and the correct tributary stream within the river system. After many years of research **Hasler** and coworkers discovered that during the smolting period the salmon **imprint** upon its olfactory system the characteristic odour of their native streams (*i.e.*, the chemicals released by the vegetation and soil in the watershed of the parent stream). So during their return journey salmon are able to discriminate between the water coming from their native stream and that from other tributaries (Box 14.4). They essentially are recognising landmarks, but because the scent of birth place flows all along the migratory route, it is not necessary for the young salmon to remember landmarks all along the outward journey.

Box 14.4.

1. Salmon's precision in homing. One of the most elaborate studies on homing behaviour of salmon has been made by **Andrew L. Pritchard, Wilbert A. Clemens** and **Russell E. Foerster** in Canada. They marked 469,326 young sockeye salmon born in a tributary of the Fraser river and they recovered nearly 11,000 of these in the same parent stream after the fishes' migration to the ocean and back. What is more, not one of the marked fish was ever found to have strayed to another stream.

2. Salmon have very sensitive chemical sense. Studies have shown that young salmon hatched in freshwater could not return to sea until their salt secreting cells had developed. Some workers believe that fish have an extremely sensitive chemical sense that enables them to perform this feat (homing). Another theory links the accelerated metabolism at spawning time with the need for oxygen which increases as they ascend the hard waters of a stream.

2. Migration of *Hilsa*. Amongst Indian fish, *Hilsa ilisha,* which is found in Bay of Bengal, provides an explicit example of anadromous migration. During breeding season for spawning it ascends the Ganga river and is seen as far off as Allahabad.

3. Migration of eel. The two species of Atlantic eels, *i.e.*, European eel (*Anguilla anguilla*) and American eel (*Anguilla rostrata*) show catadromous migration. The spawning ground and spawning time of species was discovered by **Schmidt** (1967). It was found that these fish spawn near the "Sargasso sea" in the Atlantic Ocean, south-east of Bermuda. The European eel is found toward the eastern parts.

A. Migration of European eel (*Anguilla anguilla*). The European eels are found towards the eastern region of the Atlantic Ocean and in inland waters of countries near the shores of Europe, some of which are Murman Coast of North Africa, Iceland, the Azores, Madeira and Canary Islands, the Baltic and Mediterranean countries, the Black sea, sea of Azor and the Red sea. Spawning of European eel occurs during spring seasons and eggs are laid at depth of 500-700 meters with the temperature ranging between 10-12°C. On hatching, the larvae ascend to a depth of about 200 metres, where the temperature is warmer (*i.e.,* around 20°C). These larvae retain a yolk sac and are called **protocephaline larvae.**

Protocephaline larva transforms to **leptocephalus larva** after reabsorption of yolk content. Leptocephali are confined to a depth of about 100 metres and have a flattened leaf-like, glassy-transparent body measuring about 5 cm. The skeleton remains unossified at this stage, gut has a straight tubular structure, teeth are long and pointed and eyes are large and silvery. The larvae swim by eel-like undulatory movements and is reported to show diurnal vertical migratory movements, coming towards the surface only at night.

Leptocephali migrate from the Sargasso sea to the 'nursery grounds' of coastal waters by passively drifting with the warm water currents. They reach middle of the Atlantic Ocean by their second summer and finally touch the coast of Europe by the third summer. During this period the larvae grow to attain a large size. Transparency, serpentine movements and large size help the larvae defence against predators on their long migratory route. During autumn and winter of the third year, leptocephali metamorphose to form **"mini eels"** or **'glass eels'** called **elevers.** During metamorphosis the larvae stop feeding, their body becomes cylindrical with a decrease in length and the needle-like teeth are replaced by new ones. Elevers are initially translucent but gradually develop pigmentation on the body. Elevers may remain in coastal waters for 1-2 years, blocking the mouth of rivers till they are strong enough to swim up the rivers. They now measure about 15-20 cm long and are as thick as a pencil. The male fish prefer brackish water and stay in the estuarine region where as females prefer fresh water and are seen to ascend up river. In the stronger fish, larval teeth are lost and intestine shortens and the anal aperture moves forward. Pigmentation gradually increases, transforming the eels into yellow, cylindrical fishes, called **yellow eels.** These fish spend 8-10 years on feeding and growing; they become sexually mature **silver eels** before starting the spawning migration to the sea (*i.e.,* to their breeding grounds in the Western Atlantic South of Bermuda). During maturation of eels, the following prominent nuptial changes occur: a change in colour from yellow to silvery, enlargement of eyes, development of pointed pectoral fins, shrinkage of alimentary canal and enlargement and maturation of reproductive organs. These changes trigger a movement to over 5000 to 6000 kilometers in search of more favourable conditions of salinity, depth and warmer temperature required by the gametes.

B. Migration in American eel. Interestingly, like the European eel the American eel (*A. rostrata*) is also catadromous and spawns in the 'Sargasso sea'. It is similar to the European eel in many biological aspects. The larvae called **leptocephali** drift to the American coast by water currents. In about two years they reach coastal waters (the nursery grounds) to change into **elevers** or **mini eels** before ascending the rivers. On reaching the parental habitat they feed and grow till sexual maturity is attained.

V. Factors for Fish Migration

Fish migrations are influenced by a number of physical, chemical or biological factors:

1. Physical factors. They include quality of water, bottom materials, water depths, pressure, temperature, light intensity and photoperiod, currents, tides and turbidity.

2. Chemical factors. These include salinity, pH, dissolved oxygen, types of dissolved organic substance, smell and taste of water.

3. Biological factors. These include sexual maturity, blood pressure, food, memory, effect of endocrine secretions and the related physiological changes in the body systems and behaviour (social response, biological clock). Presence or absence of predators and active competitors may also be taken as the biological factors.

VI. Theories/Hypotheses Regarding Fish Migration

Several laboratory and field studies have been conducted to solve the mystery by which a fish can select the course and subsequently maintain its direction during long migratory movements. Following theories/hypotheses have been forwarded to explain the homing/ navigation during fish migration.

1. Olfactory hypothesis. According to this hypothesis, a fish may recognize its surroundings and path of migration by detecting the specific odour from a stream or water body. This odour may depend on the bottom topography, *i.e.,* soil, plants and springs at the bottom from which the water body acquires its organic quality. Scientists have also termed this as the **stream factor.**

A.D. Hasler (1983) has supported olfactory hypothesis with **imprinting** and regards it as the sensory basis for homing. He has suggested that the salmon may learn to recognize the organic quality of water in the early days of its life cycle and this may become imprinted for the rest of its life. Thus, a fish smells its way home as would a fox hound. The paleo-cortex is, therefore, a dominant part of fish brain as it receives impulses from the olfactory tissue. This region is much less significant in non-migratory fishes as well as in humans.

2. Internal biological clocks. Experimental evidences have suggested that the fish (salmon) during its migration is able to adjust its course or orientation through internal biological clocks. These adjustments to seasons (through latitudes) as well as to the time of the day (through longitudes) are according to appearance of light.

3. Sun-compass mechanism. Evidently open-water migrations in salmons may be directed by a sun-compass mechanism. These may be directed either by the altitude of the sun or by changing diurnal azimuth of the sun or by the combination of the two. To swim in a single direction, the fish probably calculates appropriate horizontal angles to the sun during different times of the day.

4. Ground water seepage hypothesis. According to this theory which was proposed by **Harden Jones** (1980), the assembly area for spawning are identified by marine fish with the help of the chemical that enter the sea by ground water seepage.

5. Pheromones. Pheromones are suggested to play an important role in identifying the spawning area. They may prevent further spawning in a particular area after a sufficient number of eggs have already been laid.

6. Celestial reference points. The use of the celestial points in maintaining a steady course in mid waters was suggested to be useful, especially in nocturnal travel. Nocturnal direction may also be assisted by gravity, magnetic field, oceanic currents, etc., especially during moonless or starless nights.

7. Tidal transport. This mode of migration is performed by Plaice of South Bight of North sea and saves energy and problems of direction are eliminated if the right tide is caught at the right time.

8. Gradient theory. This is based on the salmon's presumed ability to follow gradients of physico-chemical characteristics of water, mainly temperature and carbon dioxide. It has been shown that salmons swim in the direction of coolest waters and that high CO_2 tension repels migrating salmons.

14.3. BIRD MIGRATION

The bird migration is a seasonal movement from one habitat to another and back again, to get the advantage of favourable conditions. It is traditional, hereditary (instinctive) and of regular occurrence at definite intervals every year. Usually, the birds migrate to the northern hemisphere in spring to breed and return to southern hemisphere in autumn to pass the winter (Box 14.5).

Fig. 14.9. Some migratory birds.

Box 14.5.

Birds, with their marvellous ability to fly and with their unparalleled mobility, have explored almost the entire globe searching for food, territory, nesting sites and other necessities for survival. Because of their high metabolism and very high energy consumption, birds must have abundant and unfailing sources of food. In geographical regions with seasonal climatic changes, ecological changes occur which may require the birds to move away if they have to survive. In colder countries the intense chill of winter closes lakes and streams and snow fall covers the land and its vegetation. This compels the birds to move to warmer regions. In countries around equator, summer or dry season may reduce the plants, insects, worms and food sources; the temperature gets very high and to escape that the birds have to emigrate to some more favourable place.

Thus, birds migrate in order to secure optimum conditions for breeding and feeding, *i.e.*, to utilise the trophic niches of both the hemispheres and to utilise the old and protected nesting places year after year.

I. Migratory Status

A migrating bird is called a **migrant.** Depending on the seasons of migration of different birds, different names are given to the migrants. These are as follows:

Fig. 14.10. Permanent resident birds at Keoladeo National Park, Bharatpur.

1. Winter visitors (N $\rightleftharpoons$ S). These include those species which move from their breeding grounds to spend their winter in a more suitable place, where food is also in plenty. For example, fieldfare, snow-bunting and redwing arrive in autumn chiefly from the north, stay throughout winter and fly northward again in spring. Greylag goose, pintail, common teal, gadwal, shoveller, common pochard and European starling are the winter visitors (migrants) visiting Keoladeo National Park, Bharatpur. The wagtails (*Motacilla* spp.) appear in India in the beginning of winter season and at its end they leave this place, migrate to Siberia to spend summer and to lay eggs.

2. Summer visitors (S $\rightleftharpoons$ N). From spring onwards, south starts getting too hot, while luxurious summer conditions prevail in the north. The birds of south leave it in spring for north to spend the summer, breed and return to south again in autumn, when the north gets cold, *e.g.*, swifts, swallows, nightingales, cuckoos, etc. The cuckoo (*Eudynamis scolopaccus*) arrives in Uttar Pradesh in the beginning of March and breeds there. During the month of August it leaves this place and migrates to South India and Ceylon to spend winter there.

3. Transient visitors or passage migrants. These are summer and winter visitors while migrating from south to north and vice versa, stop at some places for the sake of rest only. For example, gargany teals migrate from Mangolia or Siberia to Bharatpur to go South India which is their actual breeding ground. While returning they again stop over at Bharatpur.

4. Permanent residents. These birds are found in a particular area throughout the year and do not migrate from one place to another. For example, cotton teal, spotbill duck, whistling teal, barheaded goose, mallards, comb duck (Fig. 14.10) are all permanent residents of Keoladeo National Park, Bharatpur.

II. Types of Migration

1. Daily migration. Many birds make daily movements from their nest in response to environmental forces such as light, darkness, temperature, humidity and food availability. Birds may make daily migrations from their resting sites to feeding areas. Crows, house sparrows, starlings, rockery herons, etc., show this type of migration.

2. Local migration. Because of heavy rains, floods, excessive heat and cold, birds leave an area for some time and return when the crisis is over. Flowering of certain plants and ripening of seeds also stimulate some birds to migrate locally.

3. Seasonal migration. Migration of birds in response to changes in the season is called **seasonal migration.** In tropical and subtropical areas, this occurs in the beginning or end of warm season. In temperate areas, the movements are triggered by onset of winters. The birds which rest in temperate zones, go to warmer areas. Such migrations of birds are known in northern hemisphere, Europe, North America and Asia.

4. Moult migration. In most ducks, males and juvenile birds migrate short distances northward for *moulting* leaving behind females and young birds in the breeding ground.

5. Cyclic migration. Some migrations of birds are seasonal but do not occur at regular intervals. The cyclic migrations of the snowy owl in search of leming (chief food of snowy owl) in United States in winter occurs in three to five years.

6. Partial migration. In certain cases all the birds of a group of migratory birds do not leave the native land and hence are always represented by certain individuals. But these individuals are not always the same, *e.g.*, songthrus, redbreast, titmouse, finch, etc.

7. Vagrant or irregular migration. Some of the birds (such as herons) sometimes disperse for a short or long distance for the sake of food and safety. The birds can also be swept away by powerful winds and hurricanes (oceanic storms) to very long distances.

8. Altitudinal or vertical migration. The birds living at high altitudes (hills) descend at lower altitudes in winter to save themselves from the intense cold of high altitudes. They return again to high altitudes with the advent of summer. Coots of Andes mountain in Argentina move downwards; mountain quails which nest at elevation upto 9500 feet in central California mountains, migrate downward to area below 5000 feet in the fall and in the summers they go back to the hills. Violet green swallows of U.K. and willows in Siberia are also vertical migrants. The Himalayan white caped redstarts (*Chimarrhornis*) resides at a height of 8000 to 14000 feet. During winter it descends to the height of 2000 to 8000 feet.

In certain cases, such as **blue grouse** of USA, the direction of migration is reveresed with the season. It passes winter high up in the rocky mountains probably to avoid competition and predators.

9. Latitudinal or equatorial migration. The most familiar migrations are those from north to south and vice versa. A large number of Eurasian and North American birds cross the equator to spend the winter in Africa or South America. The golden plover of the Arctic tundra passes winter 12,800 km down south in the pampas of Argentina. Cuckoo and storks pass summer in northern hemisphere and winter in south.

10. Longitudinal migration. Movement of birds from east to west or vice-versa is known as longitudinal migration, *e.g.*, evening grossbeaks (finches-like birds) that nest in northern Michigan spend the winter in New England; California gulls that breed in Utah migrate westward to winter on the Pacific Coast, starlings move from east Europe or Asia to the Atlantic Coast.

Diurnal and Nocturnal Migration

On the basis of their wing powers and method of getting food, **William Brewster** divided the birds into following three categories:

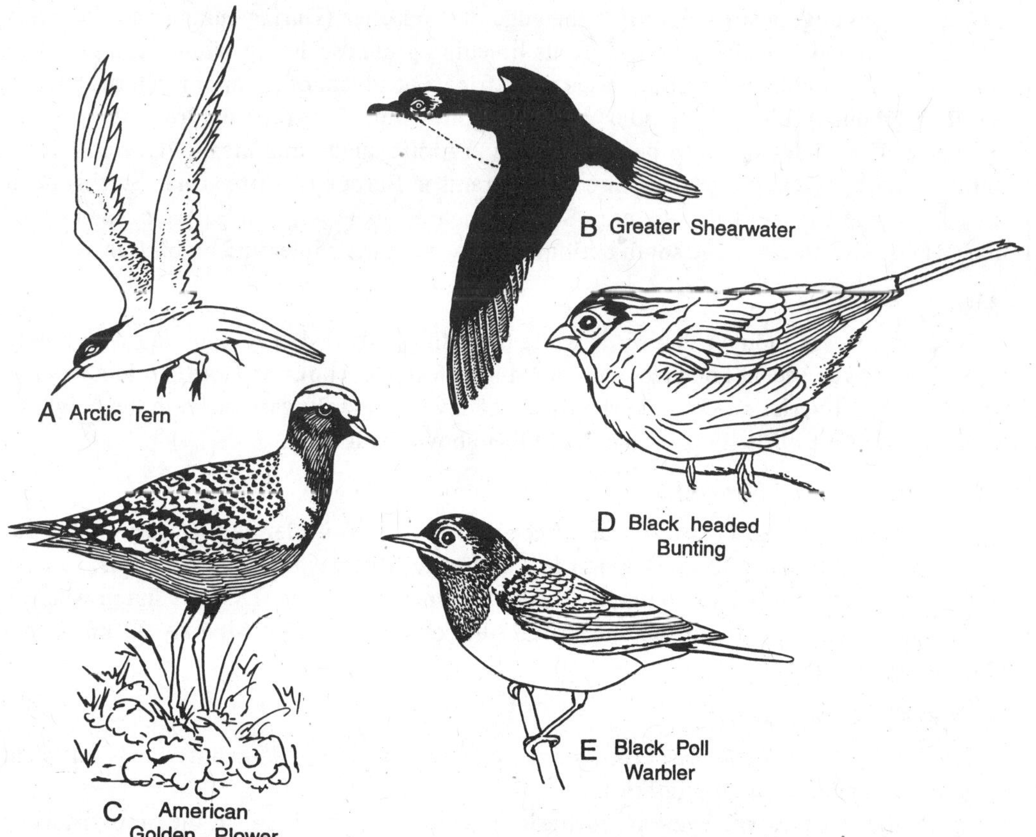

Fig. 14.11. Long distance migrant birds.

1. Night fliers. Small birds such as passerine birds (sparrows, titmice, jays, crows, etc.), cuckoos and wood-peckers.

2. Day fliers. Large birds such as **hawks, pigeons, swallows** and **robbins,** etc.

3. Few birds such as **geese** and **ducks,** migrate both by night and day.

The day fliers are usually restless and strong winged. These can feed while on wings, have least danger of enemies and can migrate to long distances. **Turkey vultures** (or Turkey buzzard—*Cathartes aura*) cover a distance of 3600 miles without a stop. Broad winged birds such as **eagle, storks** and **cranes** use thermal currents for soaring by which they spend less energy for flight.

Night fliers are usually small birds. Darkness provides them protection from large predatory birds. It also gives birds opportunity of using all the daylight hours for feeding thereby enabling them to build up sufficient energy resources for sustaining long distance flights.

Further, bird migration does not always take place by means of flight. **Penguins** (*Spheniscus*) breeds on the land of Antarctica continent. In the winter they move hundreds of kilometers by floating on packs of ice and return south in summer. They cover a greater path of the distance by *swimming* in the sea and *march* to their nesting sites. Migratory **guillemots** (*Uria*) mostly *swim* and **American coot** (*Fulica*) *walks* for kilometers across the country.

III. Range of Migration

The distance travelled by migratory birds depends upon the local conditions and the species concerned. The longest distance of about 17600 kms is covered by the **arctic tern** (*Sterna paradisea*) which migrates from north to the edge of Antarctica (During summer arctic tern is found in Canada and Greenland which are its breeding grounds; during winter it migrates to Atlantic Islands in South via South America and Africa. On advent of summer it returns back to its breeding grounds). Likewise, **golder plover** migrates from the Arctic tundra to the pampas of Argentina. The **pectoral sandpiper** that breeds in Arctic tundra migrates to its winter home in South America. The most spectacular of all migrants in Europe is **white stork.** Storks spend summer in Europe but spend the winter in South Africa. Birds of north India and of the base of Himalayas visit the places in the south periodically to avoid the winter and go back in summer.

IV. Altitude of Migration

Formerly, it was thought that birds while migrating travelled very high and with a very fast speed. However, recent knowledge obtained by **telescope, radar** and **radio telemetry** has pointed out a great variation. Some birds fly at sea level, some fly very close to the height of Mount Everest. Most birds fly less than 7,400 feet above sea level.

V. Flying Speed during Migration

Flying speeds of migratory birds have been measured by **Doppler radar.** It was found out that the speed ranged from 32 to 64 kmph in small **song birds,** while in larger birds, such as **cranes** speed varies from 40 to 96 kmph. Smaller perching birds can fly 32-59 kmph whereas **falcons, ducks** and **geese** fly at the speed of 77-96 kmph, **humming bird** fly 32 kmph and **sandpipers** fly 96 kmph.

VI. Origin of Migration

Several hypotheses have been proposed to explain origin of migration but none is sufficient to explain the phenomenon of migration:

1. Northern ancestral home hypothesis. According to this view, in the prepleistocene period birds inhabited northern hemisphere and were non-migratory. In the pleistocene, when the glacier forced its way from north to south, the birds migrated to south to save themselves from intense cold. In the following summer when the ice melted, birds returned to their nests in north. Ultimately this practice became hereditary in Lamarckian fashion.

2. Southern ancestral home hypothesis. According to this hypothesis, the bird population in the southern tropics became congested in the past, so that some of the birds migrated to the north for breeding purposes and returned to the south as soon as breeding was over. By moving northward they took advantage of longer days (16 to 18 hrs) which was necessary for gathering food, etc., during the breeding period.

3. Hypothesis of periodic response. It is understood that the tropical birds inhabited colder northern parts in the past where plenty of food was present. But these had to fly to south when the cold increased too much in winter and food became scarce due to hibernation of insects. **Woodburry** explained that by this migration, since the birds survived from the intense cold, it became the habit of the birds and ultimately in course of time, the habit was **naturally selected.**

VII. Causes of Migration

1. Shortage of food supply on the breeding ground. With the end of summer in the northern hemisphere, where the birds breed, food supply falls short due to increase in the number of birds (as a result of breeding) and secondly due to the beginning of hibernation of insects. It causes birds to migrate south.

2. Environmental factors. According to some workers fall and rise in the temperature are responsible for migration. These environmental conditions change the endocrine state and metabolic conditions of the birds in such a way as to cause migration.

3. Internal factors. Some birds migrate due to physiological changes in the gonads and sex hormones which occur with change in the season.

4. Photoperiodism. The effect of length of day or photoperiodism triggers bird migration. The day length affects pituitary and pineal body in the brain which in turn prepare birds for migration. The migratory restlessness is initiated from this stage. For example, in India, increase in day length (starting from February/March onward *induces* the Siberian birds to migrate to their breeding habitats in Siberia. On the contrary, decrease in photoperiod (starting from September-October onward) induces these Siberian birds to migrate to Indian Subcontinent.

5. Fat deposition. According to **Wolfson,** the substantial subcutaneous and visceral fat deposits play an important role in migration of the species. This fat deposition is followed by the development of migratory restlessness (zugunsuhe). Fat provides metabolic water as well as energy in route.

VIII. Navigation

Way-finding or navigation in migratory birds is done with the help of **landmarks** such as great river valleys, coastal lines, chains of oceanic islands and mountain ranges. However, a large number of bird migrants travel at night when they cannot easily make use of land marks. There it has been suggested that nocturnal (*i.e.,* night fliers) migrants take the help of **constellations of stars.** It is proposed that birds are guided by "tradition" based on experience. [During migration birds seek guidance for one year from those birds who have followed well in the previous years.]

A German ornithologist, **Gustav Kramer,** has shown that migratory birds which travel by day orient themselves by the position of sun (**Sun compass mechanism**). However, the sun compass orientation is an **instinctive behaviour** since young birds that have never migrated before make similar navigational orientations to sun when travelling independent of their parents.

IX. Advantages of Migration

1. Securing a **better climate** for living by avoiding unfavourable climatic conditions (such as intense cold, snowfall, hot summers, stormy conditions, etc.) and food shortage by migrating.
2. By alternately exploiting two different habitats for food due to migration, more birds are able to exist.
3. Geographical wandering or change in habitat provides greater **variety in birds diets.**
4. The long summer days provide birds with **long working hours** to gather food to feed young ones.
5. In their breeding grounds the migratory birds arrive in large numbers hence the **pressure of predation** is divided among a greater number of eggs and young and this results in greater individual survival.
6. Migration provides certain evolutionary benefits. A long distance migrant, is subjected to different kinds of natural selection pressures in its breeding grounds, its winter quarters and while on journey between two habitats. Indeed migratory birds have a greater range of adaptability than the resident birds.
7. Migration promotes the geographical dispersal of birds. Dispersal may isolate small populations of a species from each other. It tends to increase the rate of evolution (**Isolating mechanisms**).

X. Disadvantages of Migration

Migration is disadvantageous to the birds in the following ways:

1. A journey without rest is tiresome for the birds and most of the passive bird succumb at sea.
2. Sudden changes in weather, *i.e.*, pouring of heavy rain or snow or blowing of stormy wind sometimes perish a number of migrants.
3. The young and defenseless birds are exposed to various natural enemies (predatory birds and hunters).
4. The telegraphic wires, towers, lighthouses, etc., sometimes take a great toll of the migratory birds.

14.4. ORIENTATION

The simplest response to a stimulus is either **kinesis** (an increase in the activity of the individual) or **immobilisation** (an inhibition of the activity of the individual). **Orientation** refers to the way an organism positions itself in relation to external cues. One of the early investigator, of animal orientation, **Jacques Loeb** (1910), theorized that asymmetrical stimulation of an animal's sensory organs results in differential contraction of muscles on the opposite side of the animal until the symmetry is restored for both sense organs and muscle action. Other investigators concluded that not all animal orientation could be fitted to Loeb's sensory organ-muscle scheme, and they found that different organisms possessed different systems of orientation (**Mast, 1938**).

Later, **Fraenkel** and **Gunn** (1940) summarized the work of orientation up to that date and defined general classes of orienting reaction. **Kineses** are random movement patterns in response to stimuli in which there is no orientation of the organism's body to the source of stimulation.

The rate of movement increases with the intensity of the stimulus. **Taxes** (singular **taxis**) are directed reactions involving (in a single-stimulus situation) an orientation of the long axis of the body in line with the stimulus source. Movements toward the stimulus are **positive taxes;** movements away from the source of stimulation are **negative taxes.** For instance, planaria (*Planaria* spp.) exhibit a negative **phototaxis,** *i.e.*, movement away from a light source. Extensive studies have revealed other, often complex systems of larger-scale orientation and navigation in different organisms.

Spiders (*Araneus diadematus*) constructing webs present an interesting apportunity to investigate orientation in three-dimensional space (*e.g.*, vertical web of *A. diadematus*). Spiders that build **vertical webs** use at least three types of cues for orientation : light, gravity and the threads or lines of the web they are constructing or moving about (**Crawford,** 1984; **Eberhard,** 1988). The spiders use gravity to determine their direction within the vertical plane, light to discriminate between the sides of the web, and a spatial memory distances and directions on various lines involved in the web's construction (**Drickamer** *et al.*, 2002).

14.5. NAVIGATION

The most complex form of spatial orientation is navigation. Navigation requires not only a compass or directional sense, but also some kind of map. According to **Griffin** (1955), navigation involves following three types of orientations.

1. **Pilotage** or steering a course using familiar landmarks;
2. **Compass orientation,** the ability to head in a particular compass direction without reference of landmarks, and
3. **True navigation,** the ability to orient toward a goal such as home or breeding areas without the use of landmarks and regardless of its direction.

Thus, *true navigation* requires more than a compass, because the animal must not only determine the proper direction; it must also find its goal at the end of journey. The ways in which animals navigate between two given areas (either by homing back to the home area or migrating to a new one), sometimes thousands of kilometers apart, has long intrigued biologists. The term '*navigation*' implies two things: first a degree of familiarity with the destination or goal and secondly, a lack of familiarity with the geography of the area between the destination and the present position. **Navigation** *is thus a means of determining the direction of a familiar goal across an unfamiliar area.* As such it is a form of *goal orientation* (**Baker,** 1981) to be contrasted with **simple orientation** which is concerned solely with taking up a particular direction rather than seeking a specified destination.

Although long distance navigation is known in several taxonomic groups including insects, fish, reptiles and birds, it is with birds that most progress has been made.

Navigation Mechanisms

Animals are known to possess compass of various types, based upon features of the geophysical environment such as the magnetic field of the earth. To be sure that a particular physical feature is used as a compass, it is necessary to show that the animal can detect the phenomenon and that it can use it for orientation under natural conditions. Because of the uncontrolled variability of the outside world, testing for sensory capability is best done in the laboratory. A favorite method used by those interested in study of navigation is the **cardiac conditioning method** illustrated in Figure 14.12. By this and other methods researchers have shown that pigeons, other birds and animals are sensitive to the following stimuli: (*i*) sun compass cues; (*ii*) star compass cues; (*iii*) magnetic cues; (*iv*) gravity cues; (*v*) barometric pressure cues; (*vi*) polarised and ultraviolet light cues; (*vii*) odours.

1. Sun compass cues. One of the earliest discoveries about animal navigation was that the sun and stars (*i.e.*, celestial cues) provide important cues for orientation. At one time it was suggested that birds displaced from their home area might compare the sun's arc (*i.e.*, the apparent path traced by the sun across the sky) in the new area with that expected at home (*e.g.*, **sun arc hypothesis of Matthews,** 1953). However, now it seems clear that the *sun is used as a simple compass and that only its azimuth (i.e., direction from the observer) provides information for orientation.*

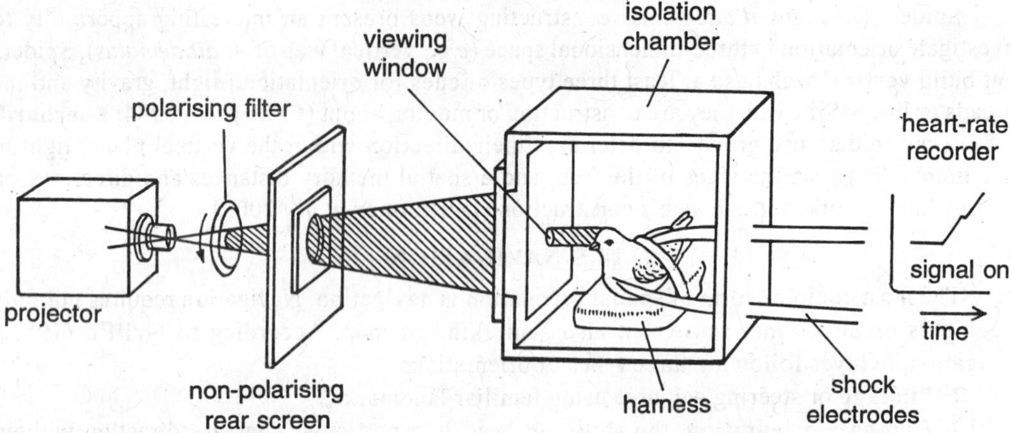

Fig. 14.12. Cardiac conditioning apparatus. It is used for testing the sensitivity of pigeons to various stimuli (polarised light in this example).

Gustav Kramer (1951) discovered the fact that pigeons can use the sun as a compass. Sun compass sense of birds were shown by Kramer's classic experiments in which the locomotory orientation of caged starlings could be altered predictably by using mirrors to '*move*' the sun. These caged starlings are deprived of all visual landmarks and only the sun and sky are visible

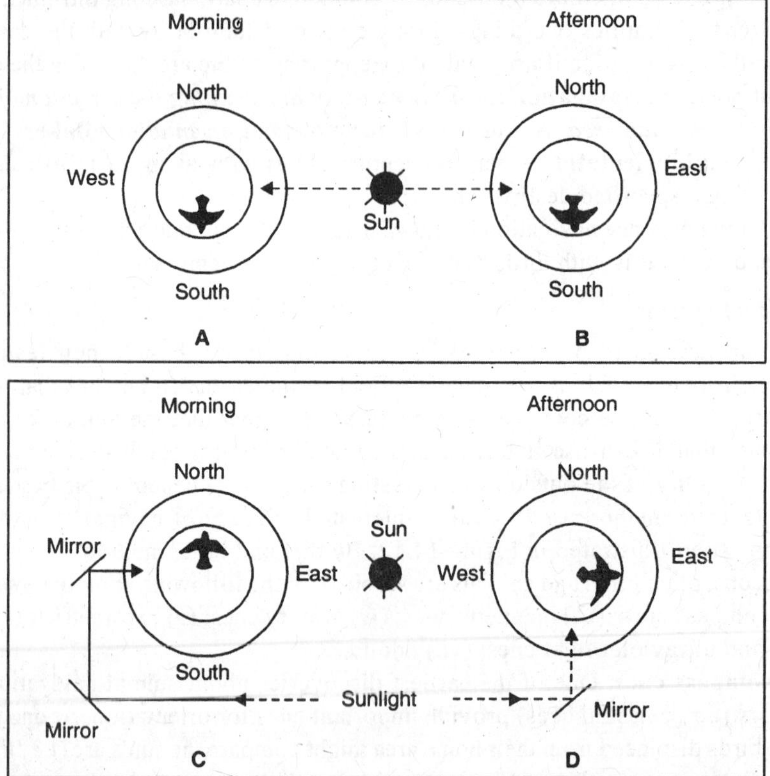

Fig. 14.13. Sun compass orientation in birds during migration. A—and B—show caged bird orienting south, keeping the sun on its left in the morning and on its right in the afternoon. C — and D — show how the effect of redirecting the sunlight by way of mirrors change the orientation, *i.e.*, to north in C and to east in D.

to them. They exhibit fluttering movements ('zugunruhe') and spend most of their time on the side of the cage oriented in the direction of the migratory pathways (Fig. 14.13). This orientation behaviour only persists when the sun is visible; if the sky is overcast or if such a circular cage is moved indoors, the bird shows no directional preference. It is possible to shift the apparent position of the sun from one side of the cage to the other by the use of mirrors. When this is done, the orientation of the bird shifts as the mirror is moved (Fig. 14.13).

Animals that use the sun navigation over long distances have to compensate for the apparent movement of the sun in the sky. During a day, the sun shifts its position from east to west, north of the equator. If a bird is headed south, the sun should be kept on its left side in the morning and on its right-side in the afternoon. This means that the migrating animals have to know what time of day it is in order to use the sun as a compass. Thus, to make use of sun compass in its orientation a bird has to contain a **clock sense** so that the bird knows if it is morning, noon or afternoon and thus where the sun is on its east to west progression through a day. A starling or a pigeon has a **biological clock** and its compass orientation can be disturbed if it is tricked into resetting the clock, as revealed by the following experiments.

1. The starling's biological clock can be reset experimentally (**Hoffman,** 1954) by keeping the bird in a lightproof room and exposing it to artificial photoperiod (Fig. 14.14).

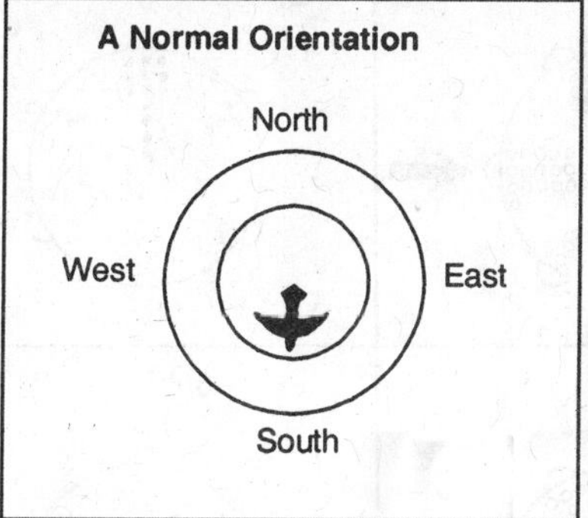

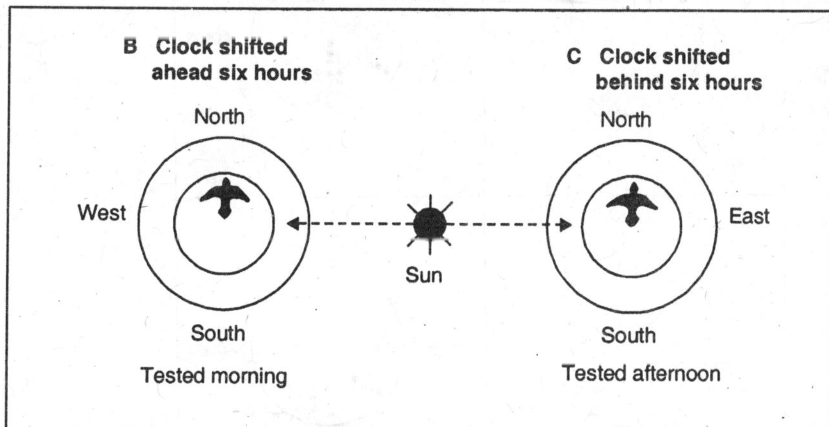

Fig. 14.14. Effect of photoperiod upon sun compass orientation. A — Normal orientation in both morning and afternoon; the bird is heading south. B — and C — The effect of artificially changing the light cycle under which the bird is kept before testing. Bird in (B) was kept in a light-day cycle running 6 hours ahead of the natural daylight schedule. When tested in the morning with sunlight, it reacted as if it was afternoon, keeping the sun on its right. C — Clock of bird shifted 6 hours behind the natural daylight schedule. When tested in afternoon, it acts as if it was morning, keeping the sun on the left.

2. Klaus Schmidt-Koening (1960) in his various experiments with homing pigeons, tested the effects of shifting the biological clock 6 hours forward, or 6 hours backward, or by 12 hours. Such a pigeon that has been trained on an artificial light/dark schedule, different from the outside, will navigate with predictable error beneath the natural sun: pigeons who think it is six hours later in the day than it is, will head off 90° anti-clockwise away from the correct direction (Fig. 14.15). Pigeons used to the northern hemisphere but moved to the southern hemisphere likewise orient with the predictable 180° error at noon.

The pigeon's sun compass has been shown to be sufficiently accurate for navigational purposes, provided the bird makes corrective measurements at intervals during their journey. However, the compass orientation of the sun, in relation to local time, is primarily of use in determining longitude. It is the sun's altitude (or azimuth) that alters with changes in the latitude of the observer. There is some evidence from operant conditioning experiments, that pigeons can make fairly accurate measurements of changes in azimuth. Pigeons are suspected to estimate the sun's altitude by measuring shadows rather than sun directly (**McDonald,** 1970).

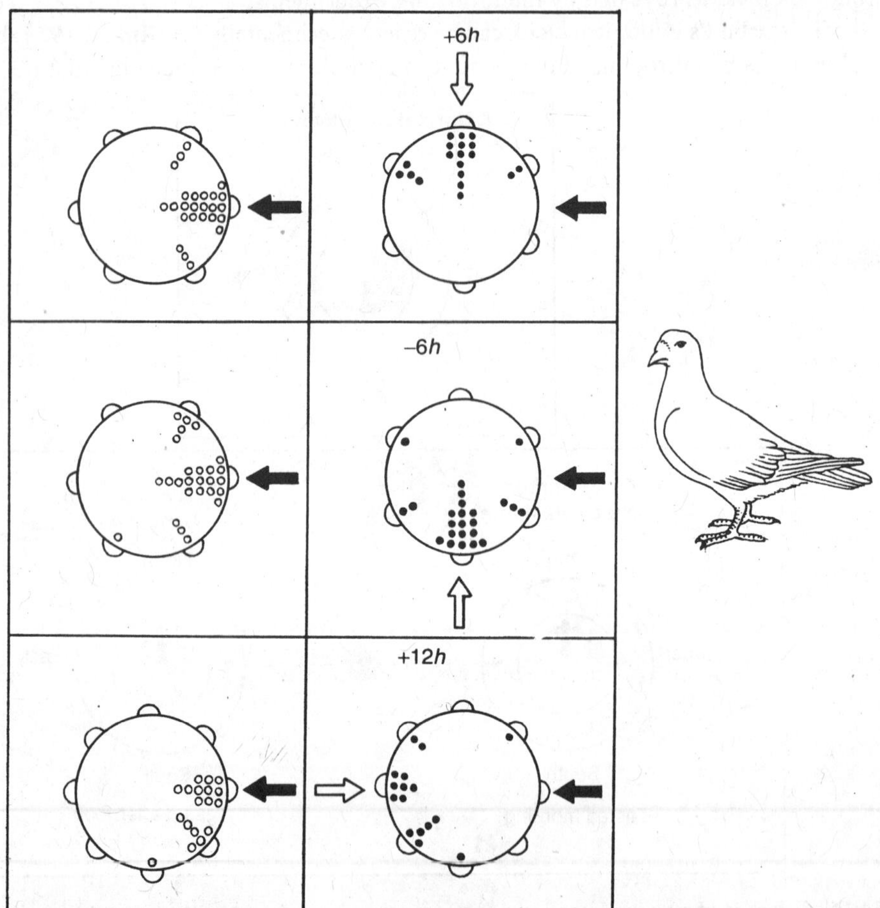

Fig. 14.15. Results of clock shift experiments with homing pigeons. The pigeons were trained in a circular cage to look for food and peck in a particular compass direction. They were tested by counting the pecks at food dishes around the periphery (each dot indicates one peck). Direction of pecking of controls are illustrated on the left. Directions of pecking by clock-shifted birds are shown on the right. The directions are those that would be expected if the pigeons used the sun as a compass, making allowance for the movements of the sun through the day.

2. Star compass cues. Like the sun, the stars also seem to provide only compass information. Unlike the sun compass, however, stellar navigation does not require time compensation. Stellar orientation has been shown in a number of animal species. For example, if songbirds are caged during the period when they normally would be migrating, they show a typical **migratory restlessness,** at night this directionality is related to the stars (**Sauer** and **Sauer** 1955). **Bellrose** (1968) released mallard ducks (*Anas platyrhynchos*) at night with small lights attached to their feet. Birds released under clear skies flew almost directly north. Those released in overcast conditions flew in all directions.

To a terrestrial observer, the stars appear as if on the inside of a spherical surface, called the **celestial sphere.** At any given time, the stars form a definite pattern on the celestial sphere, a pattern that moves so as to give impression that the celestial sphere is spinning. In reality, the earth is spinning about its polar axis. The point in the northern sky about which the celestial sphere seems to rotate is called the **north celestial pole.** A group of major stars such as **North Star** or star **Polaris** and **Big Dipper** are located very close to this pole and for practical purposes North Star can be used to determine the observer's northerly direction. The North Star provides a reliable fixed compass cue.

A classical study using indigo bunting (*Passerina cyanea*) in New York has supported the hypothesis that star patterns are used in navigation. This small bird moves between breeding sites in the eastern half of the United States to a wintering home in Central Mexico or the Caribbean islands in a series of night time flights (**Emlen,** 1975). In a series of elegant experiments using the apparatus illustrated in the Figure 14.16, in a planetarium, **Steve Emlen** (1960s to 1970s) showed that buntings used learned star patterns rather than the azimuth of individual

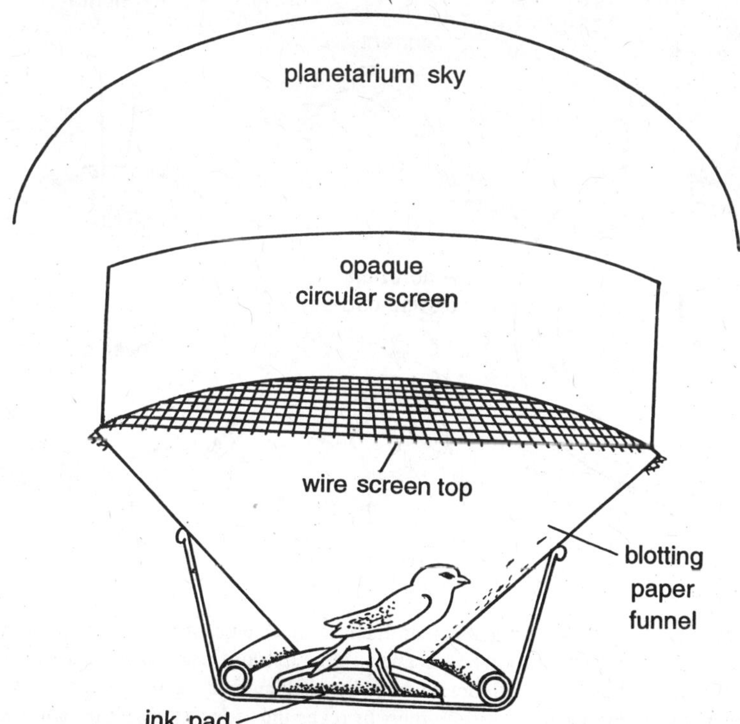

Fig. 14.16. Experimental cage for measuring migratory restlessness. When the bird (bunting) which stands on the ink pad attempts to leave the cage, it produces inky footprints on the blotting paper lining the funnel.

stars to gauge the axis of apparent rotation of the night sky. There is a critical period for **imprinting** just prior to the first autumnal migration, during which young buntings must learn the star patterns in order to be able to navigate later. However, while birds initially respond to the rotation of stellar constellation, they use it only to learn the star patterns which indicate the axis of rotation (*i.e.,* North Star and other fixed stars of north celestial pole). Thereafter, they rely on the learned pattern. The axis of rotation is merely a reference point against which star patterns are calibrated.

3. Magnetic cues. In 1882, **Vignier** suggested that the Earth's magnetic field could provide a navigational grid. Birds might be able to detect and measure the three essential components of the field: **intensity, inclination** (the angle made by the field with the horizontal) and **declination** (the angle between magnetic and true north). Since isoclines of intensity and inclination tend to cross at oblique angles, the magnetic field provides a poor navigational grid unless declination is also incorporated.

Miniature magnets have been found in bacteria, honeybees and pigeons. Honeybees can detect magnetic changes of less than 10^{-3} gauss. There is evidence that the internal clock of bees is influenced by magnetic phenomenon (**Gould,** 1980).

William Keeton (1971) showed that homing pigeons were also sensitive to magnetism. **Walcott** (1972) with pigeons and **Southern** (1972) with laughing gulls have shown that *altering the magnetic field about a experienced bird* (either by strapping a bar magnet to its back or by outfitting the subject with helmet consisting of a **Helmholtz coil** powered by a mercury battery on the birds back) will disorient the bird if it is released far from home but only on *overcast days* (Fig. 14.17). Otherwise, sun cues take precedence, a fact that originally led workers to

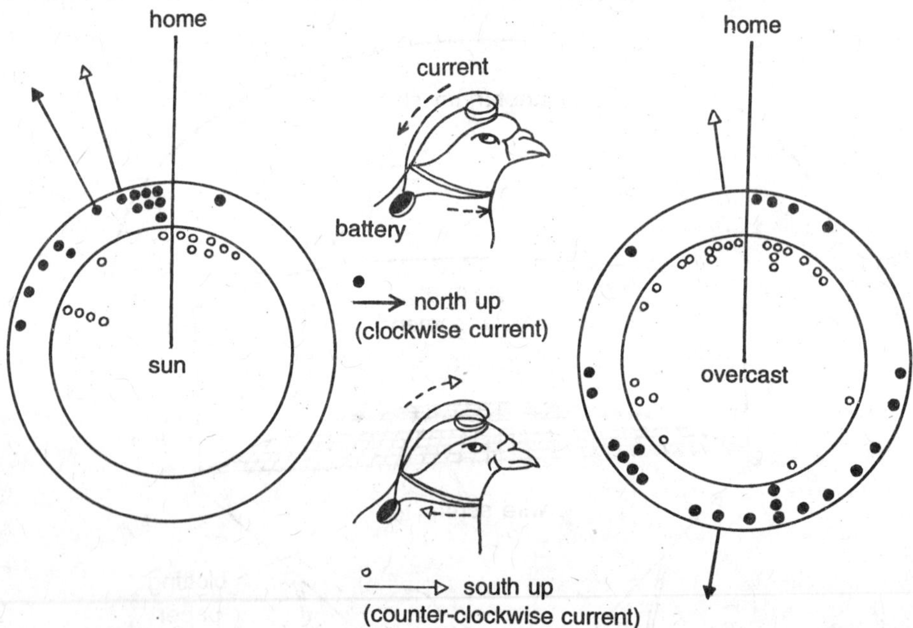

Fig. 14.17. The orientation of homing pigeons is influenced by changes in the magnetic field by Helmholtz coils. These coils above the pigeon's head and around its neck induce a relatively uniform magnetic field through its head. The coils are powered by small mercury battery on the birds back. Direction of the induced field can be reversed simply by reversing connections of the battery. The strength of the magnetic field can be varied by controlling the amount of current passing through the coils. The dots represent the angles at which pigeons flew off from the release site: the direction home is straight up; straight down represent 180° away from home. There are two magnetic conditions which do not affect orientation on sunny days (left) but do on overcast days (right).

conclude incorrectly that a bird could not detect magnetic field cues. It can, but *does not pay attention to them if the sun* is shining. But when the reliable information provided by the sun position is missing, the ability to detect the north-south running lines of magnetic forces may provide the pigeon with a compass.

4. Gravity cues. When animals show sensitivity to magnetism, they often respond to gravitational cues as well. **Larking** and **Keeton** in 1970s tested the sensitivity of pigeons to gravity by looking for the influence of the monthly gravitational cycles caused by the changing relative positions of the earth, sun and moon.

5. Barometric pressure cues. Recently pigeons have been shown to respond to pressure changes. Pigeons are known to be sensitive to pressure changes equivalent to a 10 meters, or even smaller, change in altitude. Sensitivity of this sort could help to compensate for the reduction in gravity with increased altitudes. Birds about to set off on migration seem to be able to gauge wind conditions at high altitude, taking off only when they are favourable. Similarly certain eastern American migrants fly to the east of a high pressure area in autumn but to the west, ahead of a cold front, in spring.

6. Infrasound cues. Sound with a frequency of less than 10 hertz is called **infrasound.** Humans cannot hear it, pigeons can detect infrasonic frequencies as low as 0·60 hertz. Infrasound can travel many hundreds, even thousands of kilometres, thus allowing birds to orientate to distant mountains (via the wind blowing through them) or shorelines (from the sound of breaking waves or surf). The sensitivity of pigeons to infrasound is great enough for them to detect Doppler shifts as they fly towards and then away from the sound source. In this way they could achieve the directional information necessary to orientate to the source.

Box 14.6.
Doppler effect

The change observed in the wavelength of a sonic, electromagnetic or other wave is because of relative motion between the wave source and an observer. The Doppler effect named for **Christian Johann Doppler** (1803-1853) who first described it in 1842, is of paramount importance in astronomy.

7. Polarised and ultraviolet light. In normal unpolarised light, the wavelike vibration occurs equally in all planes. In polarised light, the vibrations are greater in one plane. When unpolarised sunlight is scattered by atmospheric molecules, **polarisation** occurs, greatest for the light scattered at an angle of 90 degrees to the rays of the sun. This means there is a pattern of polarisation in the sky that changes according to the position of the sun.

Karl von Frisch (1967) discovered that the orientation of honeybees, as indicated by their communication dances depends upon the position of sun, even when the sun was clouded. He showed it was necessary for the bees to see only a small portion of blue sky and that the essential information was confined to the ultraviolet portion of the spectrum. When ultraviolet light from the sun was passed through a polarised filter, **von Frisch** found that the orientation of the bee dance changed in accordance with the angle of polarisation.

Cardiac-conditioning experiments (Fig. 14.12) clearly show that pigeons like the honeybees are able to perceive rotation in the plane of polarisation of light, but it is not known how they interpret this information.

8. Odour. While it is well-known that some animals, *i.e.,* slugs and salmon, can navigate by olfaction, the suggestion that birds are able to do so is quite surprising. Birds are usually regarded as **microsomatic** (*i.e.,* having the sense of smell feebly developed). **Papi** and coworkers (1992) have suggested that pigeons can home by scent. Pigeons might learn the directions of different odours arriving at their home loft; odour A, for instance, may come from the south, B from the east and so on. If a pigeon is displaced some distance south, odour A will be stronger. The bird therefore determines its position to be south of the home and used one of its 'compass' system to fly north.

9. Other navigational cues. Some other types of cues might also be used as navigational aids by different animals. For example, fish can navigate using carbon dioxide concentration gradients, temperature changes, sounds from other fish, geoelectric information, olfaction, magnetism, polarised light and the sun.

14.6. METHODS OF STUDY OF BIRD MIGRATION

Following methods are used to study migration in birds:

1. Banding or ringing. It is most popular method of studying bird migration (Box 14.7). Large scale ringing has enabled ornithologists to plot migratory path or routes of many bird species. There are various methods of banding migratory birds throughout the world. Ringing methods involve the use of coloured metal or plastic leg bands, systematically numbered and having the address of the branding agency to which the recovery of the banded bird may be reported. The birds are first caught or trapped alive in large number by using **mist nest** or **boom-net.** This net is thrown into air by explosives and spread out by metal weights which soars over a flock of birds.

The bird banding laboratory at Lawrel and Maryland in USA have maintained more than thirty million banding records and add a million new ones every year. In India, at Keoladeo National Park, Bharatpur extensive research on migratory birds is in progress, especially on their ecology, behaviour and migratory routes. The work is primarily being conducted by **Bombay Natural History Society (BNHS)** in collaboration with State Forest Department. *BNHS is the only authorized institution to carry out bird banding studies in India.* More than 100,000 birds have been ringed in India between 1973-1998.

Box 14.7.
Tagging of other animals

For tracking animals during homing and migration, plastic or metallic bands are placed around the legs of birds or around the wing bones of bats. These bands can be colour-coded and may have identifying number on them. The movement of fishes may be followed by capturing them and marking their fins with numbered tags. Sea turtles have been tracked using balloons attached to their shells; the balloons bob along the surface of the water as the turtle swims.

In one of the studies Spanish and Turkestan sparrows ringed at Bharatpur were recovered in Kazakhistan. By using ringing technique it has been possible to establish the migratory routes of birds, *e.g.*, the main migratory routes of birds in India run down the valleys of the Indus in the west and Brahmputra in the east. Ringing at Bharatpur has established that European swallows return not only to same locality but even to same place for nest building purposes every year, covering a distance of thousands of kilometres.

Migratory birds of Keoladeo National Park. A total of 360 species of birds have been recorded from Keoladeo National Park and a great majority of them are migratory. Most birds migrate from Himalayas, Ladakh, Tibet, Central Asia, Siberia and Eastern Europe. Some of the most commonly seen migratory bird species are **greylag goose, pintail, common teal, gadwall, shovellers, tuffed duck, crested pochard, common pochard, mallards, wigeon, common shelduck, marble teal** (Fig. 14.18). In Bharatpur, winter migration of birds starts in the August with the arrival of **garganey teals** and ends in December by the arrival of **Siberian cranes.** The return migration of most bird species starts in March.

2. By collecting dead migratory birds killed at television towers or by airport ceilometers or other human-made high towers and poles interfering with migration. (**Note:***Ceilometer* is a photoelectric device used with a mercury-vapour lamp to determine the ceiling height of clouds under all weather conditions).

3. By tape-recording or chip-counting. The bird calls through a tape recorder, with the help of a parabolic receiver, microphone and amplifier. This is used to estimate the number of migratory birds by tabulating the call notes heard at night.

4. Radio-tracking. Modern technological methods of radar and radio-tracking have largely superceded our earlier means of observing birds with binoculars. Now individual birds can be spotted on their migratory route, and their course and behaviour can be determined over longer distances with a great deal of precision. **Miniature transmitters,** weighing 2-3 g, have been developed in the USA. Easily carried by small birds, these transmitters send a signal that may then be located by

Fig. 14.18. Migratory birds of Keoladeo National Park, Bharatpur.

means of stationary direction-finding aerials or by vehicles and aeroplanes. The routes of many migratory birds have been tracked with this method. Researchers are now working on a largely automatic method of tracking birds on their migrations, with the aid of **satellites.**

5. Radar tracking. With the use of **radar tracking instruments** much knowledge has been gained about what happens to birds on their migration. In contrast with the **flight-monitoring radar instruments** used earlier, this new device detects individual birds and automatically "keeps an eye" on them. When a migratory bird passes the instrument at a greater or lesser angle, the exact range, bearing, and altitudes as well as ongoing changes in these values can be recorded or printed out, and the animal's course of flight relative to the ground can be plotted on a chart. Furthermore, the birds wing-beat frequency can be determined; this identifies or at least indicates the species and ensures that the same individual is still being tracked. Finally, the exact migratory routes of individual birds as well as other bits of information can be recorded from the natural course of events during the flight as they occur within the range of radar instrument.

6. Laboratory methods. There are some aspects of bird migration that we can study in the laboratory, where the animals live in cages or aviaries. Here we can conduct conditioning experiments or we can perform tests on birds in the state of intense activity characteristic of migratory restlessness (zugunruhe). One of the results found with conditioning experiments was that birds use the sun as a compass.

14.7. MIGRATORY RESTLESSNES

In experimenting with birds that are ready to migrate, researchers take advantage of the fact that even with the best care these animals will become restless, jumping and fluttering about in their cage as the time of migration approaches. The activity of such a bird can be recorded by an ingenious method (Fig. 14.16). The bird sits on an ink pad located inside a "funnel" or cone covered with blotting paper. It can look at the sky through the wire-mesh cage cover. While attempting to leave the cone in an upward flight direction, the bird stains the blotting paper with its footprints. The experimenter then assesses the amount of ink staining on the paper. Under certain conditions, activity during migratory restlessness becomes **directional,** usually corresponding to the species-typical direction of migration.

14.8. HOMING

Homing is a complex instinctive behaviour of migratory fish, birds and other animals. It compels an animal to return to its original locality, after taking migratory journeys. The homing instinct is seen most prominently in the salmon. Fish inhabiting smaller water bodies such as lakes, ponds and pools also have this remarkable instinct and if displayed, they can return to their home territories. **Gerking** (1959) has aptly defined homing as "*the return of animal to a place formerly occupied, instead of going to other equally probable new places*".

If pigeons, for example, are sent by a rail in closed basket to several kilometers away and there set at liberty, they will return to their loft which they can recognize without any error. Thus, they share with the rest of the migratory birds, a certain sense of orientation and homing.

QUESTIONS

Long Answer Questions

1. What do you understand by migration ? Discuss its possible reason and biological significance. (*Allahabad 1991*)
2. Write an essay on 'Migration in Animals'.
3. Define migration ? Describe this phenomenon in birds. (*Punjab 1996*)

4. Define migration. Explain the reasons, purpose and controlling factors for bird migration with suitable examples. *(Purvanchal 1999)*
5. What is migration ? Describe the various types of migration and modes of flight during migration in birds. *(Purvanchal 1993)*
6. What is migration ? Explain the reasons and purpose for bird migration with suitable examples. *(Purvanchal 1996)*
7. Write an essay on migration in birds. *(Magadh (H) 1996, 98)*
8. Give a detailed accounts of navigation.
9. Giva detailed account of migration in fishes and birds. *(Punjab 1998, 99)*

Short Answer Questions

1. What do you mean by migration of fishes ? Differentiate between catadromous and anadromous type of migration with suitatble example and their migratory habits. *(Purvanchal 1993, 95)*
2. What is migration ? Explain the reasons and purpose for fish migration with suitable examples. *(Purvanchal 1998)*
3. What do you mean by migration ? Write an essay on fish migration. *(Punjab 1996; Purvanchal 97; Rohilkhand 2000)*
4. Describe catadromous fish migration. *(Punjab 1995)*
5. Discuss the significance of ringing and radiotelemetry in bird migration. *(Purvanchal 1996)*
6. (*a*) What is migration ?
   (*b*) Name the factors stimulating birds to undertake migration.
   (*c*) Name some birds which are migratory. *(Magadh (H) 1994)*
7. Write short notes on the following :
   (*i*) Migration *(Bangalore 1993)*
   (*ii*) Migratory restlessness *(Purvanchal 1996)*
   (*iii*) Range of migration in birds *(Purvanchal 1993)*
   (*iv*) Navigation *(Purvanchal 1997)*
   (*v*) Homing instinct *(Purvanchal 1998, 99)*
   (*vi*) Factors influencing migration of fishes *(Purvanchal 1994)*
   (*vii*) Seasonal movements. *(Magadh (H) 1992)*
8. Differentiate between the following :
   1. Potamdromous migration and oceanodromous migration *(Purvanchal 1994)*
   2. Latitudinal migration and longitudinal migration. *(Purvanchul 1995)*

Very Short Answer Questions

1. Give two examples of migratory fish. *(Purvanchal 1999)*

Multiple Choice Questions

Choose the correct answer from the four alternatives given

1. The study of bird migration is known as
   (*a*) ornithology (*b*) oology
   (*c*) phenology (*d*) nidology
2. Migration of birds is chiefly controlled by
   (*a*) shelter (*b*) food urge
   (*c*) amount of light (*d*) temperature differences

3. Which of these birds builds its nest outside India?
   (*a*) crow (*b*) red-vented bulbul
   (*c*) cuckoo (*d*) wagtail
4. The largest flying bird is
   (*a*) ostrich (*b*) albatross
   (*c*) dodo (*d*) penguin
5. During migration, birds determine compass direction using
   (*a*) land marking and water bodies (*b*) water bodies and mountains
   (*c*) mountains and landmarks (*d*) celestial bodies
6. Power of young birds to return to their original ground of their parent show
   (*a*) conditioned reflex (*b*) intelligence
   (*c*) intuition (*d*) instinct
7. Anadromous fishes move
   (*a*) to freshwater from sea (*b*) from sea to estuary
   (*c*) from river to sea (*d*) from estuary to sea
8. The famous catadromous fish in India is
   (*a*) salmo (*b*) hilsa
   (*c*) eel (*d*) rohu
9. Catadromous fishes moves
   (*a*) to freshwater from sea (*b*) from sea to estuary
   (*c*) from river to sea (*d*) from estuary to sea
10. Which of the following bird is famous migratory bird in India ?
   (*a*) arctic tern (*b*) wagtail
   (*c*) maina (*d*) goose
11. Dr. Salim Ali was a famous birds man of India. He observed and described maximum number of birds. His last desire was to discover
   (*a*) himalayan mountain quail (*b*) jerdon's courser
   (*c*) indian courser (*d*) white breasted kingfisher
12. Some birds can migrate at an average speed of *(Magadh (H) 1996)*
   (*a*) 500 miles (*b*) 200 miles
   (*c*) not more than 25 miles per hour
13. For a migratory bird, its real home is the place *(Magadh (H) 1994, 96)*
   (*a*) to which it migrates (*b*) where it lays eggs
   (*c*) where it earlier was
14. Migration is a very common feature in *(Magadh (H) 1995)*
   (*a*) birds (*b*) fishes
   (*c*) both fishes and birds.
15. One of the factors that affect migration is
   (*a*) learned behaviour (*b*) instinct behaviour
   (*c*) biological (*d*) none

ANSWERS

1. (*c*); **2.** (*d*); **3.** (*c*); **4.** (*b*); **5.** (*d*); **6.** (*d*); **7.** (*a*);
8. (*a*); **9.** (*c*); **10.** (*a*); **11.** (*a*); **12.** (*c*); **13.** (*c*); **14.** (*c*);
15. (*c*).

15

Human Ethology

Human ethology is a sub-discipline of biology and psychology. It deals with behaviour of human beings especially those which have their roots in 'animal behaviour' such as feeding, drinking, social behaviour, communication, learning , intelligence, aggression, territoriality, courtship and parental care. Human beings are different from other animals and non-human primates. There are 193 living species of primates, out of these only one species- *Homo sapiens* is not covered with hair. **Dr. Desmond Morris** has rightly called them **" The Naked Ape"** (Box 15.1). According to Morris, we are *intensely vocal, acutely exploratory, over crowded apes.* He has forwarded the following nine possible reasons why humans lost hair from their body:

1. An infant chimpanzee at birth has hair on its head, its body is naked. If this condition is delayed in the chimpanzee's adult life by **neoteny,** the adult chimpanzee's hair condition would be very much like the human beings. The growing human foetus starts off like a typical mammalian baby covered with hair. Between the 6th and 8th months of gestation the human foetus is completely covered with fine hair called **lanugo,** it is shed just before birth. The neoteny explanation only provides a clue as to how the nakedness could have come to human beings.

2. When a **hunting ape** (*i.e.,* human being) given up its nomadic life and started at fixed places its fur got infested with skin parasites. By casting off its hairy coat, it helped getting rid of the parasites.

3. Another hypothesis is that hunting ape had developed a **untidy feeding habit** of eating meat with dripping blood, which is used to stick to fur coat, clogging, it paved the way to disease. Vultures tend to plung their heads and necks in carcasses and due to reason just mentioned

Box 15.1.
Dr. Desmond Morris

Doctor **Desmond Morris** made human ethology popular all over the world. He is a famous zoologist, ethologist, author, painter and film maker. Dr. Morris was born in Wiltshire in 1928 and educated at Universities of Brimingham and Oxford. He has been a zoological research worker at the University of Oxford (1954-56), Head of Granada Television's Film Unit at London's Zoo (1956-59), curator of mammals at Zoological Society at London Zoo (1959-67), and he was Director, Institute of Contemporary Arts. Dr. Desmond Morris published 48 valuable scientific papers between 1952-1967. He has written about 50 popular books between 1958 and 1999. His famous T.V. programmes are **Zoo Time** (1956-57), **Life** (1966-67), **The Human Race** (1982), **The Animal Road Show** (1987-89), **The Animal Contact** (1989), **The Animal Country** (1991-96), **The Human Animal** (1994), **Human Sexes** (1997), etc. Dr. Morris popularized Human Ethology by writing most wonderful books such as *The Naked Eye* (2000); *The Human Sexes : A Natural History of Man and Woman* (1997); *Intimate Behaviour : A Zoologist's Classic Study of Human Intimacy* (1997); *Body Talk: The Meaning of Human Gestures* (1995); *Illustrated Baby Watching* (1995); *The Human Animal " A personal View of the Human Species* (1994); *The Animal Contact : Sharing Personal View of the Human Species* (1994); *Man Watching : A Field Guide to the Human Behaviour* (1977); *Intimate Behaviour* (1972), *The Human Zoo* (1969); *The Illustrated Naked Ape: A Zoological Study of the Human Animal* (1967). Before 1967, Dr. Morris published some, books in the co-authorship of his wife **Ramona**, *e.g., Men and Snakes, Men and Apes* and *Men and Pandas.*

above may lost feathers from head and neck to keep them clean. All these explanations do not fit very well because the males of lions who have similar feeding habits but still have a mane.

4. It was **frequent fire** in the forest and around early human's dwellings that led to burning of fur then loss of it forever.

5. Another speculative theory is that before becoming the hunting ape, the original ground dwelling ape left forest and went into an **aquatic phase.** Only his head protruding from the water retained hair.

6. An argument is that loss of hair was a social feature; it developed to give a signal. Naked (hair less) patches of skin can be seen in a number of primate species used for signalling and species identification.

7. An interesting hypothesis holds that naked skin had its use in **sexual signalling.** Females prefer hairy males and vice versa. So, females lost hair and males still retain them.

8. Most commonly held explanation is physiological in nature; it regards that humans lost hair as a **cooling device.** The hunting ape chased his prey (food) which generated considerable over heating. Therefore, the **fur coat** was gradually shed.

9. The fur coat was further reduced due to covering of body with clothes which are made from plant materials or animal skin.

Human ethology can also be considered as a **connecting link** between **animal behaviour** and **human behaviour** or animals ethology and human psychology. Unlike animal ethology, human ethology is difficult to understand because it is much complex, since on one hand, animals depend more on **instincts**, humans use more of learned behaviour. Human behaviour includes lots of **pretence** also, for example, we can look gloomy at some body's funeral, whereas, in reality we are not so unhappy. Likewise, sometime we can laugh at a poor joke (of our boss or love one). Whereas, we are not so amused. That way humans have a control over their behaviour. We generally assume that we can always behave as we want in a specific manner at any given time. However, our nonverbal communications tell otherwise.

15.1. ETHOLOGICAL CONCEPTS AND HUMAN BEHAVIOUR

1. Concept of Fixed Action Pattern

The classic example of inborn human behaviour in the newborn baby reacting immediately to its mother's nipple by sucking. Similarly, smiling and frowning are also inborn actions. If we look at child born blind and deaf, they still smile and frown. In human adults, raising of eyebrow while greeting is an inborn action. Aggressive displays and appeasement actions are inborn actions. Any action of adult animal is considered inborn action, if it occurs in every human being.

Like the animals, human beings also have two kinds of behaviours: *instinctive behaviour* and *learned behaviour.* There is much behaviour that is inherent or inborn. For example, human behaviour, such as **crying, smiling, crawling, walking, sucking, swallowing, coughing, laughing, aggression, swinging of the arms, sexual activities** and **parental care behaviour**; all of these are genetically pre-wired and are produced at the arrival of proper stimulus. These actions are produced automatically and quickly without consulting the conscious brain.

2. Concept of Sign Stimulus

Like the animals, human species respond to one another by **visual releasers**. One of the first social behaviour in humans is a **smile** (a fixed action pattern). The first time you smiled, your mother, father, brother or sister smiled back to you. For the rest of your life, you will smile more frequently to a smiling face. *Big round eyes, rounded cheeks* of **infants** invoke parental behaviour in us. Similarly, sight of nice, beautiful glistening lips, protuberant breasts, round hips, hairless legs and arms, big eye lashes, arched eyebrows, rosy cheeks in females and presence of tall, dark handsome man stimulate sexual behaviour in males and females, respectively. All

TV commercials provide visual stimuli for us to buy that product. These commercials are also based on **super normal stimulation theory.** For example, when oystercatcher bird is provided with large-sized eggs, it leaves its own eggs and starts incubating super normal sized eggs, they attract the bird more. Similarly, the natural releasers that identify **masculinity** in most human societies are **tallness**, **broadness** and **wideness of shoulders, narrow hips** and **facial hair** (moustache and beard). In contrast, **shortness**, **thinness**, **hemispherical round hips, firm breasts, narrow waist, deeply placed belly-button** and **rounded face** identify **feminity**. Nice shapely lips of human female attract the male attention; if lipstick / lipgloss is applied on them, they become all the more attractive, because it now falls in the category of **super normal stimulus**. The advertisement of a tooth paste on TV always show extra sparkle on teeth, this is a super normal stimulus; it shows sparkles all around a person attracting attention of all those girls who are passing by. Top actresses advertise for soaps, for example, for the Lux and she says with all her make up on "*Lux is secret of my beauty*". This act as a super normal stimulus for a girl next door, conveys to her that soap not only cleans you but it also makes you beautiful like an actress.

Facial expressions studied by **Charles Darwin** are all example of **visual releasers** and **fix action patterns.** As mentioned earlier the **smile** is a universal gesture among humans, indicating *pleasure, appeasement* and *friendliness.* A **flying kiss** is a common gesture seen among human beings. Both of these gestures have evolved from the **fear grimace** and **pout faced lip smack** of non-human primates. In them these are sign of mild fear and a kind of "sorry". A subdominant member in a rhesus monkey when comes in direct eye contact with a dominant member at feeding or resting site gives a **fear grimace** (*i.e.,* pulling of lips back, showing the front teeth and lowering of ears and tail) and this may or may not followed by a **lip smack** (a type of flying kiss) meaning "please do not be annoyed I did not mean to bother you".

Visual releasers. We recognise babies of all mammalian species as those with heads larger than the rest of body, a big forehead, big button-like non-blinking eyes, rounded face and chubby cheeks. These **releasers** give us information about innocence, purity and lack of cunningness. Generally, we consider this combination to be very cute. These characteristics are well known by cartoon and toy manufacturers. The world famous Disney characters have been developed on this knowledge. The **Mickey mouse, Donald duck, Goofy, Tintin, Astrix, Barbie** all have been there, liked by every one universally because they incite our hidden **parental instinct.** Many types of dogs have been bred to manifest these same cute features; the **Pekingese** bread is one of the most successful attempts. Juveniles of other animals, *viz.*, cow, elephants, dogs, cats, tigers, lions, monkeys, donkeys, horses, goat, sheep, chicken, duck, etc., are so cute, because they possess some of the same visual releasers.

Auditory releasers. Sounds are also significant in our lives. **Vocalization,** *i.e.,* production of sound is **instinctive** but the language has to be learnt. Although language is important in human communication, the **tone of voice** also communicates—it communicates emotions. What is spoken is not nearly so important as the tone of voice used when speaking? Soft, whispering tones are comforting and may be uttered to lover or a child; a harsh tone conveys annoyance. We associate infants and young children with cute talks and mispronunciation. These sounds are copied by adults when speaking to infants and small children, because they carry signals of love, affection and care.

Music and musical rhythms are interesting source of human sound production. A tune or melody stimulates our pleasure center. The distinctive beat of rock music and an African chant are equally stimultating and arousing for dancing. Liking for the melody is inherent.

3. Imprinting in Humans

The concept of **imprinting** during a sensitive period proposed by **Konrad Lorenz** provides a useful framework for **mother infant bond,** also called **attachment bond** in human beings. **John Bowlby** (1969, 1973, 1980) is one of the major supporters of attachment bond theory.

Attachment is an innate human survival mechanism. It is an emotional tie that a person or animal forms with another person or animal. This tie binds these two beings together and it continues over time. Human attachments last for a lifetime. When the animals are born, they see their parents moving around them during the sensitive period, so they get attached to them. But in case of human beings it may be parents, grand parents, aunts and servants. No wonder we remember the "aaya" (nanny) of our childhood all through our life, with much fondness because of the establishment of strong attachment bond during our childhood all through our life, with much fondness because of the establishment of strong attachment bond during our early childhood called imprinting.

Attachment bond helps the infant maintain **proximity** (closeness for warmth) to its caretaker–parents, aunts, sisters, friends and servants. Babies ensure proximity to the caretaker, when frightful, they cry. This brings the caretaker, who tries to stop the crying. Through this process infants learn, in a basic way, that they *can have an effect on the behaviour of other people;* that they can have an effect on the world. Animals also do this; the young ones (*e.g.*, rhesus monkey, puppy) scream and their caretaker comes running for help or rescue.

Attachment bond provides the young child with a **secure base** from which it explores the world. As young children grow older, they gradually move away from the caretaker as they explore the environment. However, they still maintain contact with the caretaker through visual and verbal communication. Attachment bond affects personality and the individual's feelings of self reliance. Attachment bond is an indicator of the quality of care received by the infants. According to the experts, secure attachment relationships are established when parents feed the baby on time, wrap the baby in soft clothing, keep the baby in a comfortable house, these fulfill the physical needs, the baby also needs love and affection in the form of **caressing, cuddling**, **hugging**, **playing** and **giggling** from the caretakers. Money buys physical comforts and time spent with baby provides mental comfort (see **Mathur,** 2004). For normal psychological development of a baby, the parents have to provide things in a balanced way. Too much can produce fussy, indisciplined babies; too little can produce nervous babies. The strength and quality of the infant attachment bond to the caretaker is formally assessed by "**strange situation test**" by observing the way the child reacts when separated from the caretaker. This idea initiated a considerable amount of research. For example, if a child is deprived of love, affection, care and fondling during the imprinting time, it will develop behavioural or psychological disorders "**Jactatio Capitis**" is a disease caused by deprivation of enriched atmosphere to a child during its imprinting time (3 month to 1½ years of infant age). Such children will grow into social abnormal adult (who performs a public boasting or bragging).

Early childhood experiences influence sex life when this child grows into an adult. Human beings do not become functionally male and female until they are teenagers. But the sex roles of male and female develop when they are only four or five years old. Children who have unusual rearing circumstances during the critical period, *e.g.*, babies of single parent may learn the sex role of that particular one parent only. If the baby boy is reared by mother alone he may develop into feminized male. Therefore, for normal up-bringing, it is necessary that a human child is brought up by both parents.

One characteristic of imprinting is its **irreversibility**. Once the appropriate or inappropriate behaviour is learned, it is never completely unlearned or forgotten. It is well known fact that it is much easier to teach a youngster a **foreign language** than to teach a middle aged. Elders can learn a foreign language, but they are rarely proficient with the language and almost never learn to "think" in the new language, as a young child might. Therefore, more and more people are sending their young kids to learn to play a game, or play an instrument, or learn to dance. So an adoption of a child must be done at its young age.

4. Kinships and Human Social Systems

Kinships means helping related individuals in a family, or even unrelated members of a colony or tribe. This property is shared by animals also. **Sociobiology** is a branch of animal behaviour dealing with evolution of social behaviour. Kinship has been studied in genetically related animals. Trait for helping related individuals in human beings has come from animals. Among human beings where relatedness is known precisely, from **genealogies** and from **blood typing,** there is evidence that *aggression is lower between those which are closely related* and that *relatives tend to help each other at the time of any need* such as the child birth, marriage and helping a relative during fights and at other times of need (**Chagnon,** 1981). Kinship is applied to the relationship between parents and their offspring and predicts, where certainty of paternity is low, a male invest in the children of his full sisters in preference to his own sons and daughters, since on average they will be more closely related to him; cross-cultural comparisons indicated that uncles and aunts tend to care more for their nephews and nieces in which women or man mate loosely (promiscuously) (**Alexander,** 1977). This is true for animals also.

Human societies are believed to be evolved from monkey societies. Non-human primates form following kinds of social groups - (*i*) male-female and recent offsprings (*e.g.,* gibbons, marmosets, lemurs, titi) - human beings form same groups, husband - wife and kids. This grouping is called **nuclear** or **unit families.** (*ii*) One male and many females with young ones (*e.g.,* langurs, patas, howlers). Human beings too form harems, where one male has many wives. (*iii*) Many males and many females with their young ones and juveniles (*e.g.,* baboons, rhesus, saki, wolly and squirrel monkeys). Humans also form such groups as **joint families.** (*iv*) Chimpanzee follows no one type such groupings, they can be seen in male bands, female groups, male-female babies, big groups of many animals roaming around together. In modern times many human beings like to follow this pattern, no rigidity, you do what you want homosexual or heterosexual, be with someone today and next time change the partner.

5. Chemical Communication by Pheromones in Human Beings

Communication through **pheromones** is considered primitive. Animals (*e.g.,* honey bee, silk moth, dog, deer, lion, etc.) depend largely on this method, because they cannot write, they cannot speak like us, they do not have expressive eyes like ours. But it is surprising that in spite of our other developed communicative skills, we have retained this primitive animal's form of communication through pheromones.

Odours (pheromones) play significant role in human sexual behaviour. Pair formation, falling in love involves odours. We have odour preference which changes also. Before puberty human beings have preferences for sweet and fruity smells, at sexual maturity, there is shift to flowery, oily and musky smells. Humans do not have any mammalian scent glands for the production of pheromones. Instead, in them occur numerous apocrine sweat glands which secrete higher proportion of solids. These glands are concentrated in arm pits and around genitals. Hair present there act as scent traps. There are 75% more apocrine glands in the human males than the females. As it is normal body odour, it is a source of attraction. We use perfumes to add to good smell and make it more attractive.

Axial hair (hair in armpits) and **pubic hair** (hair around sex organs) seem to have no use whatsoever and have been considered remains of our ancestral fur coats. Closer examination, however; reveals that these hairly areas (including scalp) have specialized glands called **sebaceous glands.** These secretions from these glands have more proteins and oil than the waterly fluid that pass from the sweat glands. These glands are more abundant and more oily than in females. The secretions stick to hair, creating scent traps. Scent is best dispersed by current of air as humans walk or run.

The role of baby scent in human sexual behaviour has attracted much attention. In 1971, **Maclintock** maintained notes on the menstrual cycles of 135 girls of a hostel and observed synchronization in onset of their menstrual dates, the reason assigned was body scents or pheromones. A female living with a male has relatively short menstrual cycle, because the body smell of male has an effect on female's reproductive physiology. Female rhesus monkey secretes a pheromone '**copulin**" from its vagina that stimulates mounting and mating in male rhesus monkey. Similar copulins have been identified in human female vaginal secretion also. Identification of human pheromone from male sweat has been carried out by **Dr. George Dodd** of Warwick University in U.K. It is identified as **X-androstenol** (related to sex hormone) its principal function is in attracting the females during sexual attractant pheromone (androstenol).

Women are particularly sensitive to the smell of **musk.** a common pheromone of many vertebrates specially musk deer found in Himalayas. It used by males to attract females to their territories during mating season. Sensitivity to musk is strongest in sexually mature women at the time of ovulation. Women working with musk products have shorter menstrual cycles than other women, indicating that this male pheromone has some influence on women.

Pheromones have significant role in maintaining **mother-child bond.** It is common observation that a young human infant is most comfortable and quiet with its mother. This is because it starts recognizing its mothers by smell when it is six days old; by six week the baby starts **orienting** its head towards the mother's breast and that is again guided by the mother's scent.

Pheromones produced by humans are washed away each time they take bath. If they are allowed to accumulate, a strong odour can be felt known as **body odour.** We all are familiar with occasional strong smell coming from armpits and genitalia. The odour can linger after the producer has long left the scene. Body odour evolved as "**human awareness**" pheromone that perhaps helped our primitive ancestors to locate members of our own widely scattered species. This smell increased **group cohesiveness** and **facilitated breeding** among small bonds of humans.

6. Territorial Behaviour of Human Beings

Human beings are most territorial animals among the animal world. There is a constant fight for small pieces of lands between countries, cities, villages, colonies, roads or houses. The trait of keeping a territory and defending it by animals has been passed on to human beings during the course of evolution. There are parallels between the territoriality displayed by animal species and the **possession** and **protection** of property by humans.

During human evolution, territoriality had an obvious advantage. **Early humans** moved in **small bands** or groups and frequently set up temporary camps or territories. These areas became home bases for hunting extending into nearby home ranges. This area was well-protected and contained food, water and shelter. Fossil evidences suggest that human territories by the side of a water body were preferred where infants and women were secure. There were obviously competition for these good sites with other human tribes and they were defended against human or animal intrusion.

Primitive human beings marked their territory in simple ways. They used clear landmarks, such as trees, rocks, caves and shore-lines of lakes and rivers. Markings today are far more elaborate, because there are so many competitors (Box. 15.2). Today, we have become highly territorial, in some sense much more than animals. We make a **boundary wall** around our house to mark our territory. We restrict the entry by putting in **main gate.** During the day we keep it closed, during the night we lock this gate from inside. In this way, we protect our family, food, water, garden and other belongings. We often employ help in defending our family territories, *e.g.*, **guard dogs, burglar alarms, special locks, watchman** and **security guards**. We display

Box 15.2.

Research comparing modern and primitive cultures indicate that all humans, regardless of culture, display the same protective behaviour, *viz.*, **territoriality**. This behaviour increases with the growth of population. For example, the shore birds gather in large numbers for breeding in a limited space, this increases **competition** and **aggression.** To reduce this, these birds acquire territories and defend them. It is known that crowding increases territoriality. In developed countries where there are less people, they build houses but do not make boundaries (walls) around them. Whereas, we in India, because of high population pressure, defend our houses by making a boundary wall.

greeting signals to a familiar face but treat strangers with suspicion, carefully evaluating them through peepholes and partially closed doors / curtains before we approach.

Types of human territories. Desmond Morris has classified human territories as follows:

1. Tribal territory. It is a territory around a tribe, village, city or country. **Patriotism** for a country can be the cause of most dreaded wars and bloodsheds we have had (*e.g.*, Indo-Pak wars) and are still having all over the earth.

2. Family territory. Human beings are very defensive for the family territory that is their homes. Human houses are marked in many ways. One house often may look much different from another, and that in itself is a marker that identifies "**my territory**". Our markers include **fences, walls, trees, rockworks, sidewalks, advertisement of the name of the occupant and** the **address**, among many others. Our home territories are also **breeding territories**, because breeding and other sexual activities account for a large share of the behaviour performed in a home territory. The centre of the territory is the **bedroom – the nest.** We feel very secure in the privacy of our bedroom. Feeding, grooming and parental activities occur in other regions of the family territory.

3. Floating or moving territory. We take a moving territory with us, particularly when the whole family units leaves together. Families claim tables at restaurants or picnic grounds as **temporary territories.** But one of the interesting moving territory is the car.

4. Individual space. Like the animals, human beings also have **individual space,** *i.e, territory that goes with individuals everywhere.* There are several levels of the individual space : **(1) Intimate space**, space that is reserved near our bodies for intimate relations with other humans, which frequently involves physical contact. Intimate space is reserved for parents siblings, or mates. Others who attempt to intrude are discouraged and often physically expelled. **(2) Casual space.** This space extends around us about one arm length. This is the ring of privacy we reserve for ourselves while talking to friends and acquaintances of our own peer group. **(3) Social space.** This is a large space around us for interacting with familiar people at wedding receptions or music concerts.

Modern society creates many situations in which we cannot preserve or defend our personal space. We are often crowded close together in elevators and in buses, subways, metros, and trains. In these situations, we become **non-persons**, according to Dr Morris: our facial expressions become bland, our bodies are motionless, and our eyes look upward or downward at the floor. We **avoid looking** at other people in the same situations, and they do likewise. We avoid sending any social signals, because they be misinterpreted and cause defensive behaviour. In very crowded situations such as a jam-packed elevator (lift), we also pull our arms in front of us and stand as erect as possible to lesson the space we occupy. This helps us avoid contact with the person next to us.

7. Aggressive Behaviour of Human Beings

1. Dominance hierarchy. Most non-human primates follow social hierarchy with one dominant male as incharge. In gorillas, there is a big male, and he is tolerant towards other males. This basic system changed when naked ape (Human being) became coopererative hunter with a fixed home. They became territorial and defended homes. When they hunted they took help of weaker males. They follow mild hierarchy, with stronger members at top and weakers at the bottom. It started with group or tribal defence but gradually became involved in individual home defence.

2. Ritualized combats. The danger initiates fear as well as aggression. The battle can be won without shedding blood. Animal species follow this model, most of the animal species settle their disputes without causing undue damage to its members and are benefitted by this process. Animals tend to develop the **ritualized combats. Threats** and **counter threats** have replaced actual fights. Blood fights do take place but rarely when aggressive signalling and counter signalling have failed. As a result of this we can see in many animal species, elaborate threat rituals and dances.

Animals have also developed many **appeasing signals**, gestures and postures, lying down, bowing, grooming, lip smacking are some. What are our own special behaviours of **threatening** and **appeasements** ? We human beings can not frighten our rivals by raising our hair to look bigger, because of almost naked skin. We still do it in moments of anger or fear – "*all my hair rose* " – but as a signal it is of no use. But nakedness gives us the chance to send other visual colour signals. We can go "*white with rage*", "*red with anger*" or "*pale with fear*". We avoid actual fight by exhibiting aggression or appeasement by gestures such as clenched fist (*i.e.,* "mukka"), stamping of feet (Chimpanzees and gorillas also do this), slamming the hand or fist on a table and smashing objects on the wall or floor (Chimpanzee and gorillas also perform this by tearing, smashing and throwing branches and vegetation). We make **threatening facial expressions** along with **shouting**. All these are our displays to avoid an actual bloody fight. We may cover our disinterest in aggression by giving **appeasement signals.** Most animals show appeasement by lowering the eyes, ears, head tail and limb; during total submission the animal (*e.g.,* dogs) rolls on their dorsal side. Human beings also have some interesting appeasement signals. Lowering of human body can be in the form of **grovelling**, **prosterating**, **kneeling**, **bowing, curtsying**. The key signal here is lowering of the body to appease dominant individual. A friendly **hand shake** has developed from holding of hands in chimpanzees.

Further, to avoid or ignore an uncomfortable situation we indulge in **displacement activities** such as rearranging ornaments, light a cigarettes, clean the glasses of spects or of sun glasses, look at the wrist watch, pour a drink, nibble a piece of eatable, hum a song, avoid eye contact, scratch our head, bite nails, rub, pick or blow nose, touch ear lobes, scratch ear, rub palms and shake legs. Over the time naked ape has developed intellectual control over aggression. The **display of national strength** is done by exhibiting shows of **army, airforce** and **navy**. On our **republic day parade** (26th January) we Indians show off our latest artillery, planes, submarines, ships, missiles, tableaus displaying glimpses of economic growth, culture of a state or highlighting some social movement such as awakening about AIDS or Oral Polio drops. **India** advertises about its army and **Pakistan** does the same thing; we test the **nuclear bomb** or **missile** and Pakistan too tests a bomb or missile. These are all displays to avoid a real war to prevent blood shed. The **peace talks** are the part of **appeasement**. We share this instinct with animals.

Is aggression innate or learned ? It is partially determined by heredity and partially by surroundings. Aggressive behaviour is not **pathological** or **abnormal**. It is very much part of the normal behavioral patterns of almost all animal species and human beings. **Darwin** explained that stronger and healthier animals are selected for reproduction to produce better progeny. That is true for us also, therefore, this trait has been favoured during evolution. Aggressive behaviour is essential for our survival.

Sports of any country form the ritualised aggression. For example, in a football game, there is an **offensive side**, a **defensive side** and tests of strength, goals and rules that prevent serious injury to the participants. Although in many ways our behaviour is similar to that of other animals, but we are also different from all other animal species. **First,** we have group aggression. Groups of humans, as **team, tribes, battalions**, perform aggression (*i.e.*, attack) against another group of humans, a behaviour not known for other animals. **Second,** humans confrontation often ends in the death of the opponents. In terms of evolution, it is clearly maladaptive to **kill** other members of one species. It is not clear why we are different, although there has been a great deal of speculation about the cause. The most popular theories drawn on the facts that we invented **weapons**, "that we have an unusual capacity of learning, we are over-crowded, and that humans are unusually cooperative. **Cooperation** was necessary when our ancestors first became hunters; it required each member of hunting band to be specialized in job to kill a big prey.

Gestures. Human beings use many types of **gestures** to communicate pieces of information. There are (*i*) **incidental gestures**, *e.g.*, scratching, rubbing, wiping, coughing, yawning, stretching, etc; (*ii*) **expressive gestures**, *e.g.*, facial expression such as, smile, frown, fear, repulsion, pain, wonder, funny, aggression, calmposed, etc. (*iii*) **mimic gestures**, *e.g.*, putting on a good face, looking sad at a death, theatre and cinema artists, signalling for food, drink, gut firing, smoking, etc; (*iv*) **schematic symbolic gestures**, *e.g.*, cattles are represented as pair of horns by putting two fingers on the side of heads, temple tap indicating to use brain, finger rotating against temple indicate something wrong with brain, (*v*) **technical gesture**, *e.g.*, signals by policeman or fire man; (*vi*) **coded gestures**, *e.g.*, sign language of deaf and dumb, tic-tac system used on race courses, semaphore code for naval communication. It is not easy to understand all gestures made around the world because of much cultural variations; **multimessage gestures** and **gesture alternatives, hybrid gestures, relic gestures, regional signals, baton signal, yes-no signals** and **obscene signals.**

It has been suggested that some people **sleep** as **escape** from **conflict.** People who sleep more than nine to ten hours per day are more than likely involved in **displacement sleep. Yawning** or incipient sleeping is commonly exhibited by all humans when bored or frustrated. Sometimes when we are tensed we suddenly feel tired and sleepy. We tend to use the expression "we are tired of it" to describe the source of tension, we say it and we feel it. Some of human beings may smoke or drink as a displacement behaviour. If someone smoke or drink more frequently when he/she is tensed, it is possible that these activities are displacement activities for that person.

15.2. HUMAN SEXUAL BEHAVIOUR

Like animals human sexual behaviour includes following four phases: pair formation, courtship, precopulatory behaviour and copulatory behaviour. Humans and primates (*e.g.*, gorillas) also show **genital displays, homosexuality**, and **masturbation**.

1. Pair formation. Human beings have obvious **sexual dimorphism**, *i.e.*, males can be distinctly recognised from females. Dimorphic species attract mates by the physical characteristics that serves as **sexual releasers.** In humans, pair formation takes place with much rituals, *i.e.*, marriage.

(*a*) Male sexual releasers. Tall, broad, good (*i.e.*, muscular) built, well kept beard and mustache, shapely hips, good contour of front around the genitalia. Body hair, particularly bread, chest, leg and occasionally axial hair also act as a sexual releaser. Women are not usually turned on by fourteen- year old boys, who lack body hair and beard, that fullness. Men are obviously aware of the releasing effects of chest hair. They will wear unbuttoned shirt to expose chest hair.

(*b*) Female sexual releasers. Young, sexually receptive women have round, moonlike full faces, big but firm breast (Box 15.3), fuller hips, big juicy lips, slim waist, rounded thighs and women attempt to retain their youthful sexual appearance by using makeup and dressing style. Make up is commonly applied to make the face look more attractive; make up of the eyes, eyebrows, cheeks and lips adds the overall illusion. Hair-cutting and hair-styling too add attractiveness of female sex (*e.g.*, Bipasa Basu). Going a step further, elderly women may go for a facelift to get rid of from wrinkles and to recapture youthfullness.

A theory suggests that hemispherical breasts of women mimics a women's rounded buttocks (hips). It has been common in evolution for one organism to mimic the physical appearance of another or for one part of body to mimic another part. Animals may mimic rocks, twigs, leaves, or other animals, a phenomenon known as **biological mimicry.** When one part of the body mimics another part 'the phenomenon is known as **auto-mimicry.** For example, male **mandrill** baboon has a brightly coloured face with red nose, blue cheeks and yellow beard. His **genitals are auto-mimics** of his face. The penis is red, the scrotum blue, and the pubic hair yellow. The mandrill uses both facial and genital displays during courtship (see **Mathur,** 2004).

Box 15.3.
Gender Signals and Self-Mimics

Dr. D. Morris (1977) described the following terms about human sexual behaviour in his book "*Man Watching*".

1. Gender signals. These are clues that enable us to identify an individual as either male or female. In addition, they may help to emphasize masculinity or feminity in cases where the sex of the individuals is already known. Thus,

⇒ Hairless, smooth skin indicates feminity.
⇒ Round, hemispherical buttocks indicate female sex.
⇒ Firm rounded (protuberant) breasts indicate sexual maturity of human female.
⇒ Bald head top indicates age signal of human male.
⇒ Pot-belly indicates aging human male.

These, then are the natural biological Gender Signals of human species – the **hunting feature** (of human males), the **breeding feature** (of human female) and the pure display feature.

2. Body self mimicry. It occurs when one part of human body acts as a copy of some other part of the anatomy. Nose of human male is phallic mimic (*e.g.*, swollen nose tips has been seen as the glans of the penis and bulbous lobes on either side of the nose tips have been likened to scrotal-mimics. Likewise, in human female, the role of lips wearing glistening lipstick as labial - mimics has been suggested).

2. Courting. Human male being larger than women, display, **dominance** during **appeasement** and **diversionary** (recreation) behaviour. Men and women are attracted to each other by combination of sexual releasers.

Courting humans are easily identified. They are almost always those couples walking around **holding hands** (in parks, malls, window shoping at Rajiv Chok, New Delhi) or **clinching bodies** while **staring** into each other's eyes. It is obvious that some **bond** has been formed between them. Courting women often shows following postures of appeasement, such as leaning against the man, necking and thereby lowering her body, which is an invitation to mate (copulation) in animal kingdom. She may gaze into his eyes and smile. She may giggle like a child, cover her face, or suck her fingers. It will be very interesting to note that all **sucking** in human behaviour including **kissing** have developed from basic feeding response derived from suckling the breast. **Lip contact** is equally common during courtship. **Kissing** is one of the most frequent behaviour exchanging between courting humans. The kiss has evolved from juvenile food-begging responses in animals. It is common in new guinea for human mothers to chew food for their infants and to feed them **mouth-to-mouth.** In earlier times chewing of food for offspring was a necessary behaviour. Among human adults, kissing is exactly the same except that food is not exchanged.

Kissing does not always have to be lip-to-lip. In fact, lip contact with nearly any **part of the body** is commonly observed during **human courtship**. It is a very short step from kissing to **nibbling** (to bite gently) with the teeth (another feeding behaviour), which is commonly found in courting animals too (*e.g.*, lions). Often **feeding** and **food begging** are not only **symbolized** or **ritualized** ; food can actually be exchanged and eaten. Observe any courting pair in a restaurant it is commonly seen to witness that one male feeds the other by sharing bits of "roti". "pizza" or "paneer" or by putting a spoonful of ice cream into the other's mouth.

3. Precopulatory behaviour. Like the animals this part of sexual behaviour includes visual invitations, mild chasing in the form of pursuance, rubbing, sniffing, licking, sucking and genital display (*e.g.*, strip-tease shows). **Phallic gestures** are made throughout the modern world. If one studies copulation in apes, it is like to know about own species (Box. 15.4).

BOX. 15.4.

Gibbons are monogamous. In **Indian Hoolock gibbons**, found in North-East, the sex partners are attracted through pheromones, they see each other, sit and gaze, hide in the foliage, then the most interesting part of the courtship starts, *i.e.*, singing of the songs in the form of **duet song.** These are actual songs with a melody, stanza and para. Our Hindi films are full of such songs. **Chimpanzee** male and female greet each other with **hooting**, **hugging** and **back patting.** During this period females increase their play with males. The females initiates the play with slow, gentle wrestling. They also **mouth groom** the males. The female walks in front of a male, shows off her sexual swellings, the male examines the females crouches. Among *Gorillas* the adult male identifies the ready female. The female stays closer to male, may go infront of him.

4. Copulatory behaviour. Mating postures can be dorso-ventral (back to face) or ventro -ventral (face to face). During mating, a sexually aroused male penetrates his erect penis in the sufficiently lubricated vagina of sexually receptive female. After a few **thrusting** , male ejaculates semen inside the vagina of female. The male chimpanzee climbs the female from behind (ventro-dorsal) but at times mating can be ventro-ventral too. Both male and female chimpanzee exhibit **thrusting** during copulation. During coitus female often looks back at the male and makes screaming sounds. During mating both male and female gorillas thrust and make vocalizations and facial expressions.

Wild **orangutans** are solitary animals; male and female make a temporary bond only for mating. The male gives out loud calls and is identified by receptive female. They find each other, roam around in forest, foraging together. The orangutan female show **receptive behaviour** and **prospective sexual behaviour.** Prospective sexual behaviour includes play, fast brachiation, hand and hand slapping, oral genital contact, genital-genital contact and pelvic thrusting. Dorso-ventral and ventro-ventral copulation has been observed in them.

QUESTIONS

Long Answer Questions

1. Describe human sexual behaviour.
2. Write on human aggression and technology.

Short Answer Questions

1. Write short notes on the following:
   (*i*) Ritualized combats
   (*ii*) Desmond Morris
   (*iii*) Super normal stimulus
   (*iv*) Human gestures
   (*v*) Human pheromones.

Multiple Choice Questions

Choose the correct answer from the four alternative given

1. Concept of imprinting was proposed by

   (*a*) Desmond Morris (*b*) Konrad Lorenz

   (*c*) Hamilton (*d*) Wilson

2. Social grouping of gorilla is

   (*a*) one male, one female (*b*) two males, two females

   (*c*) many males, many females (*d*) one male, many females

3. Gibbon groups include

   (*a*) one male, one female (*b*) two males, two females

   (*c*) many males, many females (*d*) one male, many females

ANSWERS

1. (*b*); **2.** (*c*); **3.** (*a*).

16

Hormones, Drugs and Behaviour

16.1. EFFECTS OF HORMONES ON BEHAVIOUR

Hormones are chemical substances produced either by specialized ductless (= endocrine) glands in various parts of the body by specialized individual cells, or by neurons called **neurosecretory cells** within nervous system. In the last case, they are called **neurosecretions.** Both are carried by the circulatory system. Both are messengers to various target organs, and they influence such processes as **growth,** metabolism, water balance, and reproduction.

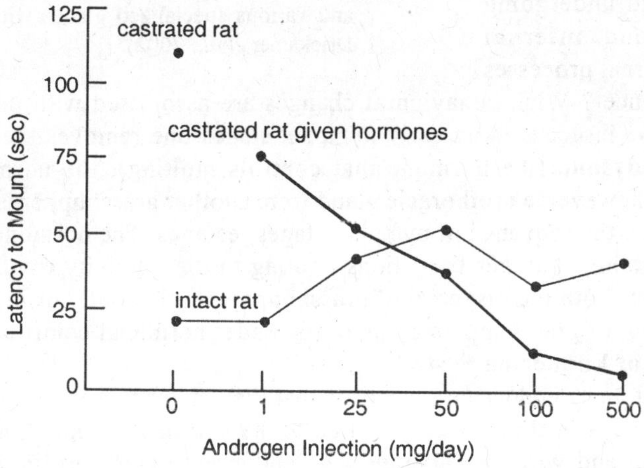

Fig. 16.1. Sexual behaviour in castrated rats. The average latency to the first mount varies in intact rats, castrated rats, and rats given daily injections of doses of synthetic androgen (after Drickamer *et al.*, 2002).

Hormones play major role in determining the behavioural activities in vertebrates. Their role in insects and other invertebrates has also been observed. Our knowledge about the effects of hormones is very old. Greek philosopher, **Aristotle** (284-322 B.C.), studied the effect of **castration** (the removal of gonads, testes or ovaries from animals) on immature male birds and found that they failed to develop sexual characters such as comb, colourful plumage, mating calls and mating behaviour. He concluded that testes were related with sexual behaviour and reproductive behaviour. Ancient Chinese associated testes in humans with the development of beard and male behaviour pattern. In 1849, **A.A. Berthold** showed that hormones are essential for sexual behaviour and reproductive processes.

Examples of Influence of Hormone on Behaviour

A male rat placed with a sexually mature female rat will mount the female within a matter of seconds and begin a sequence of behaviour that ends with copulation. A castrated rat takes much longer to initiate mounting (Fig. 16.1). Which hormonal mechanisms underlie this behaviour ? The testes are a source of androgens, hormones that affect reproductive behaviour. Injections of synthetic androgens can replace the hormone normally produced by the testes and restore mount latencies to near normal levels in castrated male rats.

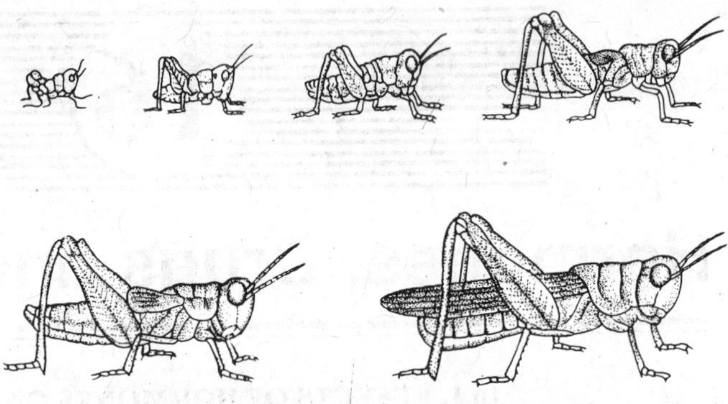

Fig. 16.2. Grasshopper development. Hoppers, young grasshoppers that hatch from eggs, proceed through a series of five to seven molts, changing in body size and in internal and external characteristics with each molt, until they attain the adult stage in several months. Molting and other changes in physiology and behaviour are controlled largely by the neuroendocrine secretions produced by the brain and various specialized glands within the insect (after Drickamer *et al.,* 2002).

Many insects, such as grasshoppers (*Melanopus* sp.), hatch from eggs and reach adult after going through a series of nymphal stages (16.2). Between the stages, insect molts, shedding its exoskeleton and undergoing other external and internal changes. What internal processes regulate this sequence ? What behavioural changes are associated with these morphological shifts ? If, when the insect is in an early nymphal stages one removes the prothoracic gland which secretes, **ecdysone** (the hormone that controls molting), the normal developmental sequence stops. If, however, a prothoracic gland from another grasshopper is transplanted, or if ecdysone is injected, the sequence of nymphal stages resumes. The hormonal changes are also correlated with certain behaviour transitions relating to diet, activity rhythms, and choice of habitat, for example. Both the dispersal of grasshoppers, individually or in large numbers and the mating behaviour of grasshoppers are partially under hormonal control.

General Features of Endocrine System

A general feature of endocrine systems is that they either involve neurons directly, as do the neurosecretory cells or they are closely tied to the nervous system by nerves that connect them with the brain and with one another. The endocrine glands are interconnected by the circulatory system. This **dual relationship** of the neuroendocrine glands and the nervous system is an important features of hormonal system in both vertebrates and invertebrates. Thus, the mechanisms that exert controlling influences on behaviour have apparently evolved in close harmony.

A neurosecretion or a hormone does not necessarily exert a direct influence on behaviour. Instead, the neurosecretion or hormone may have another endocrine gland as its target. These secretions are termed **trophic neurosecretions** or **trophic hormones** and they occur in vertebrates as well as invertebrates.

The **specificity of action** of all hormones is dependent upon the specificity of **receptor sites** at the target tissues. This is true for both peptide and steroid hormones, whether the targets are other endocrine glands or nonendocrine tissues. At any particular time, the blood-stream may be circulating many "messages" in a variety of hormones , but a hormonal effect on physiology and behaviour occurs only when the hormone makes contact with the appropriate receptor sites. Thus, the testes secrete androgens into bloodstream, and the hormones are carried throughout the body. The androgens affect seminal vesicle growth, changes in secondary sex characteristics, and behaviour through target tissues in the brain. In each instance, there are appropriate receptors for the androgens located on or inside cells of the target tissues.

Invertebrate Endocrine System

Different types of invertebrates, for example, echinoderms (*e.g.*, starfish), segmented worm (e.g., leeches), molluscs (*e.g.*, clams), and crustaceans (*e.g.*, lobsters), possess different endocrine systems. However, the role of hormones in affecting behaviour is most clearly understood for insects. For example, in a grasshopper neurosecretions or hormones that affect behaviour are released from following five major locations:

1. The **neurosecretory cells** in the brain (Fig. 16.3).

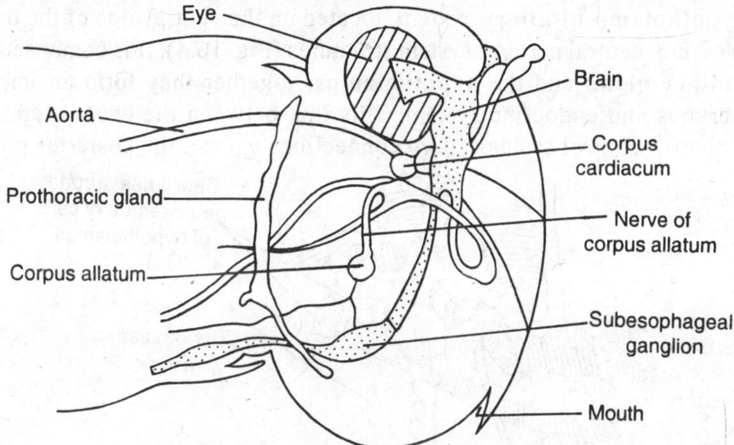

Fig. 16.3. Brain neurosecretory cells and glands of the head region in the grasshopper (after Drickamer *et al.*, 2002).

2. The paired **corpora cardiaca** just behind the brain, near the aorta.
3. The paired **corpora allata** alongside the oesophagus.
4. The elongated **prothoracic gland** at the rear of the head, near the neck membrane.
5. The male and female **gonads** in posterior body regions.

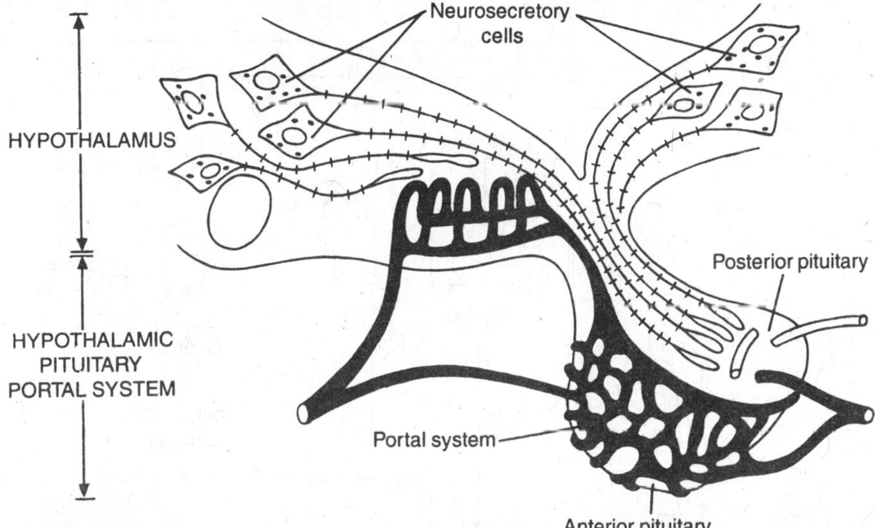

Fig. 16.4. Hypothalamus and pituitary of a generalized vertebrate neuroendocrine system. The products of various portions of the pituitary gland regulate other endocrine glands and secrete hormones that directly influence physiology and behaviour. Posterior pituitary cells are extensions of nerve cells in the hypothalamus; these cells store and release neurosecretions. Other neurosecretions released in the hypothalamus are carried by blood vessels to the anterior pituitary, when they directly influence hormone production and release (after Drickamer *et al.*, 2002).

Most insect species have similar structures, with some variations and exceptions. These five locations do not act independently : neurosecretions from the brain of many insects exert a trophic effect on the prothoracic gland which in turn produces and secretes the hormone ecdysone that controls molting. Ecdysone, in turn, controls molting from one developmental stage to the next (Fig. 16.3). Other trophic hormones influence dispersal and mating behaviour of insects.

Vertbrate Endocrine Systems

Vertebrate endocrine system includes following two major components :

1. The **hypothalamo-pituitary system,** located on the ventral side of the brain, has close connections to several central nervous system structures (Fig. 16.4). It is composed of the closely connected **pituitary gland** and the **hypothalamus;** together they form an important bridge between the nervous and endocrine system. This link between the two systems is the key to integration of the two control systems. Two connections exists : the posterior pituitary (called

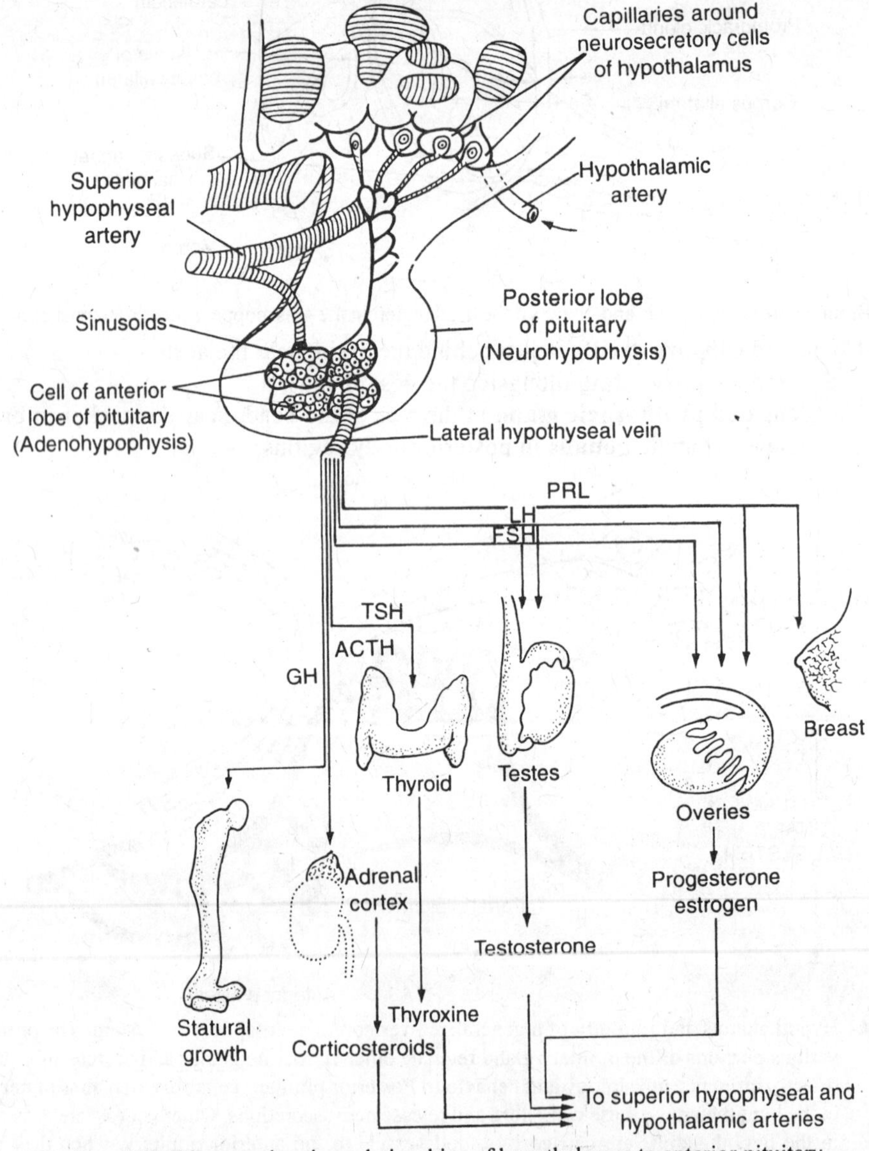

Fig. 16.5. Diagram showing relationships of hypothalamus to anterior pituitary.

neurohypophysis) is mainly composed of neurons that originate in the hypothalamus; the anterior pituitary (called **adenohypophysis**) is linked to the hypothalamus by the **hypothalamic-pituitary portal system.** The pituitary produces trophic hormones, which affect other endocrine glands, and direct-acting hormones (*e.g.,* oxytocin and vassopressin).

2. A series of endocrine glands, including the **thyroid, pineal gland, adrenal** (cortex and medulla), **pancreas** and **gonads,** are located throughout the body.

The trophic hormones released by the various parts of the pituitary are **peptide** (proteins). The peptide's basic structure consists of amino acid chains. The hormones from the adrenal glands, testes, ovaries, and placenta include **steroid hormones,** organic compounds with a common carbon atom ring-structure and varying additional side chains.

Endocrine Gland Secretions

Several pituitary gland secretions exert controlling effects on behaviour and related physiological mechanisms in vertebrates (Table 16.1). Oxytocin and vasopressin are produced by hypothalamic neurons and are stored by nerve terminals in the posterior pituitary (pars nervosa); they are then released as neurosecretions into the bloodstream. **Oxytocin** acts to stimulate uterine contractions that may facilitate sperm movements in the female genital tract after copulation and help expel the fetus during parturition. Oxytocin also has a role in **nursing behaviour ;** it stimulates the ejection of milk from the mammary glands (Fig. 16.6). **Vasopressin** influences the physiology of the kidneys; it alters urine concentration and thus helps regulate water balance. For example, excretion of highly concentrated urine and retention of body water are physiological adaptations of many desert mammals, such as camels, kangaroo rats and gerbils.

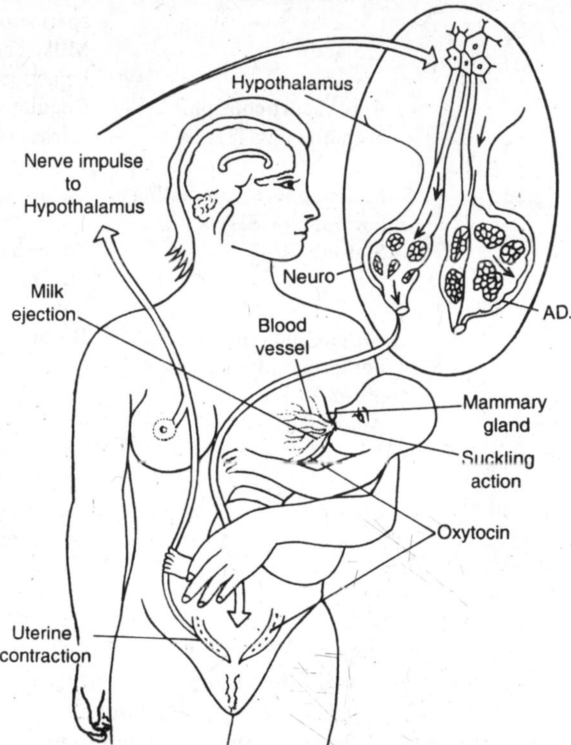

Fig. 16. 6. A hypothalamohypophyseal interaction through neurohypophysis.

The intermediates pituitary (pars intermedia) secretes a hormone important in communication. **Melanophore-stimulating hormone (MSH)** affects the concentration or dispersion of pigment granules in chromatophores, or color cells, found in many vertebrates, particularly fish, reptiles and amphibians. In the absence of MSH, pigment granules remain clumped; MSH stimulation leads to dispersion of the granules and colour change. For example, adult male three-spined stickleback fish (*Gasterosteus aculeatus*) normally have pale-coloured sides. But when two males engage in a territorial boundry display, release of MSH triggers dispersion of pigment granules, and the fish's sides take on a bright blue colour. Colour changes in vertebrates may serve as communication signals, or they may produce colouration patterns that render an animal inconspicuous against a particular background (camouflage).

Table 16.1. Sources of hormones that affect behaviour, hormones produced, and regulatory physiological effects (Source : Drickamer *et al.*, 2002).

Source	Hormone	Regulatory and physiological effects
1. Pineal gland	Melatonin	Annual reproductive cycle
2. Posterior pituitary gland	Oxytocin	Milk ejection; parturition
	Vasopressin	Water balance
3. Anterior pituitary gland	Luteinizing hormone (LH)	Corpora lutea formation; progesterone secretion; Leydig cell stimulation (androgen secretion)
	Follicle-stimulating hormone (FSH)	Follicle development (♀), ovulation (with LH and estrogen); spermatogenesis
	Prolactin	Milk secretion; parental behaviour in birds
	Adrenocorticotrophic hormone (ACTH)	Regulate adrenal glands, which release corticosteroids
4. Intermediate pituitary gland	Melanophore-stimulating hormone (MSH)	Colour change
5. Adrenal cortex	Steroids	Water balance; metabolism; electrolyte balance; blood sugar level; stress reaction
6. Adrenal medulla	Adrenaline (epinephrine), noradrenaline (norepinephrine)	Blood sugar level, stress reaction
7. Testes	Androgens	Testes development; spermatogenesis; secondary sex characteristics; sexual activity
8. Ovary and placenta	Estrogens	Uterine growth; mammary gland development; sexual activity
	Progestogens	Gestation

The anterior pituitary (pars distalis) secrete four hormones that indirectly affect behaviour; three of these are *trophic hormones* that exert their effects on other endocrine glands. In females, **follicle-stimulating hormone (FSH)** and **leuteinizing hormone (LH)** affect the cycle of egg maturation in ovaries, sexual receptivity, conception and pregnancy. In males, FSH and LH control sperm production and secretion of male hormones, or androgens. A third trophic secretion, **adrenocorticotrophic hormone (ACTH),** affects the production and secretion of adrenal cortex steroid hormones. **Prolactin,** a hormone of anterior pituitary gland, is present in birds, mammals, and fish and is important for parental behaviour. It influences milk production in mammals, crop milk accumulation in pigeons and doves, and possibly parental care behaviour in fish.

Prolactin is also found in amphibians, in certain species, it stimulates **migration** to water to breed.

The **pineal gland** within the brain secretes several hormones that are indolamines, proteins, and polypeptides. In the study of behaviour, the most important of these is probably **melatonin.** Melatonin functions in the modulation of reproduction and annual breeding activity patterns in mammals (**Reiter,** 1980), among other things. Doctors recommend it to combat jet lag and other problems in daily biological rhythms.

Both sexes produce androgens, estrogens and progestins but in different amounts. The **testes** produce androgens, primarily **testosterone,** which directly acts on tissues to produce male sexual behaviour. Testosterone is also converted to **estradiol,** which also affects sexual motivation in several species. The **ovaries** produce **estrogen** and **progesterones,** which influence mating behaviour in most mammals. Estrogen charges progesterone receptor. In addition, the ovaries release small amounts of androgens; these may be important for sexual receptivity in humans and other primates. During pregnancy, progestogens secreted by the placenta play a key role in maintaining gestation. **Adrenal** hormones are involved in maintaining the body's water balance, metabolism, and electrolyte balance. **Adrenaline** and **noradrenaline** are produced by adrenal medulla; they play important roles in emergency stress reactions.

Endocrine Gland Interactions

Endocrine glands interact with each other and with target tissue by following two methods:

1. Feedback loops. Many endocrine gland secretions are characterized by feedback loops; an example is shown in Figure 16.7. Pituitary secretion of FSH and LH is controlled by a releasing hormone gonadotrophin (GnRH) that travel via the hypothalamic-pituitary portal system from the hypothalamus. The trophic hormones FSH and LH are then carried in the blood to the testes, where they stimulate the gonadal processes of spermatogenesis in the seminiferous tubules and the production and release of testosterone from the interstitial cells (cells of Leydig).

Testosterone, in turn, is carried including the sex accessory glands (*e.g.*, the seminal vesicles, where semen production is stimulated) and the hypothalamus. Specialized hypothalamic sensory cells are part of the body's **homeostatic machinery** and continuously monitor blood levels of various chemicals, including testosterone and its metabolites. Thus, when one castrate an animal, the amount of testosterone present decreases but at the same time, releasing factors are secreted into the hypothalamic-pituitary system, travel to the anterior pituitary, and cause an increase in output of FSH and LH. This is an ethological engima that to what should then attribute any observed change in behaviour—to the lower levels of testosterone or the higher levels of FSH?

In a **negative feedback** relationship, the levels of circulating testosterone influence the secretions of hypothalamic-releasing factors to the pituitary. As the testosterone levels in the blood increases, the secretion of hypothalamic-releasing factors decreases. Conversely, as the testosterone level in the blood decreases, more releasing factors are secreted into the hypothalamic-pituitary portal system, travel to the anterior pituitary, and cause an increase in output of FSH and LH.

There are also feedback loops involving FSH, LH and the female gonadal hormones, estrogen and progesterone, during sequence of events in **estrous cycle** of a female mammal. At the start of cycle, the hypothalamus stimulates release of FSH and LH from the pituitary; FSH predominates at this time. FSH and LH stimulate the growth of follicles within the ovary and the production of estrogen hormone from the follicles. When the estrogen level in the blood reaches a peak, indicating the follicles are mature, the estrogen has a negative feedback on the pituitary, and it decreases the release of FSH. However, estrogen has a **positive feedback effect** on LH, the pituitary releases more LH when the estrogen level rises, and the LH becomes the predominantly pituitary hormone. In those mammals that ovulate spontaneously the follicle rupture and the ova are released at about the time of the transition to LH dominance. The ruptured follicles secrete estrogen and progesterone under the continuing influence of LH and prolactin. However, rising blood levels of progesterone exert a negative feedback influence on the pituitary that cause it to diminish its release of LH. In mammals, where ovulation is induced, the stimulation received during copulation triggers this sequence of hormonal events and results in the release of ova.

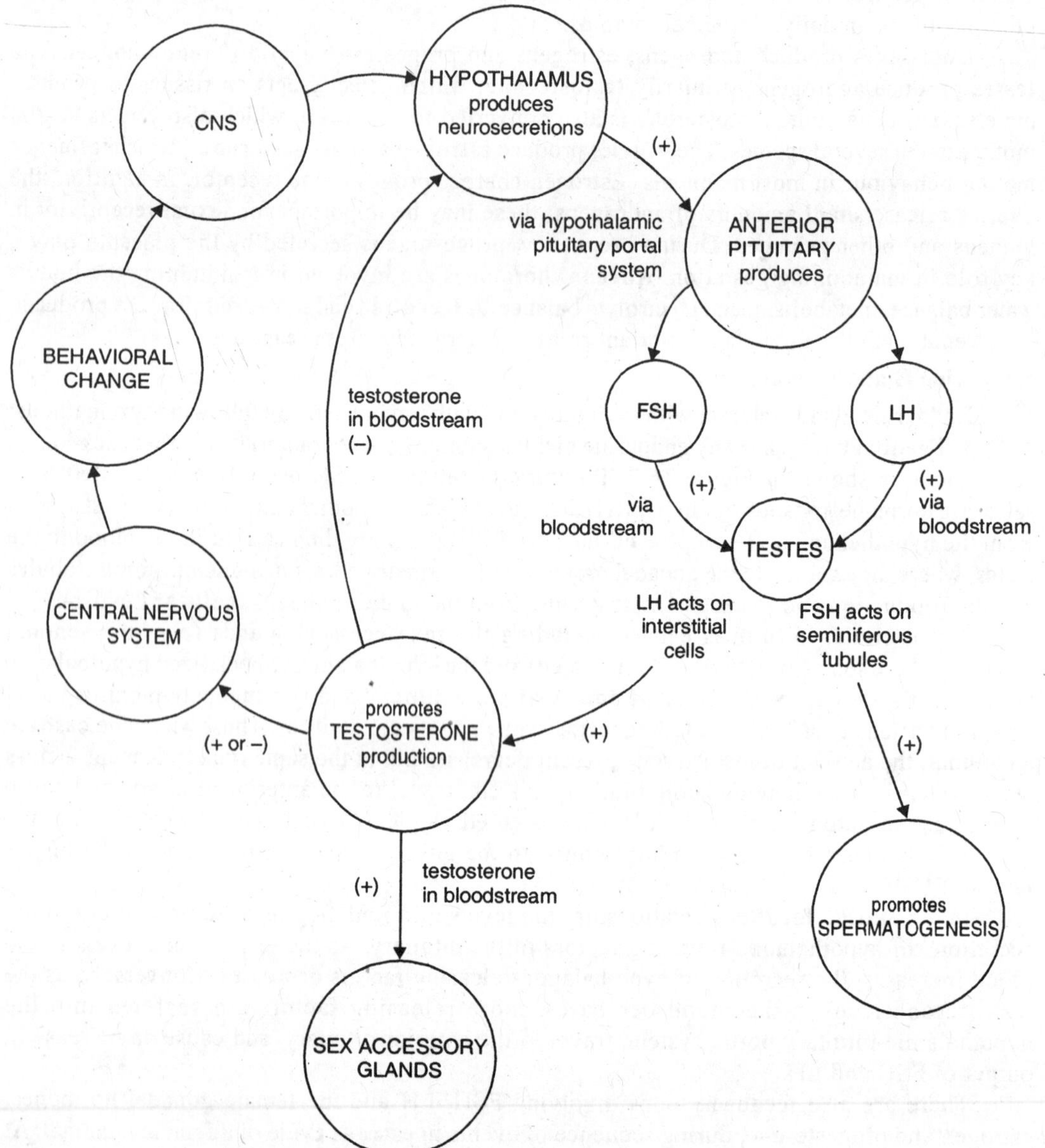

Fig. 16.7. Feedback loops in endocrine physiology of a male mammal or bird. Feedback loops are a key feature of hormone-behaviour relationship. Some of the pathways involve direct effects (*e.g.*, FSH and LH act to stimulate testes). Other pathways involve negative feedback effects (*e.g.*, testosterone in the blood-stream passing through the hypothalamus is monitored by specific receptors. High levels of testosterone may result in reduced output of the releasing factors, which affect the anterior pituitary. The (+) beside an arrow indicates a stimulatory effect and (–) indicates an inhibitor effect (after Drickamer *et al.*, 2002).

Estrogen and progesterone are also responsible for changes in behaviour. Species vary in their reproductive physiology and endocrinology but for mammals in general the **estrous phase** of the cycle, when the females are receptive to males for courtship and mating, is influenced by the two ovarian steroid hormones. The complex physiological and behavioural changes are the product of several endocrine pathways that involve positive and negative feedback systems.

Similar feedback loops exist for ACTH and adrenal steroid hormones. The feedback principle is important for a complete understanding of hormone interactions and their effects on behaviour. Feedback loops can also be influenced by factors in the environment (*e.g.*, daylength) to alter or set hormonal levels.

2. Synergism and antagonism. Two key endocrine interactions involve synergism and antagonism. Female estrogens and progesterone are often in circulation and simultaneously act together to affect sexual behaviour in vertebrates. These effects are termed **synergism.** For example, as mentioned previously, estrogen must be present to prime females to respond to progesterone. Sexual receptivity in female rats depends on the synergistic action of the two hormones (**Beach,** 1976). In contrast, some hormones, when they occur in circulation together, have **antagonistic actions;** the effects of two hormones appear to be in opposition to one another. Male pigeons initially act aggressively toward a female that is a potential mate, yet eventually their aggression is inhibited as a result of the antagonistic interaction of testosterone and progesterone.

Experimental Methods to Understand the Role of Hormones

Following techniques have been developed by investigators to explore the relationships between hormones and behaviour :

1. Extirpation or removal, of a particular endocrine glands to assess the absence of a specific hormone on behaviour.

2. Hormone replacement therapy, in which the investigator injects a specific hormone into an animal or transplants a gland from another animal to replace one previously removed by surgery.

3. Augumenting or boosting the animal's own hormone supply with exogenous hormone.

4. Antagonist or competitive chemical provision, which involves giving the animal a chemical that antagonizes a particular hormone or a chemical that competes with the hormone and blocks its receptor sites, interfering with the effectiveness of the hormone.

5. Blood transfusion to transfer the "hormonal state" of one animal to another in order to observe possible behavioural effects.

6. Bioassays to assess circulating hormone levels indirectly by measuring a secondary characteristics that is dependent on a particular hormone such as skin gland size.

7. Radioimmunoassay to measure directly circulating levels of a hormone from a blood, saliva, or faecal sample.

8. Autoradiography and cytoimmunochemistry to localize the sites where hormone uptake occurs.

9. Immunoneutralization by administring antibiotics against a hormone to effectively remove it from circulation.

Further, in most modern technique, a knockout mice provides an opportunity to explore more directly the connections between genetics and hormones influencing behaviour (**Nelson,** 1997). When a particular gene sequence has been identified, it is disabled by substituting a mutant sequence of DNA for that gene. This enables scientists to study mice that are missing a gene for example that codes for production of a particular hormone or for cell receptors that recieve particular hormonal messages (**Lydon** *et al.,* 1995). Another relatively new procedure involves implanting hormones directly into particular area of the brain. Estrogen implants in the medial pre-optic area of the brains of male rats induces them to become more maternal

(**Rosenblatt** and **Ceus,** 1998). Both of these new techniques have involved **immunochemistry.** In these procedures an **antibody** to a particular hormone is developed. It is then linked with a marker such as **fluorescent dye.** When the antibody is injected into test subject, it binds to the hormone and provides us with a picture of where that hormone is located in the body. This technique has been used for example to locate the *estrogen receptors* in specific areas of the brains of quail (**Balthazart** *et al.,* 1990).

Role of Hormones

Ethologists have divided hormonal influences on behaviour roughly into two categories :

1. Organizational effects. These effects of hormone are manifested during an organisms development. Organizational effects involve hormonal influences during critical periods in behavioural development that produce relatively permanent changes in the organism's nervous system and other tissues. For example, sex differentiation and patterns of growth for body tissues are partly under hormonal control.

2. Activational effects. In this case, hormones act as triggering influences on the expression and performance of behaviour patterns. Direct activational effects are occurring when hormone secretion or inhibition of secretion leads to a relatively rapid response. Indirect activational effects require more complex stimulations and hormone secretion sequence. **G.W. Harris** (1960) proposed a three tier linkage between neural and endocrine components with reference to the behaviour functions.

1. Central nervous system perceives and transduces external informations to adenohypophysis and to other endocrine glands via the former.

2. CNS mediates responsiveness of the endocrine axes at certain stages. For example, pituitary-adrenal and pituitary-gonadal axes become highly sensitive during injury and puberty, respectively.

3. CNS mediates between the effectors and the effecting agents. That is, it integrates sensory messages from internal system and directs the endocrine system for action.

Although cerebral cortex and limbic system have been known for coordinating many functions of the body, the hypothalamic areas have been widely investigated for their role in the different behaviour patterns.

Examples of Behaviours Controlled by Hormones

1. Hormones and sexual behaviour. Sex hormones are essential for (*i*) sex attraction and recognition of male. (*ii*) sexual interaction between two sexes or **courtship.** (*iii*) postural responses by male and female to achieve **coitus** (copulation).

For example in female rhesus monkeys, estrogen enhances the female attractiveness and female receptivity. The male copulatory behaviour, *i.e.,* erection and ejaculation are controlled by testosterone and ACTH. **Moss** and **McCann** (1973) have worked out that LH-RF (Luteinizing hormone releasing factor) from hypothalamus directly affects copulatory behaviour.

Likewise, singing in male birds to attract females of their species is controlled by male sex hormone (testosterone). **Eversole** (1941) found that immature guppies when injected testosterone exhibited normal male sexual behaviour of adult. If a chick is treated with testosterone, it starts exhibiting courtship movements. A number of vertebrate female when treated with androgens are found to perform male sexual behaviour.

2. Aggressive behaviour. Aggressive behaviour and also other social behviours are also influenced by a variety of hormones. These hormones are released from **pituitary-gonadial system** and **pituitary-adrenal system.** In red-billed weaver birds, leuteinizing hormone (LH) from pituitary controls aggressive behaviour. In mammals males are more aggressive than females because of testosterone hormone in them. Higher level of androgens in blood increases aggressive attitude in majority of vertebrates. On the contrary, the estrogens or female hormones in blood make the female more peaceful.

3. Male dominance. In social vertebrates male dominance is associated with **testosterone** hormone. When a low ranking pigeon was treated with testosterone, it rose to a higher status in dominance hierarchy. Similar results were obtained with chickens, mice and chimpanzees.

4. Territorial marking. Territorial marking was studied in mongolian gerbils by **Thiessen** (1973). Both male and female gerbils mark their territory by pheromones they secrete from their midventral sebaceous glands. Castration leads to a low level of scent marking for territory by gerbils. The activity was found to be recovered by injection of androgen or estrogen to castrated gerbils.

5. Persistence. Testosterone increases persistance in animals for a variety of tasks such as searching, discrimination and open field activities.

6. Hoarding. Food hoarding is associated with females or with female hormones. Food hoarding by male gerbils increased by castration and decreased by injecting testosterone. In female rats hoarding activity is associated with their estrous cycle.

7. Wheel running. Wang (1923) reported an association between female hormones and wheel running in nonpregnant female rats. The four days wheel running activity is correlated with the 4 days estrous cycle, fluctuating with the change in hormonal level.

8. Hormones in learning and memory. MSH or ACTH (hormone secreted by anterior lobe of pituitary gland) influences memory and learning processes. These hormones are called **neurotransmitter** or **neuromodulator.** In rats, α-MSH is more effective in eliciting behavioural activity. In addition, β-endosphin also plays important role in behaviour.

9. Adaptive behaviour and hormone. Vasopressin, an hormone of neurohypophysis, maintains adaptive behaviour of animals.

16.2. DRUGS AND BEHAVIOUR

A chemical used to balance the body functions of humans (or animals) or to treat a disease is called **drug.** The name has been derived from a Greek word *drogue* that means a herb. Drugs are also given to patients to replenish deficiencies. Drug is called **medicine** by a common human being. When a drug is prepared from chemicals in a laboratory of industry it is called a **synthetic drug,** *e.g.,* calcium, thyroxine, aspirin, etc. However, if a drug is produced from a naturally found substance it is called **natural drug,** *e.g.,* morphine, caffeine, penicillin, streptomycin. Olive oil, clove oil, cod liver oil, etc. These are expensive drugs. The most modern types of drugs are **genetically monitored drugs** which are produced by using the knowledge of **biotechnology** or **genetic engineering.** These drugs are still not easily available and are very expensive. They are being used to cure cancer, diabetes, etc.

In animals and humans the behaviour is controlled by a number of drugs. Traditionally, drugs are classified on the basis of their effects on one or other categories of responses such as instincts and learning. Drugs alter brain function in following ways:

1. By altering the general metabolism of the neurons.
2. By altering the excitability of the neuroglial membrane.
3. By blocking electrical conductance of the neuronal impulse along the neuron axon.
4. By reducing the blood supply to the neurons.
5. By altering the distribution of water and ions between the neurons and the surrounding tissue.

When the drugs influence neuronal function, changes in behaviour are observed. If these influences are widespread, the effects on behaviour tend to be general, rather than specific. These effects may be due to direct influences of the drug on brain tissue but more often indirect.

Drug addiction and Drug-dependence

A drug has three kinds of action on human body : (*i*) **beneficial,** where it cures a disease; (*ii*) **harmful**, where it causes harmful side effects, and (*iii*) **addictive** which is a drug that causes physiological or psychological dependence. Some people are unable to bear tensions and day to day problems, they are basically weak minded persons who are looking for an easy way out in life. They start smoking, drinking alcohol or coffee or taking drugs. This becomes their habit and they get **addicted.**

There are certain drugs, called **psychotropic** or **mood drugs** which act on brain and alter the behaviour, consciousness, power of thinking and perception. Persons who consume these drugs become fully dependent on them; they cannot live without these drugs. This condition is called **drug-dependence.**

It is possible to get rid of the habit-forming drugs, but, because of serious withdrawal effects, it is extremely difficult to give up those drugs on which the body becomes dependent. Deprivation of a drug to an addict person leads to nausea, vomiting, cramps, diarrhoea, painful muscle contractions and dizziness.

Types and Effects of Drugs

Addictive drugs are broadly divided into two categories :

1. Psychotrophic drugs;
2. Psychedelic drugs.

A. Psychotrophic drugs. These are mood elevators and alter the behaviour via central nervous system. These drugs are of following types :

1. Tranquillizers. These drugs depress brain activities, so are called **depressant drugs.** They give a feeling of tranquility, reduce anxiety, reduce aggression, relax mind and body, ease out stress, consequently induce sleep, *e.g.*, **phenothiazines, benzodiazepines** (Valium), **chlordizepoxide,** etc. Chlordizepoxide is used as tranquillizer and to treat withdrawal symptoms of alcoholism. This drug relives anxiety and reverse the dampening of social behaviour when the neurotransmitter serotonin is depleted or decreased, it blocks the anti-anxiety effect of chlordiazepoxide. This effect leads us to conclude that serotonin transmitter mechanism is involved in the process of anxiety and in the action of anti-anxiety drug.

Thalidomide is also a tranquillizer which was taken by pregnant mothers but were reported to be **teratogenetic** producing congenital malformations in babies such as **ectromelia** (lacking of limbs) or **phocomelia** (grossly deformed limbs) (see **Verma** and **Agarwal,** 2009).

2. Sedatives. These drugs also have depressing effect on brain. They are also called **hypnotic drugs.** Sedatives when taken by a person reduce mental activity, reduce anxiety and induce sleep, *e.g.*, **barbiturates** (sleeping pills). In prescribed amounts they act as sedatives, but in large amounts cause a dreamy state, followed by unexpected changes in mood such as abrupt crying or laughter or violence. Barbiturates are drugs for relief of anxiety Barbiturates are drugs for relief of anxiety, insomnia with the associated risk of death by overdose.

3. Opiates. They suppress brain activity, releive pain, induce drowsiness and give a feeling of well being and hallucination, *e.g.*, opium, morphine, codeine (natural), heroin, pethidine, methadone (synthetic).

Opium is highly addictive drug that consists of the dried milky juice from the seed capsules of the opium poppy obtained from incisions made in the unripe capsules of the plant (*Papaver somniferum*). Opium has a brownish yellow colour, a faint smell and a bitter and acrid taste and is taken orally or smoked. **Morphine** and **codeine** are derived from opium, while **heroin** is a synthetic compound prepared from morphine. Morphine is used to reduce the pain. Codeine has similar uses and is an ingradient in special types of medicines including some cough syrups. Heroin, commonly called **smack,** is chemically diacetylmorphine. It is so dangerous and addictive that its possession or use even for medical purposes is illegal. Certain synthetic drugs such as pethidine and methadone also produce similar effects.

4. Stimulants. Antidepressants, mood elevators, increases mental alertness, *e.g.*, caffeine (tea, coffee, chocolate), cocaine (natural), iproniazid, amphentamines (synthetic).

Ipronizaid was introduced in 1950s for the treatment of tuberculosis. It was noted that some patients experienced a sense of well-being and mood elevation soon after the start of therapy. It also showed antidepressant activity. As a matter of fact iproniazid inhibits the enzyme monoamine oxidase (MAO) involved in the degradation of neurotransmitters in the synapse. This inhibition results in accumulation of the transmitter substance in the synaptic region, increasing functional efficiency.

Another antidepressant drug, the **imiparamine** (Tofranil) inhibits the reuptake of released neurotransmitters and is assumed to have a similar function result to MAO inhibitors.

Amphetamine is a classical stimulant drug, called **pep pills,** having several effects on behaviour. Its principal action is on neurotransmitters such as noradrenaline (NA) and dopamine (DA). It stimulates both learned and unlearned behaviour in all animal species including human beings. With higher doses, streotyped motor behaviour is observed in which unlearned behaviour is repeated for long periods of time, *e.g.*, rats sniff, gnaw and lick the bars of their cages. Cats frequently show hypermobility, side-to-side looking movements and repetitive sniffing and monkeys show biting, pricking or probing of themselves. Amphetamine has marked effect on learned behaviour. It is used by human beings to heighten arousal and intellectual function. People become more active under the influence of this drug. It is supposed to be mediated by dopamine pathways. Amphetamine also releases noradrenaline. Amphetamine induces in humans a form of psychosis which may be misdiagnosed in the clinic as paranoid, schizophrenia. The drug addicts have olfactory, visual and auditory hallucination, pronounced paranoia and anorexia (want of appetite).

Cocaine. It is commonly called **coke** or **crack.** It is a narcotic drug extracted from the leaves of coca plant, *Erythroxylem coca.* It causes blockade of the dopamine transporter protein, thus, interfering with the transport of neurotransmitter dopamine. When it is taken internally, cocaine causes a temporary stimulation of nervous system and a feeling of pleasure. Later, however, the victim is seized by a feeling of great fear and may even become violent. The bad effects of cocaine addiction include lack of sleep, loss of appetite and hallucination which ultimately lead to damaged mental functions and insanity. Misuse of cocaine may also produce severe headache, convulsions or death due to cardiovascular or respiratory failure.

B. Psychedelic drugs. These drugs include LSD, Mescaline, Psilocybin, Hemp products, Tobacco, Alcohol, etc. Abuse of psychedelic drugs take a person on a "trip" to an imaginary world (**hallucination**) full of floating colours, all senses become exaggregated, light looks brighter, sounds get louder, space gets wider, distances look never ending a person feels as if he/she is flying in air.

1. LSD. This natural hallucinogen is the abbreviation of a German term D-lysergic acid diethylamide 15. It is derived from Argot fungus, *Claviceps purpurea.* This drug is extremely potent even in vary small amount. It gives a feeling of euphoria, followed by hallucinations and sense of being outside own body. Perception are vividly enhanced. Prolonged usage can destroy the personality (*i.e.,* Complete mental breakdown and suicides). LSD is highly addictive.

2. Mescaline. It is a natural hallucinatory crystalline alkaloid ($C_{11} H_{17} NO_3$) that is the chief active principle in mescal buttons which are dried tops of mescal. Mescal is a small cactus *Lophophora williamsii* with rounded stems covered with jointed tubercles that are used as a stimulants and antispasmodics among the Mexican Indians. Mescal is also the name of colourless Mexican liquor distilled especially from the central leaves of any of various fleshy-leaved agaves.

3. Psilocybin is another natural potent hallucinogen and is obtained from a mushroom *Psilocybe mexicana.* When taken intravenously, it causes euphoria and hallucinations. Large amounts of this drug lead to violence. Highly addictive chemical. Person withdraws from the society. Prone to sudden outbursts of anger.

4. Hemp products. Bhang, ganja and charas are three natural drugs which are obtained from the dried leaves and flowers of hemp plant, *Cannabis indica.* Another drug **marijuana** (White powder) is derived from another kind of hemp plant, *Cannabis sativa.* Major active component of the marijuana is **THC** (delta-9-tetrahydrocannabinol) which inhibits the enzyme adenyl cyclase, thus, lowering the concentration of the second messenger cyclic AMP (cAMP). When these drugs are taken, the pupils of the eye, dilate, blood sugar level and frequency of urination increases. These drugs are usually mixed with tobacco and smoked.

Marijuana is a mild euphoriant. Under its influence an inner calm sets in, time seems to expand. Reflexes are slowed down and perception is enhanced, may cause hallucinations in the form of visual illusions. This drug can completely take over a weak personality by making him/her an addict.

Other well known plants with hallucinogenic properties are *Atropa belladone* and *Datura.* sp.

5. Tobacco. It comes from the dried leaves of plant *Nicotiana tabacum* and *Nicotiana rustica,* which are cultivated all over the world. Tobacco contains **nicotine.** This chemical mixes in the blood very easily and acts as a tranquilizer and relaxes muscles, therefore, is habit forming. It can be smoked or chewed. Prolonged use of tobacco causes lung and mouth cancer.

6. Alcohol. The alcohol (*e.g.,* whisky and beer) is consumed as **ethyl alcohol** or **ethanol.** Alcohol is produced by fermentation of glucose by yeast and is distilled as liquors. Consumption of alcohol in small amounts gives a feeling of well being, it elevates the mood, it induces sleep. Alcohol is addictive. As the frequency and amount is increased it becomes a disease called **alcoholism.**

High levels of alcohol causes slurred speech, drowsiness, less muscular coordination, exaggregated emotions, violence. Alcohol affects the cerebrum and cerebellum of human brain. It causes gastritis, dehydration and excess urine production by decreasing the production of anti-diuretic hormone. Liver is the most adversely affected. In liver, the alcohol is converted into acetaldehyde and excess fat. This extra fat deposits in liver causing liver tissue damage which lead to a disease called **cirrhosis** and death. Severe withdrawal symptoms appear upon detoxification (no alcohol consumption).

QUESTIONS

Long Answer Questions

1. Describe in detail the role of hormone in the behaviour of animals.
2. Write how hormones of gonads, adrenal and pituitary control vertebrate behaviour.
3. Describe the hormones associated with sexual behaviour. (*Garhwal 2003*)
4. Describe experimental methods in understanding the role of hormones in animal behaviour.

Short Answer Questions

1. Give a brief account of hormonal basis of behaviour. (*Garhwal 2001*)
2. Describe the mechanisms of hormone action. (*Poorvanchal 2001*)
3. Discuss effects of hormones on aggressive behaviour.
4. Write an account of the roles of drugs in altering the behaviour.
5. Write short notes on the following :

   (*a*) Psychotropic drugs (*b*) Psychedelic drugs (*c*) Antagonistic action of hormone

   (*d*) Feedback mechanism.

Multiple Choice Questions

Choose the correct answer from the four alternatives given

1. Anger hormones are
   (*a*) adrenaline and thyroxine (*b*) noradrenaline and calcitonin
   (*c*) adrenaline and noradrenaline (*d*) thyroxine and relaxin
2. Primary producers of behaviour are
   (*a*) gonad (*b*) thyroid
   (*c*) both (*d*) none
3. Emotional and physical stress are managed by hormones produced from
   (*a*) gonads (*b*) pituitary
   (*c*) adrenal (*d*) pancreas
4. Mating in the grasshopper is under the control of
   (*a*) GH (*b*) gonads
   (*c*) corpora allata (*d*) corpora cardiaca
5. In cockroach the copulation is under control of
   (*a*) copulin (*b*) corpora allata
   (*c*) corpora cardiaca (*d*) prothoracic gland
6. Higher androgen level in blood causes
   (*a*) feminity (*b*) appeasement
   (*c*) aggression (*d*) none
7. In birds progesterone hormone is important for
   (*a*) mating (*b*) incubation
   (*c*) both (*d*) none
8. In pigeon 'crop milk' production is controlled by
   (*a*) prolactin (*b*) progesterone
   (*c*) adrenaline (*d*) not known
9. Dominance hierarchy in animals is controlled by
   (*a*) progesterone (*b*) testosterone
   (*c*) adrenaline (*d*) thyroxine
10. Barbiturate is a
    (*a*) sedative (*b*) tranquillizer
    (*c*) opiate (*d*) hallucinogen
11. Which of the following is a hallucinogen ?
    (*a*) LSD (*b*) psilocybin
    (*c*) mescaline (*d*) all of these
12. Benzo-diazepene is an
    (*a*) antidepressant (*b*) antipsychotic
    (*c*) antianxiety (*d*) scent glands
13. Methadone is a
    (*a*) opiate narcotic (*b*) sedative
    (*c*) stimulant (*d*) hallucinogen

14. Testosterone made rats aggressive when introduced into

(*a*) prosencephalon (*b*) mesencephalon

(*c*) diencephalon

15. Epinephrine is secreted by

(*a*) adrenal medulla and increases the heart rate

(*b*) adrenal medulla and decreases the heart rate

(*c*) adrenal cortex and increases the heart rate

(*d*) adrenal cortex and decreases the heart rate

ANSWERS

1. (*c*); **2.** (*a*); **3.** (*c*); **4.** (*c*); **5.** (*c*); **6.** (*c*); **7.** (*b*); **8.** (*a*);
9. (*b*); **10.** (*b*); **11.** (*d*); **12.** (*c*); **13.** (*a*); **14.** (*c*); **15.** (*a*).

Glossary

Acquisition. Gaining a response in learning paradigm (framework).

Action potential. A self-propagating wave of depolarization that moves along the membrane of a nerve or muscle.

Activational effects. Effects of hormones that act as triggering influence on the expression of a particular behaviour patterns; response is rapid, in hours or days.

Active avoidance learning. A form of operant conditioning where the animal has to act in order to avoid some noxious consequence (*e.g.*, shock).

Adaptation. 1. In evolutionary biology, any structure, physiological process, or behavioural trait that makes an animal better able to survive and reproduce compared to conspecifics. 2. The process of evolutionary change leading to the formation of such a trait.

Adaptive behaviour. Behaviour pattern that makes an organism more fit to survive and reproduce in comparison with other members of the same species.

Adaptive radiation. Evolutionary diversification that produces numerous ecologically distinct lineages from a single ancestral one, especially when this diversification occurs within a short interval of geological time.

Adaptive value. A numerical value that expresses the potential for survival and reproductive success of a particular genotype relative to other genotypes.

Adrenal glands. Paired endocrine glands, located next to the kidney in the abdomen. The adrenal *cortex* produces steroid hormones involved in water balance, glucose metabolism and electrolyte balance. The adrenal *medulla* produces adrenaline and noradrenaline, which are involved in glucose metabolism, heart rate, and blood pressure.

Aggression. Behaviour that appears to be intended to inflict noxious stimulation or destruction on another organism.

Aggressive mimicry. A technique for capturing prey in which the predator uses lures or other means to misinform the prey.

Agonistic behaviour. A collection of behaviour patterns used during conflict with a conspecific, usually indicating whether an individual is going to submit to the other animal or flight if the other does not submit.

Allee effect. The unfavourable consequences or undercrowding, as when populations fail to breed when numbers are below some critical density.

Allele. A particular form of a gene (locus) that can be distinguished from other forms (alleles) of that gene.

Alliance. A relationship between two or more individuals, in which they support each other in competitive interactions. Specifically used for dolphins in reference to long-term coalitions.

Allopatric speciation. The formation of new species by a process involving geographic barriers.

Allopatry. Population or species with nonoverlapping geographic distribution.

Altricial young. Young that are born or hatched in a relatively immature or helpless condition.

Altruism. 1. The situation when one individual acts to increase the fitness of another to the cost of its own fitness. 2. Helpful behaviour that raises the recipient's direct fitness while lowering the donor's direct fitness.

Amplitude. The maximum absolute value of a periodically varying quantity, as the peak in activity in a circadian or circannual cycle.

Analogy. Similarities due to convergent evolution.

Anisogamy. The condition in which the female gamete (ovum) is larger than the male gamete (sperm).

Anosmic. Lacking the sense of smell.

Antagonism. The condition of being an opposing principle force, or factor, as when two hormones have opposite effects on target tissues.

Aposematic coloration. Conspicuous markings of noxious animals that are easily recognizable and avoidable by a potential predator.

Appetitive behaviour. The flexible introductory phase of a behaviour sequence during which the organism is searching to obtain something to meet a need, as in seeking food, a mate stimulation of a specific type, etc.

Applied animal behaviour. A subdiscipline of animal behaviour concerned with all aspects of the behavioural biology of pets, domestic animals, in zoos and circuses, and animals in aquaria.

Arrhythmic activity. Activity that does not exhibit any clear cyclical pattern.

Aschoff's rule. Pertaining to circadian rhythm, the rule stating that when nocturel animal's are held in constant darkness, their free-running period becomes slightly shorter each day and when a diurnal animal is held in constant darkness, the free running period becomes slightly longer each day.

Axon. In a nerve cell, the extension that carries information away from the cell body.

Basilar membrane. A flat band of tissues that bears the auditory hair cells in the cochlea of the vertebrate ear; the membrane vibrates in response to pressure waves.

Bateman gradient. The degree to which the reproductive success of males increases with the number of mates they obtain.

Batesian mimics. Palatable species that evolve morphologies and behaviours similar to those of unpalatable species.

Behavioural ecology. A subdiscipline within animal behaviour that deals with ways in which animals interact with their environment and the survival value of behaviour as well as its contribution to reproductive success.

Behavioural isolating mechanisms. Differences in behaviour (usually courtship) that prevent genetic exchange between members of different populations or species.

Behaviour genetics. The study of the role that genes play in controlling behaviour.

Behaviourism. A view of the actions of animals that postulates that behaviour can be analyzed functionally in terms of stimulus and response combinations. Thus, in this view most of behaviour is a function of the experience of the organism.

Biological clock. An internal timing mechanism that involves both an internal self-sustaining pacemaker and cyclic environmental synchronizers.

Biological communication. An action by one organism (or cell) that causes a reaction from another organism (or cell) in a fashion adaptive to either one or both of the participants.

Biological rhythm. A cyclical pattern of behaviour, occurring at some regular period (*e.g.*, diurnal).

Brood parasite. An animal that exploits the parental care of individuals other than its parents.

Bruce effect. In mice, the effect of a strange male, or his odour, that causes females to abort and become receptive.

Cannibalism. The killing and eating of members of the same species.

Cannulation. The use of a hollow electrode or a fine tube to introduce a substance to a tissue (*e.g.*, an area of the brain) or to sample chemicals that are released in the tissue.

Caste. With reference to the social insects physiologically, morphologically, and behaviourally different forms of an organism that perform specialized labour in the colony.

Central nervous system (CNS). That part of the nervous system that is condensed and centrally located, for example, the brain and spinal cord on vertebrates and the brain and ganglia of insects.

Central pattern generator. A group of cells in the central nervous system that can produce the particular pattern of signals needed to cause a fully functional behavioural patterns. (or A set of neurons in the central nervous system that controls repetitive movement without sensory input).

Central place foragers. Animals that collect more than one food item at a time and bring the items to a central location for storage or feeding of offspring.

Cephalization. The evolutionary trend of the concentration of sensory and other nervous system tissue in the head region of animals.

Cerebellum. Part of the hindbrain in vertebrates; important in coordination.

Cerebrum. Part of the forebrain in vertebrates, includes the paleocortex and the neocortex.

Character. An inherited trait that is thought to be independent of other traits, used in constructing a phylogeny.

Character displacement. The process by which two closely related species interact so as to cause one or both of them to diverge evolutionarily in one or more traits.

Character state. The value of a particular taxonomic character (*e.g.*, present or absent, or a particular measurement).

Chemical synapse. The type of synapse where communication between neurons is through chemical intermediary.

Chemosreceptor. A sensory receptor that responds to a specific molecule(s).

Chromosome. A long, complex molecule of DNA, portions of which are the different units of heredity (genes).

Circadian rhythm. A biological rhythm of about a day (24 hour), length or period.

Circannual rhythm. A biological rhythm of about a year in length or period.

Cladistics. A method of reconstructing phylogenies based on the identification of monophyletic groups.

Classical (Pavlovian) conditioning. 1. A form of learning in which an automatic, or unconditioned, response (such as salivation upon detection of food stimuli) comes to be performed in reaction to a stimulus other than the one that normally elicits the behaviour (such as the sound of a bell or the shining of a light). 2. A type of learning whereby an unconditioned stimulus (US) that elicits a specific response (unconditioned response or UCR) is paired with a neutral stimulus (becomes the conditioned stimulus or CS) so that the response becomes conditioned (CR).

Cochlea. A portion of the inner ear of vertebrates that in coiled like a snail's shell and that contains the sensory structure for detecting sound.

Coefficient of relationship. The fraction of genes identical by common descent shared between two individuals.

Commonality. In learning, in inference of a phyletic, evolutionary relationship among the species exhibiting a particular type of learning.

Communal nesting. More than one female in a nest raising the young of more than one female.

Communication. The cooperative transfer as information from a singaler to a receiver .

Comparative method. A comparison of the behaviour of two or more species for the purpose of either elucidating some common aspects of the energy and evolution of behaviour or exploring the mechanisms underlying behaviour.

Comparative psychology. A branch of psychology involving the study of animals. Some comparative psychologists are more concerned with learning cognition and intelligence in human and non-human animals, while others are indistinguishable from other animal behaviourists in that they explore the causation, development, evolution, and functional aspects of behaviour in a broad range of species.

Competition. The attempt of two or more organisms to utilise the same resource.

Composite signal. A signal formed by combining two or more simpler signals.

Conditioned response (CR). The behaviour pattern that becomes conditioned during classical conditioning.

Conditioned stimulus (CS). The previously neutral stimulus, that through classical conditioning, now elicits the conditioned response.

Conflict behaviour. The actions that result when two different motivational systems are activated simultaneously within an individual.

Confusion effect. The reduced capture efficiency experienced by a predator caused by high densities or swarms of prey.

Conspecifics. Animals belonging to the same species.

Constraints. In an optimality model, the limits, both internal and external, that define the range of animal's behaviour.

Consumatory behaviour. Actions of an animal completing a behaviour sequence, as in consuming food, mating, etc.

Context. In communication, stimuli other than the signal that are impinging on the receiver that might alter the meaning of a signal.

Contiguity. The association of events in time, especially as used in classical conditioning.

Cooperation. Two or more individuals behaving to achieve a common purpose.

Cooperative breeding. A breeding system in which nonparents share in rearing the young.

Core area. The area of heaviest use within the home range.

Cost of meiosis. The reduction in numbers of alleles passed on to subsequent generations as a result of reproducing sexually compared to asexually.

Coursers. Predators that chase their prey over long distances.

Crepuscular. Daily cycles with peak activity around dusk and/or dawn.

Critical period. A discrete portion of a sensitive period during development, during which the probability for forming and reinforcing the behaviour is greatest.

Cryptic female choice. Choice of mates by females that takes place during or after copulation in cases where females mate with more than one male. Sperms of particular males is favoured in fertilized eggs.

Crystallized song. The final stage in song development, when the song variation (repertoire) has decreased and more fixed.

Currency. The resource that is being maximized in an opportunity model; usually assumed to be energy in an optimal foraging model.

Cycle. Repeating units that make up the pattern of biological rhythms.

Dendrites. The part of the nerve cell that conveys information toward the cell body; most neurons have several of them.

Depolarized. The reduction in the difference in change (potential) between the outside and inside of the membrane.

Deprivation experiments. Experimental manipulations involving removal of particular types of stimuli (*e.g.*, social, sensory, motor) to ascertain the effects later in development.

Despot. A dominant individual that controls resources.

Diapause phase. A period of dormancy, common in insect species, which occurs during the more rigorous portions of annual climatic cycle.

Diet selection model. A type of optimality model that concerns the sort of food an animal should attempt to acquire and eat.

Dilution effect. Safety in numbers ; when predators can attack only one prey at a time animals in a group are less likely to be eaten than solitary animals.

Dimorphic. Having more than one form, size, or appearance; usually referring to the difference between males and females of a species.

Diploid. Having both members of all chromosome pairs.

Direct fitness. A measure of an individual's potential to contribute genes to future generations via personal reproduction.

Discrete signal. A form of communication that is digital, or all or none; usually given at the same intensity each time, such as an alarm call.

Discrimination. The ability to detect differences between two or more stimuli and make choices based on those differences in a learning situation.

Dispersal. A more or less permanent movement of an individual from an area, such as movement of a juvenile away from its place of birth.

Displacement activity. An innate or stereotypic response to a stimulus that seems inappropriate or irrelevant to the situation.

Display. 1. A stereotyped action used as a communication signal by individuals. 2. Any behaviour pattern especially adapted in physical form or frequency to function as a social signal in communication.

Diurnal. An animal with an activity period during the light portion of the daily cycle.

Dominance hierarchy. A social ranking within a group in which some individuals give way to others, often conceding useful resources to others without a fight.

Ecolocation. The proces whereby the adults form of an insect emerges from the pupa.

Ecological niche. The range of ecological variables (*e.g.*, temperature, moisture, etc.) in which a species can exist and reproduce.

Economic defendability. Defence of a resource that yields benefits that outweigh the costs defending it.

Effector. A tissue or organ that responds to an action potential or a hormone.

Electrical synapse. The type of synapse where communication between neurons is thought electrical transmission rather than a chemical intermediary.

Electric organ discharges (EODs). Produced by some species of fish; may be used in defense, prey capture, and as a channel of intraspecific communication.

Electrophoresis. The separation of different proteins or nucleic acids within a gel matrix that is subjected to an electric field; separation is based on size and/or charge of the molecule(s).

Electroreceptors. Sensory receptors that detect electric fields.

Endocrine glands. A series of ductless glands in both invertebrates (nonchordates) and vetebrates that release hormones into the body through the blood or lymph.

Endogenous. Processes within the animal; used here with particular reference to the internal, genetically based, components of biological rhythms.

Endogenous clock mechanism. Any internal proceses that are genetically based and that play a role in setting or regulating biological rhythms.

Enrichment experiments. Experimental manipulations that involve providing organisms with particular types of stimuli (*e.g.*, social, sensory, motor) to ascertain the effects later in development.

Entrainment. The process by which a biological clock is set or reset by synchronizing with the period of some external environmental stimulus.

Epigenesis. The integrated, interactive process of behaviour development involving the genetics of the organism and its environment, including all its experiences.

Equation. With respect to comparative learning the attempts to match situations and procedures for examining learning behaviour in different species.

Estrous cycle. The period of behavioural and physiological changes from one ovulatory phase (estrus) to another.

Estrus. The period of sexual receptivity at the time of ovulation.

Ethogram. An inventory of all of the behaviour patterns of a species.

Ehtology. The study of patterns of animal behaviour in natural environments, stressing the analysis of adaptation and the evolution of patterns.

Eusocial. A social system involving reproductive division of labour (*i.e.,* castes) and cooperative rearing of young by members of previous generations.

Evolution. A change in the frequency of alleles in a population over generations. The change is caused by natural selection and/or genetic drift.

Evolutionarily stable strategic (ESS). A strategy that when employed by most individual in a population cannot be outcompeted by some alternative strategy.

Experience. All of the interactions between an organism and its environment, beginning at conception and including both external and internal influences.

Exploitative. A type of competition in which organisms passively use up limited resources; also called **scramble competition.** Contract with interference or **contest competition,** in which organisms defend or otherwise control resources.

Exploratory behaviour. The spontaneous search for and active investigation of objects, situations, or other organisms in the absence of any homeostatic need.

Explosive breeding assemblage. The formation of large groups of reproducing individuals in species that breed on only a few days each year.

Extinction. In learning, the decrease of response rate or magnitude of response with lack of reinforcement in a learning situation.

Fatigue. Loss of efficiency in the performance of a motor act when that act is repeated in rapid succession.

Feedback loop. A sequence of events in which the level of hormone of related endocrine product circulating in the blood leads to alterations in the rate of production and release of other hormones from one or more endocrine glands. Often, feedback loops involve the hypothalamic portion of the brain where specialized sensory cells monitor circulating blood levels of numerous compounds, including both hormones and products from hormone actions at various body locations.

Female defense polygyny. Males controlling access to females directly by competing with other males.

Filial imprinting. The process by which young animals form a social attachment for a particular stimulus, often the mother.

Fitness. The potential for an individual to contribute genes to future generations as a function of its adaptive traits. It may be direct, indirect or inclusive type of fitness.

Fixed action pattern (FAP). An innate behaviour pattern that is stereotyped, spontaneous, and independent of immediate control, genetically encoded, and independent of individual learning.

Flagging behaviour. Alarm signalling, as with the use of the tail.

Flehmen. A restriction of the upper lip exhibited by certain mammals soon after sniffing the anogenital region of another or while investigating freshly voided urine.

Focal-animal sampling. A technique for recording the behaviour of individual animals in which the observer concentrates on the activities of one individual for a set time period and then switches to watch another animal for a set time period.

Foci. Groups of cells in an embryo that differentiate into specific structures and organs.

Forced copulation. Rape, in which a male inseminates a female against her will and with the consequence that her fitness is reduced on average.

Founder effect. The establishment of a genetically distinct population based on a random event in which the founders of the population represent only a subset of the genetic variability in the parent population.

Free-running rhythm. The activity cycle that an animal exhibits when placed in a constant environment, its period is different from any known cyclic environmental variable.

Functional neuroanatomy. The study of the size, structure, and arrangement of cells within the nervous system, particularly the brain.

Functionalists. People who conduct studies involving attempts to discern how the mind works.

Fundamental niche. 1. The multidimensional space that an animal would occupy under optimal conditions in the environment. 2. The niche determined in the absence of competitors or other biotic interactions such as predation. 3. Total range of environmental conditions under which a species could survive.

Game theory. An optimality approach to the study of the adaptive value of traits, that focuses on the effects of the activities of other individuals of the same species in determining the fitness pay off for the attributes of interest.

Gametes. Mature haploid cells (sperm and ova) that fuse to form a zygote.

Ganglion. A distinct dense cluster of cell bodies of many neurons.

Gene. A sequence of nucleotides on a DNA molecule that is the basic unit of heredity.

Gene flow. The movement of genes from one population to another via migration or interbreeding.

Generalists. Species that occupy a wide range of habitats and eat a variety of food types.

Generalization. In learning, a principle that states if an animal has been conditioned to respond to a certain stimulus, the response will usually also occur to stimuli used during the original acquisition trials.

Generation time. The length of time between when an organism is born and when it first reproduces.

Genetic drift. The occurrence of random changes in the gene frequency in population due to chance rather than selection, gene flow, or mutation.

Genetic screening. The use of number of mutant forms of a species to test for a specific type of change or deficit, as with respect to biological clocks.

Genotype. The genetic constitution of an individual.

Glial cells. Supportive cells that are closely associated with neurons.

Gonads. Glands responsible for the production of gametes and where certain gonadal hormones are produced. These consist of the ovaries in females and the tests in males.

"Good gene" hypothesis". The argument that mate choice advances direct fitness because it enhances the survival chances of offspring of choosy individuals by providing them with genes that promote viability.

Graded signal. A form of communication that is analog, or varying in intensity or frequency, providing quantitative information about the strength of the stimulus.

Green beard effect. A mechanism for recognizing kin that does not require previous experience recognition genes enable an organism to identify and behave altruistically toward other organisms bearing that same gene.

Group selection. Selection that operates on two or more genetic lineages (groups) as units; broadly defined, this includes kin and interdemic selection.

Gustatory. Having to do with the sensation of taste, the chemoreception of molecules in solution by specialized epithelial receptor cells.

Habitat imprinting. The tendency of an animal to prefer a place to live based on early experience in that habitat.

Habitat selection. The choosing of a place to live.

Habituation. The relatively persistent waning of a response that results from repeated presentations not followed by any form of reinforcement.

Hair cell. A epithelial cell that bears modified cilia on one end; when the cilia are deflected, they change the receptor potential.

Handicap hypothesis. The hypothesis that apparently deleterious sexual "ornaments" possessd by males in many species are attractive to female, because they indicate that the males have such vigor (good genes) that they can survive even with the handicap.

Handling time. The time it takes to process a food item for consumption.

Haplodiploidy. The mode of sex determination by which males develop from haploid (unfertilized) eggs and females from diploid (fertilized) eggs, characteristic of the insect order hymenoptera.

Haploid. Containing one member of each pair of chromosomes, or half the chromosomes of a diploid cell.

Hardy-Weinberg equilibrium. The application of the binomial expansion to determine the frequency of two alleles in the absence of any evolutionary forces.

Helpers-at-the nest. Individuals other than the parents that forego reproduction to assist rearing the young of others.

Heritability in the broad sense. The amount of the variance in a trait in a population that is due to genetic factors; defined as the ratio of the variance due to genetic factors to total phenotypic variance.

Heritability in the narrow sense. A measure of the extent to which genes inherited from parents determine phenotypes; defined as the ratio of the additive genetic variance to total phenotypic variance; also called realized heritability.

Heterozygous. A condition where an individual has two different alleles of the same gene on homologous chromosomes.

Hibernation. A condition of deep sleep and reduced metabolic activity observed in some animals, particularly, during the winter months.

Hippocampus. Part of the vertebrate brain important in learning and spatial memory.

Homeostasis. A tendency toward a stable or equilibrium state with respect to the internal physiological conditions of an animal.

Home range. The area in which an animal spends most of its time.

Homing. The ability of an animal to return to its home site after being displaced.

Homology. A similarity between two structures that is due to inheritance from a common ancestor.

Homozygous. A condition where an individual has two copies of the same allele at the same locus on homologous chromosomes.

Hormones. Chemical products of ductless glands and neurosecretory cells that are carried by the circulatory system and that influence various physiological processes in the body.

Ideal free distribution. The distribution of individuals among resource patches of different quality that equalizes the net rate of gain of each individual. Assumes that organisms are free to move and have ideal knowledge about patch quality.

Imprinting. A process that occurs when an animal learns to make a particular response to only one type of animal or object. The sensory modes used for establishing such a connection can be visual, auditory, olfactory, or some combination of these, depending upon the animal.

Inbreeding depression. Reduction in offspring fitness resulting from mating among kin; caused in part by the accumulation of deleterious recessives.

Inclusive fitness. The sum of an individual's direct and indirect fitness.

Indicator traits. In reference to sexual selection, traits that provide information to members of the opposite sex about the health or fitness of the bearer.

Indirect fitness. A measure of the potential for an individual to contribute copies of its genes to future generations through its influence on the reproduction of nondescendant relatives.

Individual distance. The fixed minimum distance an animal keeps between itself and other members of its species.

Induced ovulators. Species in which copulation is necessary to effect ovulation.

Industrial melanism. The increase in the prevalence of darker-coloured morphs as a result of environmental changes caused by industrial pollution.

Innate. Behaviour that has either a fixed genetic basis or a high degree of genetic preprogramming.

Innate releasing mechanism (IRM). A neural process, triggered by the sign stimulus, which preprograms an animal for receiving the sign stimulus and mediates specific behavioural responses.

Instar. The name given to the form of an insect or arthropod between molting episodes.

Instrumental conditioning. See Operant conditioning.

Intelligence. 1. A collection of mental capacities including imagination, problem-solving ability, memory, the ability to use information gained from past experience, perceptiveness, and behavioural flexibility. 2. The process by which animals obtain information about their environment and retain and use the information to make decisions during the course of their behavioural activities.

Intention movement. The preparatory movements that an animal makes prior to a complete behaviour pattern, as with using movements before taking flight.

Interference. A form of competition in which organisms defend or otherwise control limited resources, also called contest competition.

Intermittent reinforcement. Schedules or reinforcements that involve providing rewards, not on every trial, but on usually one or four other types of schedules : fixed ratio, fixed interval, variable ratio, or variable interval.

Interneuron. A neuron that connects two or more other neurons.

Interobserver reliablity. A method for checking on the agreement between two or more people watching the same animals behave.

Irritability. The ability of a cell to undergo a change in membrane potential

Isolating mechanisms. Any physical or behavioural characteristic that prevents successful exchange of genes between members of different species or population.

Iteroparity. The production of offspring by an organism in successive bouts.

Kinesis. Random locomotion patterns in which there is no orientation of the organism's body axis to the source of stimulation.

Kin selection. The selection of genes due to individuals assisting the survival and reproduction of non-descendant relatives who possess the same genes by common descent.

Knockout gene. A genetically engineered mutant gene that has been targeted for disruption so that it no longer functions normally.

K selection. Selection favouring, characteristics that are adapted to stable, predictable environments.

Latent learning. Association made with neither immediate reinforcement or reward nor with particular behaviour evident at the time of learning. The animal may store and use such information in appropriate situations at a later time.

Learning. The relatively permanent change in behaviour or potential for behaviour that results from experience.

Lek. 1. An area used, usually consistently, for communal courtship displays. 2. An traditional display site where males gather to defend small territories that lack resources useful to females, which nevertheless visit the site to mate.

Lesion. An area of tissue that has been destroyed by an agent such as electric current or a chemical.

Life range. The larger geographic area that an animal utilizes over the course of its life.

Limbic system. A group of nuclei and brain regions in vertebrates that control emotion, and sex and feeding drives.

Local resource competition model. In female philopatric primates, females giving birth to a disproportionate number of males that will disperse and not compete with each other, or high-ranking females giving birth to a disproportionate number of daughters that are able to compete for resources effectively.

Locomotion fear dichotomy. A process during the imprinting process in young birds whereby they initially are more likely to locomote and follow a stimulus objects and later develop a fear toward objects, thus, imprinting occurs when the locomoter tendency is high and fear of objects is low.

Locus (pl. Loci). The position of a particular gene on a chromosome.

Macroevolution. Major evolutionary events occurring above the species level.

Male dominance polygyny. A mating system in which males compete and acquire dominance ranks that influence their access to females, with higher-ranking males often having greater mating activity.

Marginal value theorem. A model that predicts when an organism should cease foraging in one patch and travel to another.

Maternal condition model. In polygynous mammals, females that are high ranking, well nourished, or otherwise in good condition giving birth to a greater proportion of sons since their sons will outreproduce their daughters and vice versa for females that are in poor condition.

Mating system. The species-typical pattern of mate finding, reproduction and parenting of offspring.

Mechanoreceptor. A sensory cell that is stimulated by some mechanical force or pressure.

Menses. The period of shedding of the lining (endometrium) of the uterus and associated fluids if an ovum is not fertilized, most notably in primates.

Menstrual cycle. The period from the end of one ovulatory cycle, as demarcated by menstrual flow to the end of the next cycle in female primates.

Metacommunication. Communication about communication, where one signal changes the meaning of signals that follow.

Microevolution. Changes in gene frequencies or traits that occur in small increments.

Migration. A seasonal movement from one location to another.

Monestrous. A species in which the female is receptive for only a few days once each year.

Monogamy. A mating system in which a male and female bond for some period of time and share in the rearing of offspring.

Monophyletic origin. A group of organisms that evolved from a single ancestral type.

Monozygotic. Twins that arise from a single zygote, hence, two genetically identical individuals.

Mosaic. An organism whose tissues are made up of two or more genetically different types.

Motivation. Internal processes that arouse and direct behaviour.

Motor neuron. A neuron that synapses with a muscle membrane.

Mullerian mimics. Noxious species that resemble each other.

Muller's ratchet. The steady accumulation of mutation in a population of asexual organism over time.

Mutation. A change in the sequence of nucleotides in a gene.

Mutational analysis. An analysis that examines the effects of specific mutants for the effects on particular phenomena with respect to morphology, physiology, or behaviour.

Mutualism. Interaction between two organisms of the same or different species from which both benefit.

Natural selection. The disproportionate survival and reproductive success of organisms that possess certain alleles, as a result of the influence of those alleles.

Navigation. The process by which an animal uses various cues to determine its position in reference to a goal.

Neoteny. The persistence of juvenile traits in the adult form.

Nerve cell body. The largest part of a neuron that typically contain a nucleus.

Neuron. A nerve cell.

Neurosecretions. Hormone-like chemical substances that are produced by specialized neurons (called neurosecretory cells) that affect various physiological processes in the body.

Neurotransmitter. 1. A chemical that is released by one neuron and affects the probability that the next cell in the network will fire. 2. A chemical released by the presynaptic membrane of synapse that attaches to receptor molecules on the postsynaptic membrane and causes a change in the permeability of that membrane.

Nocturnal. Animals whose primary activity occurs during the dark portion of the daily cycle.

Nuptial gift. A food item transferred by a male to a female just prior to or during copulation.

Observability. Not all members of a group are always seen the same proportion of time. Animals of different ages, sexes, or of different dominance status may be seen more or less than other animals.

Observational learning. The tendency to perform an appropriate action or response as the result of having observed another animal's performance or similar actions in the same situation.

Olfactory. Having to do with the sense of smell, the chemoreception of molecules suspended in air.

Operant conditioning. A kind of learning, allied with trial-and-error learning, in which an action (or operant) that is rewarded becomes more frequently performed.

Operational sex ratio (OSR). The ratio of receptive males to receptive females in a population at any one period; a measure of the intensity of competition among males for mates.

Optimality theory. The theory that the attributes of organisms are optimal (*i.e.,* offer the best ratio of fitness benefits to fitness costs); a theory used to generate hypotheses about the adaptive value of characteristics, as in optimal foraging theory, which analyzes alternative foraging decisions in terms of their fitness payoffs.

Organizational effects. Refers to the effects of exposure to certain hormones during critical periods in development that affect the central nervous system and other structures, producing morphological and behavioural changes in adulthood.

Orientation. The way an organism positions itself in relation to external cues.

Oscillator. The internal mechanism that is the clock in a biological rhythm.

Outbreeding depression. A reduction in the fitness of offspring resulting from matings among individuals from different populations or demes, possibly caused by the breaking up of coadapted gene complexes.

Parental investment. The risk taken by a parent and the time and energy it invests in an existing offspring that reduces its chances of producing additional offspring in the future.

Patch model. A type of optimal foraging model that assumes that prey occurs in discrete clumps and that seeks to predict where and when organisms will forage for the prey.

Pelage. The body covering of hair in a mammal.

Perception. The analysis and interpretation of sensory information.

Perceptual world. The manner in which an organism's brain analyzes and interprets all incoming stimuli. Each species has a different view of the world, its *Umwelt.*

Period. The duration of one cycle of a biological rhythm.

Peripheral nervous system. All nerves that run to and from central nervous system.

Phase. A specific, recognizable portion of the activity cycle.

Phenotype. Any measurable aspect of an organism; an attribute or trait of an individual that arises because of a developmental interaction between heredity and the environment.

Phenotype matching. A proximate mechanism of kin discrimination in which an individual's behaviour toward another is based on how similar they are in appearance, odour, or some other phenotypic trait.

Pheromone. 1. A volatile chemical released by one individual to communicate with another. 2. A species-specific odour released by animals that influences the behaviour and/or physiology of conspecific.

Photoperiod. The number of hours of light in 24 hours.

Philopatric. Remaining near the place of birth after sexual maturation.

Phonotaxis. Orientation with respect to sound.

Photoreceptors. Sensory cells that contain photopigments and response specifically to light energy.

Phototaxis. Orientation with respect to light.

Phylogeny. The evolutionary history of a group of organisms.

Piloting. The use of familiar landmarks to find a direction or area.

Pineal gland. An endocrine gland located near the midline of the brain that produces melatonin, a hormone involved in biological rhythm, particularly in annual cycles.

Pituitary gland. The master gland (along with the hypothalamus) of the endocrine system of vertebrate animals. Located directly below the hypothalamus of the brain, the pituitary produces or releases a variety of hormones that target other endocrine glands of the body.

Plastic song. One of the stages of song development in birds during which the bird has begun to sing using the species-typical pattern, but with a certain degree of variation.

Play behaviour. Certain locomotor, social and manipulative behaviour patterns exhibited by young and some adult animals and birds.

Polarized. A description for a membrane that has a potential difference due to an unequal distribution of ions across the membrane.

Polyandry. A mating system in which some females mate with more than one male and males usually invest heavity in offspring.

Polygamy. A mating system in which both males and females mate with several members of the opposite sex.

Polygenic trait. Traits that are influenced by many genes and that are usually continuously distributed within a population.

Polygyny. A mating system in which some males mate with more than one female and females usually invest heavily in offspring. It may be of four types female defense polygyny, lek polygyny, resource defense polygyny and scramble competition polygyny.

Polygyny threshold. The point at which a female will benefit more by joining an already mated male possessing a good territory rather than an unmated male on poor territory.

Polyphyletic. Describes a group of taxa that arose from multiple ancestors.

Postzygotic isolating mechanisms. Isolating mechanisms (barriers to interbreeding) that occur after the formation of zygote.

Precocial young. Young that are born or hatched at a relatively advanced state of development and are capable of a more independent existence beginning at birth or hatching.

Preparedness. The genetically based predisposition to learn. Depending upon the learning situation, animals of a particular species may be prepared, unprepared, or contraprepared to learn.

Prezygotic isolating mechanisms. Isolating mechanisms (barriers to interbreeding) that occur before the formation of zygote.

Profitability. The ratio of energy gained to the handling time of type of food.

Promiscuity. A mating system in which there is not prolonged association between the sexes and multiple mating by at least one sex.

Proprioreceptor. Sensory receptor that relays informations about the position and movement of the body; located primarily in muscles, tendon, and the inner ear.

Protandry. A sequentially hermaphroditic species in which individuals change from males to females.

Protogyny. A sequentially hermaphroditic species in which individuals change from females to males.

Proximate causation (factor). Mechanistic explanations for how behaviour occurs, including, in particular, hormones, the nervous system and behaviour development.

Psychobiology. The study of the mechanism and function of the central nervous system from both psychological and biological perspectives.

Psychopharmacology. The study of brain's behaviour in terms of chemical psychology and psychological parameters.

Puberty. The age at which an organism can first reproduce.

Punctuated equilibria. The hypothesis that evolution occurs in relatively rapid bursts, interspersed with long periods of stasis.

Quantitative genetics. The study of continuously varying traits.

Quantitative trait locus (QTL). The set of genes that governs the quantity of a trait that is not completely determined by any one gene acting alone.

Realized niche. The multidimensional space that an animal actually occupies in the presence of competitors, predators, pathogens, and limited food.

Receiver bias. The increased sensitivity of perceptual system, to certain stimuli as a result of natural selection. These biases may then influence the evolution of communicative behaviour.

Receptive field. 1. That portion of a receptor surface that is monitored by a higher-order nerve cell. 2. The region of a sensory surface that causes the change in activity of a neuron.

Reciprocal altruism (reciprocity). The trading of altruistic acts by individuals at different times, *i.e.,* the payback to the altruist occurs some time after the receipt of the act.

Recombination. The formation of new combinations of alleles through exchange of sections of homologous chromosomes during the process of meiosis; sometimes refers to the combinations of new allele, that result from fertilization.

Reconciliations. Social interactions, such as appeasement, that take place between two or more individuals following agonistic encounters.

Red queen hypothesis. The hypothesis that sexual reproduction has evolved because the genetic variation that results is adaptive in the evolutionary "arms race" between hosts and their pathogens and parasites as well as between predators and prey.

Reflex. A simple, sterotyped behaviour in response to a specific stimulus.

Refractory period. In neurobiology, the period during which voltage gated sodium channels in a neuron cannot be depolarized to threshold polarity by the influx of sodium in adjacent channels due to hyperpolarizing effects of potassium exiting through potassium voltage gated channels in the region.

Regulatory gene. A gene that turns other genes on and off.

Reinforcement. In learning, anything that alters the probability of behaviour.

Releasing factors. Hormones or neurosecretions from the hypothalamus that travel either via the hypothalamic-pituitary portal system or along axons between the hypothalamus and the pituitary where they exert effect in terms of production and release of hormones.

Reproductive effort. The energy expended and risk taken to reproduce, measured in terms of the decrease in ability of the organism to reproduce at a later time.

Reproductive success. The number of progeny born, or surviving progeny produced by an organism.

Resource defense polygyny. Males control access to females indirectly by monopolising critical resources.

Resting potential. The unequal distribution of charges on either side of neuron membrane that is not undergoing an action potential.

Risk-sensitive foraging. A model that incorporates variation in prey encounter rates to predict where predators forage. Depending on their internal state and other factors, predators may be risk-prone, foraging where encounter rates are variable, or risk-averse, foraging where encounter rates are relatively constant.

Ritualization. Evolutionary process by which behaviour patterns become modified to serve as communication signals.

r selection. Selection favouring rapid rates of reproduction and growth, especially among species that specialize in colonizing short-lived habitats.

Runaway selection. Selection for male ornaments that happens due to the genetic correlation, and resulting positive feedback relationship, between the trait and the preference for the trait.

Salience. In the context of classical conditioning, the fact that animals can learn to make certain association, depending upon which stimuli are involved, more rapidly than other associations.

Scan sampling. A method for recording behaviour observation that revolves looking briefly at each animal at prescribed time intervals (*e.g.*, every minute or every five minutes) to record their activity at the time of the sample.

Scramble polygyny. A mating system in which males actively search for mates without overt competition.

Search image. The hypothetical mental image of a prey species that a predator forms as it improves in its ability to capture that particular species.

Semantic communication. Tc denote the use of different alarm signals to warm about different predators.

Semelparity. The production of offspring by an organisms once in its life.

Sensation. The process of transducing environmental stimuli or energy into action potentials.

Sensila. A chitinous, hollow, hairlike projection of the arthropod exoskeleton that serves as an auxiliary structure for sensory neurons.

Sensitive period. 1. The time interval when an animal can develop an imprinting attachment. 2. In behaviour development this refers to time intervals when particular events must occur for proper ontogenetic sequencing.

Sensitization. 1. Enhanced responsiveness to a repeated stimulus. 2. Strengthening of a response that was initially produced via a CS resulting from a pairing with a US and UR. 3. A stimulus priming the animal to pay particular attention to what follows.

Sensory adaptation. A process that occurs at the level of the sensory receptors and that consists of a slowing down or cessation of nerve impulses transmitted to the central nervous system.

Sensory filter. Neural circuits that selectively transmit some features of a sensory input and ignore other features.

Sensory neuron. A neuron that is modified to respond to a particular set of stimuli.

Sexual imprinting. A process by which young of many species establish an attachment to opposite sex conspecifics : the effect manifests itself later in life during the process of finding mates.

Sexual selection. Selection in relation to mating. It is composed of intrasexual competition among members of one sex (usually males) for access to the other sex and intersexual choice of members of one sex by members of the other sex (usually females).

Siblicide. The killing of one sibling by another.

Signal. Any behaviour pattern that conveys information. Patterns modified through evolution to convey information are called displays.

Signaling pheromone. A chemical signal that is quickly perceived and caused an immediate response.

Sign stimuli. Specific external stimuli that trigger stereotyped responses from conspecifics, usually called fixed action patterns.

Social organization. The species-typical pattern of relationships among all members of a group. This would include spatial distribution patterns, interindividual relationships involving dominance hierarchies or territoriality, mating systems, parenting, and dispersal.

Society. A group of individuals belonging to the same species and organized in a cooperative manner.

Sociobiology. A study that involves the application of the principles of evolution to the study of the social behaviour and social systems of animals.

Sound window. The use of freqeuncies for communication that are transmitted through the environment with little loss of strength (attenuation).

Specialists. Species that occupy a narrow range of habitats and eat a narrow range of foods.

Speciation. The process of two populations that share a common descent evolving in different ways so that they ultimately do not interbreed and are therefore different species.

Species. The basic (natural) unit of classification, consisting of groups of actually or potentially interbreeding individuals that are reproductively isolated from other such groups.

Species-typical behaviour. Actions and displays that are broadly characteristic of a species.

Sperm-competition. A situation in which one male's sperm fertilize a disproportionate number of eggs when a female copulates with more than one males.

Spontaneous ovulators. Species in which females release eggs whether they have copulated or not.

Stalk-and-rush. Predators that approach prey as closely as possible, then close with a sudden burst of speed.

Stereocilia. The short, modified cilia at the apex of a hair cell (Hair cell is the basic unit of the inner ear of vertebrates).

Strategy. Alternative course of action.

Stretch receptor. Sensory receptor that responds to stretch, found in muscle tissue, lungs, and other organs that undergo changes in positions or size.

Stridulation. The production of sound by an insect rubbing one body part against another (*e.g.*, in male crickets).

Structural gene. A gene that determines the amino acid sequence of a protein.

Sun compass. The mechanism in which animals use the sun and an internal clock that allows them to adjust for movement in the sun as they navigate.

Sympatry. Populations or species with overlapping geographic distributions.

Synapse. A junction between two neurons or between a neuron and a muscle fiber in which action potentials along the presynaptic membrane will influence the post synaptic membrane.

Synaptic processes. The threadlike extensions at the end of an axon.

Synergism. The interaction of two or more agents or forces so that their combined effect is greater than the sum of their individual effects, as when two hormones combine to affect target tissues.

Syntax. Information provided by the sequence in which signals are transmitted.

Syrinx. In birds, a part of the respiratory tract specialized for sound production.

Taxis (pt. taxes). Directed reactions to a stimulus involving an orientation of the long axis of the body in line with the stimulus source.

Territory. 1. An area occupied exclusively by an animal or group of animals that is defended. 2. An area that an animal defends against intruders.

Thermoreceptors. Sensory cells that are sensitive to changes in temperature.

Threshold. The minimum stimulus necessary to initiate an all-or-none response.

Tradition. A behaviour pattern that is passed from one generation to the next by learning.

Translocation. Moving animals from one location to another, for instance to determine whether and how soon they shift their activity cycle to match the photoperiod and/or other features in their new location.

Transplantation. 1. The movement of neural or hormonal tissue from one area of an animal to another or from one animal to another. 2. The movement of individuals or colonies from one location to another.

True language. Communication that includes the use of symbols to represent abstract objects or ideas, and syntax, where those symbols convey different message depending on their sequence.

True navigation. The ability to maintain or establish reference to a goal without the use of landmarks.

Ultimate causation (factors). Those aspects of behaviour that are concerned with why the behaviour evolved and its functional significance in an ecological context.

Ultradian rhythm. A cyclical rhythm of less than 24 hours.

Umwelt. The sensory and perceptual world of the animal, depending upon the type of sensory receptors that it possesses and the internal nervous system processes for receiving and interpreting those stimuli.

Unconditioned response (UCR). In classical conditioning, the animal's response that is initially given to the unconditioned stimulus.

Unconditioned stimulus (US). In classical conditioning, the stimulus that produces what is initially the unconditioned response.

Utility. In foraging theory, the value that a resource has to an animal.

Voltage-gated channels. Ion-specific channels that open only in response to a specific polarity across the plasma cell membrane.

Wisconsin General Test Apparatus (WGTA). An apparatus designed originally at the University of Wisconsin Primate Centre for use with various species of primates in a series of learning problems. Modifications of this apparatus have been used for a variety of animal species.

Zeitgeber. Any entraining agent that plays a role in setting or resetting an internal biological clock. Examples include sunrise or sunset.

Zugunruhe. Restlessness that an animal displays at the time of its usual migration.

SELECTED READING

Alcock, John, 1974, 1993 — *Animal Behaviour : An Evolutionary Approach,* Sinauer Associates, Inc. Publishers, Sunderland, Massachusetts.

Barnard, C.J., 1983 — *Animal Behaviour : Ecology and Evolution,* Croom Helm, London.

Begon, M., J.L. Harper and C.R. Townsend, 1996 — *Ecology,* Blackwell Science, Berlin

Brusca, R.C. and G.L. Brusca, 2002 — *Invertebrates*, Sinauer Associates, Inc. Publishers, Sunderland, Massachusetts

Drickamer, L.C., S.H. Vessey and E.M. Jakob, 2002 — *Animal Behaviour Mechanisms, Ecology and Evolution,* McGraw-Hill, Boston.

Eisner, T. and E.O. Wilson, 1975 — *Animal Behaviour : Readings from Scientific American,* W.H. Freeman and company, San Francisco.

Hinde, R.A., 1970 — *Animal Behaviour,* McGraw-Hill, New York.

Krebs, J.R. and N.B. Davis, 1993 — *An Introduction of Behavioural Ecology,* Blackwell Science : Oxford, London.

Kumar, Vinod, 1996 — *Animal Behaviour,* Himalya Publishing House, Bombay.

Manning, A. 1967, 1972 — *An Introduction to Animal Behaviour* : Addison-Wesley, Reading, Mass.

Manning, A. and M.S. Dawkins, 1998 — *An Introduction to Animal Behaviour,* Cambridge University Press, Cambridge, U.K.

Mathur, Reena, B.S Tomar and S.P. Singh, 2004 — *Evolution and Behaviour,* Rastogi Publications, Meerut.

McConnell, J.V. 1986 — *Understanding Human Behaviour,* CBS Publishing Japan Ltd., New York.

McFarland, D., 1985 — *Animal Behaviour,* Arnold-Heineman (The Pitman Press), Great Britain.

Mohanty, D., 2009 — "*After two failed tries, scientists slow and steady on Olive Ridley's trail*", Indian Express (7.2.2009).

Morris, D., 1982 — *Manwatching : A Field Guide to Human Behaviour,* Jonathan Cape Thirty Bedford Square, London.

Odum, E.P. and G.W. Barrett, 2005 — *Fundamentals of Ecology,* Thomson Books, Australia.

Reese, E.S and F.J Lighter, 1978 — *Contrasts in Behaviour,* John Wiley and Sons, New York.

Smith, R.L., 1977 — *Ecology and Field Biology,* Harper and Row, Publishers, New York.

Tinbergen, Niko, 1965 — *Animal Behaviour,* Time-Life Books: New York.

Wilson, E.O., 1975 — *Sociobiology,* Belknap press, Cambridge, Massachusetts.

INDEX

A

B

E

F

G

H

I

J

K

L

M

N

O

P

Q

R

S

T